Proceedings of the International Conference on Transformations in Engineering Education

R. Natarajan
Editor

Proceedings of the International Conference on Transformations in Engineering Education

ICTIEE 2014

Springer

Editor
R. Natarajan
Chairman, Board for IT Education
Standards (BITES)
Bangalore, Karnataka, India

ISBN 978-81-322-1930-9 ISBN 978-81-322-1931-6 (eBook)
DOI 10.1007/978-81-322-1931-6
Springer New Delhi Heidelberg New York Dordrecht London

Library of Congress Control Number: 2014950665

Printed on acid-free paper

Springer is part of Springer Science+Business Media (www.springer.com)

Preface

The International Conference on Transformations in Engineering Education (ICTIEE) was organized by the B V Bhoomaraddi College of Engineering, Hubli, India, during January 16–18, 2014. This volume includes the plenary, keynote, and contributed papers presented during the conference. The contributed papers were subjected to a detailed evaluation procedure involving a number of competent reviewers.

Currently, we are witnessing a range of transformations in engineering education both internationally and nationally. Older paradigms and practices are being rendered obsolete, and new ideas are emerging. Some of the new buzzwords surrounding engineering education include MOOCs, global engineers, outcome-based teaching and learning, good governance, academic leadership, technology-assisted learning, and research-intensive institutions.

International partnerships, networking of institutions, harmonization of quality assurance, and accreditation systems and processes are becoming the concern of institutions across the world. There is intense activity worldwide in disseminating and sharing knowledge and experience of different countries and institutions.

ICTIEE 2014 was truly international in character. There were several international speakers and foreign exhibitors of education-related products and services in addition to their Indian counterparts. A large number of students participated in the conference. The international student organization SPEED (Student Platform for Engineering Education Development) organized several activities during the conference. It was very heartening to see the active participation of the delegates in the different sessions and their networking with the speakers and the other delegates.

We are thankful to the faculty members of BVB College of Engineering, who willingly undertook a variety of responsibilities, and also to the reviewers who completed their tasks on time.

It is hoped that this proceeding volume will be very useful and of lasting value to academic leaders, faculty members, professionals, and engineering students.

Dr Krishna Vedula and Dr Ashok S. Shettar served as coeditors for this volume. I am very grateful for their support in this endeavor.

Bangalore, Karnataka, India R. Natarajan

ICTIEE 2014 Committee

Chief Patron

Prabhakar Kore

Advisors

Dr. Jose Carlos Quadrado, President, IFEES,
President, Instituto Superior de Engenharia de Lisboa (ISEL), Portugal.
Dr. R. Natarajan, Chairman, Board of IT Education Standards (BITES), India,
Former Director IIT Madras, India.
Dr. Hans J. Hoyer, Secy General, IFEES,
Executive Secretary, GEDC, Marquette University, USA.
Dr. Gururaj 'Desh' Deshpande, Trustee, Deshpande Foundation,
Life Member, MIT,
Sparta Group, Sycamore Networks, Tejas Networks, USA.
Mr. Arjun Malhotra, Chairman and CEO,
Headstrong Corporation.
Mr. H. U. Talawar, DTE,
Bangalore.

Conference Chairs

Dr. Ashok Shettar, Principal, BVB College of Engineering & Technology, Hubli – 580031 (India)

Dr. Krishna Vedula, Professor, Dean Emeritus, Umass Lowell, USA., Executive Director, IUCEE

Program Chairs

Dr. M. P. Ravindra, President, IUCEE India Society, Bangalore, India.
Dr. Bopaya Bidanda, Professor, University of Pittsburgh, USA.
Dr. Anil Sahasrabuddhe, Director, College of Engineering, Pune (India)
Dr. Ashok Kalbag, Secretary, IUCEE India Society, Mumbai, India.
Dr. Gopalkrishna Joshi, Director, CEER, BVB College of Engineering & Technology, Hubli – 580031 (India)

Publication Chair

Dr. R. Natarajan, Chairman, Board of IT Education Standards (BITES), India, Former Director IIT Madras, India.

Program Committee

Dr. Vinod K Lohani, Professor, Virginia Tech University, USA.
Prof. B. L. Desai, Vice Principal, BVB College of Engineering & Technology, Hubli – 580031 (India)
Dr. P. G. Tewari, Professor & Head, Mechanical Engineering, Dean (Academic Affairs), BVB College of Engineering & Technology, Hubli – 580031 (India)
Dr. N. H. Ayachit, Professor, Rani Chennamma University, Belgaum
Dr. Linganagouda Kulkarni, Professor CSE, PG Coordinator, BVB College of Engineering & Technology, Hubli – 580031 (India)
Prof. S. C. Padashetty, SPFU, Bangalore
Dr. A. K. Kulkarni, SPFU, Bangalore

Contents

Part II Presented Papers

Part III Presented Posters

About the Editors

Editor

R. Natarajan received his M.E. degree from the Indian Institute of Science, Bangalore, and his M.A.Sc. and Ph.D. degrees from the University of Waterloo, Canada. He has worked as a National Research Council Fellow in Canada, and as a Humboldt Research Fellow in Germany.

He served as Director of the Indian Institute of Technology, Madras, from 1995 to 2001, and as the Chairman of the All India Council for Technical Education, a statutory body of the Government of India, from 2001 to 2004. He was the Vice President of The Indian National Academy of Engineering during 2002–2006, and the Chairman of the Research Council of the Central Fuel Research Institute, Dhanbad, during 1995–2005. He is currently the Chairman of the Board for IT Education Standards of Karnataka. He is a Fellow of Indian National Academy of Engineering, Indian Society for Technical Education, National Academy of Social Sciences, and Institution of Engineers (India). He has been conferred Honorary Doctorate Degrees by University of South Australia, Jawaharlal Nehru Technological University, Kanpur University, Nagarjuna University, Purvanchal University (U.P.) and NIT, Agartala.

He has been a Member of the Editorial Boards of many prestigious journals including *Fuel*, *Fuel Science and Technology*, and *Indian Journal of Technical Education*. He is currently on the Editorial Board of the *Journal of Engineering Education*, published by the American Society for Engineering Education (ASEE).

Associate Editors

Krishna Vedula is Professor of Chemical Engineering and Dean Emeritus, Francis College of Engineering, University of Massachusetts, Lowell. Dr. Vedula is currently Founder and Executive Director of the Indo-US Collaboration for Engineering Education (IUCEE) facilitated initially by ASEE. IUCEE has the objective of improving quality and global relevance of engineering education in India and US. For the past five years, IUCEE has facilitated interactions between several hundreds of US faculty and several thousands of Indian faculty. He has also been the President of International Federation of Engineering Education Societies (IFEES) from 2010 to 2012.

Dr. Vedula is well recognized globally for his contributions to engineering education, research, administration, and outreach. He is internationally recognized for his research in processing and properties of materials for high-temperature applications, with particular emphasis on powder processing and inter-metallic compounds. He has been made a Fellow of American Society for Metals (ASM) and a Fellow of the American Society for Engineering Education (ASEE) in recognition of these professional achievements.

As dean of engineering at University of Massachusetts (1995–2003), he has demonstrated his leadership in building unique partnerships with businesses, K-12, state agencies, and other educational institutions. He was responsible for raising several millions of dollars for scholarships, endowments and facilities from private sources and for overseeing a significant improvement in the quality of students and faculty as well research and teaching facilities at the Francis College of Engineering. He is Founder of "Massachusetts STEM (Science Technology Engineering and Mathematics) Collaborative" aimed at increasing the number of students in the state interested in pursuing science and engineering. This has now evolved into a program of the State of Massachusetts.

Dr. Vedula has B.Tech. (IIT Bombay, 1967), M.S. (Drexel University, 1969) and Ph.D. (Michigan Tech University, 1980) degrees in Materials Engineering. He has 30 years of academic teaching and research experience in materials science and engineering as well as engineering administration, including 10 years as a faculty member at Case Western Reserve University, 5 years as chair of the materials science and engineering department at Iowa State University, and 8 years as dean of an engineering college at UMass Lowell. In addition he has spent 2 years managing federal research and educational programs.

Ashok S. Shettar earned his Ph.D. from the Indian Institute of Science, Bangalore (India). He has three decades of experience in Engineering Education. His conviction and systemic practice of Outcome-Based Education (OBE) have taken his Institution to greater heights in India. He has the experience of acting as resource person in workshops on outcome-based education in local, regional and national levels in India. He is an active researcher in Engineering Education. Dr. Ashok Shettar is currently the Principal of BVB College of Engineering and Technology, Hubli, India.

Part I
Invited Papers

Transformations in Engineering Education Globally

Stephanie Farrell, Hans J. Hoyer, and Duncan McKenzie Fraser

All of us are aware of the critical but exciting evolution of research universities from the 1960s until today, a metamorphosis in which universities have gone from insular "ivory towers" to key drivers of economic prosperity. This has resulted in dramatically expanded missions for engineering colleges and the program of engineering education. Leading global experts have described the global drivers of those expanding missions in terms of accessible and inexpensive global communications, global partnerships, globalization, expanding access to higher education, expanding opportunities for talent, cost control, innovation, global "grand" challenges, and leadership challenges. Recent global discussions related to the transformation of engineering education have emphasized that excellent pedagogy in engineering education can benefit greatly from the appropriate use of today's rapidly developing information and communication technology. A wide variety of approaches to the use of ICT greatly improve learning experiences by providing more personalized and effective learning environments for students. In addition, approaches such as MOOCS are rapidly making educational opportunities available to people who otherwise would not have access to college courses. Not only are MOOCS making high-quality educational courses available to an unprecedented number of students, these courses also provide a plethora of student data that helps advance learning science and provides guidance and feedback which can be used to improve student education. The widespread use of ICT in engineering

S. Farrell (✉)
Department of Chemical Engineering, Rowan University, Glassboro, NJ, USA
e-mail: Farrell@rowan.edu

H.J. Hoyer
College of Engineering, Marquette University, Milwaukee, WI, USA
e-mail: h.hoyer.ifees@gedc.info

D.M. Fraser
Department of Chemical Engineering, University of Cape Town, Cape Town, South Africa
e-mail: duncan.fraser@uct.ac.za

R. Natarajan (ed.), *Proceedings of the International Conference on Transformations in Engineering Education: ICTIEE 2014*, DOI 10.1007/978-81-322-1931-6_1,

education is in its early stages. It seems clear from the global ongoing discussions during the last couple of years that this field will continue to develop rapidly and has the potential to transform global engineering education in the relatively near future.

Strongly influenced by the US National Academy of Engineering (NAE), the moderator of the panel focused on numerous critical strategic issues confronted by the global engineering education community. These issues focused on the reality that the combination of global connectivity and accelerating change will increasingly create abrupt disruptions and that universities and colleges will need to fundamentally change their mindset to be able to respond to those changes. The fact that too many colleges are behind in the use of new technologies in their overall culture is deeply disturbing and presents a real challenge. Partnerships globally will be embedded in the engineering culture and will characterize engineering operations for many if not most of it functions. Students must be prepared to effectively engage in the global community. "Working around the world" for them will be similar to "working around the country" which has been the paradigm in the past. It has become increasingly clear that the engineer's reality is global. Engineering colleges should nurture global talent in leadership, management, and "current skills" for professionals across disciplines through targeted degree and certificate programs. They are poised to create university-wide cultures of innovation that would extend beyond their campuses to surrounding communities and beyond. They will engage multidisciplinary teams, usually led by engineering, focusing on great global challenges such as environmental degradation, climate change, and other universal issues. Shaping the university and college cultures so that they can contribute and prosper under these world drivers is the most important and interesting responsibility of university and college leadership today and engineering will play a central role in all of these strategic challenges.

1 Panel Discussion: Transformations in Engineering Education Globally

Stephanie Farrel
Department of Chemical Engineering, Rowan University, Glassboro, NJ, USA

The Corporate Member Council of the American Society for Engineering Education together with IFEES recently surveyed engineering stakeholders around the world to develop a series of competencies and characteristics necessary for a modern engineer to work in a global environment. This study and others send a clear message that we must improve the focus on fundamentals, teach more real-world thinking, increase coverage of emerging areas, teach problem-solving skills, offer more instruction on oral and written communications, instill in students an awareness of ethical, environmental, and social issues, and so on. These findings have significant overlap with those of that National Academy of Engineers [1] and National Science Foundation [2] as well as the outcomes sought by accrediting bodies such as ABET, IChemE, and Engineers Australia. In addition, we must not increase the total number of credits or time required to complete a degree. It is clear that these goals are not achieved by traditional engineering education and that a major transformation is necessary to create an engineering ecosystem that supports the development of global competencies. What is the role of teaching, course, and curriculum reform in this transformation?

In recent years, research on instructional practice and learning in engineering has led to a variety of teaching strategies that effectively enhance learning outcomes. These strategies are accessible to educators through a variety of mechanisms such as journals and conferences, workshops, webinars, and certification programs such as IGIP's ING-PAED International Engineering Educator certification. Yet most instructors continue to rely on traditional and ineffective teaching methods in the classroom. Similarly, some engineering programs have successfully reformed courses and curricula to improve learning outcomes, but faculty members are often hesitant to engage in such efforts.

Over a decade ago, Felder et al. [3] explained that the gap between the current state of knowledge and practice results from the perception and reality that good teaching is not valued in terms of career advancement. The authors made a compelling case for the need to create a positive campus climate for good teaching. Fourteen years later, we are making very gradual progress toward this goal. This panel will explore the following questions? What can be done to accelerate this progress? How can forward-thinking administrators create incentives for faculty and departments to engage in assessment and reform of their programs? How can a positive culture for excellent teaching and educational scholarship be established on campuses?

2 Panel Discussion: Transformations in Engineering Education Globally – Focus on Africa

Duncan McKenzie Fraser
Department of Chemical Engineering, University of Cape Town, Cape Town, South Africa
duncan.fraser@uct.ac.za

I will confine my discussion to the parts of Africa that I am familiar with, which means that I will focus on sub-Saharan Africa. As engineering education in Africa can only be understood within the general education context, I will deal with that first.

All levels of education in sub-Saharan Africa have undergone significant transformations since most nations in this region gained independence from the colonial powers in the 1960s. In the 1970s there was a push for universal primary education, which pulled resources away from higher education. Apart from the changes postindependence, further transformation of the education sector was forced by the structural adjustment programs of the 1980s promoted by the World Bank and the International Monetary Fund, with a further reduction of state funding. In the 1990s this led in many countries to the introduction of fees for tertiary students, which caused some unrest.

Pre-independence, most sub-Saharan countries had only one or two universities, with many of them on expansive campuses which included both student and staff housing, the exceptions being Nigeria and South Africa. There has been an increase in the number of tertiary institutions since then, and more recently quite a significant private university sector has arisen in some countries.

South Africa had a very different education trajectory, achieving independence in 1910. While its education system was always skewed along racial lines, this became entrenched under the apartheid regime in power from 1949 onwards. This regime even introduced ethnic universities for black students. A well-resourced and good quality education system for whites was in stark contrast to an under-resourced and poor quality education system for blacks. This started changing prior to the first democratic elections in 1994, and since then there has been a strong push to equalizing resources and quality right across the system.

The engineering education sector in sub-Saharan Africa has felt the poor resourcing of higher education particularly strongly. Laboratories were often provided by donor funding, but upkeep was a problem and most gradually fell into disrepair. Libraries have been mostly very inadequate, especially in the light of most students not having access to textbooks. There are almost no resources for doing research. Academic staff have been badly paid, to the point that many have had to have a second income to survive, which usually detracted from fulfilling their teaching responsibilities properly. In Nigeria around the turn of the century, the government unilaterally doubled the intake into all engineering schools, without any increase in resources, which led to a decline in standards.

Most teaching in engineering education is by chalk and talk, with no significant student engagement, even though some lecturers are keenly aware of the education issues and trends elsewhere, having largely obtained their PhDs overseas. The opening up of the Internet does provide some hope for significant changes in the future, although this is severely limited at present by the very low bandwidth in most countries.

Engineering education in South Africa is a notable exception to all the characteristics noted above, with good resources and facilities, although the approach to teaching and learning in many places is still very lecturer-focused. Note that black students were only able to study engineering in South Africa from the early 1980s onwards. In my institution there is one very significant transformation of undergraduate engineering education, based on the latest research into teaching and learning.

References

1. Vest C. Educating engineers for 2020 and beyond. Reform Eng Educ. 2006;36(2):38–44.
2. Transforming undergraduate education in engineering, Phase I: synthesizing and integrating industry perspectives; 2013 May 9–10; Workshop Report, National Science Foundation, 2013
3. Felder R, Stice J, Rugarcia A. The future of engineering education: making reform happen. Chem Eng Educ. 2000;34(3):208.

The Important Tool for the Transformation in Global Engineering Education: Mobility

Hasan Mandal

Keywords Global engineers • Mobility • Quality assurance

While global challenges (such as changes in demographic structure, demands for better living conditions, sustainability, climate change, and other changes in nutrition, water sources, health, energy, environment) are becoming a part of our daily life, the importance and the necessity of the graduates who have global attributes are becoming more and more important. Since most of these global challenges are much more related to the engineering issues, engineering education with a global perspective is needed. The competences to achieve global engineering attributes can not only be obtained from one engineering college. Many engineering schools are requesting at least a term for study abroad. In year 2020, 7 million of the students will be mobile globally, and the aim for the European Commission is that 20 % of graduates will take at least study abroad. There are many types of mobility which are as follows:

- Migratory Mobility
- Vertical (Intercycle) Mobility
- Individual Mobility for Study Period Abroad
- Short Organized Group Mobility (summer schools)
- Erasmus-Type (Exchange) Mobility (international and national levels)
- Integrated Mobility (double/joint degree)

One of the important obstacles for the mobility is the recognition of the courses which are taken from the other institutes. Although most of the engineering colleges are applying similar curriculum, many students are having a problem to transfer their knowledge and skills.

H. Mandal (✉)
Vice President (Research and International Relations), Sabanci University, Istanbul, Turkey
e-mail: hmandal@sabanciuniv.edu

R. Natarajan (ed.), *Proceedings of the International Conference on Transformations in Engineering Education: ICTIEE 2014*, DOI 10.1007/978-81-322-1931-6_2,

European approach is to improve transparency, recognition, and therefore the mobility in higher education systems by using the tools of Bologna Process. Bologna Process is simplified by the following figure.

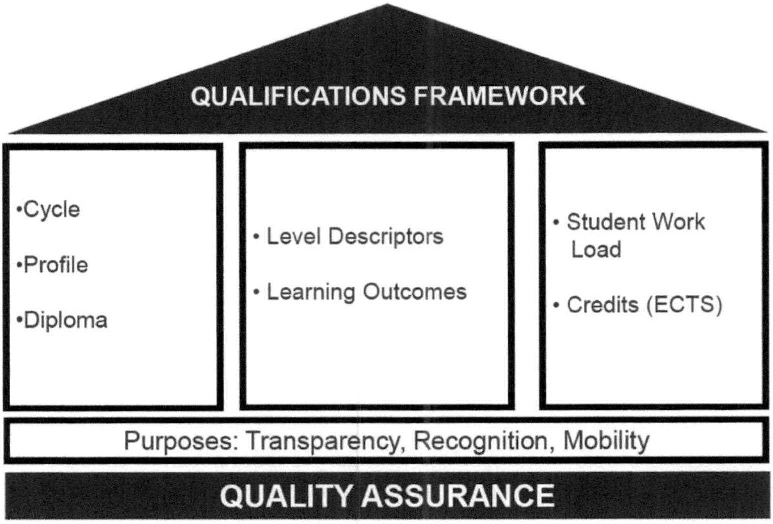

As can be seen from this figure, Qualifications Frameworks (QF) and Quality Assurance (QA) are the main pillars. The others (cycles, profiles, diplomas, level descriptors, learning outcomes, student workload, and credits) are just the tools. Therefore, the understanding QF and QA are rather important.

In this respect, the accreditation experience of engineering programs either by WA or EUR-ACE systems is giving a great advance for the transparency and the recognition and therefore the mobility of the learners in all levels.

Adopting MOOCs for Quality Engineering Education in India

Deepak B. Phatak

Abstract In India, over 5,000 engineering colleges affiliated to different universities offer conventional engineering education. Teachers in colleges do the teaching, but universities rigidly control the programme of study, syllabus, and examinations. The quality of education is a matter of concern. MOOCs (Massive Open Online Courses) permit learners to access and benefit from the teaching by renowned professors. MOOCs offer an unprecedented opportunity to revitalise education. These cause complete disintermediation of the university system, making them very affordable; however, they have several shortcomings in their present form. Students enrolling for a MOOC still have to conventionally study the subject for their degree. Complete absence of physical group activities in a classroom under a teacher's mentoring is another serious issue. Conduct of practical sessions in laboratories is an important aspect of engineering education, for which MOOCs offer no alternative.

We propose a blended MOOC model for adoption in India. It envisages acceptance of MOOC grades by a university towards its degrees. It also stipulates an important role for local teachers, who will use a 'flipped classroom' model of teaching. They will conduct group discussions and problem-solving sessions rather than mere lecturing and locally give and evaluate assignments of which the marks will be factored in the final grade. They will also conduct laboratories where needed. They will thus mentor and guide students, under their charge.

Keywords MOOCs • Blended education • Flipped classroom • Group activity • Teacher centric • Laboratories

---

This invited paper is based on an earlier keynote, delivered at Educon organised by 'Sakal group', at Istanbul in September 2013, and its subsequent versions delivered on several other occasions.

D.B. Phatak (✉)
Department of CSE, IIT Bombay, Kanwal Rekhi Building, Mumbai, 400076, India
e-mail: dbp@it.iitb.ac.in

R. Natarajan (ed.), *Proceedings of the International Conference on Transformations in Engineering Education: ICTIEE 2014*, DOI 10.1007/978-81-322-1931-6_3,

1 Growth of Engineering Education in India

India has about 5,000 engineering colleges today, a 25 times increase from just about 200 colleges 30 years ago. Annual enrolment of students in these colleges is now over 1,250,000. The number of qualified and experienced teachers is highly inadequate. As India moves to increase the GER (Gross Enrolment Ratio) in coming years, the need for quality teachers will increase, and the supply-demand gap will widen.

The quality of engineering education in Indian colleges suffers primarily because of the rigidity of our university system. A number of engineering colleges are affiliated to a university. While teaching takes place in individual colleges, it is the university which controls the syllabus and also the examinations. Teachers are required to teach strictly as per the prescribed syllabus which changes at a snail's pace, if at all. There is a fixed plan for the examinations. The papers include standard options such as 'Answer any 6 out of 10 questions'. The questions themselves are mostly descriptive, such as 'explain something', or 'describe something briefly', or 'write short notes on any two of the following', etc. Such papers are rather sweetly called 'theory' papers. Problem-solving abilities, central to engineering education, do not get adequately tested at all. As an added attraction, the questions are often repeated across examinations conducted in different years. A teacher, who teaches, has no control or even a say in these matters. Someone else is going to set the question paper, and some other one is going to evaluate the answer books. It is not uncommon to find that students routinely bunk classes and attend coaching classes or read guidebooks, which provide specialised training in passing such 'standard' examinations.

I myself studied in a similar college in the city of Indore. It was and is a good college, known for its quality faculty and students, but with a similar make-up. My own experience was that I studied 100 % syllabus in the first year. Having discovered the 'choice' in all examinations, I studied only 60 % syllabus in the second year, still securing more than 80 % marks. In the third year, I enhanced my skills to predict with reasonable accuracy, as to which questions were likely to be 'repeated' from the papers of previous years, in the next examination. I could further optimise on study time. Smarter students resorted to guidebooks and coaching classes, with more optimal outcomes.

For me, things changed drastically when I joined IIT Bombay for my MTech in 1969. My first brush with the IIT style examination was Prof. Jimmy Isaac's 'open-book' test. Time specified was 2 pm 'onwards'! Delighted, I carried six books to the exam. I was shocked to see no descriptive questions and no choice at all. There was simply a set of problems to be solved. Jimmy's style of conducting the examination was also very funny. He kept walking around the class, reading the solutions being worked out by students, occasionally slapping a student on the back, shouting 'not this way idiot, read that chapter from that book, and solve again'. I recall receiving two slaps. I returned the answer book at 6 pm. I was not the last student out, as there was one more joker still struggling. I was completely frustrated, but upon reflection later, I realised that I learned more in those 4 h than I did in a semester back home in my college.

A teacher in the IIT system has absolute control of the subject he or she teaches. Why is it that one Jimmy Isaac succeeds in motivating us to learn better, whereas like-minded teachers elsewhere are not able to do so as effectively? Why is it that the university system has not changed significantly, whereas the IIT system has thrived on continuous experimentation with the process of teaching, learning, and testing?

A large number of very good engineers come out of our engineering colleges. There are several excellent teachers in many colleges. Yet, the IITs somehow manage to produce better trained engineers. While it is true that funding available to most engineering colleges is meagre as compared to IITs, I believe that the true differentiator is the immense autonomy and its responsible and creative use, which one teacher enjoys and the other does not.

2 Role of a Teacher

To learn, to gather knowledge, and to use that knowledge for making life more comfortable, prosperous, and enjoyable are now integral parts of human endeavours. The society recognised very early that the vast amount of knowledge, gathered and accumulated over generations, must be passed to the next generation, for it to be substantially assimilated at an early age by the young humans. It is this urge to educate the young of the species quickly, which prompted humanity to create the institution of a 'teacher', charged with the responsibility of codifying, storing, disseminating, and enhancing knowledge. No other species (barring perhaps the bird Jonathan Livingston Seagull described by Richard Bach!) has created such an institution. In order to support this institution of a teacher, structures were created, which were known in India as ashram or gurukul in old times. Temples, mosques, and churches often served as places for teachers and disciples to congregate. A teacher in those days had the same stature as that of a parent. In fact, the ancient Indian tradition required a student to leave the safe abode of parents at the age of 6 or 8 years and live at the teacher's home for the next 12 years to learn from the guru. Apart from disseminating knowledge, the teacher was also responsible to inculcate values and discipline in the minds of the young. Under the mentorship of a teacher, a student would mature from a young learner into an informed and wise adult, ready to live in, and contribute to, a healthy, prosperous, and happy society.

Over the centuries, these support structures got formalised into schools, colleges, and universities. Curiously, these are now called institutions, and the original 'institution' of the teacher has now assumed the status of a paid employee of these institutions. The stature of a teacher, with the attendant autonomy, is still maintained in some places, such as most of the well-known universities, including our IIT system, and in some renowned colleges. However, in the affiliated colleges and their universities in India, the education process, particularly the engineering education, is governed by hierarchy, procedures, rules, committees, rigid syllabi, a set pattern of examination, and the like. Most teachers have lost the freedom to interact with students as they wish, to decide on the way topics of a subject ought to be discussed, and to conduct the evaluation process as they deem fit.

When a teacher endowed with autonomy teaches in a college or university, as I teach at IIT Bombay, things are different. There is a defined syllabus, which gets updated even as teaching progresses in a semester. What I cover is the operative syllabus, what I set is the examination style, and the grade I give is the final grade accepted by the Senate of the Institute. In the IIT system, two teachers teaching the same subject in different years often teach it completely different from each other, even emphasising different topics in their offerings. This is accepted and respected, because students still learn the essentials as these are always covered by both the teachers. More importantly, students understand that engineering problems are neither 'simple' nor 'standard'. It is such autonomy that permits a teacher to challenge the students' minds, to make them think and learn, and to make them better prepared to face the real world.

A teacher from an affiliated college, on the other hand, is forced to behave like an employee, doing only what the 'system' stipulates. This teacher has been led to believe that the only 'job' to be done is to deliver assigned lectures strictly as per stipulated syllabus, conveniently broken into very well-defined modules, to supervise examinations, and to evaluate answer books (of some other students) in a stereotype fashion. The task of imbuing discipline of thoughts, academic rigour, and ability to attempt solving hard problems is not his or hers any more. Inculcating values about life in the minds of students and mentoring them to become contributing citizens of the society may remain farthest from the thoughts of such teachers. I have interacted with a large number of exceptionally good teachers in such a setup of affiliated colleges. They all feel deeply stifled and tell me that their attempts to discuss real-life problem-solving in the class are often rebuffed by students as being 'out of syllabus'.

Several educational leaders in the country have observed that the general quality and preparedness of teachers in most autonomous institutes, including IITs, and in several university departments, together with their research orientation, make it possible for them to exercise such autonomy effectively. It is said that the teachers from most affiliated colleges do not always possess such qualifications, experience, and research orientation, and therefore only a rigidly defined system is able to get the best contribution from them, in our educational process. While there is some truth in these observations, making a local teacher just a cog in the wheel seriously undermines his or her ability to positively influence the learner's mind. Besides, is not every teacher, even in such rigid setting, desirous of improving oneself, wanting to try solving harder problems, and wishing to participate in meaningful research work? Has the system ever given them sufficient time and training in these aspects?

I believe that there is a way out to ensure a central role to our teachers, while ensuring compliance with global educational standards. This is possible because the technology available today permits us to free the teachers from the mundane tasks assigned hitherto and to resurrect their central position as a guide and mentor to the students under their charge.

3 Onslaught of MOOCs

It has long been predicted that the next disruptive changes due to the Web will occur in education and health care. Research interest in e-learning paradigms is actually quite old. Personalised tutoring systems have been under development for decades. Open-source educational content of high quality is now available in searchable digital formats. A knowledge seeker today has unprecedented access to quality content.

3.1 Working of Present MOOCs

The operations of any MOOC offering resemble the sequence of activities followed in a conventional teaching of that subject. A MOOC is announced by some well-known professor of an institute, through one of the companies offering MOOCs. The announcement includes the broad course content, prerequisites if any, and the course start date and duration. Learners from all over the world register online, without having to satisfy any eligibility criterion. This is what makes the course 'open' and accessible to any learner. From the start date, the course lectures can be accessed on the website. These are video-recorded lectures. Activities such as practice problems and trial quizzes are embedded in these recorded lectures. Usually, a weekly schedule is prescribed. During each week, the participants listen to the lectures, solve practice problems, and give trial quizzes. The participants are required to appear for online tests every week or fortnight. The test questions are either multiple-choice or short-answer type and are automatically graded. Scores for these weekly or fortnightly tests are accumulated and are factored into the final grade. There could be additional online examinations, similar to our conventional mid-semester and end-semester examinations, again graded automatically and counted towards final scores. Some courses require assignments (long answers) to be submitted online. Evaluation of such assignments is often done through a peer evaluation process. Automatic grading of such long answers is not yet possible. It is being attempted through specially designed machine-learning algorithms, currently under research and development.

A suitable IT infrastructure is put in place to run MOOCs. This comprises of a cluster of servers running a sophisticated server application. This cluster is connected to the Internet with a fairly large bandwidth so as to support millions of online users across the world. The software permits creation of all the learning material including lecture videos, handouts, and question banks for quiz and test examinations. It provides online registration and access control to course components through an extensive LMS (learning management system). There are server modules to permit participants to upload their assignments and conduct online examinations. A good discussion forum is provided for participants to pose questions and difficulties and to receive explanations and clarifications from the professor as also from other participants. Last but not the least, there are modules to capture and store actual use of the system by participants. Since all interaction is online, the back-end software is able to capture the entire activity by each participant.

When and how long a participant watched a lecture are known; when did a participant attempt a quiz is recorded; which answers were given against each question is also recorded. A MOOC offering is thus able to collect huge data on the academic behaviour of the participant throughout the duration of the course. This data is almost a gold mine and feeds into extensive research happening in the areas of pedagogy. The feedback is used to make subsequent offerings better.

The online examinations and quizzes are ordinarily not supervised. Most offerings instead have an 'honour code' which participants agree to. Basically, the code stipulates that a participant will attempt all examinations on one's own efforts. A certificate is issued to all participants who complete the course successfully, based on this honour code. Some courses also offer an 'audit option', which does not require participants to pass examinations. Some of the MOOC offerings are adopted by conventional universities, where the course material is from the MOOC content, but the university conducts its own examinations.

The revenue model in such offerings is very interesting. Initially, all companies offered these MOOCs free of charge. Once the credibility is established in the minds of participants, there are now attempts to charge for certification. The charges currently vary between $50 and 100 for each course. Since the examinations are not supervised, there is a limited value perceived by participants, and the number opting for such paid certificates is rather low as of now. Some offerings give a 'validated identity' certificate. What is done is that the identity of a person is initially recorded based on an appropriate identity proof. In all interaction during the course, the participant is required to use a webcam through which it is ensured that the same person is doing all the work. Additional checks, such as a 'typing profile' test, are done for such interaction. In any case, the certificates so issued seem to have a limited value proposition as of date and do not compare with the value that a participant associates with a regular grade from an established university. Employers and universities also do not consider these to be equivalent to a regular course.

3.2 Limitations of the MOOCs

Online courses and makers of e-learning artefacts sometimes seem to imply that an individual student can complete all studies in self-learning mode. There is no evidence yet to suggest so. However, the use of technology for providing quality education on a truly massive scale is quickly coming of age as seen in the last two years. MOOCs are rapidly gaining popularity. Some notable efforts by Coursera, edX, and Udacity are well documented. A Wikipedia article [1] provides a balanced view of pros and cons of MOOCs. IIT Bombay has already announced offering of MOOCs in select subjects from 2014, in collaboration with edX [2].

I am personally not convinced that the MOOC offerings in the present form alone will succeed in reshaping the existing Indian higher education system. My opinion is based on some prominent concerns. The first is that, in the MOOCs offered so far, the completion rate has been very low. A survey published by Katy Brown [3] indicates that in the online courses with registration of over 100,000 participants, the completion rate has only been between 3 and 13 %. I have tried to analyse possible reasons for such huge dropout, from the perspective of an Indian student who is already enrolled in a university and is studying in an affiliated college. The second concern is the serious disruption which the present MOOCs cause in the human interaction prevalent in the present system. I am particularly worried about the complete removal of any role for a teacher in a college, which to me is a disastrous consequence of the disintermediation of the entire university system, which MOOCs bring about in their present form. Another important concern is the lack of exposure to practical sessions conducted in engineering laboratories for the students, as a part of regular curriculum. I present my arguments on these concerns in the next subsections.

3.2.1 Dilemma Faced by Students Enrolled for a University Degree

This analysis is based on the different styles of education in the two systems of Indian engineering education – in the general university system ('A'), covering over 4,800 engineering colleges in India, as against what takes place in IITs, NITs, and similar well-known institutes, and in several university departments ('B').

Consider a student of 'A' studying a certain subject as a part of the university curriculum. The student is required to appear for and pass the examinations conducted by the university. The marks and grades obtained therein are counted towards the degree which the student eventually earns. These are vitally important for the student, since all employers look at these carefully before offering a job or before even calling someone for a job interview. Assume that the same subject is now available as a MOOC. The student may register for the MOOC offering, hoping to learn the subject better, because some renowned professor from a well-known institution 'B' is offering it.

After some initial online interaction with this MOOC, the student may find that the level of questions in the tests or assignments is far more difficult than the ones found in his or her college and university examinations. At the least, these will be significantly different from the familiar style of examination papers. As mentioned earlier, the standard university papers require long answers which can be reproduced from memory. Such questions are rarely, if at all, asked in question papers of 'B'. The 'A' paper will typically announce upfront a choice such as 'Answer any five questions'. There is no such choice in papers from 'B'. Many teachers in 'B' offer open-book examination, a practice simply non-existent in 'A', but will be used with a MOOC.

The student must prepare for own university examinations in any case. If the learning from the MOOC does not directly enable the student to prepare better for the standard university examinations, then the efforts on MOOC may appear to be a waste of time, and a guidebook will perhaps be a far better choice for the student!

3.2.2 Disintermediation of Human Interaction

Consider the difference in the engagement with students in these two models. In the conventional system, students interact daily with teacher and with each other on numerous occasions. Such group activity is a vital component of education. The learning that happens during such group interaction actually goes much beyond just the academic learning. The importance of teamwork and leadership is better understood by students through group activities. Lifelong friendships are formed during these interactions, resulting in a strong human network which often proves very useful in future professional life. Participation in games and sports, cultural activities and debating competitions, and technical festivals contributes to the total education which one receives during the college days. Indeed, there is much more to education than just the excellence in understanding and mastering the technical subjects; but, for the present discussion, let us limit our analysis to just the academic learning.

In a classroom environment, one does not learn by listening to lectures alone, but has a chance to seek explanation from the teacher or from fellow students. The interaction is largely verbal, sometimes even in one's native language. Such help is more comforting and acceptable than the discussion offered by a MOOC, where both question and answer need to be typed in a discussion forum and only in English. Group assignments and projects, where a team of students work together for several weeks to solve a rather complex problem, are considered an important learning component and are routinely practised not just in IITs and NITs but also in many colleges where some internal evaluation is permitted by the affiliating university. Most teachers additionally provide handholding support to weaker students in a class, through individual or group discussions.

All these academic interactions are facilitated by the teacher teaching that subject in the college. The importance and usefulness of organising and facilitating such group interactions is critical to education, independent of how good or bad the 'lecturing' is. A conventional MOOC has no place for any of these.

3.2.3 Practical Laboratory Sessions

Every engineering curriculum emphasises inclusion of practical sessions in laboratories. Students get to work on actual equipment and instruments, learn to make measurements of all concerned operational parameters through observations and readings in well-defined experimental setups, and understand how things work in practice. These hands-on sessions in laboratories form a crucial component of engineering education the world over. This is where students dirty their hands and learn all the real-world aspects of engineering.

In the current MOOC offerings, such practical sessions are impossible to arrange in any online environment. I have seen some online courses where video-recorded 'experiments' are included as part of recorded lectures. There are also many e-learning tools, which try to explain such experiments using animations, simulations, and video clips. All these at best complement, but do not substitute hands-on experience. Literature on MOOCs is indeed silent on this aspect, almost to the extent of having a blind spot in this regard.

I believe that practical understanding of real instruments, gadgets, and equipment forms an important component of engineering education. The infrastructure needed to learn these things exists only in established colleges and in engineering industry. It is also important to remember that practical sessions are group activities, again organised and supervised by a teacher in a college. Without these, learnings from MOOCs will not provide complete understanding of 'engineering'.

4 A Blended Model for Adopting MOOCs in India

IIT Bombay, in a unique experiment, proposes to introduce a blended model which will address the limitations of MOOCs in their present form.

4.1 Acceptance by Universities

First and foremost, it is necessary to resolve the dilemma faced by a student of an affiliated college, mentioned earlier. Should time be spent in learning from a MOOC, or is it better to stick to tricks in passing university examinations with good grades? We propose that a university should accept the grade obtained by the student in a MOOC to be the grade for an equivalent university course. This addresses the problem of duplicate efforts. In fact, the student will not have a choice, since the university will declare the course to be taught as MOOC only. This has been attempted in the USA through a notion termed SPOC, but our model has uniform applicability to a large number of universities and affiliated colleges which teach courses with a near similar syllabus, defined on the lines of a model syllabus approved by AICTE (All India Council for Technical Education). The fact that teaching system and examination patterns are similar across most of our 5,000 engineering colleges affiliated to over 300 universities will help us achieve this scale. Incidentally, the chairman of AICTE, Dr. Mantha, announced last year that AICTE may permit up to 15 % of the credits of a degree to be obtained through MOOCs. Another committee appointed by UGC is currently addressing the issues related to MOOC offering in Indian higher education system.

One major issue in the MOOCs is that the simple 'honour grade' given by a MOOC will not be acceptable. So the online examination will have to be supervised and so certified. This is already being done in some global offerings. In our model, we suggest that the teacher in the college be made responsible for such proctored online examination.

4.2 Human Interaction

The second component of our suggested approach is more important. It attempts to transform the local teacher into a true mentor and guide to students. In our model, regular classroom engagement in the college still continues, but instead of lectures, the teacher will hold discussion and problem-solving sessions, as in the 'flipped classroom'. A teacher will now be able to spend much more time with the students in these discussion sessions, helping weak students through more elaborate explanations and challenging the better performing students with harder problems. In fact, teaching will be personalised to a much greater extent. Additionally, we propose that the local teacher should give individual assignments and offer course projects to be done by small teams of students. The project activity will typically be in the second half of a semester long course, spanning 4 to 6 weeks. Each project will be offered, evaluated, and graded by the teacher, and not by the college or university administration. The score for each student in the assignments and group project will be directly factored into the final grade awarded by the MOOC. The weightage of the MOOC online score to the teacher's score is proposed to be 80:20 or 85:15. This makes the local teacher an important stakeholder in the entire process. More importantly, by restoring the autonomy of the teacher in this fashion, I believe that the stature of the local college teacher will be resurrected and will be brought closer to the one originally intended for teachers.

Some senior leaders of our education system have pointed out that there may be resistance from teachers who are used only to their current style of functioning, who may genuinely be concerned about their preparedness in assuming this new role. They will now have to face questions and difficulties posed by the students all the time. They will have to solve more difficult and complex problems. They have to innovate in assigning and guiding course projects. They now have to take responsibility of marks they award to each student out of the 15 or 20 % of the total marks. Well, with autonomy comes such responsibility!

The only way to reduce and remove this reluctance is to train these teachers extensively in their new and more important role. For those Indian universities which accept our MOOC grades, we propose to train teachers of all affiliated colleges in this approach. We know we can, because such training on large scale is precisely what we do well, through the T10KT (Train 10 Thousand(K) Teachers at a time) programme, being conducted by IIT Bombay [4] and IIT Kharagpur [5], under NMEICT (National Mission on Education using ICT). In fact, encouraged by our project review committee chaired by Prof. Nigavekar, ex-chairman of our University Grants Commission, we propose to conduct our T10KT workshops themselves through a blended MOOC – a real bootstrapping approach.

4.3 Practical Sessions

Laboratories are used to conduct practical sessions for students and form a critical component of engineering education. Such practical sessions constitute a mandatory part of several courses for an engineering degree. While the present MOOCs cannot handle this aspect at all, our blended model will be able to resolve this issue. The local teacher-in-charge of the course will also supervise and conduct these sessions. Indeed, when designed in conjunction with a common MOOC syllabus, some uniformity in quality of practical sessions across the colleges and university can be achieved.

5 Blended MOOCs from IIT Bombay

The proposed blended MOOC model is based on the experience gained in experiments done by some of us in running a flipped classroom at IIT Bombay. In the blended model, students study available digital material on their own, including viewing of pre-recorded lectures. They then attend classroom tutorials and discussion sessions and practise solving problems in groups, under the supervision of a teacher. This 'flipped' classroom model has been tried in IIT Bombay and at some other sister institutions. It was found initially that the students do not always listen to pre-recorded lectures on their own as advised and are often unprepared for classroom discussions and problem-solving sessions. This was addressed by conducting quizzes in every class. My colleague Prof. Kannan Moudgalya has nearly perfected this approach. In the picture (Fig. 1), he is using the Clicker application on Aakash tablets [6] to conduct a quiz.

A senior alumnus of the institute, Ruyintan Mehta, attended one such class during his visit to IIT Bombay. He was convinced that this approach will definitely make students more attentive. His reasoning is very simple – students are always more alert when giving examination of any kind, when marks are involved! Another colleague, Prof. Kameswari Chebrolu, has successfully conducted her course on computer networks using the flipped classroom. She reports that students want video-recorded lectures of shorter length.

When IIT Bombay decided in early 2012 to start offering MOOCs, a committee deliberated at considerable length for choosing a suitable partner from amongst the well-known alternatives and decided upon edX. Our decision to partner with edX was influenced by several factors. One is that edX is a non-profit organisation,

Fig. 1 Conduct of Clicker-based quiz using Aakash tablets

initially funded by contributions from MIT and Harvard. There is nothing wrong with for-profit companies, but our comfort level is greater otherwise. Another is MIT's OpenCourseWare history. This is in consonance with our own desire at IIT Bombay to open source as much knowledge content as possible. Last but not the least is their decision to open source the edX platform itself, for future collaborative development. This was critical because we had started planning to offer the blended MOOCs to Indian students, and a proven platform which could be modified and deployed for Indian usage will significantly reduce development efforts and time.

We have decided to offer four of our courses in the next academic year through edX. We plan to try at least one course for adoption of this blended model by multiple universities. I have held preliminary discussions with several academic leaders of Indian universities and engineering colleges, and the initial response has been very encouraging. We will probably offer such a course initially only to a limited number of students of select colleges and universities, restricting it to say about 200,000 students with involvement of about 5,000–8,000 teachers. We will expand after fine-tuning our approach from the experience. In future, it should be possible to engage all of 1,125,000 Indian students being admitted to first year of engineering degree, for example, in some core course offered as a MOOC. It is my dream that such scale in quality education is achieved within the next 2 years.

There are a large number of additional issues which need careful application of mind and foolproof solutions. I will just mention one of the technical issues in information management, which we will face: the LMS (learning management system) will have to permit representation of groups of students hierarchically arranged. One must be able to identify a student as belonging to the specific administrative node in this hierarchy and maintain all records of interaction with local teacher, scheduled submissions made locally, marks awarded by the local teacher, etc. MIS reports will have to be generated appropriately and separately for a batch of students and the associated teacher, for students of a department in a college, for the college as a single group, and for the entire university. Apart from a GUI for data entry, file uploads by local teachers will have to be permitted. It is not a difficult problem, but one which must be addressed by appropriate modification to the platform used.

There are other practical problems in implementing such blended solutions across India, where a stable and continuous availability of high-capacity Internet bandwidth to engineering colleges may be difficult to guarantee. We propose to build a distributed implementation, with a local server in each college, so that the students can continue to work in a LAN environment in case of Internet outage. IIT Bombay proposes to use a suitably modified edX platform to initiate our blended MOOC offering.

6 Conclusion

In any mass-scale adoption of technology for education, the teaching-learning process must be carefully defined and tuned. Issues of examinations, evaluation, and certification need to be addressed. The value proposition that such technology-enabled education offers must be examined from the point of view of all stakeholders – students, teachers, institutions, and employers. The transition in our education towards greater use of technology may take several years. Technology-enabled learning will continue to mature and evolve, but one thing is certain: enhancing quality of education on such a large scale, as needed in India and in several developing countries, cannot be achieved without extensive and appropriate use of technology, adopted with ingenuity and focus.

We have analysed the present MOOCs in this context and have proposed a blended model for engineering education in India. Our vision is to explore, examine, and solve pedagogical and technical issues and establish the best possible model of blended learning, for Indian education system. We sincerely hope that a successful implementation of this model will lead to its adoption in other spheres of higher education, with possible eventual use in our school education as well.

Acknowledgement I am grateful to a very large number of people, including several colleagues for their significant contributions. These contributions have come by way of their actual experiments and the wisdom which they shared with me. The successful conduct of T10KT programme by our project teams at IIT Bombay and IIT Kharagpur, supported by the National Mission of Ministry of Human Resource Development, has given us confidence to try this new approach on blended MOOCs. Since the contributions by people are too numerous to list, I will just say 'thank you all', for now.

I wish to specifically thank the Sakal foundation, organisers of Educon, where I could put together and present my initial thoughts, the constant follow-up by Prof Krishna Vedula, and Shri Ashok Kalbag for ICTIEE, which has enabled me to give a concrete form to many of my ideas for a blended MOOC implementation in India. I am grateful for the help by our project advisor Shri Prakash Vaidya in organising and proof-editing this paper.

References[1]

1. http://en.wikipedia.org/wiki/Massive_Open_Online_Course
2. https://www.edx.org/school/iitbombayx
3. http://www.katyjordan.com/MOOCproject.html
4. http://www.it.iitb.ac.in/nmeict
5. http://www.nmeict.iitkgp.ernet.in
6. http://www.akashlabs.org

[1] The concerned pages of these Web references were downloaded on 30 December 2013.

Outcome-Based Accreditation and ABET

Michael K.J. Milligan

Abstract ABET is a not-for-profit, nongovernmental accreditor of degree granting programs in applied science, computing, engineering, and engineering technology. ABET uses a peer review process to evaluate academic programs at the associate, bachelor, and master's levels. This paper discusses the evolution of the ABET accreditation system from an input-based system which focused on what was being taught, to an outcomes-based system which focuses on what is being learned.

Keywords ABET • Accreditation • Assessment • Outcomes-based • Quality assurance

ABET is a not-for-profit, nongovernmental accreditor of degree granting programs in applied science, computing, engineering, and engineering technology. ABET uses a peer review process to evaluate academic programs at the associate, bachelor, and master's levels.

ABET is a federation of 33 professional and technical societies that represent their specific professions. These member societies play several key roles within ABET, including participation on the Board of Directors, developing program criteria, recruiting program evaluators, and approving policies.

ABET accreditation is an 18–20 month process that consists of the preparation of a self-study by the program being evaluated, an on-site visit by the ABET Review Team, due process response, decision-making meeting, and official notification of accreditation action to institution/program. Programs new to ABET accreditation may be required to undergo a Readiness Review prior to officially applying for accreditation.

Accreditation provides value to numerous constituents, including students, industry, institutions, and the general public. Accreditation plays a key role in

M.K.J. Milligan Ph.D., P.E (✉)
Executive Office, ABET, 111 Market Place, Suite 1050, Baltimore, MD 21202, USA
e-mail: MMILLIGAN@abet.org

R. Natarajan (ed.), *Proceedings of the International Conference on Transformations in Engineering Education: ICTIEE 2014*, DOI 10.1007/978-81-322-1931-6_4,

society as public safety for engineering, and other regulated professions is of critical importance.

The accreditation status of a program is often used by students in selecting quality programs, as it is an indication that the programs will adequately prepare them to enter the profession and enhances their licensure/employment opportunities.

Similarly, employers also use accreditation as a guide during the hiring process, as it ensures that the graduates of accredited programs have attained the necessary knowledge and skills required to begin practicing their profession. In addition, it enhances mobility of the global workforce, especially with the increase of multinational corporations. Accreditation also provides industry an opportunity to guide the educational process by participating on the Industry Advisory groups of the programs.

Institutions with ABET-accredited programs demonstrate a commitment to quality education which helps attract strong students and faculty. ABET-accredited programs enjoy global recognition, therefore facilitating the mobility of graduates and students.

Since its inception, ABET used input-based accreditation criteria which dictated the major elements of the program, including program curricula, faculty, and facilities. However, in the 1990s, working closely with industry, ABET began developing a new set of outcomes-based criteria, called Engineering Criteria 2000 (EC2000). Unlike the input-based criteria, which focused on what was being taught, EC2000 places the emphasis on what the students learn.

As a result of ABET's collaboration with industry to identify the necessary skills required for the profession, a set of student learning outcomes, a key component of the criteria, were created. The student outcomes specify the knowledge and skills that graduates of ABET-accredited programs should possess as a result of their matriculation through the program. Programs must demonstrate that graduates have attained these student outcomes.

Assessment is a key component of the outcomes-based accreditation system. Assessment is the systematic collection, review, and use of information for the purpose of improving student learning and development. Programs must be assessed to measure if and how well student outcomes are being attained. The information gathered during the assessment process must be used to improve the quality of the program as part of the Continuous Quality Improvement (CQI) process.

CQI is the ongoing process to improve the quality of students' educational experience. It is a systematic, documented and repeatable process which assesses the performance of the program against the relevant criteria. Accreditation is part of the CQI process as it verifies that programs meet certain level of quality. In order to obtain ABET accreditation, a program must have an established and effective CQI process.

EC2000 was implemented over three accreditation cycles, including a pilot cycle. Implementing EC2000 required changes to the program evaluator and team chair training structure and content. Previously, each individual ABET member society, responsible for recruiting the program evaluators, was also responsible for training them. With the shift to EC2000, ABET's constituents expressed the need for a centralized training. As a result, the Partnership to Advance Volunteer

Excellence (PAVE) was introduced. PAVE training consists of a combination of web-based training and face-to-face training.

Initially, EC2000 was met with significant resistance from faculty, as it presented a fundamental change to the traditional process. The input-based system was perceived to be more direct and did not require a strong CQI process. However, after experienced the effect of the outcomes-based system on the quality of the programs was recognized by both faculty and ABET evaluators.

To determine and measure the impact of EC2000, in 2002 ABET commissioned the Center for the Study of Higher Education at Pennsylvania State University to conduct a study as a means of verification of the outcomes-based accreditation model. The main focus of the study was twofold. First, it aimed to determine the impact, if any, EC2000 had on student learning outcomes in ABET-accredited programs and institutions. The second focus was on determining what impact, if any, it had on organizational and educational policies and practices that may have led to improved student learning outcomes.

The three-and-a-half-year study concluded that the 2004 graduates were better prepared to enter the profession than their 1994 counterparts. The study also found that the program curricula also placed more emphasis on professional skills and that there is higher level of faculty support for continuous improvement. However, there was no clear indication that EC2000 actively encourages innovation in program delivery or content. As a result of the success of EC2000, many specialized and institutional accreditors have adopted outcomes-based accreditation globally.

ABET has been engaged globally for over 30 years. In 2007, ABET began accrediting programs outside the USA and has since accredited 365 programs. ABET is also engaged globally through its memoranda of understanding (MOU) and mutual recognition agreements (MRA).

ABET signs MOUs with other accreditors of technical education in many parts of the world. The purpose of these MOUs is to provide a platform for collaboration and exchange of information on matters relating to quality assurance and technical education. In addition, the MOUs also serve as a mechanism to assist accrediting bodies to develop their systems.

In addition to accreditation of programs and MOUs, ABET is also engaged in five mutual recognition agreements (MRAs). MRAs are agreements among mature accreditation bodies from around the world. MRAs recognize the substantial equivalence of accreditation systems and, therefore, also recognize the programs accredited by those bodies. The purpose of MRAs is to establish recognition of the educational base technical professionals receive in different countries. This is especially important with ABET's existing MRAs covering the engineering, engineering technology (2-year and 4-year programs), and computing disciplines.

Another aspect of ABET's international engagement consists of its membership in the International Federation of Engineering Education Societies (IFEES) and the Global Engineering Deans Council (GEDC). Membership in these organizations provides ABET with a platform to engage in discussions regarding technical education to better understand the needs of technical programs around the world.

ABET understands that undergoing an accreditation review requires extensive time and preparation. ABET offers various resources such as assessment workshops and webinars on a range of relevant topics. In addition, each year, the ABET Symposium is a great resource for programs interested in ABET accreditation. The Symposium offers tracks on program assessment, accreditation preparation, innovation in technical education, and program evaluator development.

While great strides have been made in accreditation, many challenges still lay ahead. As education continues to change due to, among other things, globalization and advances in technology, accreditation must also adapt. ABET and its constituents continue to work together to address some key challenges, which include the increase of distance/online education, nontraditional students, and globalization of the workforce.

Present Status and Challenges Ahead for Engineering Education: Global and National Perspectives

R. Natarajan

Abstract The present status and the challenges and opportunities ahead for engineering education, both at the national level and at the global level, are explored in this paper. There are many distinctive and unique characteristics of the new millennium which place new demands on the engineers we produce. We are living and working in a world that is characterized by volatility, uncertainty, complexity, and ambiguity, requiring innovative strategies for educating our engineers. The emerging accreditation systems and processes and the Washington Accord Graduate Attributes provide guidelines for producing global engineers ensuring global mobility. Nationally, we have several policy pronouncements over the years, which serve as drivers of change. The recent global university rankings have attracted worldwide attention of the academic leaders and other stakeholders and are serving to determine university policies and practices. We have to create research-intensive institutions in our country if we have to figure among the top of these lists. Some of the issues of concern are discussed. Professional ethics and human values are important for engineers in the practice of the profession and need to be inculcated through both curricular and extracurricular means. As for the agenda ahead, the destination is reasonably clear, and there are several national policies and reports which serve as the road map, but several hurdles and challenges have to be overcome in order for us to bridge the gap between the potential we possess and its actual attainment.

Keywords Accreditation systems • Graduate attributes • Global rankings • Autonomy • National policies • University-Corporate sector partnership • Research-led universities • Global engineers • Professional ethics

R. Natarajan (✉)
Chairman, Board for IT Education Standards (BITES), 9, UNI Building, Miller tank Bed, Vasanth Nagar, 560052 Bangalore, Karnataka, India
e-mail: prof.rnatarajan@gmail.com

R. Natarajan (ed.), *Proceedings of the International Conference on Transformations in Engineering Education: ICTIEE 2014*, DOI 10.1007/978-81-322-1931-6_5,

1 Introduction

As we navigate through the twenty-first century, we are confronted with a world characterized as a "VUCA" environment. Coined in the late 1990s, the military-derived acronym stands for the volatility, uncertainty, complexity, and ambiguity. It is also credited to Bob Johansen of the Institute for the Future (www.iftf.org). In this world, achieving success in any human endeavor requires the positive flip side of VUCA: vision, understanding, clarity, and agility (Cisco WebEx – CLO Webinar). Central to this world is the challenge of change – not only in magnitude but also in its velocity and acceleration.

Apart from change, there are several distinctive characteristics of the new millennium: population explosion in developing countries; demographic disparities; depletion of natural resources (energy, materials); environmental degradation (air, water, soil); demand for mass education; significant impact of technology on society, commerce, lifestyle, education, and entertainment; widening of disparities in society (digital divide, literacy divide, education divide, prosperity divide, technology divide, haves and have-nots); social tensions; and globalization.

We are living and working in a changing world – we are witness to worldwide changes in technology; in education and training; in work, professions, and careers; and in management and organization of institutions. There have been significant changes in the practice of engineering as a profession, driven by environmental considerations, customization, availability of sophisticated diagnostic and computational tools, a wide choice of materials, and innovation as the basis of global competitiveness.

On change, Angela Merkel recently said: "The Question is not whether we are able to change, but whether we are changing fast enough." On resistance to change, Maurice Maeterlinck, the Belgian Nobel Laureate in literature, observed: "At every crossway on the road that leads to the future, each progressive spirit is opposed by a thousand men appointed to guard the past." Educators, like the rest of us, tend to resist major change. But change we must! In this age when speed is of the essence, there is no time to waste!

Engineering education has been recognized universally as the vehicle for national development and global competitiveness. It needs a major rethink in the "flattening" world: On the challenge for engineering education, former US Secretary of Education Richard Riley has said: "We are currently preparing students for jobs that don't yet exist, using technologies that haven't yet been invented, in order to solve problems we don't even know are problems yet."

2 Some Contemporary Issues in Technical Education

2.1 Liberalization of Technical Education

This is a favorite complaint of the private sector. The implications are as follows: freedom to charge fees, freedom to select and admit students, freedom to appoint faculty and staff, freedom to collaborate with foreign institutions, freedom to expand capacity, freedom to diversify program offerings, and freedom to operate multiple shifts. Some of these are indeed reasonable, but some are not!

2.2 Autonomy

It relates to decision-making on whom to teach (students), what to teach (curriculum), who will teach (faculty), and how to assess (exams). Autonomy has multiple dimensions: academic and administrative – managerial, financial, and functional.

Some of the contemporary issues in engineering education are as follows: the disconnect between those who teach and those who learn, those who recruit and those who seek jobs, and those who frame policies and those who function within the system; theory and practice of assessment of learning; and performance on the job. How do we bridge these disconnects?

Salman Khan (of Khan Academy fame), in his recent book *One World Schoolhouse*, has revisited some fundamental assumptions of student learning: "How do people actually learn?" "Why do grown-ups sense such a disconnect between what they studied in school and what they do in the real world?" "Lessons should be paced to the individual student's needs, not to some arbitrary calendar; what we now have is *One-pace-fits-all* curricula and education."

The current buzzwords in Indian education are the following: access, affordability, collaboration, diversity, employability, equity, excellence, expansion, global engineer, inclusion, innovation, outcomes-based education, and quality.

3 National Policy Guidelines for Bringing About Changes in Higher Education: Indicative List

There have been several policy guidelines over the years:

- Periodic National Policies on Education – 1986–1990
- Science Policy Resolution (1958), Technology Policy Statement (1983), Science and Technology Policy (2003), and the recent Science, Technology and Innovation Policy
- NKC (National Knowledge Commission) Recommendations
- Yash Pal Committee Recommendations
- Twelfth Plan Document
- Periodic Recommendations of Review Committees of IITs and NITs
- The RUSA (Rashtriya Uchchatar Shiksha Abhiyan – National Higher Education Mission, January 2013)
- The Pending Bills in the Parliament

There are lessons to be learned:

- Implementation is as important as planning.
- We must learn from the success or failure of past actions.

4 Drivers of Change in Higher Education

Quality can be driven by regulations and accreditation; AICTE regulations demand fulfillment of prerequisites, such as faculty (both numbers and qualifications), infrastructure, etc.; accreditation can be a driving force for change as far as quality is concerned; the mandatory peer review mechanism enables an outside-in view of the institution. The Washington Accord and ABET, which prescribed outcomes-based teaching, learning, and assessment paradigm, are driving several changes in our technical institutions. Of course, there are other change agents, as well, for example, the founding mandate, national and state policies, perspective plans, and leadership.

5 A Summary of the Features of Indian Engineering Education

Our technical education system is characterized by several unique features: huge size, many asymmetries and divides, diversity (of many types), variable quality, frequent changes of policy, many international collaborations, many strengths and weaknesses, many opportunities and challenges, and lessons learned and not learned.

6 Some Mismatches in Our Technical Education System

- *Employer's Requirements and University Offerings*
    - The design of university curricula is principally undertaken by university faculty; though the academic bodies contain industry members, they seldom commit time.
    - The affiliated system precludes quick response to rapidly changing and emerging requirements.
    - Inadequate industry-institute interaction and communication preclude an understanding of each other's roles and requirements.
- *Education Received and Employment Sought*
    - Students' choice of engineering is based not on aptitude and a desire to pursue an engineering career, but on other considerations.
    - The general demand is for IT-related branches; the traditional branches represent only a fallback option.
    - Whatever the branch of specialization, the goal is an IT-related job; the IT employers also encourage this trend.

- *Manpower Creation and Employment Generation*
  - Neither reliable estimates nor mechanisms exist for manpower planning – unplanned quantitative expansion is justified as fulfilling aspirations of young persons to pursue engineering degrees (not necessarily careers!).
  - Many graduates are unemployed for extended periods of time, and many others are underemployed.
  - While self-employment and entrepreneurship are laudable goals, the ambience is unsuitable for promoting them.

7 The Symbiotic and Synergistic Partnership Between University and Corporate Sectors

Two words highlight the interaction between university and industry – symbiosis and synergy. Not only do the two partners depend on each other and derive mutual benefit from the partnership, but also the overall impact can be much greater when the two partners function in phase and in resonance. It is necessary to create a win-win situation for both partners.

Two of the principal stakeholders in the engineering education system, viz., the students and employers, are increasingly demanding relevance and pragmatism. The university, representing the intermediate system, has an important role to play in this venture, and the faculty must develop an appreciation for engineering practice. The universities should earn revenue through their services; otherwise they will always be tied to the apron strings of government.

Industry cannot get instant knowledge any more than university can get instant experience. We need bridges between engineering in university and technology in industry through researchers and educators with commitment. We need policy initiatives by both industry and university to create interfaces to facilitate the interaction.

8 Industrial Training

There are essentially two facets of industrial training: what is given to engineering students during their degree program before they are employed and what is given to working professionals while they are employed. The objectives, means, and strategies are as different as those between initial formal education and continuing education.

9 The Emerging Significance of Global Rankings

In the recent past, three global university rankings, viz., Shanghai Academic Ranking of World Universities (ARWU), Times Higher Education (THE), and Quacquarelli Symonds (QS), have emerged and are exerting a serious impact on the policies and practices of universities worldwide. In her 2011 book *Rankings and the Reshaping of Higher Education: the Battle for World-Class Excellence*, Ellen Hazelkorn [2] has pointed out that "Rankings demonstrate the new environment of higher education, and act as a driver of change." "The extent to which these changes are productive or useless is still controversial, but HEIs are worried about their impact on the reputation of their institution, individuals, and the country as a whole."

In as much as rankings provide a comparative assessment and evaluation of universities worldwide (in the case of global rankings) and nationally (in the case of national rankings), they provide an opportunity for benchmarking. The essence of benchmarking, according to the International Benchmarking Clearinghouse is: "the practice of being humble enough to admit that someone else is better at something, and being wise enough to try to learn how to match, and even surpass them at it."

It is important to keep in mind the limitations of these rankings. Malcolm Gladwell [1], in his article in *the New Yorker*, has explored the question of "what college rankings really tell us." Different colleges have different perceptions of their educational function, raison d'étre: For Yale, it is "to assemble the most gifted group of freshmen it can – Selectivity; Penn State sees its educational function as serving a wide range of students –Diversity." "There is no *right* answer to how much weight a ranking system should give to these two competing values. It's a matter of which educational model you value more."

10 Creating Research-Led Universities in India

Mashelkar, in his recent Book *Reinventing India* [3], draws attention to widely accepted examples of globally recognized WCUs: Harvard, MIT, Oxford, Cambridge, Stanford, Caltech, Princeton, etc. He asks: "Do we have any in India? "The answer is no. "Can we create them?" he asks next. Yes is the answer. He asserts that it is crucial to create "a great intellectual environment."

He has identified five key factors prevalent in WCUs:

1. Absolutely uncompromising pursuit of excellence in both teaching and research
2. Continuous thrust on not only "working" at the frontiers of research but "creating" new frontiers or doing S&T that will "lead" and not "follow"
3. An uncompromising attitude towards selection of both faculty and students – going for the very best and that too on a globally competitive basis. Rigor to be employed strictly in the promotion and retention of the most talented and accomplished faculty
4. Undying commitment to the true institutional autonomy without any political interference whatsoever
5. Full understanding of the route to "Lakshmi" through "Saraswathi," not only understanding the wealth creation potential of knowledge but also creating an "ecosystem" in which this can happen most effectively

10.1 What Do We Need to Do to Create World-Class Research-Led Universities in India?

Mashelkar has given the following recipe for creating WCUs in India:

1. "Political will and ambition and massive funding; China is a shining example.
2. An understanding that in future, the private sector will have to play a critical role in higher education. Entry of foreign universities of great prestige (not all and sundry) has also to be viewed with an open mind.
3. We need to put explicit demand on S&T. This demand has to be created by both industry and society. The 'i' in industry must stand for 'innovation' and not for 'imitation' or 'inhibition.'"

We need to recognize that there is nothing like "intellectual democracy." He quotes Lotka's law: "1 % of population usually carries 90 % of intellectual energy." We must benefit by throwing open the doors of our academic institutions to foreigners.

In the special social circumstances that we have, we must recognize the importance of an astute balance of "expansion, inclusion, and excellence." Balancing "expansion, inclusion, and excellence" in the Indian context requires great will, skill, and innovation, he says.

11 Issues in Research Promotion in our Technical Institutions [4–6]

- *How do you promote R&D in a nonresidential institution*?
  The faculty and students (may) have to travel long distances, both ways, and have little time or energy to devote to R&D, particularly experimental work in labs. The only recourse may be to pursue computational or theoretical R&D, with much work done at home, after office hours, and on weekends. This requires commitment and dedication.
- *How do you promote R&D in a UG institution*?
  The faculty and students have no exposure to R&D.
- *How do you promote R&D in a nonmetro institution?*
  Typically, there are no industries or R&D labs in the vicinity, and opportunities for interaction with such organizations are nonexistent.
- *How do you promote R&D in an institution with very few faculty members having PhD?*
  There is no possibility of peer-to-peer interaction in R&D matters.
- *How do you promote R&D in an institution with heavy teaching loads – because of faculty shortages, excessive admissions, and temporary faculty working on contract basis?*
  R&D requires concentration, and sufficient time must be available for pursuing R&D.

12 Some Issues of National Concern [7, 8]

Here is a list of some current issues of national concern:

I. National Issues

- Regional imbalance in quantitative expansion
- Small proportion of institutions of quality; there are no Indian institutions in the top 100 world-class university rankings.
- Even within states, institutions are concentrated in the vicinity of metros/cities; there is not much demand in rural institutions.
- Manpower assessment and planning are largely absent, at both the national and state levels

II. Institution Related

- Self-financing institutions depend solely on student fees, and hence fees are high.
- Most of the institutions are only engaged in teaching; there is not much research, consultancy, and industry interaction.
- While institutions of excellence resist quantitative expansion, the majority of institutions clamor for more branches, more seats.
- There is a tremendous scarcity of academic leadership, at both the institutional and departmental level. Academic administrations largely learn on the job – and learning involves making mistakes.
- Commercialization (for-profit institutions).

III. Industry Related

- Industry does not value PG qualification.
- Employment generation is not commensurate with outturn of graduates.

IV. Faculty and student related

- Lack of adequate and well-qualified faculty.
- A large proportion of faculties do not possess PhD/MTech qualification.
- We don't produce enough PhDs/MTechs annually.
- PhD/MTech intake capacity is small (while there has been a rush for quantitative expansion at the UG level, there is no corresponding enthusiasm for such expansion at the PG level, for obvious reasons).
- Graduates are not attracted to the teaching profession, and hence, there is not much demand for PG qualification.
- The QIP capacity is not significant.
- In the prevalent affiliating system, the role of the teacher is relegated to a minor one; curricula remain un-updated for long periods; while international and global practices employ innovative ET and IT tools to improve teaching effectiveness, we are still laboring with outdated teaching techniques and systems.
- Student projects are largely inappropriate, ineffective, and unimaginative.

13 Globalization and Global Engineers

As far as globalization is concerned, there are differences in perceptions of developed and developing countries: In the matter of economy, globalization means favorable trading opportunities and expanded markets for developed countries, while for developing countries, it means deregulation, enhanced privatization, and currency integration. In the matter of education, for developed countries, globalization means enhanced markets for educational products, processes, and services, while for developing countries, it means study opportunities for those who can afford it and competition to local institutions. In the matter of employment, for developed countries, globalization leads to erosion of jobs and entails competition from low-wage workforce from developing countries, while for developing countries, it enables offshore jobs and provides opportunities for short-term employment abroad.

The ASEE CMC (Corporate Member Council) Special Interest Group for International Engineering Education has identified over the past two years the attributes of a global engineer: "What skills and experiences will today's engineering students need to develop, while in school and throughout their careers, to successfully compete in today's and tomorrow's global workplace?"

The study has identified eight attributes: engineering science fundamentals; engineering; context in which engineering is practiced; communication; teamwork and leadership; flexibility; curiosity and desire to learn, for life (show initiative, inquire, and learn); and ethical standards and professionalism.

14 Instruction in Professional Ethics and Human Values

More and more engineering educators and professionals are realizing that just as students require technological knowledge and skills before going out into the world of work, they also need instruction in professional ethics and human values. Because engineering, just as law and medicine, has a significant interface with, and impact on society, engineers have ethical obligations to the public at large. Engineers also have ethical obligations to their employers, clients, and the profession. Ethical issues and problems arise in situations involving conflicts of interest, responsibility for public health and safety, trade secrets and proprietary information, gifts from contractors and others, honesty in research and testing, etc.

Learning about ethical issues in a classroom environment has the advantage that students have the opportunity "to hear different perspectives, allowing them as a group to generate solutions that might never have occurred to them individually. It also gives the students a sense of ethical unity, and confidence that they need to stand up for principles in the work world."

While there is no quick-and-easy formula for solving ethical problems, there are some tools to help in the process. The 1990 Conference on Ethics in Engineering Education has devised a seven-step procedure for moral decision-making.

The steps include: recognition and definition of the ethical problem, investigation of the facts, formulation and evaluation of alternatives, gathering support, and negotiating with others. It is cautioned that because handling ethical problems is never a cut-and-dry process, these steps should be viewed more as a checklist than as an algorithm.

Despite increased interest in engineering ethics, many engineering professors feel that there is no room for the topic in their already-packed syllabi. Currently, engineering schools have found two basic ways to accommodate ethics instruction in their curricula:

1. By introducing freestanding courses in ethics
2. By integrating ethics across the curriculum

15 The Gap Between Extant Potential and Performance

Particularly with reference to India, we often bemoan the conversion of the extant potential to performance – in several sectors: education, economy, and GNP. In this context, the following observation of Mahatma Gandhi is very significant:

> The difference between what we do and what we are capable of doing would suffice to solve most of the world's problems.

16 The Road Ahead

The destination is clear – we aspire to be globally competitive and have academically reputed systems and processes. As for the road map, there are several national policies and opportunities for benchmarking with world-class universities. But the road is strewn with many impediments and obstacles – lack of commitment, delays, policy paralysis, complacency, and many other usual suspects. In this age of rapid change, speed is of essence; we must start acting now!

17 Concluding Remarks

This paper has dealt with a wide range of international and national perspectives on technical education. The issues discussed have both long-term and short-term implications. The imperatives of globalization and internationalization demand prompt and purposeful responses and proactive initiatives if we want to achieve competitiveness, productivity, and national prosperity.

References

1. Gladwell M. The order of things: what college rankings really tell us. New Yorker. Feb 14, 2011.
2. Hazelkorn E. Rankings and the reshaping of higher education: the battle for world-class excellence. Basingstoke: Palgrave Macmillan; 2011.
3. Mashelkar RA. Reinventing India. Pune: Sahyadri Parkashan; 2012.
4. Natarajan R. The importance of research in technical education. AICTE Newsl 2013;March.
5. Natarajan R, Anandakrishnan M, editors. XXI century priorities in engineering education. New Delhi: Macmillan India; 2010.
6. Natarajan R, Misra A. A journey of the growth of engineering education in India. In: Guptara P, editor. Where did we go wrong? The role of 200 years of modern education (1813–2013) in India's triumphs and tragedies. (to be published).
7. Natarajan R, Vedula K. Engineering education in India – current status and future directions, Invited paper. Chin Acad Eng. 2010.
8. Natarajan R, Shetty NR, editors. Multiple stakeholder perspectives of higher education. New Delhi: Macmillan India; 2012.

Innovation and Entrepreneurship in Engineering Education

Gururaj 'Desh' Deshpande

Abstract The twenty-first century poses several grand challenges: clean water, clean air, climate change, energy, biodiversity, and sustainability. We need to *invent*, *innovate*, and *entrepreneurially* implement solutions to solve the mega problems. The solutions to these problems will not be solved by a select few. Billions all over the world will have to contextualize the solutions to their problems and implement them. As we democratize innovation and entrepreneurship, the twenty-first-century Engineering Colleges will play a central role in inventing, innovating, and educating *entrepreneurially* students who can solve problems. We need Engineering Colleges all over the world to be involved in the process.

Keywords Innovation • Entrepreneurship

The twenty-first century poses several grand challenges: clean water, clean air, climate change, energy, biodiversity, and sustainability. We need to *invent, innovate*, and *entrepreneurially* implement solutions to solve the mega problems. The solutions to these problems will not be solved by a select few. Billions all over the world will have to contextualize the solutions to their problems and implement them. As we democratize innovation and entrepreneurship, the twenty-first-century Engineering Colleges will play a central role in inventing, innovating, and educating *entrepreneurially* students who can solve problems. We need Engineering Colleges all over the world to be involved in the process.

G.D. Deshpande (✉)
Sparta Group

Sycamore Networks

Tejas Networks, USA

MIT, Cambridge, MA, USA
e-mail: desh@spartagroupllc.com

R. Natarajan (ed.), *Proceedings of the International Conference on Transformations in Engineering Education: ICTIEE 2014*, DOI 10.1007/978-81-322-1931-6_6,

After a short tenure in teaching at universities, Desh has pursued an entrepreneurial career since 1980. Thirteen years ago, when he joined the MIT as a Board Member in Cambridge, Massachusetts, he got reconnected back to the academic world. Together with other like-minded individuals at MIT, he has been experimenting with ways in which to make the innovation at MIT have a bigger economic and societal impact. After the initial success of the approach at MIT, my wife and I started a similar effort in India targeted at fostering innovation in Indian universities to make a difference in social entrepreneurship. The effort in India is now eight years old. This paper summarizes the results of these two efforts to encourage discussion on how such approaches can be used to further the Engineering Education in the twenty-first century.

1 The Effort at MIT

MIT with its culture of "mens et manus" (mind and hand) already excels in making the technological innovation useful to the world. However, there is an opportunity to make it better. The "Deshpande Center for Technological Innovation" was set up at MIT in 2002 with funding of $20 million. The Kauffman Foundation did an extensive study of this center and the von Liebig Center at the University of California, San Diego. They found the models to be successful in promoting academic innovation to directly address real-world problems. Lesa Mitchell, Vice President of Advancing Innovation at Kauffman Foundation, is facilitating similar centers at other universities. The following are some of the lessons from this effort.

1.1 Insight

Researchers who work on applied research always think about how their ideas can impact the world, both economically and socially. There is no lack of desire on the part of the researcher to see the impact. The idea has to be directed towards solving a burning problem in the world to have an impact. You cannot mandate innovation. However, bringing the knowledge of what the world needs to the innovator will help the innovator make choices that increase the probability of impact.

Research at universities now is where the engineering practice was in industry a few decades ago; an engineer designed the product and the salesman then went looking to sell the product. Today in industry, an engineer only starts designing the product after fully understanding what the customer needs. However, in the current practice of research, the researcher innovates and patents the idea, and then the technology licensing offices try to find applications for the patents. The center at MIT has found that bringing the practitioners and the researchers together early on changes the culture of innovation. There is a lot of give and take between what is possible and what is worth solving to come up with an innovation that can have impact. Injecting relevance early in the process of innovation increases the probability of that innovation having a bigger impact on the world. The faculty members fully embrace this idea. In fact, MIT has created a new course called I-Team which brings engineering and MBA students together to explore how to target ideas at appropriate markets. This has been a very popular course.

Researchers, in the campus environment, are idea generators. When a new idea comes along, the researcher is not only excited by the elegance and novelty of the idea but is also excited about where it can be useful. A few months down the road, the researcher will have ten more ideas that have sprung up from the original idea. Left to himself or herself, a researcher will choose to pursue one of those ideas that make sense to his or her own environment. However, if the researcher is connected with practitioners, he would have the benefit of relevance to pick an idea that has a better chance of creating a bigger impact.

1.2 Process

The center achieves its mission through several approaches: grant program, catalyst program, innovation teams (I-Teams), and events. Faculty members apply for grants to the center twice a year. The funding from the center enables the faculty and their students to pursue exciting new avenues of research on novel technologies that could have a significant impact. These grants are selected by a panel of faculty and business leaders and are selected based on potential for impact, technical merit, team considerations, and time frame. Ignition Grants of $50,000 are awarded to fund proof-of-concept explorations, and Innovation Program Grants are awarded in the range of $50,000–250,000 to build on existing innovations at MIT and bring them closer to commercial viability. The objective of the funding is to nurture ideas with market potential and reduce the uncertainty around them so that an external party would invest in the technology. This could occur through various means, such as a VC-funded start-up or licensing by a company. In addition to the funding, the grants bring with them publicity, mentoring, and connections with the business community. Volunteers from the business community are central to achieving the center's mission of helping MIT innovators achieve market impact. Catalysts are a highly vetted group of individuals with experience relevant to innovation, technology commercialization, and entrepreneurship. Catalysts provide individual contribution to the center and do not represent any company interests in their role as catalysts. Catalysts are chosen based on the following qualifications:

- Experience in commercializing early stage technologies and/or mentoring researchers and entrepreneurs and industry expertise.
- Willingness to proactively provide assistance to MIT research teams.

- Willingness to abide by the time commitment, confidentiality, and conflict of interest guidelines.
- Commitment to the interests of MIT researchers and the center. The I-Teams (innovation teams) program provides an action-based learning experience for graduate students where students evaluate the market potential for research projects being conducted at MIT and develop "go-to-market" strategies.

The center hosts a variety of events to bring together minds from the MIT and business communities. The IdeaStream Symposium, held each spring, is our largest event. The symposium is intended to showcase new MIT technology, educate the business community about leading-edge new technologies, and facilitate connections between VCs, entrepreneurs, industry, and MIT innovators. These symposia are by invitation only. The center also collaborates with other programs on and off campus to promote a variety of events to enhance innovation within the community.

2 Results

So far the center has reviewed close to 500 proposals submitted by the faculty. The center has supported about 100 projects with about $10 million in grants. The grants have resulted in 30 startups that have raised over $450 million in capital. Over 60 faculty and 500 students have participated in the program, and a new course has been designed to capture the process of taking the innovation to the market. The process can be summarized by three actions: select, connect, and direct. Active participation of the business community in all three activities is essential for success: selecting appropriate research to fund, connecting the innovator to the marketplace, and directing them when they need help.

3 The Effort in India

India is a vast country and several educational institutions have come up over the last 60 years. There are institutions like the Indian Institutes of Technology, All India Medical School, national law schools, Indian Institutes of Management, and Indian Institutes of Public Health which have international recognition. They get their strength from being able to select a few thousand from millions and holding smart students together in a campus environment.

The faculty members at these institutions are dedicated, but lack the research infrastructure. This is improving, but has a long way to go. India has to deal with two issues: millions of people and low affordability. If innovation is to have an impact, Indian universities have to bring new solutions at very affordable prices to millions. Indian scientists have shown promise; the Indian space program has shown results with modest investments.

Indian industry understands the opportunity; for example, Telecom companies add 10 million cell phones a month that can be recharged incrementally 2 cents at a time and Tata Motors sells a $2,000 Nano car. India graduates approximately 400,000 engineering undergraduates from approximately 3,000 colleges. We picked BVB Engineering College, a college in a small town called Hubli, to see how we can bring innovation to this institution. The college has 4,000 engineering students. The students are bright and the teachers are dedicated. However, the students are taught to study and do well in exams. Students spend their time preparing themselves to provide canned answers to questions posed at the exams at the end of the academic year. There is a total lack of innovation in that education system. Students walk around with hundreds of problems all around them, but they do not know that they have the ability to solve them.

In order to connect the students and the faculty to the problems in the surrounding area, we built a center for social entrepreneurship in the campus. For the last eight years, we have been funding approximately 70 programs by NGOs (nongovernmental organizations) in the areas of education, agriculture, livelihood, and health. The enthusiastic participation by the students to get involved in these programs has been overwhelming. The participation started from the engineering campus and now has spread to all the surrounding colleges. Last year 10,000 students conducted 2,500 projects. The enthusiasm of the students has now spread to the young faculty. They are bringing technological as well as other innovations to projects.

Problems exist in this world because people do not see a solution. Therefore, you need innovation to solve even the simple problems. The innovation always comes from a fresh perspective. The Social Entrepreneurship Center provides that fresh perspective to the students and young faculty by bringing their bright minds and the problems together. The results of this effort have been excellent. For example, a kitchen was built in the city that serves midday meals to 185,000 school children every day. The kitchen uses good management tools and technology to maintain very high-quality standards and has managed to serve nutritious meals that the kids love for 12 cents a meal. After the program was optimized in this city, it is now duplicated in other parts of India. This program currently serves 1 million meals to school children every day. By using a similar model as the center at MIT, difficult social problems are being solved in India by connecting the academic innovation of universities to the needs of the real world.

4 Conclusions

The twenty-first century poses several grand challenges: clean water, clean air, climate change, energy, biodiversity, and sustainability. The universities can play a central role in coming up with solutions to these problems. The solutions will need "Eminent Technological Innovations" and thousands of innovative ways to localize the eminent innovations. The universities of the twenty-first century have to create educational programs and an echo system that injects a lot more innovation and entrepreneurship into their institutions to remain relevant to the world.

Attributes of Engineers and Engineering Education for the Twenty-First-Century World

P.E. Seeram Ramakrishna

Abstract The twenty-first century is witnessing ubiquitous technology; workforce economically active beyond normal retirement age, learning lifelong, interacting, and mobile globally; businesses with markets and supply chains spanning the world; growing demand and enrollments in professional engineering programs; diverse engineering education providers; worldwide comparison of engineering schools; and international recognition of accredited engineering programs and mobility of students. The surge in demand for professional qualifications in emerging nations led to the scenario of undergraduate engineering education being akin to liberal arts education with diverse career options. Recent decades also saw widening of engineering field covering many new disciplines which flourished at the interfaces with other disciplines. Yet in every country, employers are able to find only a fraction of graduates with requisite knowledge, skills, abilities, and attitudes suitable for employment. In other words there is a need to rethink the way engineers are trained and to align the curriculum to the needs of the twenty-first-century world. Sharing of best practices and continual improvement in all aspects of engineering education are desired for preparing engineers for the twenty-first century.

Keywords Engineers • Engineering education • Graduate education • Researchers • Innovation

---

P.E.S. Ramakrishna, FREng (✉)
Global Observatory for Learning, IFEES, GEDC and AAET, National University of Singapore, Singapore, Singapore
e-mail: seeram.rk@gmail.com

R. Natarajan (ed.), *Proceedings of the International Conference on Transformations in Engineering Education: ICTIEE 2014*, DOI 10.1007/978-81-322-1931-6_7,

1 Twenty-First-Century World

Forecasting is often fraught; nevertheless, a best attempt must be made as the foresight enables engineering education providers to understand what they need to do now to better prepare for the future. In the end how the future will pan out depends on what they do now.

During the past 15 years the world saw economic expansion, financial crises, and unrest in certain regions. The following 85 years will likely witness them repeating for varied reasons which include emerging geopolitical conditions, population growth, internal and external migrations, further urbanization, competition for natural resources, changing climate, and energy mixes of nations. Longevity coupled with improved healthcare and birth rates suggests that the active world population will rise, contributing to continued economic expansion. The number of centenarians is projected to grow to 15 million from the current half a million. Future generations will be healthier, motivated, and economically active for periods longer than the current generations, upon formal tertiary education. In other words, the economically active life span upon graduation from a university will be in the range of 50–60 years. It will no longer be unusual for people over 65 to be working full time. Holding more than ten jobs in a working career will be the new normal compared to the current average of two to three jobs during the career of an individual. According to *The Atlantic* (theatlantic.com), the US National Bureau of Economic Research scientists analyzed the data from the lives of great inventors and Nobel Prize winners in recent decades and inferred that most people achieve big breakthroughs in their careers when they are in their late 30s [4]. Innovators have been peeking slightly later in life in part because today's scientists have more to learn than their predecessors did. The same trend has been observed with poets and painters. More women will seek higher education and pursue careers. Moreover, the future generations are likely to experience rapid technological innovations that will shape living conditions and businesses. Hence, the education and skills they acquire at the tertiary institutions and lifelong learning (learn, unlearn, and relearn) are far more important than ever before. Countries and universities need to provide for growing need for continuing education. Environmentally friendly, autonomous transportation will be ubiquitous. Air, water, and soil are likely to become cleaner with greater attention to the cradle-to-grave cycle of manufactured products. Bio-factories for medicines, nutritious food, and environmentally benign products will be the new normal. Advances in technologies and pursuit of sustainable development slant energy mixes of nations toward renewables and lower resource consumption per capita. Widening income inequalities will likely erode social values. The proportion of single, two-person, or three-person families will increase around the world. As the families become nuclear, future generations will rely on public systems (universities, community organizations, cultural organizations, welfare groups, religious organizations, etc.) to develop character and acquire social values and interpersonal and life skills, which were acquired naturally in respective family settings in the case of earlier generations. The world will be increasingly multi-polar with diverse geopolitical influencers. Innovation nodes will be globally distributed and interconnected. High-quality educated workforce, scientific research capacity and capabilities, and an ecosystem conducive of innovate products, services, and governance are key enablers of competitiveness of countries.

2 Universities and Engineering Education Providers

Global comparison of thousands of universities around the world emerged in the beginning of the twenty-first century. The world ranking lists were generated by comparing each university as a whole with others. More recently, the universities are compared at the individual discipline level. Increasingly universities and engineering schools are pressured to move up in the rankings. Institutions around the world are chasing the top few universities, which take fewer students and primarily are research universities. This is leading to isomorphism of institutions and losing the sight of the broader mission, i.e., advancing social mobility and facilitating social engineering. It is important to recognize that the students have diverse needs, and societies need diverse education pathways. Efforts should be aimed at developing incoming students to their full potential and evaluate the education providers based on the extent of value added. The twenty-first-century world needs university education to facilitate students' disciplinary competence as well as social conscience and compassion. It is more important to differentiate education offerings and continually improve the education quality by benchmarking with peers and aspirant institutions meaningfully. Quality assurance or accreditation with appropriate benchmarking serves better purpose than the mere rankings (see Table 1).

Table 1 Comparison of rankings and accreditation

	Rankings	Accreditation
Assessor	Market-oriented organizations (QS, THE, ARWU, etc.)	External peers
Performance review	Based on databases	Comprehensive – internal review, self-evaluation report, benchmarking, site visit, and audit
Outcome	Ranking lists	Accredited or not accredited
Criteria	One dimensional; based on ranking criteria and weights	Multi-dimensional; based on evidence and demonstration of outcomes
Contents	Well received by the media, public, and students	Difficult to comprehend
Purpose	Business motive	Quality assurance and enhancement of standards; social accountability
Perception	Independent; benchmarking tool for national and global comparisons	Limited by peers' perception of performance
Period	Annually	E.g., once in 5 years
Effect	Better students, staff, resources, and influence	Mobility of graduates; resources

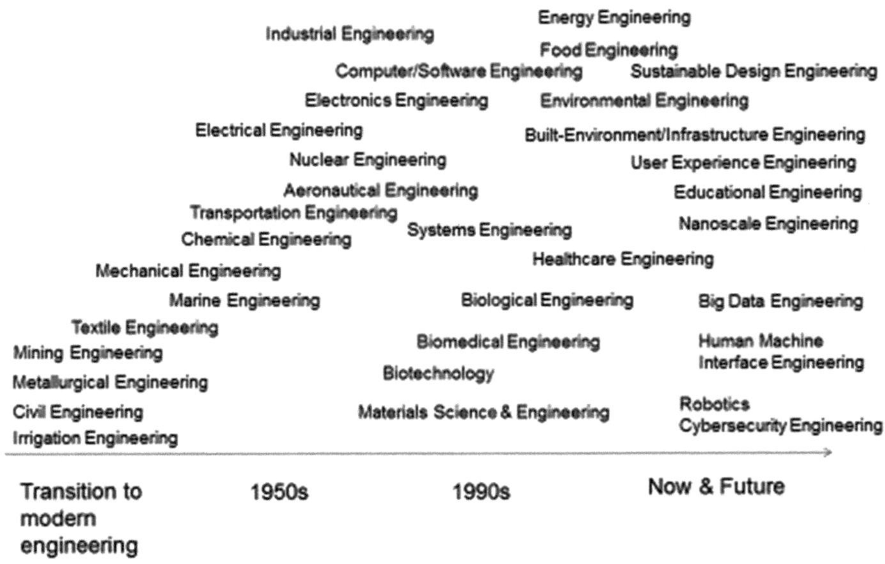

Schematic 1 Dynamic nature of engineering field

Universities are to find ways and means to facilitate an interdisciplinary learning environment via flexible curriculum which is necessary for the future generations.

There are about 6,000 engineering education providers worldwide enrolling millions of students. Majority of them came into existence over the last 50 years. They are diverse in terms of nature of institution (public, semipublic, private; teaching, research, and research intensive), size (boutique, medium, and large enrollments), management structures, funding mechanisms, range of programs, location, mission, vision, mode of delivery (full time, part time, online, mixed mode), level of resources, quality of students, quality of faculty members, flexibility of programs, national and international partnerships, etc. The engineering field itself is not static over the centuries (see Schematic 1), and it has several new disciplines to address the needs and expectations of industries, businesses, economies, and nations.

For the aforementioned reasons, the engineering education providers has a pivotal role in enhancing (a) relevance of engineering education to the changing businesses, economies, and societies; (b) teaching abilities of academics; and (c) tailoring of curricula and pedagogical approaches to the future engineers' needs and expectations. Emerging countries have the largest proportion of engineering education providers, programs, and students, yet they lack engineering education research centers and programs. In emerging countries many engineering education providers are experiencing shortage of adequately qualified faculty members. Hence, they are employing (a) graduates without Ph.D. degree and (b) Ph.D. degree holders with strong research credentials as faculty members. In both scenarios these faculty members lack the pedagogical skills of teaching students and deeper understanding of engineering education. There is a need for training future faculty members with

Table 2 Examples of training of engineering educators

Institution	Degrees	Notes
Virginia Tech Department of Engineering Education	Certificate	Engineering Education Graduate Certificate
University of Michigan Engineering School: Center for Research on Learning and Teaching (CRLT)	Certificate	Certificate in Engineering Education Research
University of Georgia Faculty of Engineering: Engineering Education Research Cluster	Ph.D. Masters	M.S. and Ph.D. in Engineering with Area of Emphasis in Engineering Education Research
Texas A&M University College of Engineering	Ph.D.	Ph.D. in Interdisciplinary Engineering
Purdue University School of Engineering Education	Ph.D.	Ph.D. in Engineering Education
Carnegie Mellon University Engineering School: Program in Interdisciplinary Educational Research (PIER)	Ph.D.	Ph.D. in Interdisciplinary Educational Research via Mechanical Engineering program
Arizona State University Ira A. Fulton School of Engineering	Ph.D. Masters	Example: Ph.D. Mechanical Engineering with concentration in Engineering Education
Aalborg University (Denmark) UNESCO Chair in PBL in Engineering	Ph.D. Masters	PBL in Engineering Education

engineering education knowledge and skills. In this regard future engineering faculty members are encouraged to take a concentration or minor in engineering education while doing Ph.D. in engineering research (e.g., see Table 2).

2.1 Attributes of Engineers

The world's leading businesses are global and integrate supply chain across the continents. They source finances and talent worldwide and customize products to the markets they serve, as opposed to the past practice of standardized products made in one place and sold worldwide. This has been possible due to the availability of educated workforce, investments, and innovation capabilities in various regions of the world; globalization of trade, finance, and talent; and availability of modern transportation, information, and communication technologies (ICT). These enterprises require engineers with strong communication skills, open to diverse approaches, and who can lead multicultural teams. Such engineers are termed as "global engineers." However, currently most of the engineering graduates are employed by the businesses and industries whose operations are local and regional, not global. These industries need engineers with strong technical skills relevant to the local market and societal conditions and connections to sustain and grow businesses. Perhaps we may choose to refer to them as "mass engineers." What is common between the two types of engineers? They must be proficient in ICT skills as well as problem-solving skills with real-world experience. They need to be innovative, relevant, persistent, and lifelong learners to keep their jobs in the twenty-first-century world. Both will become engineer-leaders with experience and sustained achievements in respective ways [3].

Recently, the American Society for Engineering Education (ASEE) produced a report entitled *Transforming Undergraduate Education in Engineering*, with an aim to develop a clear understanding of the qualities engineering graduates should possess and to promote changes in curricula, pedagogy, and academic culture needed to instill those qualities in the coming generation of engineers. Key knowledge, skills, and abilities desired of future engineers include (a) engineering science fundamentals, (b) engineering, (c) context in which engineering is practiced, (d) communication, (e) teamwork and leadership, (f) flexibility, (g) curiosity and desire to learn for life, and (h) ethical standards and professionalism.

Accreditation bodies with motivation to enable mobility of engineers and working professionals via international recognition of qualifications have documented the desired outcomes of engineering education. For example, the Accreditation Board for Engineering and Technology (ABET) [5] lists various competencies, namely, (a) ability to apply knowledge of math, science, and engineering; (b) ability to design and conduct experiments and analyze data; (c) ability to design a system component or process; (d) ability to function on multidisciplinary teams; (e) ability to identify, solve, and formulate engineering problems; (f) understanding of professional and ethical responsibilities; (g) ability to communicate effectively; (h) understand the impact of engineering solutions in a global and societal context; (i) lifelong learning; (j) knowledge of contemporary issues; and (k) ability to use techniques, skills, and engineering tools necessary for engineering practice. Similarly, Washington Accord (WA) member nations identify certain abilities expected of a graduate from an accredited program. They are grouped under broad headings, namely, academic education, knowledge of engineering sciences, problem analysis, design and development of solutions, investigation, modern tool usage, individual and teamwork, communication, the engineers and society, ethics, environment and sustainability, project management and finance, and lifelong learning.

2.2 Undergraduate Engineering Education

Recent decades saw tremendous expansion of engineering enrollments in several countries. This led to a wide variety of engineering education providers, students, engineering disciplines, curricula, and pedagogical and delivery approaches. The unbridled expansion led to (a) compromise in the quality, rigor, and standards of education; (b) graduates without jobs that fit their training; and (c) education unable to inspire students who seek personal fulfillment and motivated to contribute to the society. The real-world demands on engineering graduates and engineering

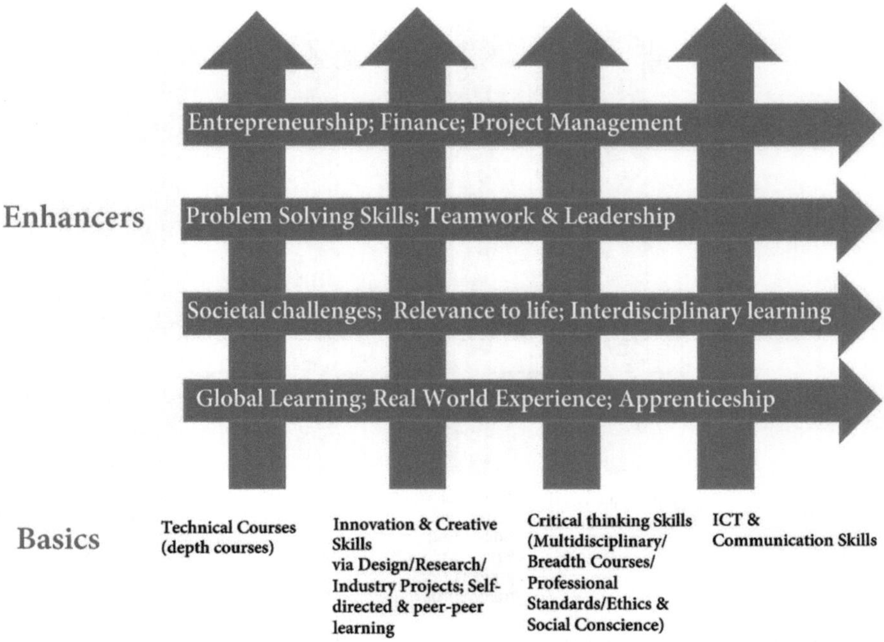

Schematic 2 Key features of training undergraduates

education have drifted apart. Engineering education should be updated and fine-tuned to nurture engineers with necessary skills and knowledge as shown in the schematic [1, 2]. Transformation in engineering education is needed in order to enhance students' interest and to improve relevance of engineering education to the respective societies (see Schematic 2). Experimental learning involving societal challenges, hands-on approach, internships in industry, teamwork, internships abroad, and entrepreneurship opportunities are identified to enhance students' interest in engineering.

3 Training Graduates and Researchers (Training the Trainers)

In our technology ubiquitous world, youth dream of scientific research-led innovation career as a worthy option. They desire to help people and society by developing new innovations. Hence, they are showing interest in graduate education and research. Every year about 20 % of university undergraduates upon graduation pursue this path either in home country or overseas. Their number is growing around the world. During the graduate studies and research training, they hope to equip themselves with right knowledge, skills, and experience. Key question on their

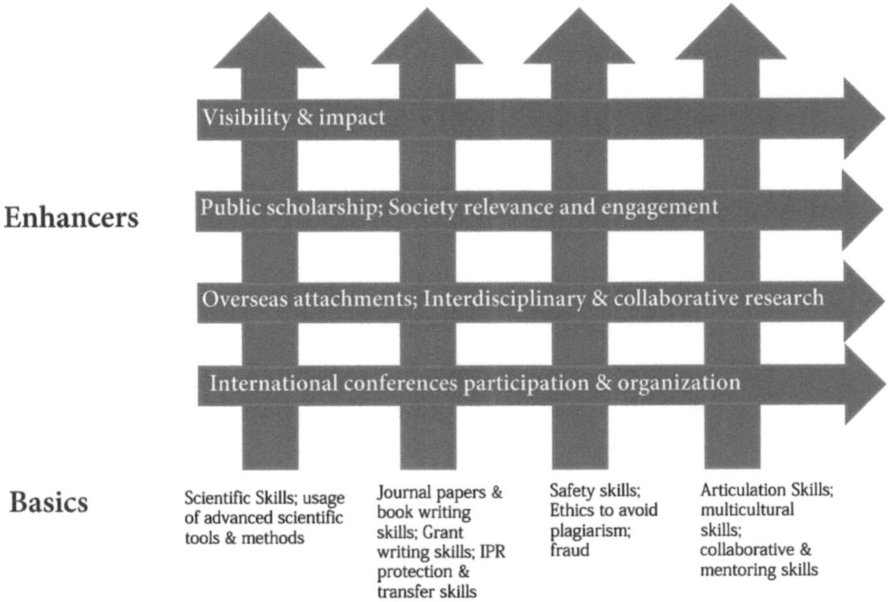

Schematic 3 Key features of training graduates and researchers

minds is how best to prepare for the research-led innovation careers in a globalized competitive, cooperative world.

Most nations and universities focus their attention on the undergraduate education and relatively less or no attention is given to the training of quality graduates and young researchers. Graduate training is also equally important as a good proportion of them will become academics who will train and inspire future generations. Many of them will also become chief scientists, chief technology officers, and chief information officers in diverse companies around the world. They contribute to the organizations and support their chief executive officers in foresight, strategies, vision, and new innovative products. Along the way some of them will become entrepreneurs by setting up start-up companies based on their own experience and research-led innovations. Hence, from these diverse considerations, there is a need for training quality graduates and junior researchers in a more holistic way and future ready, and what should that be?

Universities have been paying attention to the students' abilities to use modern tools and methods and advanced scientific knowledge and skills imparted in specialized domains. However, solving many challenging problems such as clean water, energy, transportation and environment, healthy life and affordable healthcare, and ubiquitous security and comfort require integrative approaches. Most interesting and useful innovations in recent years are a result of multidisciplinary and interdisciplinary efforts. Hence, the students should be given opportunities to collaborate and interact with other disciplines. For sustaining research-led innovation career which typically spans over four decades, an individual needs to raise funds regularly in a competitive and resources-limited world. This requires effective communication and marketing skills to convince diverse investors and stakeholders while maintaining ethics and research integrity and respecting intellectual property rights of others. In order to develop and implement innovative solutions to the societal challenges, often researchers must be able to work with others from different cultures, languages, norms, and safety standards. Hence, the graduate researchers need exposure to overseas research cultures and practices and intellectual property protection and transfer processes. A quality program should have various aspects indicated in Schematic 3, which will enable graduates and researchers to be most effective in society engagement and relevance. More importantly such a holistic approach will enable them to pursue passion to the full extent and inspire the generations to come [6, 7].

4 Conclusions

Engineering education is at the crossroads. There is need for (a) training engineering educators, (b) developing flexible and customized curriculum to suit students' needs and expectations, (c) updating ranking and accreditation criteria and standards to enhance the quality and social impact, (d) customized engineering education strategies appropriate for an institution and country, and (e) thought leadership in engineering education. Sharing of best practices around the world and continual improvement in all aspects of engineering education have never been this important in the history of engineering profession.

References

1. http://www.europeanbusinessreview.com/?p=9889
2. http://news.nus.edu.sg/images/resources/news/2014/2014-05/2014-05-02/CAREER-tabla-2may-p7.pdf
3. http://serve.me.nus.edu.sg/seeram_ramakrishna/index/GEDC2013_Chicago_Conf_Rpt.pdf
4. www.theatlantic.com
5. www.abet.org
6. http://engineeringeducationlist.pbworks.com/w/page/27610307/Engineering%20Education%20Departments%20and%20Programs%20(Graduate)
7. http://en.wikipedia.org/wiki/Engineering_education_research

Engineers Without Borders

Bopaya Bidanda

Keywords Globalization • Engineering education • Technological future

The trend towards globalization affects engineering educators on three interrelated dimensions: a dramatically new landscape of world economy, the next generation of engineering job, and the education of the "next gen" engineer.

It is critical that the next generation of engineers be globally competent. This is a broad outcome with multiple nuances, implementation methodologies, and assessment metrics that will be detailed in the next few paragraphs. Our mission is to enhance engineering education by enabling the "next gen" engineer to be equally at "home" in practicing engineering in Hubli, Hannover, Hokkaido, Helsinki, or Hackensack. This aligns with Jack Welch (former Chairman and CEO of General Electric) who believes that "Globalization has changed us into a company that searches the world, not just to sell or to source, but to find intellectual capital – the world's best talents and greatest ideas." These words by the visionary Jack Welch aptly summarize the phenomenon of globalization that has swept over the world. It has practically influenced every aspect of human life – culture, politics and economy, and especially engineering education.

1 Stages of Globalization and the Global Economy

Globalization has evolved tremendously over the ages. In his book, World Is Flat: A Brief History of the Twenty-First Century, the journalist Thomas L. Friedman details the three stages of globalization as follows:

B. Bidanda (✉)
Department of Industrial Engineering, University of Pittsburgh, Pittsburgh, PA, USA
e-mail: bidanda@pitt.edu

R. Natarajan (ed.), *Proceedings of the International Conference on Transformations in Engineering Education: ICTIEE 2014*, DOI 10.1007/978-81-322-1931-6_8,

1. Globalization 1.0 (1492–1800) – In this period, countries began to recognize the importance of spreading their wings to outside their homeland and looking for opportunities abroad. These years were characterized by countries trying to globalize for resources and imperial conquest. This period laid the foundation stone for the much richer global exchange that was to happen in the forthcoming eras.
2. Globalization 2.0 (1800–2000) – This was the era when globalization started to really mature. Unlike the earlier stage, where only some imperialistic countries were reaping the benefits, this stage provided opportunities for the corporate world to globalize. Multinational companies grew and concepts such as global outsourcing and supply chains emerged.
3. Globalization 3.0 (2000 onwards) – Globalization now not only encompasses trade but also exchange of talent and knowledge. In today's world, individuals are able to globalize. Companies search the entire world for talented engineers. This has led to the rise of what can be called a "global economy" where the economies of individual nations, companies, and individuals are deeply interconnected.

The global economic phenomenon has led to a sea of change in the nature of business and jobs. At big corporations, global sourcing of products and people has become the norm. Factors that have enabled this remarkable growth in globalization include:

Ease of accessibility and collaboration: Today's world has become much smaller in the sense that airfares have been stable in absolute terms over the past two decades and actually plummeted in real terms. Telecommunication costs have also dramatically dropped, leading to cheaper and easier collaboration between companies. But perhaps the most crucial factor has been the rise of the Internet. This global behemoth of interconnected computers has led to a tremendous increase in the ease of communication and information exchange. Helsinki and Hubli are just a click away!

Government policies: Having realized the boost that foreign investment can provide to their economy, governments have started providing incentives to firms that bring business to their country. These incentives include faster and easier licensing, attractive tax rates, and allowing free currency flow.

International alliances: Regional alliances like the European Union, NAFTA, Association of Southeast Asian Nations (ASEAN), and Asia Pacific Economic Cooperation Forum (APEC) help tremendously to facilitate collaboration and interaction between their member nations on various fronts like trade, culture, and technology by removing bottlenecks.

Talent pool in low-wage countries: There is an abundance of talented and highly educated engineers in the emerging nations and beyond – in diverse countries such as India, Brazil, China, Russia, Eastern Europe, Middle East, etc.

2 Educating the Next-Generation Engineer

In tomorrow's engineering workplace, competent technical skills are a necessary but not sufficient condition. However, in addition to the above-stated skills, it is necessary that an engineer possess the ability to work in the dynamic, multicultural teams that are present in global corporations. Also it is necessary to blend high level of technical skills with superior communication skills. So today's engineering curricula must incorporate the following techniques to educate tomorrows "next gen" engineer:

1. Add new and relevant areas to the engineering curriculum relevant to the local workplace. Engineering curriculum (or at least, technical electives) must be evaluated and refined every 2 years.
2. Engage student in exciting, team-based, authentic experiences starting from freshmen year.
3. Establish interdisciplinary teams for projects. Academic institutions are among the few entities that divide engineers into silos by discipline. In the corporate workplace, every engineer irrespective of discipline is a problem solver.
4. Offer students opportunities for leadership in multiple areas.
5. Offer exposure to cross-cultural working environment. Emphasize to students that it is a global marketplace today. Students must be offered opportunities to experience the international workplace while in college. Technical electives and humanities courses that focus on globalization, diversity, and a cross-cultural environment must be offered to students.
6. Engineering courses must utilize global resources such as the Internet, educational modules, etc. as part of their curriculum.

3 Conclusion

The interplay of global factors will lead to significant changes in the structure, delivery mode, and content of engineering education. Successful engineering institutions are ones that will embrace these changes and allow them to produce globally competent engineers.

Pedagogy Training and Certification for Faculty

Michael E. Auer

Abstract Never has the speed of development in the area of engineering been as accelerated as it is today, as we observe the enormous and driven growth of the area of engineering. Today's tendencies require concerted new efforts in engineering education – or in other words, the importance of pedagogy in the field of engineering is growing enormously. These changes strongly demand new didactic and pedagogic paradigms. The International Society of Engineering Education (IGIP) offers to contribute to the relevance and pedagogical aspects related to developing educational concepts in engineering education.

Keywords Engineering education • Training • Pedagogy • Certification • Competences • IGIP

1 IGIP and Engineering Education

IGIP has a more than 40-year tradition of contributing to engineering education, and its members and many activists have contributed to making IGIP a leading global engineering association. IGIP presently has a worldwide membership of about 1,750 members (individual, affiliate, institutional). More than 1,300 professionals all over the globe at this moment bear the title of "International Engineering Educator – ING.PAED.IGIP." IGIP also works in good partnerships with international associations as IFEES, ASEE, IEEE Education Society, SEFI, IUCEE, and IELA, to name just a few.

M.E. Auer (✉)
International Society of Engineering Pedagogy (IGIP), Hubli, Karnataka, India
e-mail: president@igip.org

R. Natarajan (ed.), *Proceedings of the International Conference on Transformations in Engineering Education: ICTIEE 2014*, DOI 10.1007/978-81-322-1931-6_9,

The aims of the International Society for Engineering Education (IGIP) are as follows:

- To improve teaching methods in technical subjects
- To develop practice-oriented curricula that correspond to the needs of students and employers
- To encourage the use of media in technical teaching
- To integrate languages and the humanities in engineering education
- To foster management training for engineers
- To promote environmental awareness
- To support the development of engineering education in developing countries

It is important to consider that humankind has never faced such a rapidly changing and dynamic global environment which demands so much of engineers as we are witnessing today.

And as our environment changes, it is imperative we better learn to adapt, which requires us to question and, when necessary, be open to changes regarding our educational systems, pedagogy, and all our methods and processes in engineering education.

Never before have the challenges in education and pedagogy been as challenging as today. Never has so much been demanded of engineers. But what exactly is engineering?

> Engineering is the discipline, art and profession of acquiring and applying scientific, mathematical, economic, social, and practical knowledge to design and build structures, machines, devices, systems, materials and processes that safely realize a solution to the needs of society [1].

A short definition of engineering might be "Exploiting basic principles of science to develop useful tools, objects and processes for society." This means that engineering is the link between science and society, which can include almost anything that people come into contact with or experience in real life. The concept of engineering existed long before recorded history and has evolved from fundamental inventions, such as the lever, wheel, and pulley, to the complex examples of engineering we witness today. But today, there are two actual tendencies:

First, we can observe an enormous (and accelerated) growth of the area of engineering.

Besides the traditional fields, new engineering disciplines have been created and more are in the process of creation. Some examples are as follows:

- Software engineering
- Data engineering
- Information engineering
- Medical engineering
- Neuroengineering
- Gene engineering
- Social requirement engineering, etc.

And new tasks requiring new competencies within traditional engineering disciplines have grown in number and complexity:

- Online engineering
- Remote engineering
- Virtual engineering
- Reverse engineering

On the other hand, we can observe a terrific decrease of the life cycles of technical (or engineering) products (and processes or technologies too!) [2, 3].

For example, how many years did it take for the following products to reach a market audience of 50 million?

- Radio 38 years
- TV 13 years
- Internet 4 years
- iPod 3 years
- Facebook 2 years
- Tablet PC (iPad) 1 year

The field of engineering has never suffered such reduced times to bring their innovations from concept to market. Competition in the field of technology is now measured in weeks.

Thirdly, engineers tend to spend the majority of their working week (around 60 %) engaged in activities which involve interaction with others (meetings, supervision, writing reports, etc.), and only around 40 % is devoted to technical engineering activity [4].

This shows, for example, the NACE's Job Outlook 2012 survey.

Survey participants rated "ability to work in a team structure" and "ability to verbally communicate with persons inside and outside the organization" as the two most important candidate qualities [5].

These realities require a concerted effort to evolve engineering education into what today's reality is demanding of practicing engineers. In other words, many traditional educational models and practices are no longer functional. For this reason, the importance of pedagogy is growing at an enormous pace. The need to innovate and apply new paradigms to the teaching-learning process is an absolute necessity [6].

2 New Questions of Today's and Future Engineering Education

All these trends result in new questions and the resulting need to evolve educational practices, especially in engineering pedagogy. Some of these important questions to consider include the following:

- What learning approaches have to be used to effectively respond to these changes?
- What are the pedagogies that provide the most effective learning experiences for engineering students of the twenty-first century?

- What learning skills in engineering education need to be developed and how can engineering teachers succeed in guiding their students to achieve them?
- What pedagogical approaches have been found to support the different phases of the present lifelong learning continuum, or is more research necessary?
- What are the approaches that enable competence in leadership skills in a multicultural working environment, and what is the best way for these competencies to be delivered?
- Ambient technology is becoming a reality. What does ambient learning in engineering education look like? How can it be designed, delivered, and assessed?
- How can engineering education support individualized and personalized learning to compensate for individual differences (learning styles, learning strategies, learning preferences, field dependency, etc.)?

These are some of the reasons why the relevance and importance of engineering pedagogy is growing so enormously.

Logically the European Commission stated in its report on improving the quality of teaching and learning in Europe's higher education institutions (JUNE 2013): "There is no law of human nature that decrees that a good researcher is automatically a good teacher, or that a first class honors student in biochemistry with a brilliant PhD will automatically be a good teacher of biochemistry."

The recommendation is "All staff teaching in higher education institutions in 2020 should have received certified pedagogical training. Continuous professional education as teachers should become a requirement for teachers in the higher education sector."

3 IGIP's International Engineering Educator Title

IGIP has established a prototype curriculum for engineering pedagogy which is already used in several countries. In contrast to ABET or ENAEE, IGIP is not an accreditation body for engineering curricula. By passing the curriculum as proposed by IGIP in any accredited or other institution worldwide, IGIP states that a given engineering educator with an ING.PAED.IGIP title has all the competencies needed to teach to the highest standards with the best available teaching technologies. Interested engineers can continue their education in accordance with the IGIP Curriculum and obtain a diploma that will provide the knowledge and skills necessary for engineers to become better teachers. IGIP, worldwide, already has 46 approved educational centers and more than 1,300 approved "International Engineering Educators" (ING.PAED.IGIP).

The IGIP model's point of departure is that individual engineering lecturers initiate and are responsible for teaching and learning concepts that train engineers and technicians. The quality and success of engineering studies are decisively influenced by teacher competencies in the area of pedagogy as pedagogical skills represent a network of knowledge and skills that transmit knowledge and experience, much like Web 3.0. For this reason, technology and educational practice must go hand in hand when we are dealing with the education of engineers.

Engineering educators expand their typical engineering subject competence by acquiring teaching and learning skills in theoretical and practical coursework corresponding to the objectives of the ING.PAED.IGIP curriculum.

The ING.PAED.IGIP is a registered program which certifies a certain educational level for teachers, trainers, or instructors. Any engineering educator who passes the curriculum at any IGIP accredited training center for International Engineering Education or other institution worldwide, and whose education, training, and professional experience meet IGIP standards, may apply to be registered as an "International Engineering Educator ING.PAED.IGIP".

The qualification profile of a specialized engineering pedagogue is based on two pillars [7, 8]:

- Engineering qualifications which were earned through a recognized and/or accredited engineering study program plus relevant professional experience
- Educational qualifications in engineering pedagogy acquired in the course of a comprehensive educational program

The goal of IGIP accreditation is to insure that graduates of the accredited engineering pedagogical programs are well prepared to perform their teaching duties in engineering subjects and meet the criteria required to become International Engineering Educators, ING.PAED.IGIP. Another goal is to promote quality assurance, quality improvement, and modernization of engineering pedagogy programs and to create public awareness of the high quality of the IGIP program for engineering pedagogues. Accreditation is a voluntary process which educational institutions must apply for through the responsible IGIP national monitoring committees.

The accreditation criteria defined by IGIP for a program for engineering educators are as follows:

- Organization of the program
- Entrance requirements for the first year students
- Skills/abilities of the graduates
- Engineering pedagogical curriculum
- Lecturers and professors
- Institutional resources
- Quality control and feedback

4 Competences in Engineering Pedagogy

An "ideal" teacher with a technical background should acquire the necessary professional competences of an engineering educator. These general professional competences consist of two main groups:

- Technical expertise
- Specific engineering pedagogical competencies

Educational theory offers different lists of competences [9]. The IGIP concept of engineering educational competences is to be summarized as follows:

- Pedagogical, psychological, and ethical competences
- Didactical skills and evaluative competences
- Organizational (managerial) competencies
- Oral and written communication skills and social competences
- Reflective and developmental competences

The IGIP Recommendations for Engineering Pedagogy Studies (in short IGIP Prototype Curriculum) are described in detail in [10].

Interested institutions and engineers, teachers, and students are welcome to contact one of the 23 IGIP national monitoring committees or the IGIP headquarters in Austria.

5 Conclusions

Technical university teaching has often been perceived as a poor cousin to research. Few technical universities require any specific technical teacher education for their academic staff. Interestingly, this is the only level of learning where academic staff receives no teacher training. Yet teaching is an art that, at least to some degree, can be taught, if the institutions of higher learning support it as an important element. The International Society for Engineering Education (IGIP) is working to assure that graduates of accredited engineering pedagogical programs are well prepared to perform their teaching duties in engineering subjects and meet the criteria for IGIP registration as International Engineering Educators, ING.PAED.IGIP. IGIP's ultimate goal, however, is to promote quality assurance and quality improvement and modernize engineering pedagogy programs and educational practices. Its intention is also to create public awareness of the high quality of engineering pedagogue programs.

References

1. http://en.wikipedia.org/wiki/Engineering. Last visited 5 Feb 2014.
2. Toru Iiyoshi, Vijay Kumar MS. MIT.
3. Alexandrov V. E-learning and Web 3.0. In: Proceedings of the ICL2009 conference, 23–25 Sept 2009, Villach.
4. http://en.wikipedia.org/wiki/Engineering. Last visited 5 Feb 2014.
5. Albrecht D, Koskinen T. http://openeducationeuropa.eu/en/elearning_paper. Last visited 5 Feb 2014.
6. Feifan Ye. Proceedings of the IEEE EDUCON2011 conference, 4–6 Apr 2011, Amman.

7. Dobrovska D. Pedagogická a psychologická příprava učitelů odborných předmětů. Praha: ISV; 2004. 196 pp. ISBN 80-86642-33-X.
8. Vanecek D. Progressive education methods. In: New trends in engineering pedagogy. Workshop of the International Society of Engineering Education IGIP. Praha: ČVUT; 2006. ISBN 80-01-03459-3.
9. Turek I, Hrmo R. Kľúčové kompetencie I. Bratislava: Slovenská technická univerzita; 2003. ISBN 80-227-1881-5.
10. http://www.igip.org/. Last visited 5 Feb 2014.

Transforming Engineering Education: The Role of Engineering Educators in Making Meaningful Change

Khairiyah Mohd Yusof

1 Summary

Transformation of engineering education is taking place all over the world to address the needs of the twenty-first century. Gone are the days when engineering graduates will automatically gain employment; the world now does not need just any engineer – the world needs good engineers who are well prepared to take on the challenges of this century. Realizing these challenges, what is our role as engineering academics in this transformation? As individuals, how can an academic teaching engineering students transform engineering education? While we may feel powerless and doubt that what we do as individuals can make a difference, it is important to remember that transformation can take place at different levels: individual or grassroots, departmental or faculty, and institution and nationwide levels. While change can start from the top-down, sustaining it from the bottom-up is just as crucial.

We need to remember that meaningful change always starts small, from individuals at the microlevel. In my own experience, the seeds of transformation started when I began reading about others' efforts to improve engineering education in the USA through reading the Chemical Engineering Education Journal back in the early 1990s. In addition to increasing my knowledge, reading the papers in the journal also made me realize that efforts in engineering education should be written and shared with others.

In the early 2000s, when I started to change the way I teach to help students in my class learn better, the university management was willing to support me to go to conferences and workshops in Malaysia or Singapore, provided that I train other academic staff when I afterwards. Well, the best way to learn is to indeed implement what we learned and teach others. Providing training to other academic staff allowed

K. Mohd Yusof, Ph.D. (✉)
Centre for Engineering Education, Universiti Teknologi Malaysia,
Johor Bahru, Johor, Malaysia
e-mail: khairiyah@cheme.utm.my

R. Natarajan (ed.), *Proceedings of the International Conference on Transformations in Engineering Education: ICTIEE 2014*, DOI 10.1007/978-81-322-1931-6_10,

me to bring about change at my own institution but also at the national level. Organizing and conducting workshops, conferences and seminars to bring in experts in engineering education to learn from, and to provide a platform for sharing and exchanging ideas also allowed us to develop a community of practice among engineering educators in Malaysia and Asia. To sustain the transformation, it is important to be a reflective practitioner through conducting research and writing about them. Becoming a reflective practitioner through conducting research on my own practice and to answer the questions that I have about engineering education to enhance its quality. Writing further pushed me to read about what others are doing, and discover educational principles that I can apply to become the foundation of good practices. Thus, writing about my practice and research in engineering education allowed me to be more critical to improve my implementation and share with others not only in Malaysia but also from throughout the world. This is the key to playing our part in meaningfully transforming engineering education as academics – we need to read and learn – to be scholarly and knowledgeable, start the change with our classes or programs, write about those efforts, and when given the chance, train others to become champions of change at the grassroots level.

When Malaysia became a provisional member of the Washington Accord in 2003, and finally a full member in 2009, although it was initially difficult, those who could see the positive impact of outcome-based education learned and became champions to support the transformation. This transformation is still ongoing, and efforts are being made to ensure that they are made through scholarly, meaningful, evidence-based approaches. We need to realize that the outcome-based approach is not just about preparing documentation – the outcomes that we want our graduates to achieve are the attributes that we would like our graduating engineers to have to be good engineers in the twenty first century. This is why we need to bring about this transformations in our classrooms. Through all these experiences, we in Asia need to adapt this philosophy to our own culture and settings, which is why developing local champions is crucial. The time has come for us to also publish and share these experiences with the whole world to enrich and advance the knowledge in the area from the Asian perspective.

Part II
Presented Papers

Application Based-Approach of Teaching Digital Signal Processing to a Large Class in the Context of an Affiliated System

Vineeta P. Gejji

Abstract This paper presents a teaching method that effectively integrates multimedia resources, software and chalk and board teaching. This method enhances the learning of fundamental concepts of digital signal processing (DSP) in a large class. It follows the application-based approach of introducing an audio or video application. The concepts linked to the application are demonstrated using software. The math associated with the concept is worked on the board. This method helps the student to understand the difficult concepts of DSP. It is effective in a large class in an affiliated system. In an affiliated system, the curriculum is prescribed by the university. The teacher has little or no flexibility in the framing of the curriculum. The content, the number of hours and the pattern of the questions are all dictated by the university. This method has proved effective within these constraints.

Keywords Digital signal processing • Convolution • Application-based teaching

1 Introduction

The learning of DSP involves both the understanding of various theoretical concepts and the practical development and implementation of these concepts [1]. DSP is a core subject prescribed for the undergraduate course. The teaching content of the course includes theory teaching and practice teaching, which are set separately but have close connection with each other [2]. The amount of mathematics involved intimidates the student. The student has to be introduced to the applications which will ignite the interest of the student. Link the application to the concept and then the maths involved has to be introduced. The curriculum for DSP for undergraduates includes basic composition of digital signal systems, discrete Fourier transform,

V.P. Gejji (✉)
Department of E&C, Gogte Institute of Technology, Udyambag, Belgaum, India
e-mail: vpgejji@git.edu

R. Natarajan (ed.), *Proceedings of the International Conference on Transformations in Engineering Education: ICTIEE 2014*, DOI 10.1007/978-81-322-1931-6_11,

digital filters and digital filter structures. The general opinion of the student is that too much mathematical analysis and mathematical formulary are included in the curriculum. The general method followed in teaching is to resort to Power point presentation and blackboard. Continuous mathematical derivations without understanding the applications many a times put off the student, and the student loses interest.

2 Application-Based Teaching

Many DSP concepts can be demonstrated by examples which involve a great deal of computation. A list of some of the concepts is as follows: convolution, filtering, quantization effects, etc. The curriculum begins with discrete Fourier transform (DFT). DFT is derived from discrete-time Fourier transform expression. The continuous and discrete Fourier transform are covered in Signals and Systems. The flow of the topics is as follows: DFT, properties of DFT, Fast Fourier Transform, Infinite Impulse Response filter and Finite Impulse Response filters and filter structures. If the topics are linked to a project with each block of the project demonstrating the various topics of the curriculum, it is easier for the student to comprehend what is being taught.

2.1 Convolution

One of the concepts frequently used in DSP is convolution. The response of any system to a new input can be calculated by convoluting the impulse response with the new input. Convolution is expressed as given by (1):

$$y(t) = x(t) * h(t) \tag{1}$$

where y (t) is the response, x (t) the new input and h (t) the impulse response of the system. Convolution can be demonstrated by recording the impulse response of a closed stadium. The input x (t) could be the voice of a person and could be recorded in a closed room. The two signals are convoluted using MATLAB. The resulting signal when played will appear as though the person is speaking in the stadium [3]. The convolution of different signals can be worked on the board. Working by hand will help in reinforcing the mathematics involved in the calculation of convolution.

2.2 Speech Processing System

To link the various models of the chapter, a project on speech processing or video processing is picked up. The whole sequence of operation of a speech processing system could be as shown in Fig. 1.

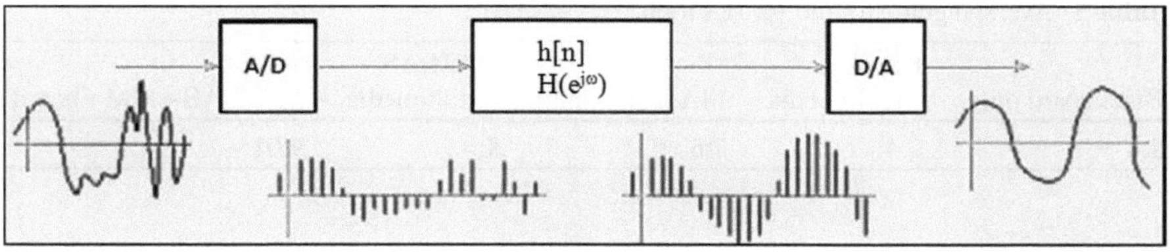

Fig. 1 A speech processing block

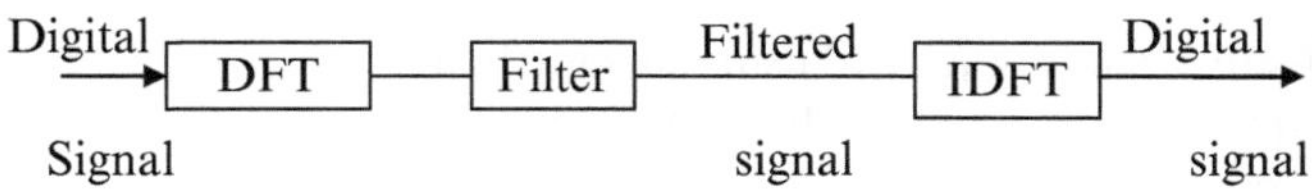

Fig. 2 DSP block

In turn, the DSP block can be expanded as shown in Fig. 2.

A speech signal is recorded in the time domain. The signal is sampled and transformed to a frequency domain signal. The transformed signal is filtered and then the inverse DFT is obtained. Each block of the speech processing system is linked to a topic in the curriculum.

The recorded audio signal is played before processing and after processing. The audio signal can be given to a MATLAB program to demonstrate the sampling process. The DFT of the samples can be worked in MATLAB. This will explain the concept behind the audio signal frequency domain characteristics. The derivation of the DFT signal from the DTFT signal can be worked on the board. Students can be asked to solve the problems related to DFT and its properties by hand. Derivations are understood better if they are worked on the board step by step. The active involvement of the student is also necessary.

The DFT samples can be given to different types of filters to demonstrate the filtering effect. The student can observe the variations in the audio clip when passed through different filters. The implementation of the filter can be done by hand and verified by MATLAB.

The advantage of the application-based learning is that the student can relate the real world example to the maths behind it. In conventional teaching methods, the maths is delinked from the application making the student wonder as to where the mathematics is used.

3 Assessment of Effectiveness

This method was adapted to a class of 95 students. The students were taught Signals and Systems of which digital signal processing is an extension, with the conventional method. To the same set of students, digital signal processing was taught with the application-based method. They were given a questionnaire to answer about the

Table 1 Average grade for the various tools

Blackboard only	Board +multimedia	MATLAB only	MATLAB +multimedia	MATLAB + MM + board
8.4	9.4	7.6	8	9.01

effectiveness of the method from their perspective. The first three questions related to the method, the fourth was the level of motivation this method provided with. The questions asked were as follows:

- How effective is teaching DSP using this method?
- Does animation help you visualise the concept?
- How good is the flow of the topic?
- Were you motivated to explore?

They were also given choices of black board, black board plus MATLAB, MATLAB only, MATLAB plus multimedia and MATLAB plus multimedia plus black board as various tools used to make them understand the basic concepts of DSP. They were asked to rate each of the combination of tools which helped them in understanding the concepts better. The rating was on a scale of 1–10, with 1 being the minimum and 10 being the maximum.

The students' feedback was positive. Some of them are listed as follows:

- Interesting lessons
- Could understand the difficult concepts
- Maths made so interesting

As for the tools used, the average of the rating given by the students was taken. It is as shown in Table 1.

Since the method was tested on the same set of students, the feedback carries more weight.

4 Conclusion

A method for teaching a large class in an affiliated system has been presented. It is based on the successful integration of black board, multimedia resources and MATLAB. This method has been tested on a class of 95 students in an affiliated system. The feedback from the students has been encouraging.

References

1. Salous S. Computer based and dedicated hardware projects for the teaching of digital signal processing. In: Teaching of DSP IEE colloquium. London; 1995. pp. 3/1–3/5.

2. Jun Zhu, Yan-Jun Hu. Application of project teaching method in digital signal processing. In: Proceedings of the 7th international conference on computer science and education (ICCSE 2012). Melbourne, 14–17 Jul 2012. pp. 1789–91.
3. Perrins E. Fun with convolution and linear time invariant systems. In: Proceedings of the 2010 midwest section conference of the American society for Engineering Education. 2010.

Keyword-Based Search and Ranking in NPTEL Lecture Videos

Manpreet Kaur and Harish Kumar

Abstract Information retrieval is one of the most important technologies of present times due to vast availability of information. Lots of systems with varying information are available through the Internet. Useful information can be retrieved from these systems. The Web is a crucial application available on the Internet. It provides a medium to study and learn various subjects in the education domain. Large amount of E-content is also available on internet sites like NPTEL, OCW, OCC, CORE, UNIVERSIA, TOCWC, KOCWC, RuOCW, BigBlueButton and YouTube. NPTEL has covered more than 600 courses which include recorded video lectures. There is no search engine available providing a mechanism to search content within these videos. All these monolithic lecture videos are around of an hour's duration. Learners face difficulty when searching a keyword-relevant content in these videos. A keyword-based system to search within a single lecture video or across multiple videos is required to reduce the time to use this content. It will be useful to have an efficient information retrieval system helping many students to search the NPTEL content. In this study, a search and ranking system is proposed for the video content of NPTEL lectures.

Keywords Information retrieval • Lecture videos • Key frame

1 Introduction

The search and rank ability of an information system is to find and rank information of interest with respect to user's need to be fulfilled by navigating various possible options in a repository so that usability and effectiveness of the system in terms of

---

M. Kaur (✉) • H. Kumar
Computer Science and Engineering Branch, UIET, Panjab University, Chandigarh, India
e-mail: manpreet327@yahoo.com; harishk@pu.ac.in

R. Natarajan (ed.), *Proceedings of the International Conference on Transformations in Engineering Education: ICTIEE 2014*, DOI 10.1007/978-81-322-1931-6_12,

precision and recall value remain high. Many universities and other institutes use the Internet to offer learning material containing thousands of recorded video classroom lecturers maintained in their repositories [1]. To browse through the entire repository for finding content related to the interest of the learner is a cumbersome job. If the learner is not familiar with the area like the first-time learner, then this problem gets more exacerbated. The learner may not be able to find the exact videos related to the relevant topic to see/listen. Hence, there is a need of a keyword query-based system which provides a relevant link portion in a video after going through the repository.

The rest of the paper is organized as follows. Present techniques used for searching in lecture videos are discussed in Sect. 2. The problems in current techniques are summarized in Sect. 3, followed by design and implementation of a system which are presented in Sect. 4. Then in Sect. 5, results and overall performance of system is discussed followed by a conclusion.

2 Literature Survey

This section outlines various techniques for searching in recoded lecture videos and some well-known video lecture repositories.

2.1 Recorded Lecture Video Search Techniques

Various techniques can be used for searching the lecture videos. These techniques work on one of the following two basic methods [2]:

Using metadata and browsing key frame: Many openly available video search systems allow search entirely on video metadata like date, author, video title, summary, author, genre, etc. Some systems provide key frame or storyboard previews of selected videos to assist the learner to preview the lecture video [3, 4]. Key frames are single images or video frames derived from shots as representatives of complete shots. These frames are basically used for content-based video management. Open video project [5] and Internet archives [6] are based on metadata search techniques.

Key frame matching: This technique requires the user to find set of images or key frame supporting the information requirement of the user. Content-based image retrieval (CBIR) search is done on the collection of images to find the video key frame from the video library [7]. In Fischlar-TRECVid-2004 [8, 9] video retrieval system, the user queries as text search terms or a set of images executed against closed captions of videos. Key frame-based technique is suitable when the user has a visual component which is clearly defined.

Table 1 Lecture video repositories comparison

Repository	Search	Navigation feature
NPTEL	No	No
OCW	Content	Speech navigation
OCC	Courses	By language or by institution
CORE	By keyword	No
UNIVERSIA	By keywords	No
TOCWC	No	No
KOCWC	No	No
RuOCW	No	No
BigBlueButton	No	No

2.2 Existing Video Lecture Repositories

National Program on Technology Enhanced Learning (NPTEL [10]), MIT's OpenCourseWare (OCW [11]), OpenCourseWare Consortium (OCC [12]), China Open Resources in Education (CORE [13]), UNIVERSIA [14], Taiwan OpenCourseWare Consortium (TOCWC [15]), Korea OCW Consortium (KOCWC [16]), Russian OpenCourseWare (RuOCW [17]), Open Source Web Conferencing (BigBlueButton [18]) are some of the main lecture video repositories. These repositories provide limited search and navigational features as shown in Table 1. NPTEL, a project funded by the Ministry of Human Resource Development, Govt of India, does not provide lecture video search facility. The only facility to assist the learner is by browsing through the directories/List. It has approximately 12,000 videos of different lectures in about 25 disciplines. The size of repository is around 4 terabytes. Learners have to select a course manually and then go through a list of titles of all lectures in that course, to find which video might have the desired explanation. Then the user has to manually browse through the videos to identify the portion of interest.

3 Problem Definition

NPTEL repository is used by a number of institutes and learners across India. It does not provide any search mechanism to find out relevant content of interest. The aim of this work is to find out a list of lecture videos from NPTEL repository where keywords provided by the learner are spoken. It should play the video from the point of utterance of that particular keyword.

4 System Architecture and Implementation

A video search engine can use a number of features of video data to search. The overall system has a number of components. The data flow between different components of search engine created for NPTEL is shown in Fig. 1:

Fig. 1 Video search engine

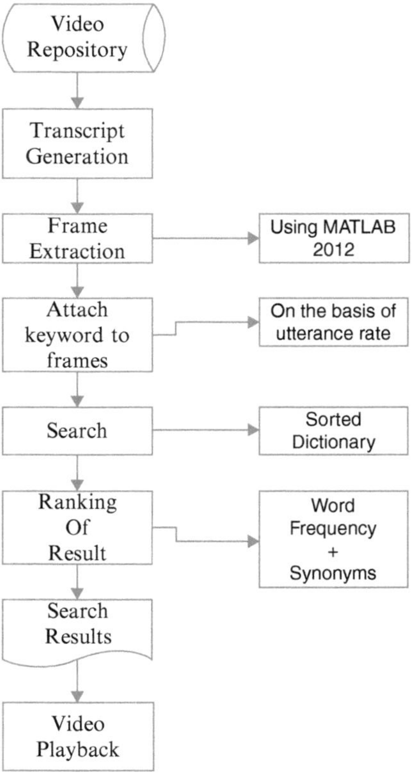

(a) *Video Repository*: When it comes to spoken content, the dialogues between speakers can be spontaneous or scripted. Complexity of the problem varies with the number of speakers and other audio elements (background score, noise, etc.) involved. The genre of the video also plays an important role from the user's perspective of searched information. Taking all factors in consideration, in this work a repository of single-speaker lecture videos has been collected from NPTEL content. 20 lecture videos of 'Data Structure and Algorithm' subject, each video of approximately 1 h, consisting of 20 viewing hours are collected for the purpose. The size of collection is approximately 5 gigabytes. Videos are downloaded from NPTEL [10].

(b) *Transcript Generation*: Transcription is a representation of content in written form. The source can either be utterances (speech) or pre-existing text. The text of video lectures has been generated through manual processes. A collection of 20 text documents of the respective videos is generated manually by listening to each video. The collection served as the dataset for whole system. The transcript is generated from the words spoken by an expert delivering lectures. Stop words are the terms that appear frequently in the text like 'a', 'the', 'is', etc. Stop words does not give any meaningful view about the repository. Hence, in transcripts 48 % of stop words are removed for fast search.

(c) *Converting Video into Frame Images*: For extracting frames of the video, MATLAB-2012 software is used. It can be used to develop algorithm, analyse

Search

VideoName	VideoPath	FrameNumber	KeyWords
Introduction to DataStructures	~/files/lec01.mp4	29	intoduction
Introduction to DataStructures	~/files/lec01.mp4	58	data structure lecture
Introduction to DataStructures	~/files/lec01.mp4	87	welcome
Introduction to DataStructures	~/files/lec01.mp4	116	data structure
Introduction to DataStructures	~/files/lec01.mp4	145	algorithms
Introduction to DataStructures	~/files/lec01.mp4	174	going learn
Introduction to DataStructures	~/files/lec01.mp4	203	today basic
Introduction to DataStructures	~/files/lec01.mp4	232	terminology regarding data
Introduction to DataStructures	~/files/lec01.mp4	261	structures notations
Introduction to DataStructures	~/files/lec01.mp4	290	
Introduction to DataStructures	~/files/lec01.mp4	319	using rest
Introduction to DataStructures	~/files/lec01.mp4	348	course
Introduction to DataStructures	~/files/lec01.mp4	377	first begin
Introduction to DataStructures	~/files/lec01.mp4	406	simple definition
Introduction to DataStructures	~/files/lec01.mp4	435	algorithm
Introduction to DataStructures	~/files/lec01.mp4	464	
Introduction to DataStructures	~/files/lec01.mp4	493	outline
Introduction to DataStructures	~/files/lec01.mp4	522	step program

Fig. 2 Word annotated with respective frame

data and create models and applications. In MATLAB ‘mmreader’ class is used to read video data from multimedia files. It is used to make frames of videos. MP4 video frame rate is 30 frames per second. So for each 1 h video, approximately 108,000 frames are extracted and stored.

(d) *Attach Keyword to Frames*: After converting videos into frame images and generating transcript, frames are annotated. The method used to annotate frames is based on utterance rate so that time alignment of words with frames can be accurately captured. Utterance rate is the speed at which a speaker speaks. On an average, a person speaks 168 words per minute which amounts to 3 words per second [19]. Annotation is done by attaching words spoken by the speaker. Since, 30 frames per second are generated, so these 30 frames are attached with each set of 3 words. Figure 2 shows the sample of words annotated with frames.

(e) *Search*: Search is carried out for a user query. The system searches for the query term in the index and returns the results after ranking them according to score. The search list is a ranked list of videos in decreasing order of score, which have the query term. Figure 3 shows search result for a term ‘tree’.

(f) *Ranking*: A user may not be interested in all the results returned in response to the query. But, the most relevant result may fulfil the user’s requirements. Scoring of result or ranked retrieval is performed to achieve this aim. Ranking is on the basis of term frequency and synonyms of the word to be searched. After searching a particular word and its synonyms from a transcript of all the videos, score is calculated. Equation (1) helps in finding the precision value and

Search tree

	VideoName	VideoPath	SearchedString	VideoScore	FrameNumber
Click 2 Play Video	AVL trees	~/files/lec07.mp4	tree	0.903225806451613	54
Click 2 Play Video	trees	~/files/lec08.mp4	tree	0.387096774193548	372
Click 2 Play Video	insertion in red black tree	~/files/lec09.mp4	tree	0.387096774193548	1767
Click 2 Play Video	Tree walk/traversal	~/files/lec06.mp4	tree	0.17741935483871	726
Click 2 Play Video	disk based data structure	~/files/lec10.mp4	tree	0.145161290322581	1428
Click 2 Play Video	Introduction to DataStructure	~/files/lec01.mp4	tree	0	0
Click 2 Play Video	stacks	~/files/lec02.mp4	tree	0	0
Click 2 Play Video	queues and linked list	~/files/lec03.mp4	tree	0	0
Click 2 Play Video	dictionaries	~/files/lec04.mp4	tree	0	0
Click 2 Play Video	hashing	~/files/lec05.mp4	tree	0	0
Click 2 Play Video	priority queues	~/files/lec11.mp4	tree	0	0

Fig. 3 Search result for term 'tree'

Eq. (2) helps in finding the recall value. With the help of Eqs. (1) and (2), score is calculated as in Eq. (3). The video with the highest score will be at the top, and other videos are arranged in decreasing order of their score:

$$\mathrm{A} = \frac{(\text{Total Relevant Words})}{(\text{Total shown Results})} \tag{1}$$

$$\mathrm{B} = \frac{(\text{Total Relevant Words})}{(\text{Total Words in Database of full Repository})} \tag{2}$$

$$\text{Score} = \frac{(\mathrm{A} * \mathrm{B})}{(\mathrm{A} + \mathrm{B})} \tag{3}$$

(g) *Video Playback*: A video containing the searched term can be played back from the frame attached with that particular term. Key frame that contains word is displayed as thumbnail. On clicking the thumbnail, the video playback starts from the instant where the searched term is spoken in the video. The frame rate of an MP4 video is 30 frames per second. For exact playback the exact frame number is required which is calculated using the speaking rate of the speaker. For playing videos, Microsoft Silverlight API is used.

5 Results

(a) *Frame Gap*: Frame gap is the difference between the actual video frame number in which the word is spoken and the frame number from which the video playback is started in the output. There is some difference or frame gap observed in

Table 2 Frame gap for different query terms

Video title	Query keywords	Actual frame #	Played frame #	Frame gap
Introduction to data structure	Data structure	56	58	–2
Introduction to data structure	Algorithm	115	145	–30
Introduction to data structure	Array	1,087	1,015	72
Introduction to data structure	Running time	7,597	7,568	29
Introduction to data structure	Pseudo code	13,230	13,227	3
Introduction to data structure	Notation	287	243	44
Introduction to data structure	Insertion sort	25,650	25,633	17
Stacks	Algorithm	5,340	5,308	32
Stacks	Java virtual machine	690	638	52
Stacks	Abstract data type	270	261	9
Stacks	Stack	61	116	–55
Stacks	Dynamic set	6,750	6,742	8
Stacks	Interface exception	300	275	25
Stacks	Data structure	5,260	5,249	11

Table 3 Frame statistical descriptive measures

Mean	15.35
Standard deviation	32.23
Variance	1,038.86
Maximum	31.07
Minimum	0.01

the method used to attach frame to words. In some cases the gap is positive, signalling a frame hit, that is, the video started from an earlier point than when the query term is spoken, while in other cases, the frame gap is negative, signalling a frame miss, that is, the video start playing from a later point of time. Fourteen queries shown in Table 2 have been considered for this experiment, and frame gap is calculated.

(b) *Descriptive Statistics*: Descriptive statistics quantitatively describes the main features of a collection. Some measures used commonly to describe data are mean, maximum, minimum, standard deviation and variance. The mean of data is the average of the values. It is the measure of the centre of distribution. The variance and standard deviation are both measures of the spread of the distribution about the mean. Standard deviation shows how much variation exists from the mean value which in our case is 32.23. Descriptive measures calculated for the frame gap is shown in Table 3.

In case of normal distribution of data, standard deviation should be near to mean. But in this data, there is a large difference between mean and standard deviation because of positive and negative frame gaps. To overcome this problem, frame gap is to be minimized; this can only be done by annotating a frame with an accurate word. The percentage of deviation of the actual frame has been calculated to find how far the searched results from the ideal are (actual frame number). Maximum deviation is 31.07 and minimum deviation

Fig. 4 Partial data for ROC curve

1	Query	Score	Result
2	AVL trees	0.9	1
3	Data structure	0.57142	1
4	Hashing	1.33	1
5	Dictionaries	1.714	1
6	Asymptotic analysis	1	1
7	Running time	1.75	1
8	Java virtual machine	1.51	1
9	Stack	1.9	1
10	Binary search	0.75	1
11	Algorithm	1.85	1
12	Programming	1.53	1
13	Pseudo code	1.6	1
14	Searching	1.714	1
15	Traversal	2	1
16	Data type	2	1
17	Sorting	2	1
18	Indexing	1.2	1

is 0.01 in absolute value, showing that the results are not too far from actual frame and the results can be counted as fairly good.

(c) *Receiver Operating Characteristics (ROC)*: To check the quality of the result, Receiver Operating Characteristic metric is used. For each threshold, two values are calculated, True Positive (TP) Ratio (the number of those words which are actually in the transcript, but not detected by the algorithm) and the False Positive (FP) Ratio (the number of those words which are not present in the transcript, but have been detected by the algorithm). The main aim is to minimize the false positive rate and increase the true positive rate:

Sensitivity is the ability to find positive results. It is defined by equation:

$$\mathrm{Se} = \frac{\mathrm{TP}}{(\mathrm{TP} + \mathrm{FN})}$$

Specificity is the ability to find negative results. It is also called selectivity:

$$\mathrm{Sp} = \frac{\mathrm{TN}}{(\mathrm{TN} + \mathrm{FP})}$$

Some data for ROC curve is shown in Fig. 4. Using this data ROC curve is generated.

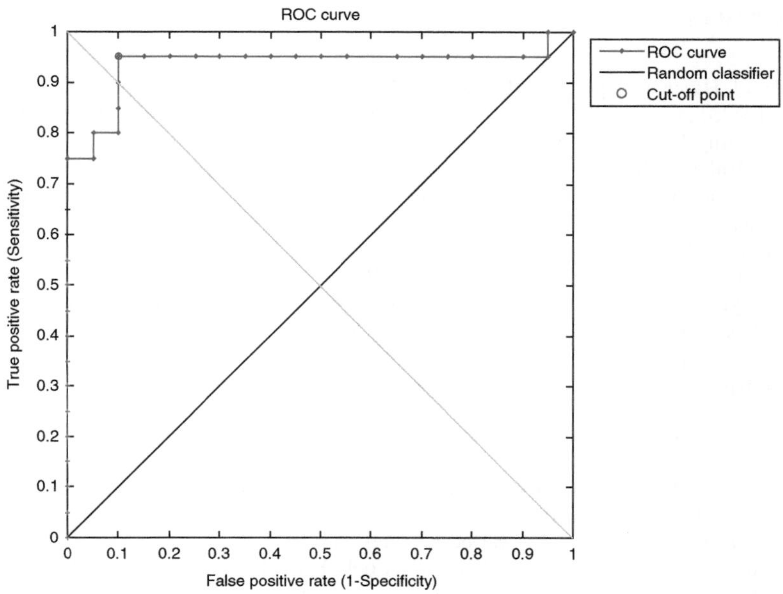

Fig. 5 ROC curve

Figure 5 shows that during query phase the accuracy is 95 %. The area under ROC curve indicates the accuracy. If the area is 1, then it represents the perfect test and an area of .5 or less is useless test. Accuracy of the test can also be considered as point system like a range of .90–1 as excellent, .08–.09 as good, .70–.80 as fair, etc. The closer the curve is to the left and top borders, the more it is accurate. A 45-degree diagonal is a random classifier; if the curve is near diagonal, then the system is less accurate. The blue circle in Fig. 5 is the cut-off point which indicates that the accuracy of this lecture video retrieval system is 95 %.

6 Conclusion

The goal of video retrieval is to retrieve video that best matches the query. Information retrieval is a method in which the user is able to convert his need for information into an actual list of videos. In this paper, a model for information retrieval based on keywords distribution in a video has been proposed. This system returns, as result, a list of videos, in which the keyword entered by the learner is spoken. Videos are ranked according to the keyword frequency. In each video of retrieved list, frame number is also mentioned where the word is spoken for the first time. Link for each video is given; upon clicking the link, the video starts playing from the mentioned frame number. Frame gap is used for analysis of the system. It is found that the system gives the results with 95 % accuracy. The display of frame related to the spoken word reduces the considerable effort on the part of the user to search through the video.

References

1. Glass J, Hazen T, Cyphers S, Malioutov I, Huynh D, Barzilay R. Recent progress in the MIT spoken lecture processing project. In: Proceedings of the 8th INTERSPEECH-2007, Belgium, 27–31 Aug 2007, pp. 2553–56.
2. Kamabathula VK, Iyer S. Automated tagging to enable fine-grained browsing of lecture videos. In: Proceedings of the IEEE international conference on technology for education, Chennai, 14–16 Jul 2011, pp. 96–102.
3. Smeaton AF. Techniques used and open challenges to the analysis, indexing and retrieval of digital videos. J Inf Syst. 2007;32(4):545–59.
4. Aradhye H, Toderici G, Yagnik J. Video2text: learning to annotate video content. In: Proceedings of the IEEE international conference on data mining workshops, Pisa, 15–19 Dec 2008, pp. 144–51.
5. The Open Video Project: a shared video collection. http://www.open-video.org/. Seen on 31 Aug 2013.
6. Internet Archive: Moving image archive. http://www.internet.archieve/details/movies/. Seen on 31 Aug 2013.
7. Hopfgartner F, Jose J. A news video retrieval framework for the study of implicit relevance feedback. In: Proceedings of the 2nd international workshop on semantic media adaption and personalization, Uxbridge, 17–18 Dec 2007, pp. 233–36.
8. O'Connor N, Lee H, Smeaton AF, Jones GJF, Cooke E, Le Borgne H. Fischlar-TRECVid-2004: combined text and image based searching of video archives. In: Proceedings of the IEEE international symposium on circuit and systems, Island of Kos. 21–24 May 2006, p. 4.
9. Lee H, Smeaton AF, O'Connor NE, Smyth B. User evaluation of F'ischlar-news: an automatic broadcast news delivery system. ACM Trans Inf Syst. 2006;24(2):145–89.
10. National Program on Technology Enhanced Learning (NPTEL). http://www.nptel.iitm.ac.in. Seen on 31 Aug 2013.
11. MIT's Open Course Ware (OCW). http://ocw.mit.edu. Last seen on 31 Aug 2013.
12. Open Courseware Consortium (OCC). http://www.ocwconsortium.org. Last seen on 31 Aug 2013.
13. China Open Resources in Education (CORE). http://www.core.org.cn. Last seen on 31 Aug 2013.
14. UNIVERSIA. http://ocw.universia.net. Last seen on 31 Aug 2013.
15. Taiwan Open Courseware Consortium (TOCWC). http://tocwc.nctu.edu.tw. Last seen on 31 Aug 2013.
16. Korea OCW Consortium (KOCWC). http://www.kocwc.org. Last seen on 31 Aug 2013.
17. Russian Open Courseware (RuOCW). http://ruocw.org
18. Open source web conferencing system developed primarily for distance education. http://www.bigbluebutton.org. Last seen on 31 Aug 2013.
19. Yuan J, Liberman M, Cieri C. Towards an integrated understanding of speaking rate in conversation. In: Proceedings of the 7th international conference on spoken language processing (INTERSPEECH-2006), Pittsburgh, 17–21 Sept 2006.

Mini Projects: A New Concept of Transformation of Teaching-Learning Process

Hanmant S. Jadhav and Savita N. Patil

Abstract The developing market economy of our country makes greater demands for engineering education. On the other hand, the training of student's basic theories, concepts, knowledge, and basic skills must be strengthened; also the cultivation of innovative spirits and practical abilities should be carried out. It is required that while broadening specialist field of engineering education, basic theories and fundamental qualities should be emphasized. This paper presents culture of learning by doing through incorporating mini project concept in small group of student at first year engineering. It results to develop problem-solving and presentation skills, excellent team spirit, increase research skills, self-confidence, attitudinal change, and sense of achievement in the students. An example of mini project is also presented.

Keywords Mini project • Experiential learning

1 Introduction

The mini project is an assignment given in a group of 3–5 students, designed to help students to develop practical ability, hands-on experience, and knowledge about practical tool/techniques in order to solve real-life problems to the industry, academic institutions, and community. An interdisciplinary approach and knowledge sharing culture will help in evolving more innovations. A culture of learning by doing something different makes a big difference to come out with new ideas [1, 2].

H.S. Jadhav (✉) • S.N. Patil
Rajarambapu Institute of Technology, Rajaramnagar, Urun Islampur, Maharashtra 415414, India
e-mail: hanmant.jadhav@ritindia.edu

R. Natarajan (ed.), *Proceedings of the International Conference on Transformations in Engineering Education: ICTIEE 2014*, DOI 10.1007/978-81-322-1931-6_13,

2 Scenario at RIT

Rajarambapu Institute of Technology, Rajaramnagar (RIT), established in 1983, private TEQIP funded an autonomous institute in Western Maharashtra offering various B.Tech and M.Tech programs in various streams of engineering and technology. The institute is committed to offer excellent academic environment conducive to overall development of students, faculty, as well as society.

3 Mini Projects

3.1 Establishment

The RIT model of Autonomy focuses on "Experiential learning which believes in learning by doing" [3]. This is achieved through hands-on experience, industrial assignments, mini projects, use of latest software, and live problem solving. RIT proactively initiated steps to transform teaching-learning process to make learning a joyful experience for students.

3.2 Objectives

Mini projects have been incorporated in the curriculum of first year engineering (F.Y. B.Tech) in order to enhance and utilize the high potential of the students and build positive attitude, which will help them in their social and technical life ahead. The mini project is designed to help students develop practical ability and knowledge about practical tools/techniques in order to solve real-life problems related to the industry, academic institutions, and community. The course mini project is one that involves practical work for understanding and solving problems in the field of technology [3, 4].

Opportunities: [5]

- Engineering will be central to addressing global grand challenges.
- Experiential learning.
- Students need more than disciplinary knowledge to succeed: teamwork, communication, customer awareness, project management, leadership, ethics, societal context, and professionalism.
- Have insight of interdisciplinary projects.
- Community involvement.
- Entrepreneurship development.

3.3 Implementation in Institute

The mini project was introduced in first year B.Tech curriculum and was introduced with a 1-day workshop to explain its concept and applications with following guidelines.

Process of implementation and evaluation of mini projects:

1. Each division (66 students) will have a separate mini project coordinator with 2 h workload.
2. Mini project will be a team work, consisting of a minimum of 3 to maximum of 5 students.
3. Topic selection should have applicability, social or technical; selection and finalization will be through project guide/supervisor.
4. Preparation of small project report as per guidelines.
5. Project group must provide complete solution to the selected problem with conceptual clarity.
6. The project will be evaluated by respective branch HOD, project guide, and senior faculty in the department.
7. The mini projects should be presented before the committee headed by dean academic/R&D, which shall evaluate for 100 marks.

3.4 Present Status of Mini Projects

3.4.1 Evaluation

Mini project is evaluated on the basis of novelty, fulfillment of project objective, experiential learning, knowledge, practical utility, cost-effectiveness, and presentation.

Projects undertaken, titles: some of them are listed below.

Sr. No	Mini projects
Automobile engineering	
1	Car parking system model
2	Modification in automotive safety belt
3	Pedal power generation
4	New life for plastic: as a fuel
5	Transparent I.C. engine model
6	Design and development of farm grass thresher
7	Development of electrically operated blackboard cleaning device
Electrical engineering	
1	Development of single line diagram in RIT campus
2	Multiuser mobile charger

(continued)

Sr. No	Mini projects
3	Clap switch on-off lamp
4	Fire alarm using IC-555 timer
5	Solar panel-based battery charger
Computer science and engineering	
1	Web site design for CSE dept library
2	Web site for F.Y. B.Tech
3	Movie making and video tutorial using open source tools
Information technology	
1	Creating web site using moodle for awareness among new comers
2	Numbering system using C language
3	Assembly and disassembly of computer system
4	Gas detection system
Electronics and telecommunication	
1	Crystal radio
2	E-Jacket
3	Small windmill for generation of power
4	Short-circuit protection in D.C. low-voltage system
5	Electronic lock system
Civil engineering	
1	Construction of windmill
2	Design of green building
3	Earthquake-resisting structures
4	Soil mapping
5	Automatic jaggery plant
6	Seismometer
Mechanical engineering	
1	Lift mechanism to bookshop
2	Tunnel light control
3	Sugar factory data analysis for electricity consumption
4	Magnet gas saver

3.5 Sample Example

1. Department: Information Technology

   Title of mini project: Creating web site using moodle for awareness among new comers

   No of students: 05 (F.Y. B.Tech)

   Description of the project: This web site is created by open source tool weekly.com. This web site provides help for first year students who are new to the college. It provides study material, the map of college, and laboratory information of all departments in campus of RIT.

Conclusions

The students have learned open source tools to create web site. They very well learned to use moodle source.

2. Department: Automobile Engineering

Title of mini project: Development of electrically operated blackboard cleaning device

No of students: 05 (F.Y. B.Tech)

Description of the project: This project aims to provide solution to the trouble caused by the chalk dust while cleaning the blackboard in the classroom. The chalk dust adversely affects the health of the instructor and nearby students in the class. A solution in the form of electrically operated blackboard cleaner, which uses wiper system mechanism, allows cleaning the board by staying away from the blackboard. The validation in terms of effective cleaning of board and the time it takes to clean the board has been made and recorded. An attempt has been made to limit the cost by using already available vehicle wiper mechanism. With some modifications, it has potential to be used on large-scale bases in rural schools and colleges to get rid of chalk dust.

4 Benefits

4.1 Gains to Students

- To help them develop the art of thinking, learning, problem-solving, and presentation skills, this will be a constant asset for them throughout their professional career.
- Excellent team spirit is inculcated which is necessary for smooth functioning of organization.
- Self-development, self-confidence through problem-solving ability, attitudinal change, and sense of achievement.
- To generate innovation and research skill set among scholars. Students tend to become creative and not remain parrot-like mugging up.
- Students will be able to write project reports, analyze case studies, give presentations, show their abilities, and make decisions based on the findings.
- Students will be able to plan and organize tasks, able to work as a team, and able to do more challenging tasks.

4.2 Gains to Faculty Advisor

Faculty advisor through mini project supervisors gets an opportunity to contribute concretely towards the goals and missions of the institute. Many creative models, teaching techniques, and learning resources have been developed through mini projects. It has led to attitudinal change and HRD leading to knowledge enhancement.

4.3 Gains to Institute

For any institute, goodwill and satisfaction of students is vital. And it is possible only when high-quality inputs to academics are provided. Academics are not limited to classroom teaching, it goes beyond that, often called experiential learning.

5 Conclusion

Students learn a lot of many things through mini projects and get opportunities to express themselves. Not only the students are the beneficiary, but other stakeholders like faculty, community, and the institution also benefit in the overall process. Hence in the wider perspective, these types of activities should be promoted in technical institutions of centers offering professional education.

References

1. Indian Society for Technical Education. News letter. 2010;XXX(2). Monthly- September. p. 3.
2. Indian Society for Technical Education. News letter. 2011; XXXI(11). Monthly- November. p. 5.
3. Naik BM. Colleges/universities need research parks to make education globally competitive (2013).
4. Naik BM. Innovative world class universities and colleges to nurture leaders for tomorrow (2013).
5. Patil SR, Thombare DG, Kulkarni SS. Positive influence of association of technical students organizations with professional bodies on academics: a case study of SAE RIT. Indian J Techn Educ. 2011;34(2):67–75. April-June.

Success Story of Industry Institution Collaboration for Enhancing Teaching–Learning Experience

P. Srinivasa Pai and Niranjan N. Chiplunkar

Abstract Collaboration between industry and academia can happen at various levels. It is for mutual benefit and can make meaningful contribution to the society. Interaction can happen for various reasons including curriculum design and development, training and skill development, basic and applied research, technology development and transfer. This paper discusses the efforts made by NMAM Institute of Technology, Nitte, an autonomous Institution to utilize this interaction in order to cover some of these areas and thereby enhance the teaching–learning experience of students and teachers and contribute to overall growth and development of the Institution.

Keywords Institution • Industry • Interaction • Teaching–learning

1 Introduction

Technical education is critical to the development of a nation. Industrial development has been attributed to the contributions from technical manpower. In the context of this globalized economy, where the focus is on knowledge, students must develop necessary skills such as familiarity with cutting-edge technology, problem solving, critical thinking and collaboration [1]. In spite of the recent global recession, the need for skilled manpower is increasing. There has been a slowdown in the recruitment process by industries, but in the near future as the global economy recovers, there are chances of improvement in the recruitment by industries.

The industry–academia interaction has undergone a paradigm shift, ever since its inception. The phenomenon of globalization has brought about a significant change in the higher education system. UNESCO has said that compatibility between higher

P.S. Pai (✉) • N.N. Chiplunkar
NMAM Institute of Technology, Nitte 574 110, Karnataka, India
e-mail: srinivasapai@rediffmail.com

R. Natarajan (ed.), *Proceedings of the International Conference on Transformations in Engineering Education: ICTIEE 2014*, DOI 10.1007/978-81-322-1931-6_14,

education policy and industrial development policy is crucial for the techno-economic well-being of a nation. Industries have played a limited role in the development of technical education. It includes to a limited extent involvement in formulating curriculum at the universities by including a few people from industries in the board of studies, encouraging students for industrial training, limited consultancy projects offered to the faculty, etc. [2]. Advances in science and technology are another factor, which has increased the need for this interaction. Also industries are looking at graduates who are readily employable with lesser time spent on their training. Thus, the importance of this interaction need not be overemphasized.

In this regard, the author's Institution has taken up several initiatives for starting these interactions and also to take it to higher levels in different departments. The initial experience of these interactions has been very encouraging and has been mutually beneficial. This paper tries to discuss some of the initiatives taken by different departments and suggests some measures to strengthen it further.

2 Need for Academia–Industry Interaction

The technical manpower required by the industries is provided by the Institutions. In this context, a need has been felt to have this interaction at different levels. In the initial years of the growth and development of technical education, this interaction has been restricted mainly to the placement of students. The number of students enrolling in technical Institutions is growing by the year. Rapid developments are happening in the area of technology, which is transforming every possible domain. Thus there is a need for strong partnerships between academic institutions and the industry. These will help educate and prepare students to be future-ready and accelerate innovations in the field of science and technology propelling inventions and discovery of new materials, products and processes, resulting breakthroughs that would help build new industries. There are immense possibilities of linkages in several areas including placements, curriculum redesign, teacher reorientation, affiliated science and technology parks and joint research, taking the outcome of research to the market. Accordingly AICTE, New Delhi, in collaboration with the Confederation of Indian Industry (CII) conducted a survey of engineering Institutions in the country in 2012, with assistance from PricewaterhouseCoopers Pvt. Ltd. The findings have been very encouraging, with more than 63 % of the Institutions (other than the IITs) having different levels of interaction with industries [3].

3 Issues Affecting This Interaction

Some of the major issues affecting this interaction include the 'compartmentalization' of thinking by both the industries and institutions. Industries feel their job is only to recruit students and they need them ready to be placed on the job immediately, whereas Institutions focus only on providing the students with basic technical knowledge, as per the curriculum, without bothering about its effectiveness and utility. According to PharmaQuest, some of the causes for poor and not so effective interactions are:

- Lack of interest from both sides.
- Curriculum outdated and not job oriented.
- University examination system puts emphasis on the reproduction of memorized facts, rather than any learning.
- Obsolete lab facilities do not attract industry to interact with the institute.
- Lack of awareness in Institutions to have this interaction at different levels and no proper support from the management/government.
- Rigidity of rules and regulations and lack of autonomy.
- Goals and objectives of education system and the industrial system do not match fully.
- Both have not realized the potential of this interaction and its mutual benefits [4].

4 Levels of Academia–Industry Interaction

Industry–academia interaction has been going on at different levels for the past several years and includes industrial extension services, cooperative education, equipment donation, exchange of personnel and collaborative research programmes. And these interactions have benefited the industry, institute and society in general. According to Pai et al. [2], this interaction can occur at three levels, which can be identified as:

- Level 1, low or primitive level – This includes students visiting industries for training or to get some exposure through industrial visits and industries visiting institutes to recruit people to fulfil the shortage of skilled, technical manpower. Other interactions that can happen at this level are encouraging students to undertake their project work on live industrial problems by supporting them in terms of providing facilities, donating old equipments to the institute for their use and also for enhancing the knowledge of their acquired theoretical skills.
- Level 2, medium level – This is a higher level of interaction, which mainly involves encouraging teachers to visit industries to get a feel of the theoretical subjects they teach and undergo some kind of training. Experts from industries can be invited for some lectures, seminars for students, in order to share their expertise and knowledge with the students. Continuing education programmes (CEP) for people from industries could be organized by Institutions, in order to help them gain up-to-date knowledge on theoretical concepts.
- Level 3, high level – This can be considered as the highest level of interaction, wherein industries and institutes are involved in each other's activities at various levels in order to increase and improve the collaboration for developing the skilled, technical manpower. There are different avenues for interaction at this level

which include curriculum development, research and development, upgradation of knowledge and skills on a continuous basis, producing patents, lifelong learning, encouragement of innovation and conversion of ideas into technologically viable products or solution and involvement through representation of both groups at the government policy level in order to evolve a proper policy framework which is conducive for these interactions.

5 Initiatives for These Interaction and Associated Issues

NMAMIT, which is an autonomous Institution under the affiliating university Visvesvaraya Technological University, Belgaum, has lot of scope for industry–institute interaction. The interactions are happening at the department level and are helping the Institution in improving the teaching–learning process and provide a different experience to the teachers and students. Some areas where interactions are happening include curriculum design and development, introduction of new industry-focused electives, faculty and student training, projects for students, additional courses and certifications for students to increase their employability and basic and applied R&D.

This section presents the different initiatives taken by some of the departments:

1. Computer Science and Information Science Department
   (a) EMC Corporation – The department has established partnership with EMC Corporation, USA, in the area of Storage Technology. This partnership started with the introduction of an elective course on Storage Technology. Further 'Train the Trainer' training course was conducted for the faculty of the department. To further this interaction, an International Conference and Symposium on Storage Technologies was conducted in December 2008, in which EMC's Global Vice President for Education Services participated. EMC has provided NAS boxes for students and faculty to work on storage-related areas. Students of this department write certification programmes, which are beneficial to their future career, for which necessary assistance is provided by EMC. Thus, this partnership has provided 'early move advantage' to the students, thereby providing an edge over other students, be it in training or placements. In fact NMAMIT has been one of the earliest partners with EMC and thus has gained an early advantage.
   (b) Infosys – Campus Connect is an industry–academia partnership initiative which started in 2004. NMAMIT was one of the Institutions which took the early lead and became a Campus Connect Partner, almost 8 years ago. Some of the initiatives undertaken under this include offering an industry-focused elective 'Essentials of IT', which is to be compulsorily taken by the Infosys placed non-IT students but which can be taken by other students also. In addition to this, Infosys has designed the curriculum of an elective called 'Industrial Aircraft Design', which has been offered to the mechanical engineering students, which is designed in two parts and offered in VIth

and VIIth semester, which has been appreciated by the students and which is taken by them almost regularly. Faculty who are supposed to handle this elective are rigorously trained at its training centre at Mysore. Thus, this initiative has been a 'win-win' situation helping the Institution in obtaining necessary learning resources and training skills, which prepare the faculty to train the students and make them industry-ready and to help the industry in getting 'employable graduates', reducing the time spent on their training and making them immediately available to be put on job.

(c) INTEL – Through Foundation for Innovative and Collaborative Education (FICE), it has started Intel Embedded Lab using ATOM processor. The necessary infrastructure was created by the company and has also provided the necessary funding and kits, which helps in training students and also in their project work. The content of the Embedded system course, which was already being offered, was suitably modified to suit the new lab requirements. Two students from the Institution underwent their training at IISc, Bangalore. The students working in this lab have presented their research work at the Intel Asia Conference held at Malaysia and also at the Intel India Conference recently held at Goa.

(d) SPAN Infosystems – The engineers of this company have designed a course entitled 'Essentials of IT Industry' and have been offering it as an open elective. The conduction of classes, evaluation and grading of performance of the students is done entirely by the engineers of the company.

(e) IBM – IBM India has set up its Centre of Excellence (COE) in the Institution, and every year about 80–100 students get certified in IBM technologies without any additional fee. Recently the Institution and IBM signed a memorandum of understanding (MOU), according to which all the Computer Science and Engineering undergraduate students have to undergo specialized credit-based courses starting from Ist to VIIth semester in one of the following specializations, namely, Cloud Computing, Healthcare Informatics and Business Analytics and Optimization.

(f) NVIDIA has set up a CUDA development centre in the Institution, through which interested students get trained on CPU–GPU programming.

(g) The Institution is a Microsoft Ed-vantage programme member at platinum level and is in the process of setting up a Microsoft Innovation Centre.

(h) Three faculty members underwent training for 15 days at Robosoft Technologies, Udupi, on mobile application development. These faculty are now handling this subject. This is a lab-oriented course wherein the experts from Robosoft Technologies assist the trained faculty members in evaluating the student's projects done in this area of specialization.

2. Electronics and Communication Department - WIPRO – WIPRO offers a training programme called Mission 10X, and NMAMIT took the initiative and provided necessary support for the conduct of this training programme for all the faculty of the Institution. During the year 2012–2013, Mission10X collaborated with 25 engineering colleges across India and established Mission10X Technology Learning Centres (MTLC) that house the UTLPs. The inaugura-

tion of all the 25 centres was completed in the month of March 2013, and NMAMIT is one of them. This lab is housed in the department and has Embedded ARM processors and offers project work to undergraduate students from E&C and Computer and Information Science students.

3. MCA department – The MCA department has been very active with regard to industry–institute interaction. For the last several years, it has been actively engaged with several industries like HP Technologies, SonicWALL, Sourcebits, etc. Recently it has started taking real-time industry projects, with deadlines. Regular monitoring of the progress is done through teleconferencing. TTH Ahmedabad has opened their development centre in the campus, in which the students work on projects under supervision of their employees and the department faculty.
4. Electrical and Electronics department – The department has been actively involving itself with industries, with a few faculty working part time in industries, which is an added advantage. Last year the department worked on a sponsored project from Digital Circuits, Bangalore, regarding the development of a maximum power point tracking for solar systems.
5. Mechanical department – The department has been active in research with many sponsored research projects, with its outcomes being relevant to industries working in areas like MEMS, machining, condition monitoring, alternative fuels for IC engines, etc. Many projects are taken up by students, which are beneficial to local industries like food, bakers and confectioners, agriculture, irrigation, etc.
6. Biotechnology department – The department has the highest funding from several funding agencies in the Institution and has been collaborating with several industries with regard to research, internship to students and framing of curriculum for new PG courses, with people from leading industries being members of several advisory bodies and board of studies.
7. Civil department – The department provides consultancy and testing services to several government and private sector organizations and government departments in the areas of soil stabilization, remote sensing and GIS applications, water resources management, ground improvement techniques, etc. It has collaborated with several organizations including M/s CADD centre in Mangalore, in providing training to students and helping in placements.

5.1 Issues Identified

Institutes must design their curriculum in line with industry requirements, at least by offering elective subjects. By this more and more industries may come forward to assist the institutes in delivering the subject contents at the institute premise itself. In a non autonomous institute, it is quite difficult to do these changes, whereas in case of autonomous institutes, these changes can be introduced very easily and newer electives can be offered as and when required in the curriculum. As far as other levels of interaction are concerned, concrete outcomes need to be shown in terms of successful implementation of projects or consultancy offered and improved student performance in projects and placements. There should be fruitful interactions with faculty resulting in meaningful outcomes at both ends. By this, interactions at different levels can be further strengthened for mutual benefits.

6 Benefits Obtained and Looking Forward

The current status of industry–academia interaction at NMAMIT is very encouraging and shows a lot of potential for further interaction. Based on the levels of interaction discussed previously, NMAMIT has had interaction at all the three levels, which shows the maturity of these interactions. The efforts have helped the Institution and interacting industries in terms of providing current curriculum to the students, providing additional training and skills to enhance their employability, providing additional training to the faculty and supporting their R&D activities and providing readily employable students to the industries. This interaction has further strengthened the efforts of the faculty in getting significant research funding from government funding agencies and also from many leading industries and organizations. There has been a significant increase in publications in leading international journal and conference proceedings. Additional benefit of these interactions has been in community development, helping local students and industries, from the knowledge gained through these interactions. Significant intellectual property has been generated both by the institution and industries helping in the development of future technology.

The departments have to pursue this more rigorously by actively involving the students and faculty. The management of the Institution has been supporting these efforts by providing necessary infrastructural facilities and financial support for the initiatives and rewarding the faculty and students by ensuring a flexible and motivating work environment. Some more support from the management will definitely strengthen these initiatives and help in reaping the true benefits of this interaction. Implementation models of leading universities of the world can be considered as benchmarks to achieve excellence in this interaction.

7 Conclusion

This paper provides an overview about the current status of industry partnerships with different departments for promoting effective teaching and learning experience to students of NMAMIT. The efforts have been mutually rewarding to both the Institution and interacting industries, with significant improvements in curriculum, training inputs, project work, R&D outputs, publications/patents and placements, to both students and faculty. Overall there has been increase in intellectual outputs, which will contribute to the overall growth and development of the Institution.

References

1. Chandran DPK, Kooi SY, Harizan MHM, Kooi CC, Hoy TT. Success story of collaboration between intel and Malaysian universities to establish and enhance teaching and research in electronic packaging. Proceedings of 34th International Electronic Manufacturing Technology Conference; 2010.
2. Pai SP, Bhat KS, Sriram NS, Samaga BS. Increased industry-academia interaction and its relevance for technological innovation in the new globalized world. Souvenir of 37th Annual ISTE Convention & National Seminar on Excellence in Technical Education through Innovation at MIT, Manipal; 2007 Dec 17–19. p. 44.
3. AICTE-CII Survey report; 2012 Nov 8. www.ritindia.edu/aicte-cii-pwc-survey-report.pdf
4. www.pharmaquest.weebly.com/uploads/9/.../industry_institute_interaction.ppt

Overlapped Concepts Pedagogy for Advanced Computer Architecture

Shampa Chakraverty

Abstract Advanced Computer Architecture is a core subject of the B.Tech course in Computer Engineering in Delhi University, taught during the sixth semester. The subject is a conundrum of concepts derived from various subdisciplines of computer science such as computer organization and architecture, networks, parallel processing, operating systems and algorithms. The pedagogical method adopted in the classroom to impart knowledge on this subject is therefore as vital as is the content, its delivery and the use of practical tools. This paper analyses the impact of using the proposed overlapped concepts methodology for teaching ACA, on students' learning experience. The aim is to enhance the understanding of complex scenarios involving multiple concepts, evaluate alternative solutions and design systems – an essential graduate attribute. The analysis indicates that the technique of seeding relevant fragments of a set of related concepts within the larger context of a base concept currently being focused on sharpens the cognitive ability to identify linkages between concepts, helps build complex scenarios and triggers ideation.

Keywords Computer Engineering education • Computer Architecture • Overlapped concepts • Pedagogical methods

1 Introduction

With the inclusion of multiple processor cores within the same chip, parallel processing has become an integral part of desktop computing. It has added a whole new dimension to the way modern applications are being developed. This architectural advancement, coupled with different flavours of distributed computing such as

S. Chakraverty (✉)
Netaji Subhas Institute of Technology, New Delhi, India
e-mail: shampa@ieee.org

R. Natarajan (ed.), *Proceedings of the International Conference on Transformations in Engineering Education: ICTIEE 2014*, DOI 10.1007/978-81-322-1931-6_15,

cluster, grid and cloud computing, has spurred a renewed interest among students to grasp the theoretical intricacies of Advanced Computer Architectures (ACA). The first four graduate attributes in outcome-based education underline the importance of investigating complex problems, analyse various aspects and design viable solutions [1]. In this paper, we propose and evaluate the overlapped concepts methodology for the classroom teaching of ACA with the objective of raising the levels of these attributes.

2 Related Work and Motivation

Etienne Riviera illustrates the use of SPLAY, a distributed system evaluation framework to spur hands-on experience of students [2]. Orsega et al. report that the learning rates achieved by using a data structure learning tool Sketchmate with feedback are better than that achieved by using the conventional paper-and-pencil method [3]. Richard Vuduc et al. report on how exit surveys have helped them in redesigning the undergraduate entry-level course on high-performance computing [4]. In [5], Francisco Javier de Hoyos describes a wide spectrum of computing platforms, development toolkits, algorithms and teaching methods that aid in teaching distributed systems.

The appropriate classroom pedagogical aspect of teaching ACA has not yet been adequately investigated. ACA emanates from the confluence of several core computer science concepts including computer organization, networks, task scheduling, compiler-driven and hardware-enabled dynamic code optimizations, parallel algorithms and operating systems. There is an urgent need to mull upon the methods that can be adopted to highlight the complex relationships between the numerous parameters that govern the overall working of high-performance computer systems. Some books give an overview of several concepts right in the first chapter and then expand each one of them in subsequent chapters such as in [6]. However, this approach is largely infeasible for a subject rife with weighty interconnected concepts such as ACA. If a potpourri of high-level concepts is given right in the beginning, it may lead to the students' interest drying up at the initial stages itself.

The limitation of books can be overcome by adopting an appropriate pedagogical technique for classroom teaching. Models that use variations of overlapping concepts have been used successfully by educators for designing curriculums [7]. We develop a model based on overlapped concepts for improving the effectiveness of the teaching learning process in ACA.

3 Overlapped Concepts Pedagogy

We adopted the following set of techniques under the umbrella of overlapped concepts methodology to teach the subject of ACA to VI semester students of B.E. Computer Engineering at Netaji Subhas Institute of Technology, Delhi University, during the even semester of two consecutive years 2012 and 2013.

3.1 Overlapped Concepts

The subject is a constellation of concepts. Each concept is a collection of artefacts such as models, examples, algorithms and explanations. One specific major concept is expounded upon as the main theme during any given time frame, providing a firm ground upon which related concepts can germinate. Concepts are patterned in an overlapped manner. A relevant portion of another related concept, which is scheduled to be explained in detail later during the course, is deliberately brought into the purview and explained enough to highlight its relevance to the current theme. This was done by carefully constructing problem scenarios involving multiple concepts. The aim is to establish the interrelationships between concepts. More than one interrelated concept may be brought in within the scope of discussion in this manner. The newly introduced topics are revisited later when they in turn are treated as the central theme. The interrelationships that were revealed earlier are emphasized once again at that time. Simultaneously, fresh conceptual connections are introduced. In this manner a chain of overlapping concepts link together a series of classroom discussions.

3.2 Scenario Building

The technique requires careful planning to construct scenarios that entail consideration of related quality issues and the application of multiple concepts to solve them. To illustrate, let us take a case study involving two architectures: vector processors and SIMD processors. They represent parallelism along orthogonal dimensions; while vector processing taps temporal parallelism, SIMD processors use spatial parallelism. Both have common grounds in that they handle array processing. They also work together in tandem to create new parallel architectures such as an array of vector processors. Coming into play here is the role of compiler optimization for nested arrays. In an endeavour to speed up execution of nested loops, should inner loop be vectorized and outer loop be parallelized? Or is it more beneficial to parallelize the inner loop and vectorize the outer one? A solution to this issue should also take into consideration a proper tuning between hardware and software and a well-planned data organization in the memory subsystem. Students are encouraged to build alternative solutions for different applications. These scenarios paint a complete picture where students understand how different architectural approaches spanning related concepts can solve the same problem for different situations and applications.

3.3 Quiz Rotary

A small time duration of 5–10 min is allocated at the beginning of each and every class to conduct an oral quiz. The quiz included simple and direct questions meant to reenforce the linkages between the concepts that had been discussed in previous classes. Instead of the traditional approach whereby oral quiz questions are thrown open to the entire class, students are given a chance to answer questions on a turn by turn basis. This allows a few students to be quizzed during each period due to time constraints. However, over a period of time, everyone gets an equal opportunity to provide or to refine answers. We generally followed this pattern interspersed with some open questions, thereby allowing everyone to be involved.

4 Evaluation

Table 1 below shows the questionnaire prepared for the purpose of our survey.

The questionnaire was designed to elicit response from students on different aspects of their learning experience during the course. The questionnaire and student responses are reproduced in Table 1. It includes seven psychometric questions based on five-point Likert scales, one multiple-choice question and one descriptive question seeking the responder's overall learning experience. The questionnaire was given to the students during the last class of the semester. Forty-four students in the batch of 2012 and 54 students in the batch of 2013 participated in the feedback exercise to give a total of 98 responses that were consolidated.

In response to Q1, two thirds of the students agreed that an appropriate pedagogical method has a significant impact on learning a subject like ACA. This sets a positive tone for evaluating its impact on different aspects of the learning process. The response to Q2 shows that 63 % of the students felt that their quality of learning improved as a result of the overlapped concepts pedagogy. However, about one fourth of the students remained neutral on this aspect, while 12 % of the students felt a reduced quality of learning. What emerged from a perusal of the students' descriptive answers is that as the course progressed, things became clearer, everything fell in place and students were able to absorb from a bigger arena of interrelated concepts with greater ease. The main cause of setbacks in learning in some cases arose when students missed classes at times and found it very difficult to remain in sync with the gradual progression from one concept to another.

Q3 probes an important aspect of the proposed method: its impact on lateral and creative thinking. The response is very encouraging: it shows that 16 % of the students who found that it gave an impetus to their creative thinking frequently, 35 % of the students expressed that this happened often and 37 % chose 'sometimes'. An examination of the descriptive answers to Q9 revealed that the complex scenarios presented to the students for solving pushed them to apply alternative solutions creatively and weigh the consequences of each. 12 % of the students in categories 1 (never) and 2 (rarely) found the simultaneous discussion of several related concepts difficult to grasp.

Q4 concerns itself with the trade-off between the breadth and the depth of coverage that is possible with overlapped concepts. Its response question reflects the flip side of this approach – a major 73 % of the students felt that the depth of coverage for individual topics in the classroom was inadequate as a result of the time devoted to

Table 1 Questionnaire on impact of overlapped concepts pedagogy and response

Option-1	Option-2	Option-3	Option-4	Option-5	Parameters
(Q1) To what extent do you think the pedagogical style for teaching the subject of ACA is at all relevant?					Median: 4 Mode: 4
Completely irrelevant: 4 %	Somewhat irrelevant: 8 %	Neither irrelevant nor relevant: 20 %	Quite relevant: 55 %	Completely relevant: 13 %	Interquartile range: 1
(Q2) To what extent did the overlapped and interwoven discussion of topics affect the quality of your learning of the subject ACA? It:					Median: 4 Mode: 4
Reduced significantly: 2 %	Reduced: 11 %	Neither reduced nor increased: 24 %	Increased: 48 %	Increased markedly: 15 %	Interquartile range: 1
(Q3) How often do you think that the concurrent discussion of related topics help in lateral and creative thinking?					Median: 4 Mode: 3
Never: 5 %	Rarely: 7 %	Sometimes: 37 %	Often: 35 %	Frequently: 16 %	Interquartile range: 1
(Q4) To what extent do you agree with the statement: the overlapped discussion of topics clarified the linkages between topics at the cost of understanding the depth of coverage of individual topics?					Median: 4 Mode: 4
Strongly disagree: 3 %	Disagree: 11 %	Neither agree nor disagree: 13 %	Agree: 44 %	Strongly agree: 29 %	Interquartile range: 2
(Q5) To what extent did the regular oral quizzes conducted in the class strengthen your understanding of the linkages between concepts?					Median: 3 Mode: 4
Did not strengthen at all: 5 %	Strengthen slightly: 22 %	Strengthen moderately: 28 %	Strengthen appreciably: 35 %	Strengthen fully: 10 %	Interquartile range: 2
(Q6) How often did the overlapped concepts method propel you to do exploratory study besides consulting textbooks?					Median: 3 Mode: 3
Never: 8 %	Rarely: 18 %	Sometimes: 35 %	Often: 31 %	Regularly: 8 %	Interquartile range: 2
(Q7) How often did you feel the need to engage in peer discussions in order to solve the complex and multi-concept parallel processing scenarios?					Median: 4 Mode: 4
Never: 3 %	Rarely: 5 %	Sometimes: 22 %	Often: 43 %	Regularly: 27 %	Interquartile range: 2
(Q8) What should be the number of related concepts discussed at a time a. Strictly sequential, one concept at a time: 15 % b. Two concepts at a time: 30 % c. Three or more topics at a time: 17 % d. As the course progresses, the number of topics can be increased gradually: 38 %					
(Q9) Give a brief write up about your overall learning experience during the ACA course					

clarifying the linkages between concepts. Q5 collates students' response to an important support activity of our scheme: oral quiz. Its response shows that 45 % of the students felt that regular oral quiz sessions conducted at the beginning of each period significantly strengthened their understanding of the linkages between concepts either fully or appreciably. Another 28 % stated that their understanding improved moderately as a result of these. Many students qualified this in their descriptive answers. Regular quizzes primarily helped them to remain attentive in class and motivated them to revise the correlations between related concepts.

Q6 probes whether the students were motivated to carry out exploratory study to further their understanding. An interquartile range of 2 reflects a wide spread in the response. 35 % of the students referred extra sources only sometimes, and 39 % of the students studied from alternative sources regularly or often, while 26 % of the students did so rarely or never. The next question, Q7 investigates how often the students engaged in peer group discussions in or outside class in order to solve complex scenarios in parallel processing. The results reflect a very encouraging aspect of this method – the boost to collaborative learning. As many as 70 % of the students said that it led them to discuss among peers regularly or often. Even among the remaining students, 22 % did engage in peer discussions sometimes. This instructor has been teaching the subject for quite a while and found the level of discussion between teacher and students as well as among students themselves in the class to be greater and more enthusiastic than earlier. Q8 seeks to find out how many concepts can be comfortably taken up concurrently. Interestingly, the mode category (with 38 % students) represented the approach to start with few topics and gradually increase the number of topics as the course progresses. A relatively small proportion of students (15 %) still advocated the traditional sequential treatment of individual concepts. Among others, most students (30 %) were comfortable with a maximum of two concepts taken concurrently.

5 Conclusion

At this point it is pertinent to revisit some of the responses and look for improvements.

5.1 Channel Curiosity

Many students reported getting distracted in the initial stages of the course due to the inclusion of multiple concepts. This initial discomfiture can indeed act as a catalyst to questioning and triggers curiosity – an essential component of learning. If the teacher capitalizes on this initial trigger by encouraging students to field queries and engage in peer discussions, a whole new process of enjoyable learning can ensue.

5.2 *Tackle the Depth Aspect*

The survey brings to surface a negative effect of this teaching method and raises a question on how to cope with the depth angle. The boost to self-learning and collaborative learning that this method triggers can be tapped to induce students to study specific topics in-depth. We conducted a successful experiment in which competing teams were made to prepare technical cum business proposals to solve specific problems in parallel processing. This endeavour saw enthusiastic participation from both toppers and relatively weaker students alike.

To conclude, the overlapped concepts method provides fertile ground for creative ideation. Mastery of the subject of ACA requires a sound understanding of the synergy between apparently disparate topics that is difficult to capture in books. By seeding new concepts within the context of base concepts, students are motivated to construct complex scenarios and seek solutions through peer discussions. Conducting oral quiz on a rotatory basis plays a significant role in helping students revise new concepts and remain alert and well prepared in class. The proposed pedagogical approach thus initiates a high-quality exploratory and collaborative learning process.

References

1. Vaithilingam CA, Rajparthiban R. A dynamic approach to outcomes based education in engineering curriculum. Proceedings of the IETEC'11 Conference, Kaula Lampur, Malasia; 2011.
2. Etienne Rivi'ere. Simplifying hands-on teaching of distributed algorithms with SPLAY. Edu-Par'12: Second NSF/TCPP Workshop on Parallel and Distributed Computing Education, Shanghai; 2012 May.
3. Orsega MC, Vander BT, Skinner CH. Two experiments using learning rate to evaluate an experimenter developed tool for splay trees. Proceedings of the 42nd ACM technical symposium on Computer science education- SIGCSE'11, Dallas, Texas, USA; 2011 Mar 9–12.
4. Vuduc R, Czechowski K, Chandramowlishwaran A, Choi JW. Courses in high-performance computing for scientists and engineers. Proceedings of NSF/TCPP Workshop on Parallel and Distributed Computing Education (EduPar), Shanghai, China; 2012 May.
5. de Hoyos FJ. Learning about distributed systems. Available at: http://openaccess.uoc.edu/webapps/o2/bitstream/10609/2285/1/Learning%20About%20Distributed%20Systems.pdf (2010).
6. Kernighan BW, Ritchie DM. C programming language. 2nd edn. Upper Saddle River: Prentice Hall Software Series; 2012.
7. Fogarty R. Ten ways to integrate curricula. http://www.ascd.org/ASCD/pdf/journals/ed_lead/el_199110_fogarty.pdf (1991 Oct).

Learning Made Easy By LabVIEW™ Software Tool

Tanuja R. Patil

Abstract Today's global world of education requires that current learning practices have to be revolutionized to challenge the traditional system of learning. There is a passion for the establishment of intelligent environment to radically transform our society to enable the different kinds of learners. Intelligent learning environment refers to a category of educational software in which the learner is 'put' into a problem-solving situation. This learning environment is quite different from traditional courseware which is based on sequential teaching and evaluation based on sequence of questions and answers.

This paper focuses on showcasing the use of modern tool, LabVIEW™, that can help to make a difference to teaching and learning experiences, by enriching the learning process. LabVIEW™ tool provides a quantitative, time-based simulation environment as well as real-time implementations, where students formulate a solution to a given problem and get feedback on how well a solution works. They can 'see' (visualize) the things happening and understand the concepts better. The simulation is a realistic visualization and animation of a mathematical problem. As we know, a visual type of learning provides powerful results for better learning, and the use of this tool has proven better results. In this paper, the use of this tool in teaching the course Virtual Instrumentation and concepts of digital signal processing C++ has been discussed. Better learning by students has reflected in their performance in exams as well as projects. This lays a foundation for active learning and smart class approach.

Keywords LabVIEW™ • Active learning • Virtual Instrumentation

T.R. Patil (✉)
I.T. Department, BVB College of Engineering and Technology,
Vidya Nagar, Hubli, Karnataka 580031, India
e-mail: tanuja_p@bvb.edu

R. Natarajan (ed.), *Proceedings of the International Conference on Transformations in Engineering Education: ICTIEE 2014*, DOI 10.1007/978-81-322-1931-6_16,

1 Introduction: Active Learning

Today's global world of education is learner-centric as learners are of different kinds. Learning is most effective when it takes place as a collaborative rather than an isolated activity and when it takes place in a context relevant to the learner.

Gardner describes eight different kinds of intelligence:

1. *Visual*: good with art and design
2. *Linguistic*: good with words
3. *Logical*: good with numbers and math
4. *Bodily*: good at action, movement and sports
5. *Musical*: good with music, tone and rhythm
6. *Interpersonal*: good at communicating with others
7. *Intrapersonal*: good at self-reflection
8. *Naturalistic*: good at appreciating the world and nature [4]

Conventional blackboard teaching may help linguistic and logical type of learners which covers approximately 25 % of learners. Ambient learning environment has to be created for active learning from all different kinds of learners, where the learning content must be adapted to the dynamic nature of their learning needs. Advances in recent software tools have facilitated learners for active learning in the classroom and also to continue their learning outside their classrooms, when and where they desire. The main objective of ambient learning is to provide a pragmatic, easy-to-use personal learning environment that is being adapted to individual needs and to high-quality learning material. Learners are provided with contextualized learning resources – a SMART classroom approach.

SMART classroom facilitates visual and kinaesthetic type of learners, thus covering almost 80 % of learners, developing creativity, analytic thinking, intuitive thinking, reflective thinking and experimentation – the key foundations for productivity and economic growth. It also helps in lesson creation, classroom management and student assessment. It reduces the time in setting up classroom technology, enabling greater focus on teaching and learning. Here, we have discussed how SMART classroom approach and the use of LabVIEW™ software tool have helped in the enhancement of learning. This approach has been used in teaching 'Virtual Instrumentation', a three-credit course for 8th semester Instrumentation Technology students. The course projects designed for this course and the projects that have won prizes show that this tool can be effectively used in teaching the concepts of digital signal processing, C++, Control Engg., Automation in process control and many other relevant courses [1–3].

2 What Is LabVIEW™?

LabVIEW™ (Laboratory Virtual Instrument Engineering Workbench) is a graphical programming environment, which is a powerful and versatile analysis and instrumentation software system for measurement and automation. LabVIEW™ consists of two main windows:

1. Front panel – Consists of controls and indicators used to design the front panel as per user requirement.
2. Block diagram – The block diagram consists of executable icons called nodes and wires and the programmes are called Virtual Instruments (VI). The block diagram is the source code for the VI. LabVIEW™ contains a comprehensive set of VIs and functions for acquiring, analysing, displaying and storing data as well as tools to help for troubleshooting.

3 LabVIEW™ Features

Hardware Implementation LabVIEW™ is used for simulation as well as real-time implementation (with the help of interfacing cards) unlike MATLAB which is more preferred only for simulation.

Drag and Drop Functions It has got thousands of drag and drop, highly interactive built-in library functions which help for easy creation and modification of VIs.

Testing and Verification It is a powerful tool which helps for design, simulation and analysis of a system/component before the real-time implementation of the same.

Flexibility and Scalability Engineers and scientists have needs and requirements that can change rapidly. LabVIEW™ provides extensible solutions to design open framework that seamlessly integrates software and hardware.

Multiplatform LabVIEW™ works on various platforms like Mac OS, Sun Solaris and Linux.

Visualization Capabilities LabVIEW™ includes a wide array of built-in visualization tools to present data such as charts, graphs and 2D and 3D visualization. Any changes in data presentation, such as colours, graph types and size can be easily done.

Connectivity and Instrument Control There are ready-to-use libraries for integrating stand-alone instruments, data acquisition devices, motion control and vision product, GPIB and RS232 devices and even PLCs to build a complete measurement and automation solution [5].

Applications of LabVIEW™ Engineers and scientists use the LabVIEW™ graphical system design as a platform to meet a wide range of application challenges.

Data Acquisition High-speed data acquisition from any sensor on any bus is possible.

Analysis and Signal Processing It can be used to perform advanced analysis and signal processing, display data on custom user interfaces, log data and generate reports and automate data collection.

Control We can easily implement PID and advanced control for process control applications, using multisim and interface with LabVIEW™, and implement mechatronics and machine design with application-specific tools [5].

4 SMART Classroom

Here the classroom used was the laboratory, equipped with personal computers and loaded with NI LabVIEW™ software. Initially the concepts were taught theoretically and later students implemented VIs with LabVIEW™ software. They could see the results then and there itself which created great interest in learning. The human brain is divided into two parts, i.e. the left brain and right brain. The left brain responds to logic and is more suitable for people with logical type of intelligence. The right brain responds for pictures, music and colours, thus enabling visual type of learners. LabVIEW™ software helps in the visualization of concepts which will be stored in subconscious mind, and it helps to REGISTER the things permanently in memory. 3-R technique is used here, i.e. registration, retain and recall. This concept makes LabVIEW™ stronger as it helps in good conceptualization and good memory [3].

Student assessment was done by giving course projects which was assessed for 40 % of internal evaluation. A team of four students had been formed and course projects were given as assignment. It was interesting for the students to carry out these projects and also interesting for the faculty to evaluate.

5 Results

We have discussed that changing teaching methods based on these concepts had an impact on creating interest in learning, student achievement and motivation. The following are the course projects done by students where learning was easy, especially in the following subjects [6].

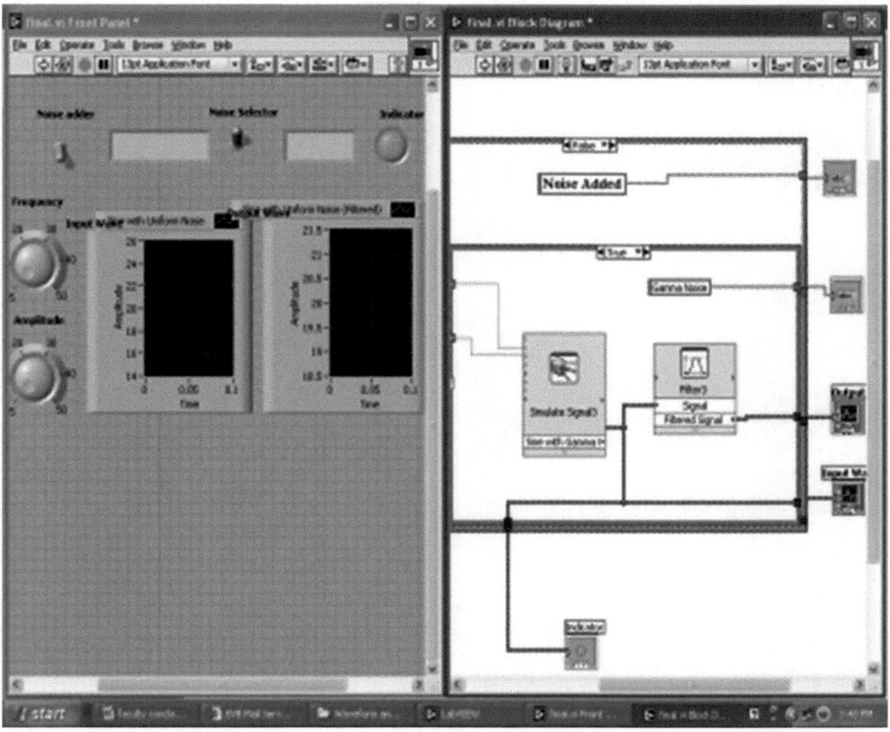

Fig. 1 Waveform analyse

5.1 *Concepts of Virtual Instrumentation*

Front panel consists of control palette where numerous controls and indicators are present. Front panel of any measuring device can be designed quickly and easily using these drag and drop icons. Here virtual measurement of external parameters with the help of sensors was discussed. The concepts of signal generators, voltmeter, CRO and waveform analysers were taught. Displaying single data and multiple data on waveform graph and waveform charts was discussed. All these 'Virtual Instruments' created appeared like real instrument panels. Real-time implementation is also possible just with the help of a suitable interfacing card unlike other software where dumping of the code is needed. Figure 1 shows the front panel (left side) and block diagram (right side) of a waveform generator and analyser, where we can see that GUI creation is very easy. And there is an easy replacement and resizing of various types of knobs, displays, indicators, etc.

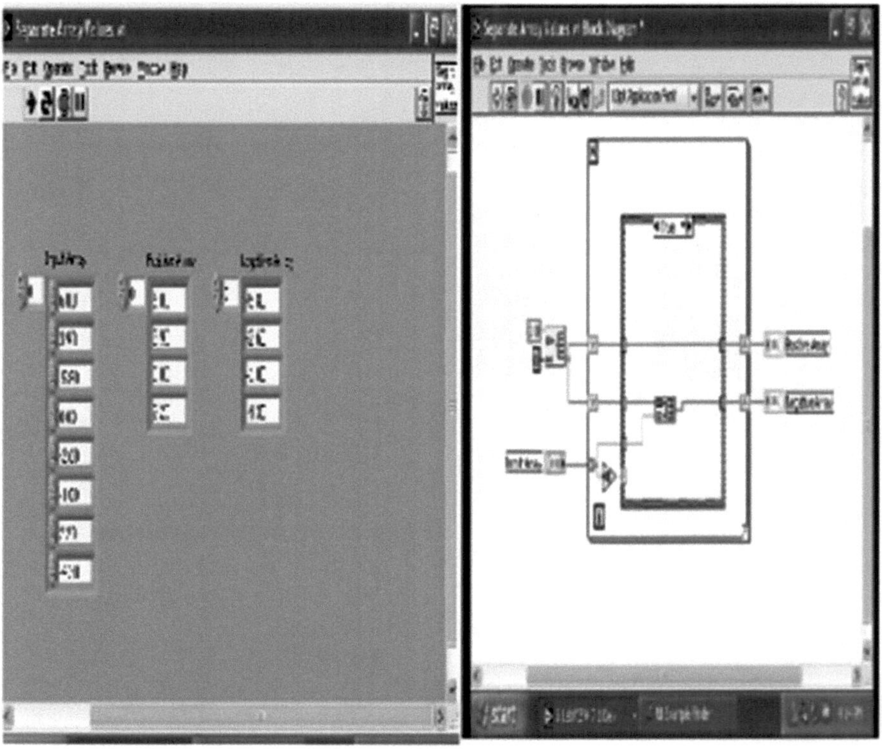

Fig. 2 Array separation

5.2 Concepts of C++

Here, the concepts like loops (for loop, while loop), arrays (1D and 2D arrays) and structures (case, sequence, event) were taught effectively. Figure 2 shows 'separation of positive and negative arrays', where we can see the use of arrays on front panel (left side) and use of case structure and for loop on block diagram (right side) where it clearly shows how easy it is to visualize arrays, loops and structures.

5.3 Concepts of Automation

Interfacing physical parameters with the help of buses and creating GUIs to visualize industrial scenario and controlling the parameters with the help of actuators were taught. 3D representation of process flow is also possible with the help of 3D graphs. Figure 3 shows GUI created for 'greenhouse automation'. This figure shows the use of comparators, various controls and indicators to measure and display parameter values like moisture, humidity and light intensity.

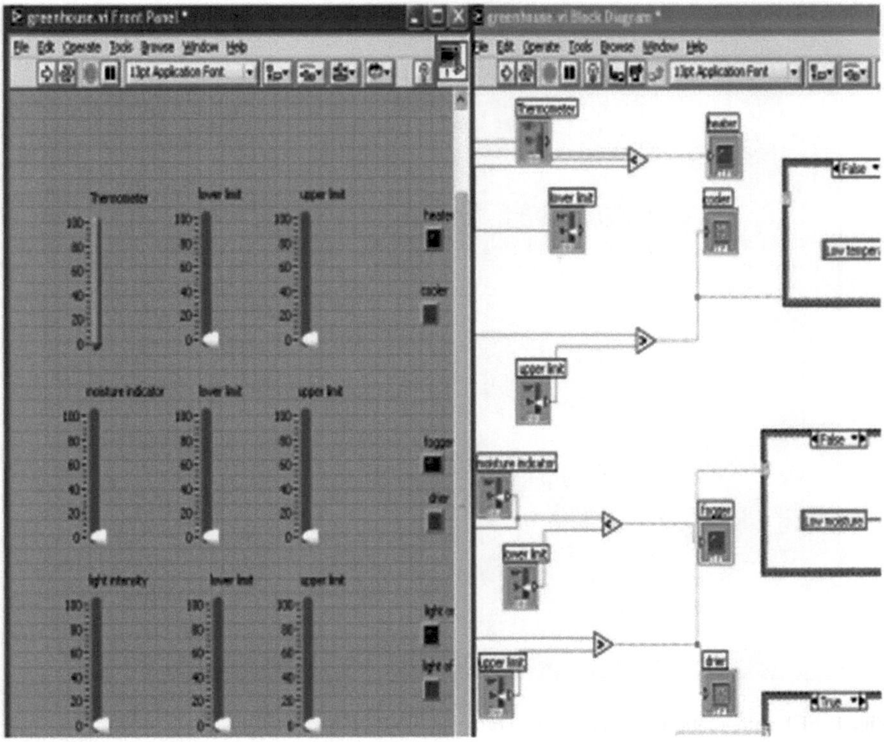

Fig. 3 Greenhouse automation

5.4 Concepts of Digital Signal Processing

Here the sampling theorem, reconstruction of the signal, computation of DFT and FFT and use of filters to remove noise were discussed. Figure 4 shows an example to compute 8-point FFT.

5.5 Animation

It is very easy and interesting to create animation. Figure 5 shows GUI created for animating the items in a mall upon the request of a user. The 3D view of these items using 3D graphs created much more interest among students (Fig. 6).

5.6 Control Engineering

The concepts of transient response of any order control systems, digital control systems and state-space analysis can be taught very effectively with the help of control simulation tool.

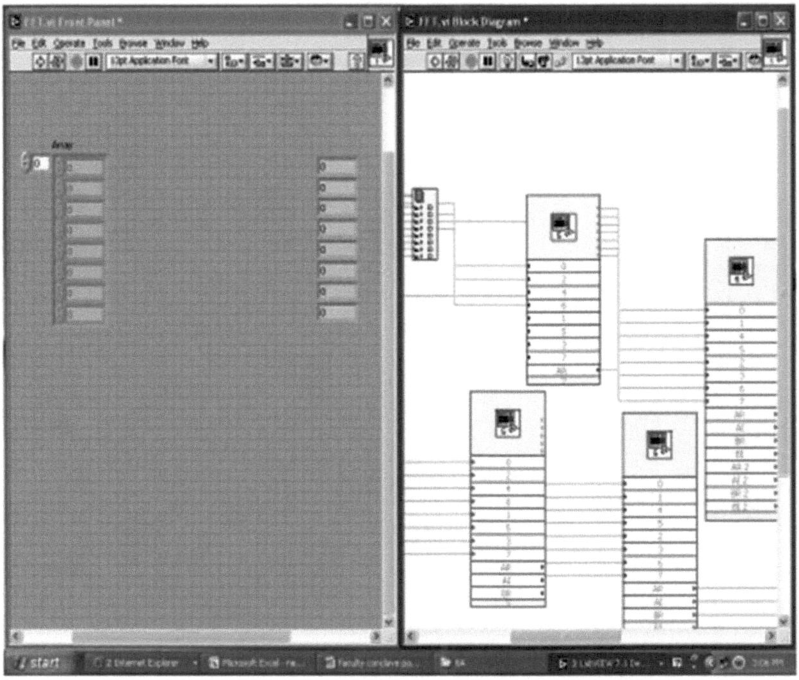

Fig. 4 FFT computation (Source: Course projects, BVBCET, Hubli)

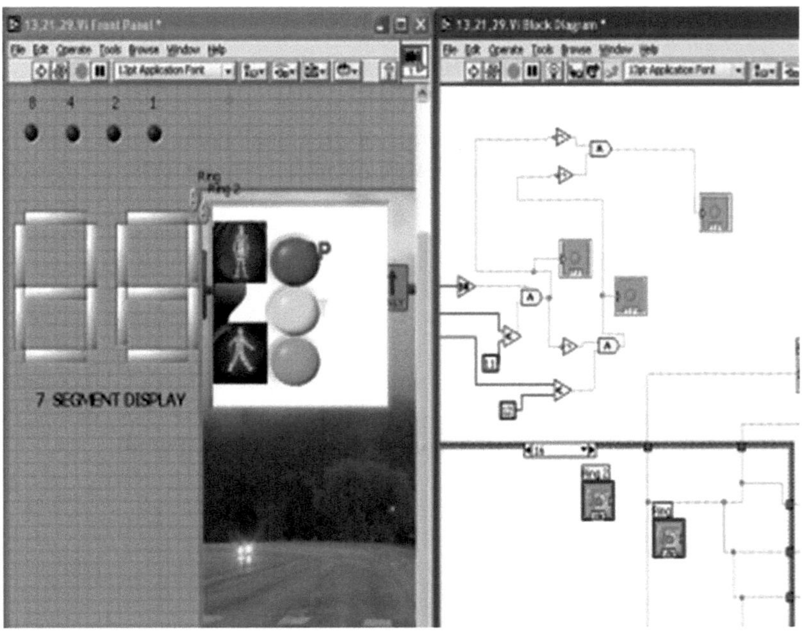

Fig. 5 Traffic control system (Source: Course projects, BVBCET, Hubli)

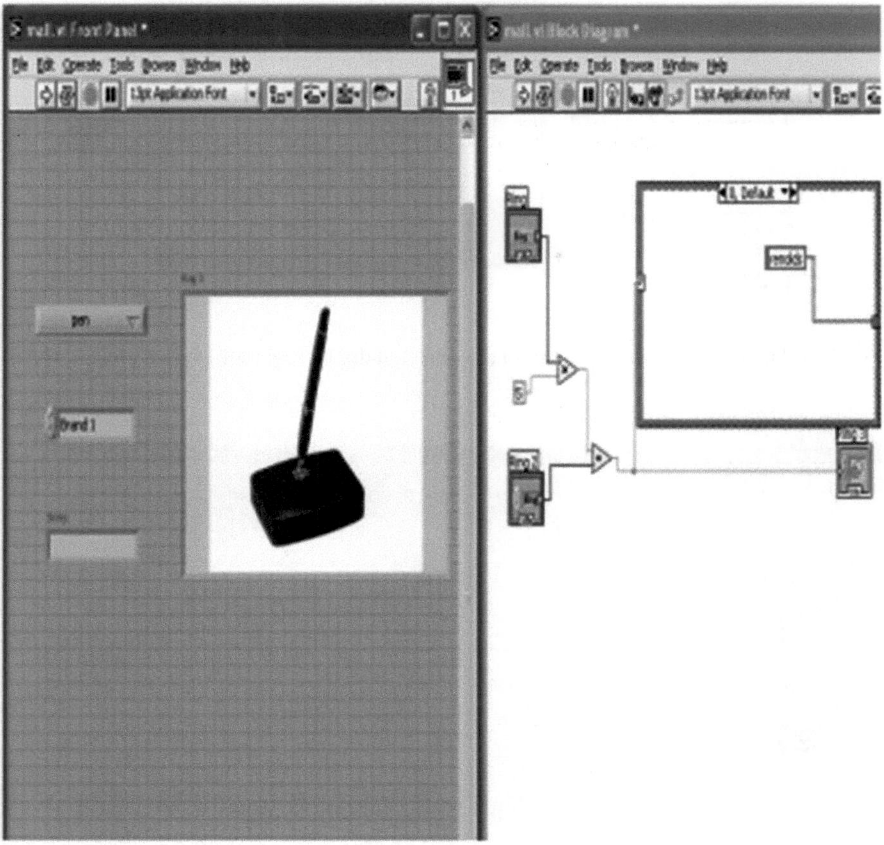

Fig. 6 Animation (Source: Course projects, BVBCET, Hubli)

5.7 GUI Creation

Figure 5 shows GUI creation for traffic control system, where we can find the use of timers to set the time (block diagram), use of 7-segment displays on front panel and use of picture ring to insert any picture on front panel.

6 Capstone Projects

LabVIEW™ was also used in final-year capstone projects. Students employed LabVIEW™ in a project titled 'Simulation of Haptic Surface' (Fig. 7) which won second prize in a national-level project competition, Jed-I project challenge held on June 13–14, 2013 at IISc, Bangalore.

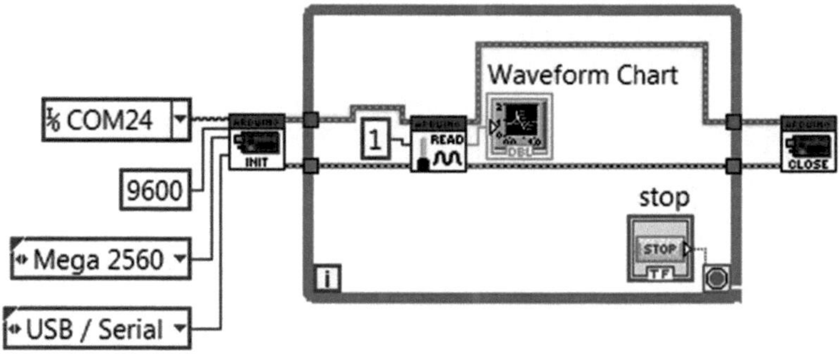

Fig. 7 Arduino interface (Source: Capstone projects, guided by Prof Nalini Iyer BVBCET, Hubli)

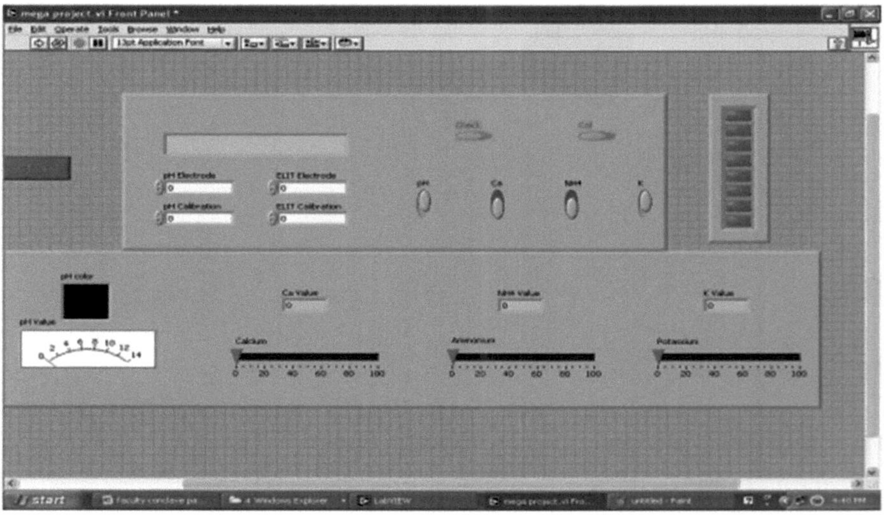

Fig. 8 Soil parameter analyser (Source: Capstone projects, guided by Prof Nalini Iyer BVBCET, Hubli)

This project aimed at creating virtual environment. Here haptic techniques are applied to simulate tactile sensation of walking on different floor surfaces. The system to create this virtual reality includes force sensors, Arduino microcontroller and LabVIEW™ software. Figure 8 shows the front panel diagram where LabVIEW™ is used to interface Arduino microcontroller with the help of Arduino-LabVIEW™ toolkit provided in LabVIEW™ 2012. As per the opinion of students, it was very easy to programme different tasks that are performed in parallel by means of multi-threading. It was possible because two or more parallel while loops could be drawn simultaneously. This is of great benefit for test system automation, where it is possible to run processes like test sequencing, data recording and hardware interfacing in parallel.

Figure 8 shows another project titled 'Soil Parameter Analyser' which won 'People's Choice Award' in the national-level project competition, Jed-I project challenge held on June 13–14, 2013 at IISc, Bangalore. This shows the front panel to measure PH, calcium, potassium and ammonium content in the soil. All these parameters were measured with the help of electrodes and interfaced to LabVIEW™ and actual implementation was done.

7 Conclusion

In this paper, we have presented how active learning can happen with the help of modern tools and technology. SMART classroom approach and use of modern tools created great interest among students. They learnt many engineering concepts with the help of LabVIEW™ software tool. Visualization of concepts helped for proper registration of the things in mind. The things once registered properly in mind cannot be forgotten. Thus, it helps for good memory also. With the help of tools like LabVIEW™, concept teaching and interfacing external parameters with PC with the help of standard buses, GUI creation has become simplified. The course projects given for students helped in creating interest for learning and motivation. Evaluation was also simplified. Hence we can conclude that providing ambient learning environment to suit different kinds of learners eases the task of learning.

Teaching-learning process becomes interesting and easy with the help of modern technology. Further scope of this topic will be to include the use of this tool in various relevant subjects.

References

1. Anderson LW, Krathwohl DR, editors. A taxonomy for learning, teaching and assessing: a revision of bloom's taxonomy of educational objectives. New York: Longman; 2001.
2. Bloom B. Taxonomy of educational objectives: the classification of educational goals: Handbook I, cognitive domain. New York: Longmans/Green; 1956.
3. Coffield F, Moseley D, Hall E, Ecclestone K. Learning styles and pedagogy in post-16 learning. A systematic and critical review. In: Popescu E, Trigano P, Badica C, editors. Proceedings of the 7th IEEE International Conference on Advanced Learning Technologies, IEEE Computer Society Press; 2004. http://www.lsda.org.uk/files/PDF/1543.pdf
4. Gardner H. Frames of mind: theory of multiple intelligences. New York: Basic Books; 1985.
5. Jovitha J. Virtual instrumentation using LabVIEW. New Delhi: PHI Learning Pvt. Ltd; 2010. ISBN-978-81-203-4030-5.
6. Labview Software Tool. www.ni.com/labview

Contemporary Access to Engineering Education and Research Using Interactive Knowledge and Modeling Techniques

Mahantesh M. Kodabagi and Geeta R. Kameri

Abstract For any knowledge-intensive undertaking (such as a stream or branch), it is critical to frame its birth and growth to understand where the discipline stands and what innovative endeavors lead to the creative accomplishments currently witnessed in its knowledge products. In this research paper, we describe the research and development of a knowledge platform called Interactive Knowledge and Modeling Techniques for Engineering Education Research. This research paper mainly describes with the explicit goal to provide a mechanism to better understand the emerging field of engineering education research and, more importantly, provide members of the engineering education research community with tools and infrastructure which allow them to understand the structure and networks of knowledge within the community at any given time. Using a theoretical model that combines ultra-large-scale data mining techniques, network mapping algorithms, and time-series analysis of knowledge product evolution, we attempt to characterize and provide insights into the topology of the networks and collaborations within engineering education research.

Keywords Computation • Data management • Engineering education research

M.M. Kodabagi (✉)
Dr. D.Y. Patil School of MCA, Lohegoan, Pune 412105, Maharashtra, India
e-mail: kodabagi.mahantesh@gmail.com

G.R. Kameri
Dr. D.Y. Patil School of Engineering and Technology,
Lohegoan, Pune 412105, Maharashtra, India
e-mail: g_kameri@yahoo.com

R. Natarajan (ed.), *Proceedings of the International Conference on Transformations in Engineering Education: ICTIEE 2014*, DOI 10.1007/978-81-322-1931-6_17,

1 Introduction

In today's competitive economy, success is increasingly attracted by knowledge and intellectual investment. To be innovative, an organization [or community] has to be adept at exploiting existing knowledge as well as exploring new ways of producing knowledge. To do so, a community must have a holistic, deep, and accessible understanding of what it knows. An informed and innovative future depends on an acute awareness of the past to avoid repeating mistakes and nonproductive paths. Engineering education has recently undergone a resurgence and reorientation that mirrors growing recognition of the unique challenges faced by both engineering educators and learners in the twenty-first century. The dramatic expansion of engineering education over the past decade has led the field to a critical juncture that demands new tools and methods to enable the community to expand and build on prior work. In this paper, we address this challenge by describing the development and deployment of an interactive knowledge platform. The engineering education community has a vision of improving and innovating how engineers are trained and preparing to make them more competitive in the global economy. To pursue this goal, the community has coalesced around several relevant initiatives such as those that have produced *The Engineer of 2020* and the draft report on *Engineering Education for the Global Economy*. The National Science Board report entitled *Moving Forward to Improve Engineering Education* explicitly points to the need for "expanding research and data collection related to engineering education." Yet we do not have a holistic view of how engineering education research can help transform engineering teaching and learning to cultivate the engineer of the future. This challenge is further compounded when one considers the international state of the field, with researchers in many different countries and regions often undertaking similar research on engineering education and professional practice. In short, we do not have in-depth insights into where we stand, nor do we know how we got here. Extant literature in engineering education5 and numerous other disciplines including learning sciences and cyber infrastructure have called for radically rethinking education research to include large-scale data and collaborations. The important question for rapidly evolving fields such as engineering education is: How do we know when large-scale research collaborations are happening? Also, how do we know that research utilizing large datasets attracts a large number of researchers to utilize these datasets? Can we take a data-driven approach to clearly point out trends in research productivity and collaboration? Information retrieval research (e.g., search engines) often helps address such problems by improving the aggregation of data and focusing on what any given document is about (i.e., word-level content analysis). However, for scientific communication, it is equally important to know who writes the document and how the document is positioned in the process of knowledge emergence. Improving access to such information demands different types of analytic tools.

Traditionally, analyzing ultra-large-scale academic data has been the domain of a few computer scientists and engineers. It requires computational techniques to acquire and manage data, analyze large-scale networks, and identify trends and patterns. To allow a broader range of researchers, educators, and other stakeholders in engineering education research community to drive the exploration of the problem space, the data gateway must not only handle the underlying computational components but also provide intuitive navigation, insightful representations, and a user-friendly interface. We attempt to characterize and provide the type of insights required by the community by utilizing ultra-scale knowledge product such as publications in journals and conferences in engineering education, the National Science Foundation grant proposals, and articles published by the National Academy of Engineering.

2 Existing Approaches for Differentiating a Research Domain Using Data and Virtualization

Engineering education researchers have long produced assessed papers that provide overviews for a variant of topics. These papers are usually written by domain experts in a comprehensive and elaborated manner and aim to review recent development within a certain scope based on related studies. Assessed papers often explore a broad range of literature concerning a topic, recognize contributions of relevant studies over a specific period, synthesize them to chart a literature roadmap, and envision the future development. In engineering education research, by applying keyword analysis to the papers in Journal of Engineering Education (JEE) from 1993 to 2002, Wankat9,10 revealed the community's interest in *teaching*, *design*, and *computer*, while *ABET* and *assessment* showed upraising trend during the second half of the decade. In sum, given the great demand of understanding the birth and growth of engineering education research, no previous study has comprehensively covered a broad range of knowledge products in engineering education research. Nor did any project attempt to construct a highly interactive platform that allows researchers to explore the field in a visual and intuitive way.

3 Methodology

3.1 Data Sources

One main data source for this project is observations and proceedings from ten international conferences targeted by *Journal of Engineering Education* and *European Journal of Engineering Education* for special AGCEER panel discussions on engineering education research. These conferences represent regional diversity as well as a focus on engineering education and provide data in the form of proceedings, lists of participants, detailed notes, and ethnographic observations. However, data for the present analysis is mainly taken from other journals and conference proceedings, including *International Journal of Engineering Education*,

European Journal of Engineering Education, and *SEFI Annual Conference Proceedings*. For each, papers from 2005 to 2007 that meet our research criteria were included in our publication database and analyzed (with the exception of SEFI 2006, which we have not yet been able to obtain).

3.2 Data Analysis Methods

We begin our comparative publication analysis by first identifying systematic research publications for our dataset. We exclude purely descriptive papers, such as those that discuss the development of modules, labs, courses, and/or curricula. To do so, we started with these six criteria from *Scientific Research in Education* [1].

3.3 Data Collection and Management

Engineering education research aims at archiving ultra-scale knowledge products in engineering education. To achieve this goal, the data server acquires metadata and full texts (when available) of academic articles relevant to engineering education from online publication data sources such as IEEEXplore, Web of Science, and EBSCO.

3.4 Data-Centered Calculation Components

Based on the data collected, the computational server aims to support the presentation layer by running corresponding algorithms. To support composite search with multiple constraints such as author and publication time, we develop a sophisticated search engine that is tuned to provide short response time. The data source dimension describes the nature of the data and how it was gathered in the reported research. Most research papers will have at least one data source. The categories considered are naturally occurring data (Nat), research-specific data (Res), reflection (Ref), and software (SW).

4 Research Activity by Country

Country-of-origin information for all 280 qualifying articles in our dataset is presented in Table 1. As indicated, 118 papers (or 42 % of all papers) included one or more authors affiliated with US institutions. Of these papers, the vast majority

Table 1 Number of qualifying research papers by country, 2005—2007

Country (1)	No. of papers	Country	No. of papers	Country	No. of papers
United States	118	Israel	5	Brazil	1
United Kingdom	24	Honk Kong	4	Chile	1
Spain	18	Portugal	4	Czech Republic	1
Turkey	16	India	3	Japan	1
Netherlands	13	Norway	3	Kuwait	1
Australia	12	Mexico	3	Malaysia	1
Germany	10	Slovakia	3	Oman	1
Sweden	9	Taiwan	2	Palestine	1
South Africa	8	Belgium	2	Poland	1
Finland	7	Greece	2	Romania	1
Canada	6	Latvia	2	Slovenia	1
Denmark	6	Lebnon	2	Ukraine	1
France	5	Russia	2		
				Total–all data	301 (2)
				Total EU only	112

1. Shaded cells indicate European Union (EU) member countries
2. Total is larger than total papers (n = 280) due to double counting of multi authored papers

(106) were exclusively authored or co-authored by US-based authors. Authors or co-authors currently affiliated with member countries of the European Union (EU) were listed on 112 (or 40 % of all) qualifying articles. The top three author locations in the EU were the United Kingdom, Spain, and the Netherlands.

5 Keyword Analysis

The challenges of keyword analysis make it difficult to identify leading areas of engineering education research, especially by region and subject. For example, in some contexts assessment may refer to outcomes-based evaluations of courses and curricula, while in others it can mean techniques for grading student work. Others have noted the wide range of meanings associated with terms like “competency” and “skill” [2]. However, our preliminary analysis reveals eight major areas, in roughly decreasing order of keyword prevalence:

1. *Curricula*, including development, evaluation, and reform at all levels (courses, programs, etc.)
2. *Assessment*, including outcomes based and competency oriented
3. *Engineering design*, including teaching and learning design-related skills, processes, tools, etc.

Engineering Education Research Differences from engineering research

- More difficult to generalize, e.g., between cultures, laws, surroundings, backgrounds
- Applied to people/ social sciences
- Quantitative & qualitative
- Human oriented – softer
- Results & future results more difficult to measure and verify
- More difficult to execute the actions of the process because of unpredictable human interactions
- Ethical consideration
- Depends on research question
- Research needs to be carefully designed from the beginning
- Results are not always quantifiable
- People and bias issues
- Ensuring you have a representative sample (lots of conversations about N)
- Outcomes of research in eng ed is improving the quality of eng ed in terms of pedagogy, epistemology
- More adaptable
- More future oriented
- Need a lot more exposure, e.g., need to go to more sites
- Results are more open to interpretation
- It is credible as opposed to verifiable/repeatable
- Demand should drive an investment of time and effort in research, e.g, faculty demanding new teaching approaches, ABET, NAE, employer, advisory board

Fig. 1 Discloses about engineering education research differences from engineering research

4. *Collaborative, cooperative,* and *team-based learning*
5. *Problem-, project-, and inquiry-based learning*
6. *Educational technologies*, including for teaching, learning, distance education, etc.
7. *Diversity*, particularly with respect to gender and ethnicity
8. *Laboratories,* developing and innovating in/for engineering education (Fig. 1)

6 Conclusion

In this research paper, we describe the design and implementations of a data-intensive knowledge platform, while it aims to document and present the evolution of engineering education. We collect, index, and allow sense making of a large collection of data through intuitive and user-friendly interfaces. Researchers and learners alike can easily explore the problem space through a web browser without technical expertise on data mining, social network analysis, or time-series analysis. We apply topic modeling techniques to the data to understand the emergence and growth of research topics within the community. Researchers, educators, and other stakeholders in the engineering education research community can visually identify potential collaborators, research patterns, topic trends, and highly related articles and also start to provide unique insights about the topology of the networks within engineering education research. It shows that the content and knowledge that rest within the networks formed by researchers are the fundamental mechanisms through which practices and methods unique to the field of engineering education and research (EER) can propagate. Promoting appropriate models for collaboration not only enhances this type of information sharing, it can also link researchers to new study sites, connect researchers to one another (including across disciplines and preferred methodologies), and expand the generalizability of research findings. Hence, our research results will be translated into concrete implications for the design of future interventions such as workshops. More specifically, the results will include a list of specific topics and regions to pursue for targeted collaborative research and capacity-building activities. Then, models for collaboration will suggest design principles for these activities. For example, if educators in one country have developed innovative learning spaces such as labs, it would be important for workshops to meet in that country so visitors could experience these spaces firsthand.

References

1. Shavelson and Towne (2002) http://web.stanford.edu/dept/SUSE/SEAL/Reports_Papers/other_papers/Scientific%20Culture_ER.pdf
2. Lemaitre D, Le Prat R, De Graaf E, Bot L. Editorial: focusing on competence. Eur J Eng Educ. 2006;31(1):45–53.

Puzzle-Based Learning: A Joyful Learning Strategy in Automotive Engineering Education

Sanjay R. Kumbhar, Sanjay T. Satpute, and Sushma S. Kulkarni

Abstract The various changes to the automobile systems and the integration of the systems have made more challenging for becoming a successful automobile engineer. Various reforms are taking place in the engineering education at the national and international levels. To stand in the competitive age, the engineering institutes are supposed to get accredited by NBA New Delhi and in the coming days to also get accredited by ABET – an International Accreditation Board. Wipro Technologies have started a giant project entitled "Mission10X," through which they have targeted to train 25000 engineering faculties. The main motto is to make teaching–learning process more effective and interesting by various means. Using puzzles is one of them. It has been well documented that crossword puzzles are an effective teaching tool. Not to mention that they are enjoyable to complete. Puzzles can be solved individually or in a group. Solving puzzles in a group will also encourage the students for group discussions and will develop their intrapersonal and interpersonal skills. In this paper, the puzzle-based learning approach has been discussed with its advantages, especially for Automobile Engineering Education.

Keywords Educational puzzles • Puzzle-based learning • Automotive puzzles

S.R. Kumbhar (✉) • S.T. Satpute
Rajarambapu Institute of Technology, Rajaramnagar, Urun Islampur, Maharashtra 415414, India
e-mail: sanjay.kumbhar@ritindia.edu

S.S. Kulkarni
Director, Civil Department, Rajarambapu Institute of Technology, Rajaramnagar, Urun Islampur, Maharashtra 415414, India
e-mail: sushma.kulkarni@ritindia.edu

R. Natarajan (ed.), *Proceedings of the International Conference on Transformations in Engineering Education: ICTIEE 2014*, DOI 10.1007/978-81-322-1931-6_18,

1 Introduction

Over the years, many students have had a teacher from whom it was difficult to learn. This difficulty may have been related to a lack of student interest in the curriculum, or it could have been that the subject was taught in a manner that did not correspond with the student's preference for learning. According to Gardner, educators tend to teach the way they were taught. Moreover, Jonassen identified that a strong relationship exists between a teacher's learning style and his or her preferred teaching style. Unfortunately, there is not a "one size fits all" approach to teaching and or learning. Thus, this creates a problem that requires attention [1, 3].

Many teachers have used puzzles for teaching purposes, and the puzzle-based learning approach has a tradition that exceeds 60 years [2]. Historians found the first mathematical puzzles in Sumerian texts from circa 2500 BC. However, some of the best evidence for the puzzle-based learning approach can be found in the works of Alcuin, an English scholar born around 732 AD, whose main work, *Problems to Sharpen the Young*, included more than 50 puzzles. Some 1,200 years later, one of Alcuin's puzzles – the "river crossing problem" – is still used in artificial intelligence textbooks to educate computer science students [3].

2 Puzzle-Based Learning Approach

Puzzle solving is a much more active type of learning and will engage students with the material more than passive types of review techniques do. Crossword puzzles also have the advantage of appealing to different learning styles. Visual learners often have strong puzzle-solving skills and feel great satisfaction when they complete one. Auditory learners enjoy step-by-step reasoning, so they also benefit from the sequential steps of completing a crossword. Even kinesthetic learners enjoy the multitask strategies required to solve a crossword. Finally, crossword puzzles have the benefit of being customizable to study content. Puzzle creation software and websites are abundant, and easy to use, so teachers can create curriculum-specific crosswords with little trouble.

The phrase puzzle-based learning is taken from the title of Michalewicz and Michalewicz, although it continues a long tradition within the mathematics, science, and engineering communities. The puzzle-based learning approach aims to encourage engineering students to think about how they frame and solve problems not encountered at the end of some textbook chapter. The main goal is to motivate students while increasing their mathematical awareness and problem-solving skills by discussing a variety of puzzles and their solution strategies.

The ultimate goal of puzzle-based learning is to lay a foundation for students to be effective problem solvers in the real world. At the highest level, problem solving in the real world calls into play three categories of skills: dealing with the vagaries of uncertain and changing conditions, harnessing domain-specific knowledge and methods, and critical thinking and applying general problem-solving strategies. These

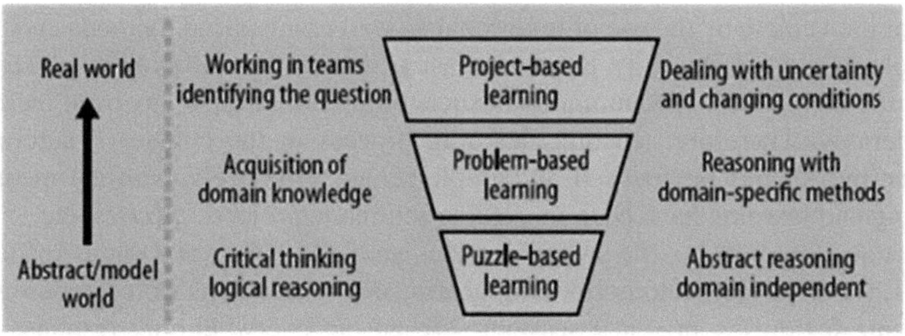

Fig. 1 Learning and skills category layers

three skill categories are captured in the three forms of learning as Fig. 1 depicts. In this continuum, each layer of skills builds upon the layers below it [2, 4–17].

Puzzle-based learning focuses on domain-independent, transferable skills. In addition, we aim to foster introspection and reflection on the personal problem-solving process: What was I thinking? What is the solution? Why did I not see it? Both problem-based and project-based learning are well established methodologies. By our description, problem-based learning requires significant domain knowledge [2, 18, 19]. The main reasons behind most students' enthusiasm for the puzzle-based learning approach are [1]:

- Puzzles are educational, as they illustrate many useful (and powerful) problem-solving rules in a very entertaining way.
- Puzzles are engaging and thought-provoking.
- Contrary to many textbook problems, puzzles are not attached to any chapter (as is the case with real-world problems).
- It is possible to talk about different techniques (e.g., simulation, optimization), disciplines (e.g., probability, statistics), or application areas (e.g., scheduling, finance) and illustrate their significance by discussing a few simple puzzles. At the same time, the students are aware that many conclusions are applicable to the broader context of solving real-world problems.

3 Automobile Engineering Education

Nowadays, Automobile Engineering is one of the most challenging engineering careers. With the increased demand and passion on vehicles, automobile engineers have many opportunities in India and abroad. The automobile engineering study deals with designing a vehicle, manufacturing new products/new vehicle, repairing, and servicing vehicles. Automobile engineer should be innovative and dedicated to their work. The future automotive engineers can be specialized in the alternative areas such as aerodynamic, alternative fuels – nonconventional and renewable fuels – the chassis, the electronics, the emissions, the ergonomic, the manufacture, the materials, the creation of fast prototypes, the security or the management, etc. The responsibilities of an automobile engineer consist of maintaining the greater level of the vehicle by the use of traditional as well as advanced methods and of the technology outpost. Recently, due to the fast growth of the automobile industries in India, the demand for the automotive expert professionals also has been increased considerably. Therefore, teaching–learning process in the course of automotive engineering should be focused and implemented effectively. Various means of teaching and learning have been in use like seminars, projects, puzzles, etc.

This section contains the sample puzzles gathered from the various resources. At RIT, for third year Automobile Engineering course, students were encouraged to solve puzzles during practical sessions. We got an overwhelming response from students while solving puzzles. Some of the points noted are:

1. Students showed enthusiasm while solving puzzles.
2. Students actively participated in solving puzzles, which shows their group discussion ability.
3. Solving puzzles showed the students' critical thinking ability.

Exercising this solving of puzzles showed us that interpersonal and intrapersonal skills of students can be encouraged and developed through this new teaching–learning approach. After conducting actual hands-on experiments, students enjoyed solving the puzzles.

The vast majority of puzzles on various topics have already been included in many books. Some of the important topics are: Automotive Engines and its Various Systems; Automotive Electrical and Electronic Systems; Conventional and Alternative Fuels; Automotive Emissions and Control Technologies; Advanced Automotive Tools and Equipments; Automotive Manufacturing, etc.

Some online softwares are also available free of cost for creating puzzles on topic of our interest. Some of the websites are www.crossword**puzzle**games.com, www.eclipsecrossword.com, www.jigsawplanet.com, etc.

For example, take the following sample puzzles.
Puzzle No. 1. Crossword puzzle: Basic automotive parts (Fig. 2)

Solution (Fig. 3)

3.1 Extensions and Commentary

This puzzle is something about identification of automotive parts by their functions. The students will try to correlate the functions of each part and will solve this crossword puzzle very interestingly.

Puzzle No. 2. Word search puzzle: Battery

Find the words in the grid. Words can go across, down, and in two diagonals (Fig. 4).

Solution (Fig. 5)

Across

1 a gear system that drives both axels at the same time, but allows them to rotate at different speeds when turning a corner

4 a piece of metal turned on a wheel to keep it on the vechicle

5 a device that uses gears and torque conversion to change the ratio between the engine RPM and driving wheel RPM

7 a device containing one of more cells that produce electricity through electrochemical action

9 ethylene glycol of other liquid used in a cooling system

10 protective device that interrupts current flow if an overload condition is in a circuit

11 a way to start a car with a weak battery with the use of cables

12 a generator in which current is changed to direct current

Down

2 a system that sprays gas directly into the cylinders

3 a rod used to measure the amount of liquid in a system

6 an electric motor which uses a geardrive to crank the engine

8 a pleated paper element, used to trap dirt, dust, etc from entering the inside of the engine

9 shafts used to transmit driving force to the wheels

Fig. 2 Crossword puzzle: basic automotive parts [20]

Fig. 3 Answer of crossword puzzle: basic automotive parts [20]

H	Y	D	R	O	G	E	N	R	K	W	F	V	H	L	T	L	E	K	C	D
M	W	B	T	E	L	E	C	T	R	O	L	Y	T	E	D	E	W	R	O	Q
L	Z	X	Z	B	A	L	L	O	Y	V	F	M	P	R	R	J	G	E	L	J
C	E	T	W	H	Y	D	R	O	M	E	T	E	R	F	Q	N	M	C	D	H
G	V	A	N	O	K	S	Y	N	B	V	P	R	E	T	I	F	N	H	C	V
A	M	V	D	W	V	M	E	H	K	Z	Z	C	M	L	V	R	T	A	R	K
S	R	W	W	P	K	E	T	D	J	W	N	L	C	B	E	W	Q	R	A	B
S	E	Q	M	P	E	V	R	N	I	A	P	Y	H	T	G	T	N	G	N	L
I	S	K	T	K	R	R	D	C	N	M	C	T	A	Q	D	N	P	I	K	H
N	E	M	Z	N	R	V	O	E	H	P	E	F	J	I	W	P	A	N	I	R
G	R	W	L	V	M	P	T	X	E	A	L	N	C	V	K	R	S	G	N	Q
K	V	R	R	G	P	N	A	E	I	U	R	A	T	H	K	S	T	Z	G	R
C	E	R	P	M	I	K	D	R	S	D	C	G	Y	C	X	E	I	G	A	G
Y	C	R	L	A	H	P	R	R	T	I	E	B	I	R	H	P	N	D	M	Y
Z	A	H	M	V	Z	E	A	R	R	I	C	Z	H	N	M	A	G	R	P	J
V	P	L	R	R	W	E	D	U	M	F	T	P	D	L	G	R	M	L	S	N
N	A	Q	L	O	L	Z	H	K	P	C	Q	I	K	T	V	A	H	B	K	N
N	C	H	P	N	K	P	F	M	M	D	T	Y	O	N	Y	T	W	B	E	B
Z	I	L	F	Q	L	K	W	Q	M	V	H	J	D	N	G	O	L	R	C	R
L	T	T	G	U	Z	L	L	K	F	W	N	D	H	M	S	R	M	H	R	F
T	Y	N	S	T	N	C	K	R	C	K	Q	V	L	V	O	L	T	A	G	E

Alloy
Cold cranking amps
Deep cycling
Electrolyte
Gassing
Hydrogen
Hydrometer
Lead peroxide
Lead sulfate
Maintenance free
Overcharging
Partitions
Pasting
Power
Recharging
Reserve capacity
Sediment chamber
Separator
Sulphuric acid
Voltage

Fig. 4 Word search puzzle: battery [21, 22]

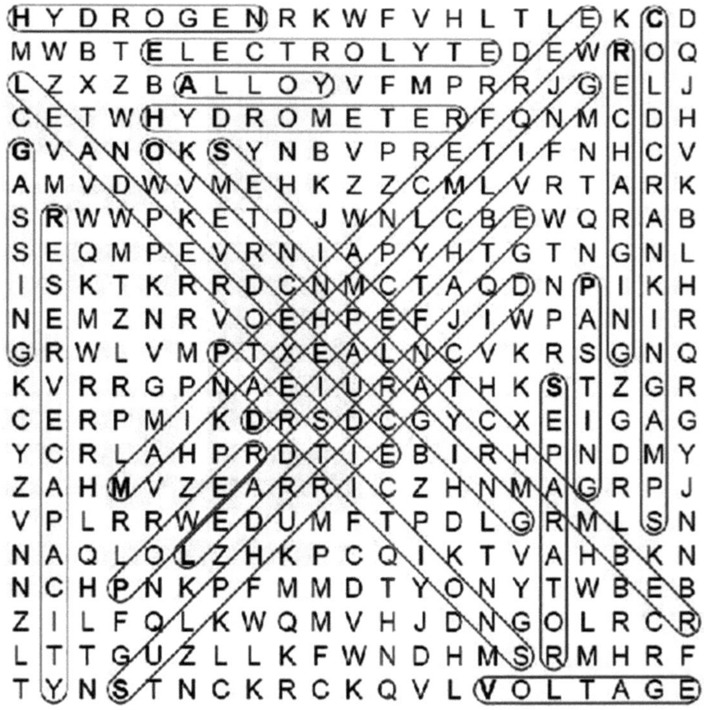

Fig. 5 Answer, word search puzzle: battery [21, 22]

3.2 *Extensions and Commentary*

This puzzle is something about identification of automotive battery terminologies. The students will try to find out the correct terminologies by searching the words in the puzzle. This activity would increase their critical searching ability. The students solve such word search puzzles very interestingly.

4 Conclusion

Crossword or word search-type puzzles can play a major role in attracting students to automotive engineering programs and can be used in talks to graduate students during practical sessions or open-day events. These can be a factor that helps retain and motivate students for joyful learning. Above all, the various types of puzzles are responsible for developing critical thinking and problem-solving skills. Through these puzzles, the ability of students to solve industry and business problems can be enhanced. Many real-world problems can be perceived as large-scale puzzles. The ultimate goal is to motivate students for joyful learning and to increase their problem-solving skills by discussing a variety of puzzles with their solution. This strategy of puzzle-based learning will really increase the effectiveness of teaching–learning process of Automobile Engineering education.

References

1. Threeton MD, Walter RA. Automotive technology student learning styles and their implications for faculty. J Ind Teach Educ. 2009;46(3):7–13.
2. Falkner N, Sooriamurthi R, Michalewicz Z. Puzzle-based learning for engineering and computer science. Computer (Computing Practices), IEEE Computer Society, 0018-9162/10/$26.00, (2010).
3. Threeton MD, Walter RA, Clark RW, Ewing JC. Automotive technology student learning styles and modes of grasping and transforming experience: Association for Career and Technical Education Research (ACTER). CTE Research and Professional Development Conference; 2010 Nov 30–Dec 2; Las Vegas Hilton.
4. http://myschoolofthought.com/2013/07/12/crossword-puzzles-for-learning/
5. http://myschoolofthought.com/tag/crossword/
6. http://vocabulary.co.il/blog/learning_vocabulary/teaching-with-crossword-puzzles/
7. http://www.scribd.com/doc/29472908/Teaching-With-Crossword-Puzzles
8. http://www.vocabulary.co.il/article.php
9. http://researcherteacher.blogspot.com/
10. http://edutechwiki.unige.ch/en/Puzzle-based_learning
11. http://cs.adelaide.edu.au/~zbyszek/Papers/PBL-IEEE.pdf
12. http://www.decisionsciences.org/decisionline/Vol42/42_5/dsi-dl42_5_instructinno.asp
13. http://nicole-weaver.com/2011/02/14/educational-value-of-puzzles/
14. http://nicole-weaver.com/2011/02/
15. http://www.andrew.cmu.edu/user/sraja/papers/2010-ieee-computer-pbl.pdf

16. http://www.visualisationproject.eu/materials/visual%20crosswords.pdf
17. http://easynewmexicanrecipes.com/soc.php?nul=printable-custom-crossword-puzzles
18. Michalewicz Z, Falkner N, Sooriamurthi R. Puzzle-based learning: An introduction to critical thinking and problem solving. *Decision*; Oct 2011.
19. Melero J, Hernández-Leo D, Blat J. Towards the support of scaffolding in customizable puzzle-based learning games. ICCSA-11; 2011; Santander.
20. Michelle's Automotive Crosswords.
21. Puzzles by Tony and Lara Martin. www.wordsearchmaker.com
22. http://www.jameshalderman.com

Improving Classroom Dynamics Through Teaching–Learning (Academic) Audit

Martand T. Telsang and Sushma S. Kulkarni

Abstract Beyond the physical arrangement of a classroom, a psychological environment is also created, based on the interaction of key players in the classroom, namely, students and teachers, and their interaction which creates class dynamics. Moos's work, which has permeated the literature on classroom environment, is based on three essential areas of classroom environment: (1) relationship dimension, which focuses on the interpersonal relationships between students and students and the teacher in a classroom; (2) personal development dimension, which centers on individual characteristics of the classroom member; and (3) system maintenance and change dimension which includes attributes such as classroom control and order as well as responsiveness to change. The process of academic audit captures the classroom dynamics in an environment of ease and comfort. Classroom dynamics here refers to delivery effectiveness, learning experience of students with different learning preferences and styles, and factors that hinder the effectiveness of both teaching and learning. A detailed audit process is laid down including the guidelines to auditors and postaudit counseling to faculty. The audit is designed to capture the five important dimensions of a teacher, namely, subject knowledge, preparation, communication, class control and concern for students, and opportunity to interact. The audit process is going to map the individual teacher and the course on a ten-point scale with appropriate weights for the five dimensions. Weightages for dimensions are decided based on the opinions of the senior faculty and academicians. Based on the scores, the faculty competency enhancement is

M.T. Telsang (✉)
Mechanical Department, Rajarambapu Institute of Technology, Rajaramnagar, Urun Islampur, Maharashtra 415414, India
e-mail: martand.telsang@ritindia.edu

S.S. Kulkarni
Director, Civil Department, Rajarambapu Institute of Technology, Rajaramnagar, Urun Islampur, Maharashtra 415414, India
e-mail: sushma.kulkarni@ritindia.edu

R. Natarajan (ed.), *Proceedings of the International Conference on Transformations in Engineering Education: ICTIEE 2014*, DOI 10.1007/978-81-322-1931-6_19,

planned and appropriate actions are initiated to train the faculty in the specified dimensions. The outcome of the audit is evident in terms of improved learning outcomes, enhancement of teaching deliveries and competency, collaborative learning, and good academic environment.

Keywords Room dynamics • Teaching dimensions • Teaching index • Academic audit • Audit procedure

1 Introduction

It is no longer enough to simply transmit information that students memorize and store for future use. Education today must focus on helping students learn how to learn, so they can manage the demands of changing information, technologies, jobs, and social conditions. In fact, a growing body of research suggests that students learn more deeply and perform better on complex tasks if they have the opportunity to engage in more "authentic" learning: projects and activities that require them to employ subject knowledge to solve real-world problems. Beyond the physical arrangement of a classroom, a psychological environment is also created, based on the interaction of key players in the classroom, namely, students and teachers which creates class dynamics. Moos's work, which has permeated the literature on classroom environment, is based on three essential areas of classroom environment: (1) relationship dimension, which focuses on the interpersonal relationships between students and students and the teacher in a classroom; (2) personal development dimension, which centers on individual characteristics of the classroom member; and (3) system maintenance and change dimension which includes attributes such as classroom control and order as well as responsiveness to change.

Teaching–Learning (Academic) Audit systematizes an institutions approach to quality by focusing on a body of content that must be considered before an analysis can be accepted or completed. This body of content is the focal areas of quality work: (1) learning objectives, (2) curriculum, (3) teaching and learning methods, (4) student learning assessments, and (5) quality assurance. Classroom environment encompasses a broad range of educational concepts, including the physical setting, the psychological environment created through social contexts, and the numerous instructional components related to teacher characteristics and behaviors.

Overcrowded facilities, too many students in certain classes, and lack of teachers' assistants are three major issues cited as potentially creating problems due to increased stress levels of students and increased teacher-reported incidences of behavioral problems. These increased stress levels and behavior problems found in larger classrooms are frequently accompanied by lower levels of academic achievement.

2 Psychological Classroom Environment

Many teachers equate student engagement and on-task behavior with classroom participation, typically a top concern for teachers. Researchers support teachers' intuition of a difference in the participation style of the different genders. Whereas girls are more likely to participate as part of the relational responsibility they feel toward the teacher, boys tend to respond more often if they feel the class is interesting and less often if the class is perceived as boring—indicating that for these students, teachers may be equally responsible for the participation level and learning.

Helen Patrick and colleagues found that there is a strong, positive relationship between students' level of motivation and engagement and their perceptions of the classroom environment as being socially supportive. The perception of a climate of mutual respect is required in order for students to increase their use of effective study strategies and increase feelings of confidence about their ability to successfully complete assignments. Furthermore, when students perceive that they receive emotional support and encouragement from their teachers and academic support from their peers, they are more likely to be on-task in the classroom and use self-regulated strategies. Classroom rules and procedures should be introduced early in the school year and consequences should be enforced consistently across students and throughout the school year. Research has shown that routine and fairness have a positive impact on behavior as well as academic quality. It has been found that teachers who run respectful classrooms are in turn more respected by their students, and students believe that these teachers also hold higher learning expectations. Teachers are encouraged to focus more on the learning task than on the outcome or grade assigned at the end of the task, although this becomes much more difficult if the emphasis in education is placed on accountability and high-stakes testing.

3 Teaching–Learning (Academic) Audit

The study classroom environment has been widespread across nearly all subspecializations of educational psychology. Researchers are interested in relationships between environment constructs and multiple outcomes, including learning, engagement, motivation, social relationships, and group dynamics. Early researchers recognized that behavior is a function of people's personal characteristics and their environment.

Academic audit is one of the three main types of higher education quality evaluation in use today around the world. (The other two are accreditation and subject-level assessment.) David Dill describes it this way: In contrast to accreditation, program review, or student assessment initiatives, [academic] audits look deeply into the heart of the academic enterprise. They test whether institutions and their faculties in fact honor their public responsibility to monitor academic standards and improve student learning.

According to Dill, the reviewers generally agree that academic audits have made improving teaching and learning an institutional priority; facilitated active discussion and cooperation within academic units by means of improving teaching and learning; helped clarify responsibility for improving teaching and learning at the academic unit, faculty, and institutional level; and provided information on best practices within and across institutions [2, 3]. Moreover, audit focuses on "education quality work" (EQW, to be defined below), which is emerging as the key element of institutional quality programs. External agencies can evaluate EQW more easily than education quality itself. Panel selection and training appear easier, cycle times can be shorter, and institutional diversity is more easily respected than in other forms of evaluation."Education quality work" (EQW) means the activities of faculty, academic leaders, and oversight bodies that are aimed at improving and assuring education quality. It should empower and stimulate faculty to continuously improve teaching and learning and help academic leaders and others to discharge their oversight responsibilities without micromanagement. EQW should not be confused with teaching itself [1].

4 Design and Implementation of Academic Audit: Case Study

4.1 Institute Profile

Rajarambapu Institute of Technology is established as self-financed private engineering college in the year 1983 with an objective of imparting quality engineering education to aspirants in general and students from surrounding rural area in particular. Currently, the institute offers seven undergraduate and six postgraduate programs engineering and three departments offer research programs leading to PhD. All the eligible programs are accredited by the NBA. The institute has successfully implemented TEQIP phase I and selected for TEQIP phase II, one among the four self-financing institutes selected all over India.

4.2 Need to Implement Academic Audit

Many of the private technical institutions which have come up in recent past, especially in small towns, are facing acute shortage of faculty with required qualifications and requisite skills and aptitude for teaching and research. It is difficult for these institutions to attract best talent and retain them. The institute referred above devised its own strategy to develop faculty by developing its own model to recruit faculty at entry level and plan career growth for each one, extending opportunity to enhance qualifications, and upgrading on continuous basis their capabilities, both technical and administrative, through planned training. This strategy calls for continuously evolving innovative practices and new systems to improve the quality of teaching–learning process, students' learning, and competency of faculty and staff to meet the ever-increasing expectations of stakeholders.

Conventional feedback system which is in practice in majority of the institutions is an indirect method of capturing students voice which has an inherent advantage in terms of easy to implement and no interference from the faculty. As compared to this, the academic audit focuses on issues related to teaching–learning with an intervention of an experienced academician as an auditor. The auditors visualize multiple facets of teaching–learning and come out with clear observations and action plans.

The purpose of the audit is to improve the effectiveness of the delivery of the faculty to enhance the students' learning in terms of clarity of concepts, application of concepts for problem solving, and grasp of the subjects to secure good grades in examination. The audit aims at bridging the gap between teaching and learning through a proper communication and feedback system.

4.3 Audit Procedure

4.3.1 Selection of Students for Participation in Audit

Ten representative students from each class are included in the audit process, which forms heterogeneous group representing typical student mix in a class. Monitor of the class will identify the students and ensure that they will be made available to the academic auditors nominated for the class for interaction and giving feedback of teaching–learning process.

4.3.2 Appointment of Auditors

Experienced faculty members having a good standing as a teacher will act as auditors. The auditors should be able to establish a good rapport and create conducive environment for the students to interact, express themselves, and critically comment without any bias and fear about the classroom teaching–learning process. The detailed schedule of the audit including the assigned class, subject teachers, and student's names will be made available to the auditors.

4.3.3 Audit Process

The auditors are expected to use their expertise and experience to have a good grasp of the classroom dynamics and should be able to assess the classroom teaching and learning progress. The focus should be on whether the learning of the students is progressing in the right direction and at the same time whether teacher is making good attempt to address the learning of the entire class. During the process of interaction, the students should be given ample opportunity to express their feedback without any inhibitions and fear of after effects. During the interaction, the effort of the auditors should be fact-finding through evidences rather than prejudges and perception.

4.3.4 Audit Venue

The audit is supposed to be conducted at a place preferred by the auditors. Only care should be taken to choose the venue other than home department of the students. This helps to maintain the identity of students confidential.

4.3.5 Duration of Interaction with Students

The audit committee can have a maximum of two meetings per audit of a class (as they are required to have feedback of five or six subjects). Each meeting should have a maximum of one-hour duration. The time and venue can be decided by auditors.

4.3.6 Confidentiality of Feedback

The proceedings of the audit meeting should not be made public and also the feedback should be given only in a format for which guidelines are given.

4.3.7 Audit Report

The audit feedback should be communicated in writing to the academic audit coordinator in a prescribed format for the individual faculty, within a time period of one week after the completion of audit. Any delay in sending the report renders the whole effort ineffective and fails to serve the purpose of the audit.

Dean Academic/Respective Head of the department is required to review the audit reports after receiving from the co-coordinator and write special remarks with guidelines to improve the performance of the faculty in areas suggested. Those who have a satisfactory performance should also be communicated with appreciation. The actions to be initiated are as follows:

1. Communicate the performance report in person to each faculty.
2. Suggest the strategy to improve the areas suggested during the remaining period of this semester and ask teacher to prepare a concrete plan of action to bring improvement.
3. Support and motivate the faculty to implement action plan and assess the progress.
4. Have an individual counseling meeting with faculty and ask faculty to come out with the lecture notes, teaching plans, etc.

5 Outcomes and Discussions

The Institute is practicing academic audit since 2010 with a sufficient awareness and training to auditors. In spite of initial teething problems during implementation, the audit model is tending toward maturity with a substantial cooperation from both faculty and students. The average scores of faculty and average teaching index and percentage of faculty below the average are represented in Table 1, Fig.1, and Table 2. The average teaching index computed on the basis of weighted average scores of five dimensions has increased by 20 % in 2012–2013 compared to 2010–2011 due to individual faculty efforts to improve the weak areas. The results indicate that planning and delivery, class control, and concern for students are the issues still to be addressed, as they are below the average of institutional average teaching dimension. In order to improve these areas, the institution has taken up many initiatives like preparing outcome-based course plan, articulating the teaching methodology based on the students learning styles, improving communication through toast

Table 1 Average values of teaching dimensions

			Average scores of all faculty		
S.No	Teaching dimension	Max. weight	2010–2011	2011–2012	2012–2013
1	Subject knowledge	3	2.14	2.25	2.43
2	Preparation and delivery	3	2.07	2.30	2.55
3	Class control	1	0.60	0.70	0.80
4	Communication	1.5	0.90	1.00	1.10
5	Concern for students	1.5	1.02	1.12	1.20
Average teaching index		10	6.73	7.37	8.08

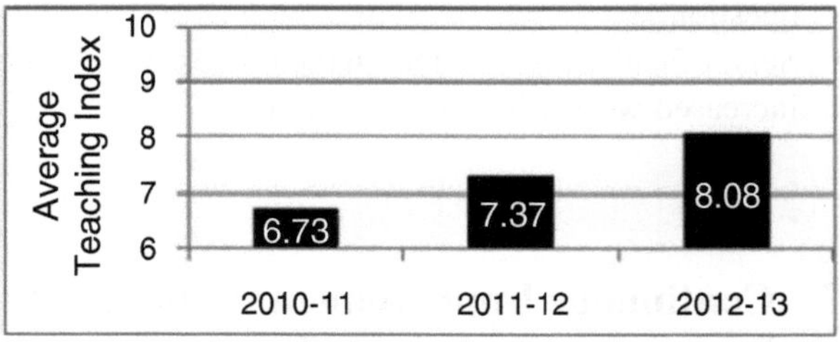

Fig. 1 Average teaching index of faculty on a 10-point scale

Table 2 Percentage of faculty below average teaching index

Year	Avg. teaching index	% faculty below avg. teaching index
2010–2011	6.73	32
2011–2012	7.37	23
2012–2013	8.08	18

master club for faculty, and also personal counseling to faculty to look into the problems of individual students and open and free communication between students and faculty.

The faculty below the institute average mainly constitute new faculty who have just started their career and pedagogical sessions are arranged to this group to raise their competency level as a teacher. Special workshops are arranged to work with them and orient them to teaching profession.

In order to encourage and reward the well-performing faculty, a reward system is being introduced, because of which they are motivated to improve the overall effectiveness of their delivery. Thus, the analysis gives us a clear understanding of areas to improve and devise training program to improve the performance. It is an ongoing exercise to achieve excellence in teaching–learning process. Academic audit is not a mere evaluation and grading of a teacher. It helps to identify the gaps in learning process of students and teaching effectiveness and sets a stage for continuous improvement. The mandatory requirement for the success of this process is the trust and respect for the students' feedback and integrity of an auditor.

6 Challenges Faced During Implementation

The challenges and obstacles faced during implementation are listed as follows:

- Students are reluctant to openly share their feedback and opinions about the teacher as there is a fear of revealing their identity to faculty.
- Faculty perceived it as a tool for management to assess their classroom performance and adverse results affect their career in the institute.
- As always, faculty assume that students' capability to give correct feedback is in question.
- Senior faculty involvement in the process was limited and always complained of increased work without many gains.

7 Continuous Improvement in the Audit Process and Sustainability

The above challenges in the audit process made the conviction strong to implement the system and improve the system continuously based on the feedback received from all stake holders. The objectives of the audit are shared with both students and faculty to seek their full acceptance and cooperation. This increases the chances of success and sustainability and whole-hearted participation of both faculty and students to make this practice unique and meaningful.

8 Conclusions

The process of academic audit captures the classroom dynamics in an environment of ease and comfort. A detailed process is laid down including the guidelines to auditors and postaudit counseling to faculty. The audit is designed to capture the five important dimensions of a teacher, namely, subject knowledge, preparation, communication and concern for students, and opportunity to interact. The audit process is going to map the individual teacher and the course on a ten-point scale with appropriate weights for the five dimensions. Based on the scores, the faculty competency enhancement is planned and appropriate actions are initiated to train the faculty in the areas. The outcome of the audit is evident in terms of improved learning outcomes, enhancement of teaching deliveries and competency, collaborative learning, and good ambience for academic environment. Continuous improvement should be the way of life as what works today may not work tomorrow. We should continuously strive to achieve excellence in what we do.

References

1. Massy WF, Graham S, Short PM. Academic quality work: a handbook for improvement. Bolton: Anker; 2007.
2. Dill DD. Quality by design: toward a framework for academic quality management. In: Higher education: handbook of theory and research. New York: Agathon Press, Inc; 1992. p. 37–83.
3. Dill DD, Massy WF, Williams PR, Cook CM. Accreditation and academic quality assurance: can we get there from here? Change. 1995;28(5):16.

IT Education and Team-Based Learning

Nathaniel G. Martin and Pradeep Waychal

Abstract The basis of the Indian Information Technology and Information Technology Enabled Services (IT and ITES) is changing from contract programming and business process outsourcing (BPO) to full life cycle development and end-to-end process ownership. These developments require professionals with better entrepreneurial and innovative skill sets. The skills depend largely on intrinsic motivation that should be fostered in colleges, if not earlier. As per self-determination theory (SDT), intrinsic motivation relies on a sense of competence, relatedness, and autonomy. We hypothesize that teamwork is one of the potent techniques for addressing the three. The paper uses ethnography to compare two case studies at an IT department at a college that had teamwork as an integral part. It also covers limitations and future scope for the study.

Keywords Team-based learning • Intrinsic motivation • Innovation and entrepreneurship

1 Introduction

India's success in Information Technology and Information Technology Enabled Services (IT-ITES) has been exemplary. It accounts for 25 % of the national export and reports 101 B USD turnover in 2012, up from merely 4 %, and 50 M USD in 1998 [1]. In the last few years, the industry is being challenged due to the declining growth rate and significant changes in work distribution: besides extending cost

N.G. Martin • P. Waychal (✉)
Department of Computer Engineering and Information Technology,
College of Engineering, Pune, Shivajinagar, Pune 411005, Maharashtra, India
e-mail: Pradeep.waychal@gmail.com

R. Natarajan (ed.), *Proceedings of the International Conference on Transformations in Engineering Education: ICTIEE 2014*, DOI 10.1007/978-81-322-1931-6_20,

advantage, industry is compelled to offer consulting, system integration services and initiate nonlinear play through products, platforms, software assets, and solution accelerators [2]. Besides, India is endowed with impressive demographic dividend, as compared to many developed economies, and can take up more business. She can do so and continue her streak of success by working on her education system. The system requires developing innovation and entrepreneurial competencies that depend on intrinsic motivation, which can be reinforced through team-based learning (TBL) [3].

The major contribution of the paper is to bring out variants of TBL and show that all of them offer huge benefits to IT education which will help the industry to keep pace with the current realities.

The paper details out background and discusses two reasonably dissimilar courses that followed different TBL approaches but returned with impressive learning. The approaches include variations in forming teams and monitoring and rewarding their performances. The paper also covers an ethnographic analysis of the students of the two courses.

2 Background

The Indian software industry has to come out of the cozy cost arbitrage-based outsourcing model and move to models based on innovation and entrepreneurship—developing novel solutions to problem. It offers three benefits over outsourcing. First, the revenue can grow in a nonlinear way. Next, it can be applied to the complex problems faced by the nation. Finally, self-reliance can help isolate national economy from international events.

2.1 Innovation and Entrepreneurship

Innovation is fresh thinking that delivers value to customers. While fresh thinking is easier, delivering value is tougher and requires entrepreneurial skills. Amabile [4] defines entrepreneurial creativity as "The implementation of novel, useful ideas to establish a new business or new program to deliver products or services." That is, entrepreneurship is innovation in the context of business—it requires not only a solution to the problem at hand but the ability to embed the solution in a business model. While entrepreneurship and innovation can aid any organization, the real payoff is in creating new enterprises or industries.

2.2 Entrepreneurship and Intrinsic Motivation

Entrepreneurship requires a different type of motivation—intrinsic motivation. It is "behaviors done in the absence of external impetus that are intrinsically interesting and enjoyable." This sense of intrinsic interest and joy is critical to pursue entrepreneurial activities.

Amabile [4] points out "no amount of skill in the domain or in methods of creative thinking can compensate for lack of intrinsic motivation to perform an activity. Without intrinsic motivation, an individual will either not perform the activity at all, or will do it in a way that simply satisfied the extrinsic goals. But … a highly intrinsically motivated individual is likely to draw skills from other domains or apply great effort to acquiring the necessary skills in the target domain." In a sense, it is more important to foster intrinsic motivation than to teach any particular skills.

Ryan and Deci [5] have developed self-determination theory (SDT) to explain the source of intrinsic motivation. It explains numerous experiments showing that both reward and punishment reduce motivation. SDT posits three basic psychological needs required for intrinsic motivation:

1. Autonomy: the feeling that one is working under one's own volition
2. Competency: the feeling that one is capable of achieving success
3. Relatedness: the feeling that what one is doing involves others

Though the traditional educational environment is detrimental for developing intrinsic motivation, Niemiec and Ryan [6] describe a theory of education based on SDT designed to increase internal motivation. This theory involves changing sources of motivation from external sources such as grades to internal sources such as a sense of connectedness to the class or the subject. The goal is to move away from external regulation where rewards and punishments are the motivation for activity toward integrated regulation where the student not only finds value or importance in the activity but builds identifications with other aspects of the self.

2.3 *Intrinsic Motivation and Teamwork*

The current education system inhibits intrinsic motivation. Benware and Deci [7] found that college students showed better conceptual learning when expecting to teach to other students than when expecting a test. Koestner et al. [8] found that students who were given controlling limits showed less intrinsic motivation than those who were given more autonomy.

The prevalent education system involves a routine process of teaching through lectures and conducting traditional examinations until the students graduate. This system decreases the students' sense of autonomy because the lectures and reading are controlled entirely by the teacher. Tests decrease students' sense of competence because they must have questions that not everyone can answer. The feeling of competence is further reduced by a perceived gap between the syllabi taught in colleges and the knowledge and skills required in the work environment. The feeling of relatedness is damaged by high-stakes competition and lack of collaboration. Students are either graded on the curve resulting in a zero-sum game or given marks to decide ranks. In either case, students compete with each other. This discourages communication between students by penalizing them for their classmates' success, taking a heavy toll on students' ability to work in teams.

So, can engineering education develop teamwork resulting in better intrinsic motivation that in turn can foster innovation and entrepreneurship needed to keep pace with current realities?

2.4 *Intrinsic Motivation of Students*

We have hypothesized that using team-based learning increases intrinsic motivation. Allowing students to choose and focus on their projects reduces their attention to grades and other external motivators, increasing a sense of autonomy. In addition, the team provides a community that can increase their sense of relatedness. Providing easily accessible materials through the Internet and providing them with a community with which they can discuss their questions increases their sense of competence. By increasing competence, relatedness, and autonomy, this type of class increases intrinsic motivation.

Further, based on our experience in global software industry, we strongly perceive need of developing team skills in today's graduates. Further, we have to move away from the earlier hierarchical or "hero"-based teams to optimizing teams that contribute equitably.

3 Case Studies

In the Odd Semester (July–October) of 2013, the authors each taught a class: human–computer interaction (HCI) and digital systems (DS). Both of these classes organized students into teams to complete projects. Though we designed the classes independently, our goal was the same—to teach entrepreneurship, intrinsic motivation, and teamwork.

3.1 *Course Similarities*

The courses were similar in that both authors tried to maximize diversity in their teams. Also, the IT infrastructure was the same for both classes.

3.2 *Diverse Team*

In HCI, students were assigned based on knowledge of English—the professor has a heavy American accent and he was worried student would have difficulty understanding him—and programming experience. In DS, students were assigned to teams randomly because the faculty perceived that there are so many differences among students that only randomness can ensure better diversity.

3.3 IT Infrastructure

Both classes used Moodle as their course learning management system. All the students had access to Internet round the clock.

3.4 Course Differences

Though the courses were both in the same department, the differences were striking.

3.5 Work Assignment

HCI—a theory course, intended to give the students understanding of user interface design and development—weekly assignments were given in each class corresponding to the reading and lecture. Teams worked on these assignments during class and then continued on their own time. The assignments were submitted using Moodle. Toward the middle of the class, the teams were asked to work as a startup company to choose a product and design the user interface for it. The exercises were graded and posted on line and the results were added into the "House Cup" totals. A scrum master for each team was identified.

DS was a theory and practical course, intended to give the students a hands-on experience with digital hardware in addition to academic understanding. Initial theory classes covered the basic elements of the digital system such as gates, multiplexer, flip-flops, and counters. Each student proposed an idea for a real-life application that can be developed using those elements. The teams were formed using a random algorithm. In the first team meeting, each team member presented his idea. The teams chose one of the ideas, a hybrid of some of the ideas, or an entirely new idea for their project. The faculty guided the team's choice to ensure that it was neither too small nor too big and then limited mentoring to encourage students to find answers on their own. The team management was completely left to the students. They were permitted but not encouraged to escalate team issues to the faculty.

3.6 Project Evaluation and Grading

Though the HCI project was not included in the student's grade, the cumulative results of the teams were presented each class as a competition called the "House Cup." The results from the team's work were displayed in Moodle along with the instructor's comments. This allowed students to see the mistakes and successes of other teams. The students' grade for the course was determined by their performance on two midterm exams and one final exam.

Table 1 Differences in the two courses

Attribute	HCI	DS
Organization	Theory	Lab in support to theory
Teams	Scrum master in each team	No hierarchy
Assignments	Weekly assignments	Single semester project
Environment	Lecture classroom	Lab classroom
Grading	Tests	Project functionality and working
Evaluation	Informal competition	Grade
Class size	86	180 total; 24 per batch
Meeting time	1 h/3 times a week	3 h/1 time a week

In DS, the students' grades for the course were based on their projects. The factors considered were comprehensiveness and novelty of functionality, optimization and variety in the design, and successful working of the application and its presentation. The working and presentation were evaluated once between the faculty and individual teams and second time by different faculty members at a poster show. The peer assessment using constant sum scale was used to decide individual contribution and grade.

3.7 Environment

The physical environment for DS was much more conducive to teamwork. DS was taught in the lab class where lab benches surrounded the classroom. Each student attended one 3 h class once a week. There were about 180 students in the entire class with about 24 in each lab class. Toward the end, they found scheduled lab time inadequate and had to meet outside the lab hours.

The HCI class was taught in a large lecture room that was filled with students. There were 86 students in the class, which was taught three times a week for an hour. There was little room and time for the students to group themselves into their teams. So they had to find opportunities to meet each other.

In both cases, students seem to have found the time to meet with each other outside class as needed. In both classes, they coordinated their activities through email and SMS. They met in their hostels, in the canteens, or in empty classrooms after regular hours. There were some difficulties meeting with students who lived off campus, and there was some difficulty for boys and girls to work together since they live in different hostels. However, none of the teams found these problems insurmountable. All the differences are tabulated as below (Table 1).

3.8 Analysis of Outcomes

Despite the difference in the courses, the results were surprisingly similar.

3.9 Methods

We performed a work practice study of the students in the classes to gather information about their experience. Work practice is a discipline in which ethnographic techniques are used to understand the way people work [9]. It is a qualitative research technique in which the ethnographer engages the subject in situ. It was developed at Xerox PARC in the late 1970s from Harold Garfinkel's Ethnomethodology [10].

Here, one of the researchers, who has experience of conducting ethnographic studies, interviewed a sample of the groups and analyzed the results for common statements.

4 Results

Through our study, we found that students enjoyed the classes and learned more. In particular, common reports were:

- The team project allowed them to work on much larger projects than they would have been able to otherwise giving them experience with projects of a more realistic size than is usually possible in a semester long course.
- Though the project was larger, the individual students did less work than they would have had to on a smaller project because the work was spread among the team.
- Though the students felt they worked less, they felt they learned more because they learned from the experience of their teammates in addition to their own experience.
- Working on the projects was less intimidating than working alone because their teammates were available when they ran into trouble, whereas the teacher was less available.
- The social interaction within the team around technical problem helped them make new friends.

In other words, the students report that these classes support the three needs as described in SDT. It gave them a sense of competence, increased their sense of autonomy, and increased relatedness.

Their sense of competence was increased because they suffered less stress from time pressure even though they accomplished, together, a larger project. Moreover, the stress of being stuck disappeared because students could turn to members of their team to get either quick answers or confirmation that they are not the only ones who are stuck.

Their sense of autonomy was increased by the control they had over the content of the projects and by de-emphasis of the relationship between the project and the grade. In HCI, the professor did not apply the grade of the team assignments to individuals' grades in the class; instead the grades were posted as a "House Cup" contest. In DS, the students knew that their grade was dependent on their projects, but did not care much about it.

Their sense of relatedness was increased by the social interaction within the team. This social interaction increased their sense of competence (they had someone close that they could discuss problems with) and their sense of relatedness (they made new friends on their teams). Indeed the connections students made in their teams helped them integrate into the college. Some of the students were direct second year students—that is, they joined the college in their second year after diploma—and they found that the team provided them with a connection to the college.

5 Conclusion

The further development of the Indian software industry will require an increasing focus on innovation and entrepreneurship. It requires intrinsic motivation, which tends to decrease in traditional classrooms. SDT suggests that TBL can be used to foster motivation. Two very different classes were taught using TBL techniques. Based on the ethnographic study, we conclude that both the classes succeeded in increasing the intrinsic motivation of the students in the class. We will require some time to see if this has developed the required entrepreneurial skills.

5.1 Further Study

We find these results encouraging and see scope for further study. We have only done a qualitative study. We are gathering questionnaires from the students and plan to apply quantitative analyses to those answers. We believe these results will confirm that the students feel that they have learned more from the class. We also need to determine whether the student's feeling translate into improved behavior in subsequent classes and, more importantly, when they move to the working world. Further, every team may not be equally effective. It will be useful to study the team dynamics and research ways to steer them toward success. One can apply the approach to other engineering disciplines with the requisite changes and reap the benefits.

References

1. Indian IT-BPO Industry. NASSCOM. Retrieved 15 Dec 2012.
2. The IT BPM Sector in India – strategic review 2013. NASSCOM; 2013.
3. Michaelsen LK, Knight AB, Fink LD. Team-based learning: a transformative use of small groups. Westport: Greenwood Publishing Group; 2002.
4. Amabile TM. Creativity and innovation in organizations. Cambridge, MA: Harvard Business School Tech, Rep. 9-396-239, 5 Jan 1996.

5. Ryan RM, Deci EL. Intrinsic and extrinsic motivations: classic definitions and new directions. Contemp Educ Psychol. 2000;25:54–67.
6. Neimiec CP, Ryan RM. Autonomy, competence, and relatedness in the classroom: applying self-determination theory to educational practice. Theory Res Educ. 2009;7(2):133–44.
7. Benware CA, Deci EL. Quality of learning with an active versus passive motivation set. Am Educ Res J. 1984;21:755–65.
8. Kostner R, Ryan RM, Bernieri F, Holk K. Setting limits on children's behavior: the differential effects of controlling versus informational styles on intrinsic motivation and creativity. J Pers. 1984;52:233–48.
9. Szymanski MH, Whalen J. Making work visible: ethnographically grounded case studied of work practice. Cambridge: Cambridge University Press; 2011.
10. Garfinkel H. Studied in ethnomethodology. Englewood Cliffs: Prentice-Hall; 1967.

Academic Information Management System Using Open Source Programming Tools and Technologies

Prakashgoud Patil

Abstract Open source refers to a computer program in which the source code is available to the general public for use and/or modification from its original design. Open source tools may be viable alternatives to popular closed-source applications and some open source tools offer features or performance benefits that surpass their commercial counterparts (http://www.webopedia.com/TERM/O/open_source.html, http://en.wikipedia.org/wiki/Open-source_software). We at BVBCET (BV Bhoomaraddi College of Engineering and Technology, Hubli) developed a wide range of Software Applications using Open Source Programming Tools and Technologies to meet the requirements of managing the Academic Information for autonomous engineering colleges. The applications developed using open source technologies include (i) Student Information Management System (SIMS), (ii) Student Assessment Management and Examination System (SAMS), and (iii) Examination Results (eResults). Brief overviews of these applications along with the benefits are presented in this paper. These tools were found to be very useful for managing and administering the information in an autonomous engineering college and well appreciated and used by the stakeholders for the last few years.

Keywords SIMS – Student information management system • SAMS – Student assessment management and examination system • Open source programming languages and technologies • BVB College of Engineering and Technologies (BVBCET)

P. Patil (✉)
Department of MCA, BVB College of Engineering and Technology,
Vidya Nagar, Hubli, Karnataka 580031, India
e-mail: prpatilji@gmail.com

R. Natarajan (ed.), *Proceedings of the International Conference on Transformations in Engineering Education: ICTIEE 2014*, DOI 10.1007/978-81-322-1931-6_21,

1 Introduction

Open source technologies offer dynamic applications for real estate tailored to meet business development needs. Open source tools are usually available at no charge under a license defined by the Open Source Initiative. Open source tools may be viable alternatives to popular closed-source applications, and some open source tools offer features or performance benefits that surpass their commercial counterparts. The term "open source tools" is synonymous with open source utility and similar to open source applications.

Our objectives were to develop software which could be user friendly and meet the requirements of engineering colleges using open source and open standard. In our work, we used open source technology – LAMP (Linux, Apache, MySQL, and PHP) to develop a wide range of products required for the management of information at BVBCET.

2 Architecture and Principal Components of Our System

LAMP is the acronym used to describe the common open source server architecture which comprises Linux, Apache, MySQL, and PHP.

2.1 Linux: The Server's Operating System

The Linux operating system provides the platform for secure and reliable operation of the web server. Access to the file system is governed by file permissions and enforced by the operating system. The operating system also provides interfaces to the networks and enforces process and user security [1].

2.2 Apache: The Web Server Component

Apache is generally recognized as the world's most popular Web server (HTTPserver). Originally designed for UNIX environments, the Apache Web server has been ported to Windows and other network operating systems. The name "Apache" derives from the word "patchy" which the Apache developers used to describe early versions of their software. The Apache Web server provides a full range of Web server features, including CGI, SSL, and virtual domains. Apache also supports plug-in modules for extensibility [2].

2.3 MySQL: A Relational Database

The MySQL database has become the world's most popular open source database because of its high performance, high reliability, and ease of use. It is also the database of choice for a new generation of applications built on the LAMP stack (Linux, Apache, MySQL, PHP/Perl/Python). MySQL runs on more than 20 platforms including Linux, Windows, Mac OS, Solaris, IBM AIX, giving you the kind of flexibility that puts you in control [3].

2.4 PHP: The Application Layer

PHP is a general-purpose server-side scripting language originally designed for Web development to produce dynamic Web pages. It is one of the first developed server-side scripting languages to be embedded into an HTML source document rather than calling an external file to process data. The code is interpreted by a Web server with a PHP processor module that generates the resulting Web page. PHP can be deployed on most Web servers and also as a stand-alone shell on almost every operating system and platform free of charge. The PHP scripts will be used to store and retrieve information and to render the HTML page layouts for transmission back to the user's browser by the Apache web server.

Figure 1 illustrates the principal components in the architecture of our system. LAMP is singularly focused towards Web applications. The architecture is very straightforward, as illustrated in Fig. 1. Linux forwards HTTP connections to Apache, which serves static content directly from the Linux kernel. Dynamic pages are forwarded by Apache to PHP, which runs the PHP code to design the page. Database queries are sent to MySQL through PHP. Administration is commonly handled through phpMyAdmin tool [4].

3 Products Developed and Its Use at BVBCET Using Open Source Programming Tools and Technologies

The various applications have been developed to meet the requirements of Student Information Management, Assessment, and Examination and placed in use. The systems developed are running for the last few years without any bugs and serving the needs of our college. The applications are user friendly, reliable, cost effective and well appreciated by all the stakeholders. Our products are being used by various institutes run by K.L.E. Society including BVBCET and other institutes from within and outside Karnataka. The applications developed at BVBCET are highly accepted and

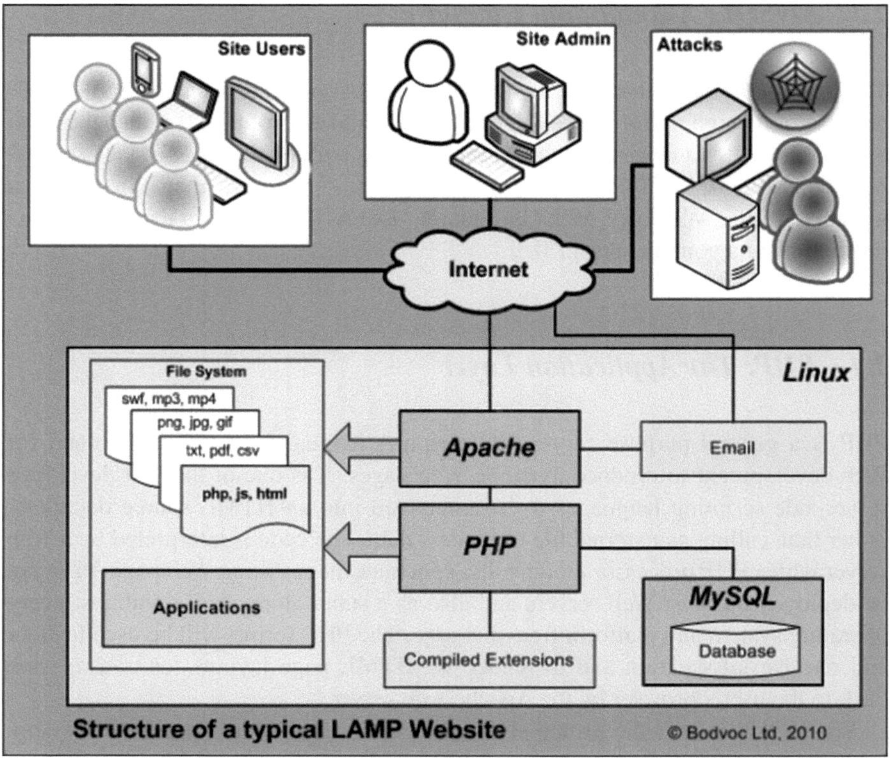

Fig. 1 Architecture and components of the system

appreciated by the clients because of a user-friendly interface, error free, seamlessly scalable, platform independent and low maintenance and easy to use and implement. Our products adapt to changes very quickly. The applications are role based and only authenticated users can login to the system. UserIDs and Passwords are provided to all the users including the staff and students. The user's authentication is done by the LDAP (Lightweight Directory Access Protocol). LDAP is an application protocol for accessing and maintaining distributed directory information services over an Internet Protocol (IP) network server. Users with little computer knowledge can use our software products very easily. The brief description of the products along with their features and few snapshots is presented in the following section.

3.1 Student Information Management System (SIMS)

SIMS is an application for engineering colleges for Managing Students and Academic Information. The main goal of this application is to streamline administrative tasks; manage student information college-wide; improve college to home

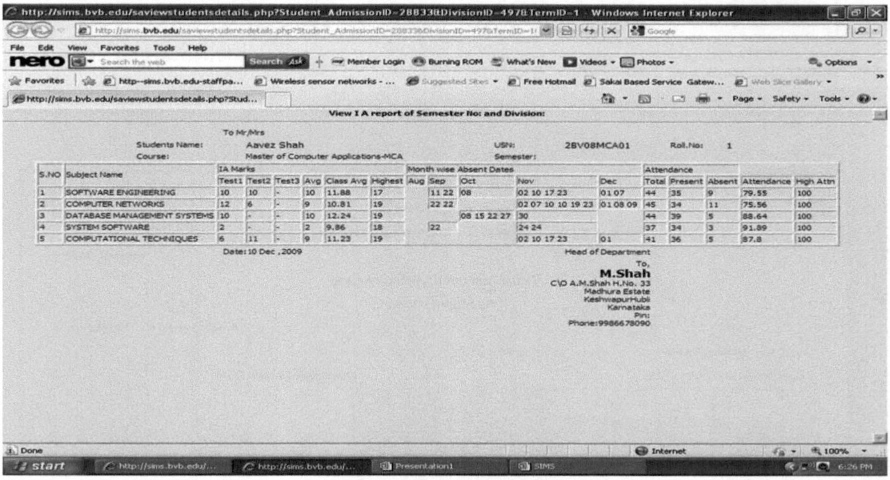

View I A report of Semester No: and Division:

To Mr/Mrs

Students Name: Aavez Shah USN: 2BV08MCA01 Roll.No: 1

Course: Master of Computer Applications-MCA Semester:

S.NO	Subject Name	IA Marks						Month wise Absent Dates					Attendance				
		Test1	Test2	Test3	Avg	Class Avg	Highest	Aug	Sep	Oct	Nov	Dec	Total	Present	Absent	Attendance	High Attn
1	SOFTWARE ENGINEERING	10	10	-	10	11.88	17		11 22	08	02 10 17 23	01 07	44	35	9	79.55	100
2	COMPUTER NETWORKS	12	6	-	9	10.81	19		22 22		02 07 10 10 19 23	01 08 09	45	34	11	75.56	100
3	DATABASE MANAGEMENT SYSTEMS	10	-	-	10	12.24	19			08 15 22 27	30		44	39	5	88.64	100
4	SYSTEM SOFTWARE	2	-	-	2	9.86	18		22		24 24		37	34	3	91.89	100
5	COMPUTATIONAL TECHNIQUES	6	11	-	9	11.23	19				02 10 17 23	01	41	36	5	87.8	100

Date: 10 Dec ,2009

Head of Department

To,
M.Shah
C\O A.M.Shah H.No. 33
Madhura Estate
KeshwapurHubli
Karnataka
Pin:
Phone:9986678090

Fig. 2 Students IA and attendance report

communication; allow college to easily make information available online for students, parents, and staff; and allow delivery via the Internet or over campus network to provide the solution that best meets college needs.

The various functionalities performed by each module are as follows:

Administrator Module: This module provides facility to create, edit, delete, and manage the following information such as term, programs, department, course, and other basic information required by the system such as qualifications, states, caste, etc.

Student Admission: This module handles student admission and other tasks such as new student registration, modify admission details and generate admission reports – branch wise, category wise. The module facilitates for division allotment (Div A, B, C, etc.) to students and generation of student identity cards. The snapshot in Fig. 2 shows the students IA and attendance report generated by the application.

Staff/Faculty Information: This module handles registration of new faculty and maintenance of faculty profiles, staff attendance, internal assessment reports, news and information, and student activities.

Term Management Module: This module handles creation of new terms, subject and subject combinations, student promotions, roll number allocation, student allotments to division and batches, subject allotment to faculty, attendance and internal assessment, parent reports, and student counseling. The snapshot shown in Fig. 3 shows the students attendance report.

Student Feedback Module: With the help of this module, college can create feedback questionnaire. The purpose of this feedback questionnaire is to gather information on students learning experience, as well as student's responses to the course and teacher(s). The information provided by the students will be useful to

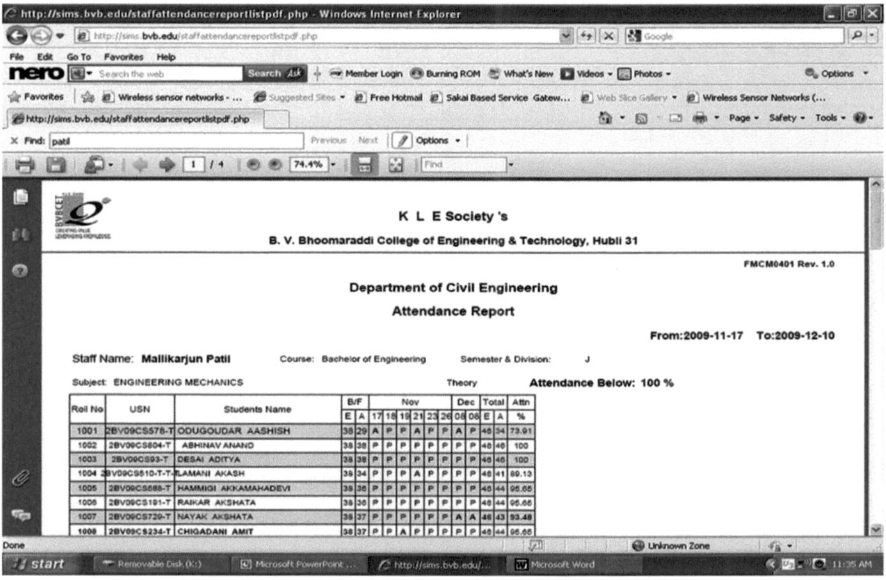

Fig. 3 Students attendance report

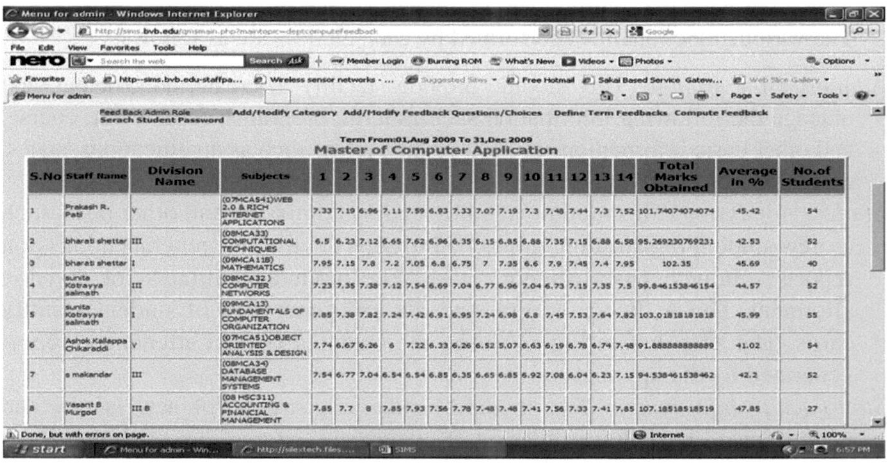

Fig. 4 Student feedback report

teacher(s) as well as to the institute in the ongoing efforts to enhance the quality of education. The data provided by the students will be treated as confidential. The student feedback is strictly anonymous – the identity of all students who provide feedback is protected. The feedback results will be computed using the total scores from the responses from all students and will be released to the

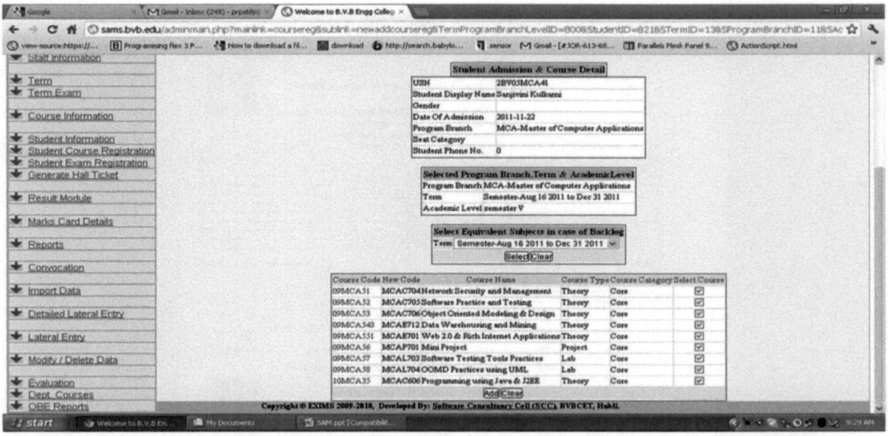

Fig. 5 Student exam registration

departments and teachers. Figure 4 shows the student feedback report generated by the application.

The SIMS became very popular and it is capable of delivering all the associated academic tasks and even it provides the facility and access to parents to keep a bird's view of their ward's progress in the classroom.

3.2 *Student Assessment Management and Examination System (SAMS)*

The main objectives of SAMS' application are to automate the process of exam conduction and evaluation. The main functionalities addressed by this system are student examination registration, exam hall ticket generation, capturing CIE marks from the departments for the courses offered, online marks entry for SEE lab examination, generate course wise candidate list for SEE – semester end examination, management of question paper templates, schemes and solutions, conduct of examination and encoding answer papers using barcodes, answer paper packet issued to evaluators for first/second/third round evaluation, entry of student answer paper marks by the evaluator through examiners login, finalization of marks based on the marks obtained in various evaluation rounds and the policy matter, and exporting final marks to eResults module for the calculation of CGPA, SGPA, and publishing exam results. The screen shot of SAMS applications describing the various features is shown in Figs. 5 and 6.

K L E Society's
B. V. Bhoomaraddi College of Engineering & Technology, Hubli 31
Examination Section(Autonomous)
ADMISSION TICKET FOR SEMESTER END EXAMINATIONS
Program: MCA-Master of Computer Applications
Examinations: Odd Semester January 2012 Examination

VTU No: 2BV09MCA31 CANDIDATE NAME: PRIYANKA B ANGADI

DETAILS OF THE COURSES APPLIED (Office Copy)

S.No	Sem	Course Code	Category/Type	Credits	Course Title
1	V	MCAC704	Core/Theory	4	Network Security and Management
2	V	MCAC705	Core/Theory	4	Software Practice and Testing
3	V	MCAC706	Core/Theory	4	Object Oriented Modeling & Design
4	V	MCAE712	Core/Theory	4	Data Warehousing and Mining
5	V	MCAE701	Core/Theory	4	Web 2.0 & Rich Internet Applications
6	V	MCAP701	Core/Project	1.5	Mini Project
7	V	MCAL703	Core/Lab	1	Software Testing Tools Practices
8	V	MCAL704	Core/Lab	1.5	OOMD Practices using UML
			Total Credits :	24	

Total fee paid in Words Total fee: RS.
Challan No: Date:
I declare that the information given is correct.If found incorrect my application can be rejected
Address for Correspondence with Phone No,
Basavaraj.Angadi

Gadag-582101 Karnataka
Ph: 8372,0 Signature of the Candidate
EmailID:

cut here cut here cut here cut here cut here

K L E Society's
B. V. Bhoomaraddi College of Engineering & Technology, Hubli 31
Examination Section(Autonomous)
ADMISSION TICKET FOR SEMESTER END EXAMINATIONS
Program: MCA-Master of Computer Applications
Examinations: Odd Semester January 2012 Examination

Fig. 6 Student hall ticket

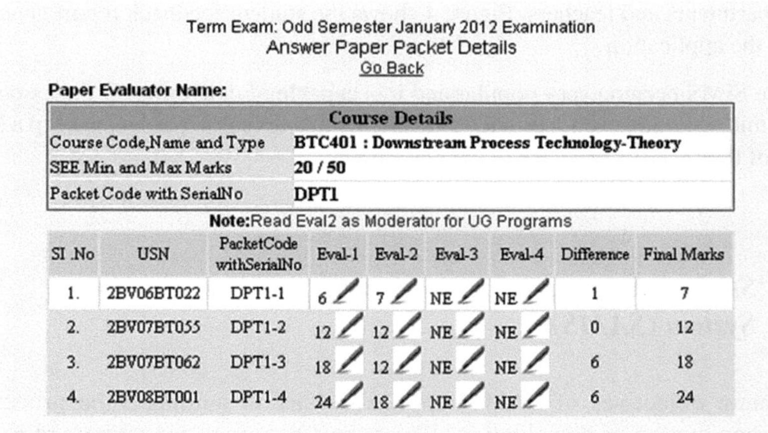

Term Exam: Odd Semester January 2012 Examination
Answer Paper Packet Details
Go Back

Paper Evaluator Name:

Course Details	
Course Code,Name and Type	BTC401 : Downstream Process Technology-Theory
SEE Min and Max Marks	20 / 50
Packet Code with SerialNo	DPT1

Note: Read Eval2 as Moderator for UG Programs

Sl .No	USN	PacketCode withSerialNo	Eval-1	Eval-2	Eval-3	Eval-4	Difference	Final Marks
1.	2BV06BT022	DPT1-1	6	7	NE	NE	1	7
2.	2BV07BT055	DPT1-2	12	12	NE	NE	0	12
3.	2BV07BT062	DPT1-3	18	12	NE	NE	6	18
4.	2BV08BT001	DPT1-4	24	18	NE	NE	6	24

Fig. 7 Three rounds of evaluation and final marks

3.3 *Examination Results (eResults)*

eResults module computes grades for the marks scored by the students in various courses registered for the examination. It takes care of grace marks policy. It computes SGPA and CGPA for the results. The eResults modules have the functionalities to generate various kinds of reports such as examination result sheet, generation of individual student's provisional grade cards, and grade card printing. Students can view their results and examination history using the user ID and password provided to them. The various kinds of statistical reports can be generated from this module. Figures 7 and 8 show the outputs generated by this module.

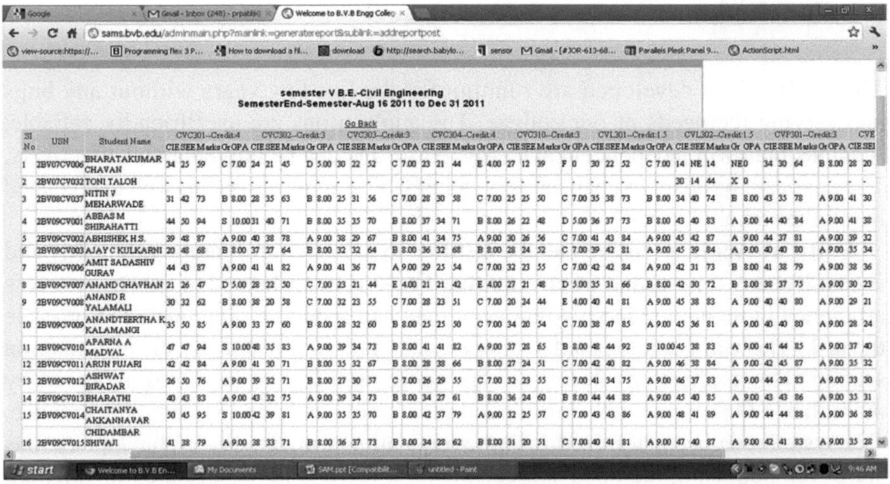

Fig. 8 Exam result sheet

4 Customer Feedback

The above said applications received good appreciation from the user because of the following features:

- Applications are web based, easily accessed from any system connected to LAN using browser.
- Offers an easy-to-use interface for any normal user.
- Designed to support large amounts of data and simultaneous access by a number of users.
- The applications are role based and system allows only authorized users to access functionalities based on their roles.

The applications developed by us are well appreciated by our beloved Principal Dr. Ashok Shettar and staff members of our institute. The email content of Principal appreciation about this work is presented below for the purpose of support and reference.

Dear Prakash,

My heartfelt thanks to you and your team for making almost impossible task into a possible one. I just visited our valuation center and it was great to see your software running smoothly all over. I think it was really hard task considering the limited time and resources you had, but you have made it happen. Once again thanks and congratulations!

Regards
Dr. Ashok Shettar
Principal-BVBCET, Hubli

5 Conclusion

The applications developed are running for the last few years without any bugs and serving the needs of our college. The applications are user friendly, reliable, scalable, platform independent and easy to use, and can be implemented effectively. The applications adapt to changes very quickly. The applications are well appreciated by all the stakeholders.

Software Applications helped the institute to manage information efficiently and access it at any time as and when required. These systems helped to improve process efficiency in managing large volumes of data and reducing time taken to complete the task. These applications led to quicker declaration of results by the examination section. The applications developed are being used in five institutes and used by more than 6,000 users with a fairly good satisfaction and good system performance. The system can be customized for other educational institutes. As the applications are developed using the open source software, there is no cost related to software licensing.

Acknowledgment The author would like to thank Dr. Ashok Shettar, Principal, BVBCET for his support, guidance and encouragement, and for an opportunity provided for such a wonderful work. The author also thanks the team members of Software Consultancy Cell, BVBCET for their hard work and sincere effort.

References

1. www.linux.org
2. www.apache.org
3. www.mysql.com
4. www.php.net

Designing Curriculum of IT: A Journey

Anupama K. Ingale, Savita P. Patil, and Sandeep A. Thorat

Abstract Rajarambapu Institute of Technology (RIT) is established in 1983. RIT has gained autonomy in 2011–2012. With the advent of autonomy, we take an opportunity to design curriculum and introduce innovative evaluation and assessment methods to ensure better learning. This paper discusses about curriculum development process of Information Technology Department. As a first step of curriculum design, process curriculum and cross curriculum competencies required are identified. According to the identified requirements, PEOs and POs are defined and then IT syllabus structure is formed based on the key areas. To support thinking competency, new assessment strategies are adopted such as programming test, mini project, quiz, demo, presentations, case studies, etc.

Keywords PEOs • POs • Innovative assessment methods • Information technology

1 Introduction

Exponential growth of engineering institutions in India, although provide opportunity for young aspirants to opt for engineering career, poses enormous challenges to develop competent engineering graduates with knowledge, skills, and ethical standards for global careers.

Rajarambapu Institute of Technology is established with the Vision of "Transformation of young minds into competent engineers to face global challenges" and Mission is "To offer the state of art technical education programmes to shape promising engineers with requisite skills, knowledge, research aptitude, values and

A.K. Ingale (✉) • S.P. Patil • S.A. Thorat
Rajarambapu Institute of Technology, Rajaramnagar, Urun Islampur,
Maharashtra 415414, India
e-mail: anupama.ingale@ritindia.edu; savita.patil@ritindia.edu; sandip.thorat@ritindia.edu

R. Natarajan (ed.), *Proceedings of the International Conference on Transformations in Engineering Education: ICTIEE 2014*, DOI 10.1007/978-81-322-1931-6_22,

ethics ensuring rewarding careers" [1]. RIT has gained autonomy in 2011–2012 which we look toward as great opportunity to design curriculum sensitive to needs of business and industries, implement the curriculum through innovative teaching-learning methods, and introduce innovative evaluation and assessment methods to ensure better learning. The Academic Council of RIT is the highest academic decision-making body which sets and maintains high academic standards at par with the best national and international institutes.

The Information Technology Department is established in year 2001. The department works with Vision: "Towards Excellence in IT Education and Research," and Mission: "To offer the highest quality of education in Information Technology leading to successful career in Industries, Research and Entrepreneurship." Our goals are to establish interactions with local and global IT industries and research institutes, to provide students exposure to latest technologies in IT industries, and to excel in employment and research opportunities. To achieve our goals, the department has signed MOUs with TCS, Wipro, Infosys, and Kahuna; CSI student chapter is started to motivate professional growth of the students.

The rest of this paper is divided into three sections. Section 2 discusses about the PEOs and POs of IT department and the key areas considered for curriculum design. Section 3 discusses innovative assessment strategies adopted by the department. Section 4 is about the other activities conducted in the department.

2 An Overview of IT Curriculum

Curriculum design should promote on developing cross curriculum (thinking competence, communication competence, and social competence) and curriculum competence [2] among students. These are the cognitive components of problem solving. In curriculum, emphasis should be on knowledge which focuses on the key concepts, models, and principles essential for providing understanding of the topic being taught and meeting the learning outcomes [3, 4]. PEOs and POs are designed according to the norms of AICTE [5] and to meet curriculum requirements.

Programme Educational Objectives (PEOs):

1. To provide students with a sound foundation in Information Technology theory and practices to analyze, formulate, and solve social and engineering problems
2. To develop an ability to design algorithms, implement programs, and deploy software
3. To develop an ability to analyze, interpret, and communicate data using Information Technology
4. To provide an excellent environment for developing professional skills comprising ethical attitude, effective communication, leadership, and teamwork

Table 1 Stream and possible weightage

Sr. no.	Name of stream	Possible weightage (%)
1	Basics of computer science	10–15
2	Electronics and computer hardware	10–15
3	Design and analysis techniques	35–40
4	Problem-solving techniques	5–10
5	Computer networking	15–20
6	Topics in computer systems	15–20
8	Professional ethics and responsibilities	5–10
9	Advanced topics	5–10

Programme Outcomes (POs):

(a) Graduates will demonstrate knowledge of mathematics, science, and engineering.
(b) Graduates will be able to identify, analyze, and formulate engineering and social problems using IT.
(c) Graduates will be able to analyze and interpret data for modeling, designing, and developing software.
(d) Graduates will have an ability to design and develop software systems of varying complexity to meet the desired needs.
(e) Graduates will be able to make use of modern tools to evaluate and implement computer-based systems.
(f) Graduates will be able to communicate information efficiently, reliably, and securely using computer networks.
(g) Graduates will be able to succeed in competitive examinations such as GATE, GRE, and various professional certifications in IT.
(h) Graduates will exhibit professional skills and ethical responsibilities.
(i) Graduates will be able to communicate effectively in both verbal and written form.
(j) Graduates will be able to work effectively in teams and have enduring learning skills.
(k) Graduates will be able to analyze the impact of engineering solutions on individuals, society, and environment.

The objective of Curriculum design is to emphasize on courses that enable students to apply their technical knowledge to face a variety of technological challenges. Hence the following job profiles were considered as core requirements for curriculum design: (a) Software Engineer (Developer, Tester, Analyst, Architect, Integration, Designer, etc.), (b) Database Administrator, (c) IT Manager (Process, Project, Enterprise and Resource, Information Services, and Business), (d) Network and Communication Engineer, (e) Research Fellow Education, and (f) Higher Studies. Each stream is identified and assigned weightage based on the needs of students and industries as in Table 1. Considering the streams and its weightages, courses are identified and syllabus structure is finalized in BOS meeting.

3 Unique Features of IT Curriculum

Keeping the perspective of employability of students and knowledge gain, some innovative components are added in the curriculum. These components focus on advanced knowledge and overall development of the students.

3.1 Professional Skills Development (PSD)

As per the feedback from companies students require effective communication skills. Professional Skills Development Courses are introduced in three semesters to enhance their communication skills. Globarena software is used as an interactive learning software where students can learn and attempt online tests. A student can achieve a level of competency required for industry by adapting this course.

3.1.1 Professional Training

As the institute is in rural area, students need exposure to outside world; the "Professional Training" course aims to give opportunity to interact with industry and gain the experience of industry environment. As a part of this course, each student needs to complete a professional training during vacation. The "successful completion of training" will be assessed on recognized certification or through project work.

3.2 Programming Course

As a developer, students are expected to have new programming skills to stay relevant in the fast-paced tech world. Hence the programming languages that are in demand by employers are identified and introduced in syllabus. Some of them are C, C++, ASP.NET, JSP, and PHP. Taking into consideration the importance of this course, an innovative assessment strategy has been adopted, discussed in Sect. 4.3.

3.3 Community Services

The department is not only focusing on technical growth but also on the social responsibilities of the students which makes them better citizens. Hence community services are introduced to make students aware of community and its needs. Students are asked to perform communal activities such as training school children from rural area and so on.

3.4 Latest Knowledge in IT Technologies

As we know technologies are changing rapidly, students are expected to adapt to new technologies to compete in the IT field. So the department has introduced a new core subject consisting introduction to new technologies evolved in past 3 years and the subject is revised once in 2 years.

3.5 Mobile Application Development

With the growing number of Smartphone and tablets, companies are looking at how they can use mobile tools. There is a need for skilled developers and designers who are familiar with the mobile application process. In this view, mobile application development course is introduced to the students where they will learn core skills in mobile design and development.

3.6 The Free and Open Source Software (FOSS)

FOSS model provides interesting tools and processes with which students can create exchange, share, and exploit software and knowledge efficiently and effectively. Advantages such as increasing interoperability, reducing costs, enabling localization, and reducing piracy/copyright infringements are compelling reasons for adopting FOSS. FOSS focuses on PHP, PERL, MySQL, Linux, and Apache.

3.7 Mini Project

In this course students develop understanding of underlying theory through working on mini project. Students are encouraged to come up with innovative thoughts or develop the projects to solve simple real-world problems. For example a team of students has to visit the shops and collect their requirement to develop a software which satisfies their needs, and provides maintenance facility. This will improve their ability to work in teams.

4 Innovative Assessment Methodologies

The important considerations when planning assessments are that it should meet course outcomes, awareness of its impact on students, establishing criteria for student assessment, and establishing fair grading procedure to minimize the marker bias [3, 4]. According to rules and regulation each semester has three evaluation schemes, in semester evaluation (ISE), mid semester evaluation (MSE), and end semester evaluation (ESE) of 20 %, 30 %, and 50 %, respectively. Each course may have two or three types of assessments in ISE. They are planned well before and dates are displayed at the commencement of the semester.

4.1 *Quiz*

Students are evaluated based on their concept understanding and applying knowledge to solve a given problem in a smart way. Quiz is launched using moodle server. It helps them to prepare for competitive exams such as GATE, GRE, and technical tests for campus interview.

4.2 *Presentation*

As per the feedback from industries, students should improve their soft skills; hence, presentation helps them to improve their presentation skills and encourage self-learning.

4.3 *Programming Test*

Programming skills are required for being a good developer and a new evaluation strategy is adopted. Question paper is divided into two parts consisting of objective and programming. During the programming test a team of experts will evaluate the students' programs. So far we have applied this strategy for C++ and Java and found significant improvements in students' programming skills. The result analysis for Java course during the year 2013 is as follows: (Table 2, Fig. 1)

Table 2 Result analysis of Java course

MSE (total 91 students, out of 30 marks)					
0–5	6–10	11–15	16–20	21–25	26–30
07	05	27	32	12	08

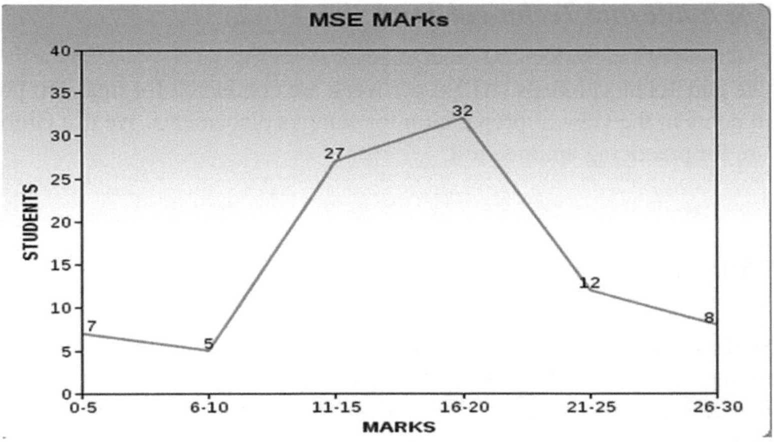

Fig. 1 Java result graph

4.4 *Demo*

To develop interest in a subject, a topic related to curriculum is given to students where they demonstrate concept practically in labs which allows them to think and analyze topic and gain in-depth knowledge. For example in operating systems, group of students were asked to study UNIX commands and show the working of the commands practically.

4.5 *Case Study*

Students are asked to choose the topic relevant to the curriculum and study which includes different tools, applications, devices, etc., and prepare a report and present it. For example in the subject Computer Network and Security, different security related tools such as Maltigo, Aircrack, Keepass, sniffit, Snort, etc., were taken as case study.

5 Other Practices for Strengthening Academic Culture

5.1 *Programming Challenge Activity*

To increase the competitive spirit among the students, this activity was started. Under this weekly different technical and nontechnical problems are given to students, so that they will get better realization of real-world problems. Attractive prizes are given to winner of the challenge by the department to encourage their participation.

5.2 *Aptitude and Technical Tests*

Aptitude and Technical Tests (ATTs) per week are conducted for final and prefinal year students in the view of preparation for campus placements. We use Globarena software for practicing aptitude test.

5.3 *Student Association*

The Students Association is formed in the department with the name "ELITE" (Emerging Leaders of Information Technology) which works to strengthen communication, effectively address student concerns, and promote a culture of academic excellence. Under this, some Social Activities, Institute Level Events, Department Level Events are conducted. To increase students' participation and to motivate, awards such as Best student of the year, Best Programmer of the year, and Best IDEA are given.

5.4 *Quality Circle*

Quality Circle is an informal group of employees who voluntarily meet together on a regular basis to identify, define, analyze, and solve work related problems. Every year a new problem statement in the department is identified and worked on it. This year we have identified Problem Statement as Difficulty in Understanding Subjects.

5.5 *CSI (Computer Society of India)*

RIT is a member of CSI and also we have a student branch of CSI for CSE/IT departments. We organized different events under student chapter such as PRO-YUDDHA'11, TECHNOSPHERE which includes project competition and on the spot paper presentation at the university level. Also technical seminars and workshops are arranged.

5.6 *Competitive Exam Cell*

Aim of the competitive exam cell is to encourage students to go for higher studies. This year 45 students are registered for GATE and few are interested in PUSs. During vacation crash courses are arranged for students. To encourage students mock tests are conducted on every Saturday.

5.7 *Toast Master*

The importance of the English language cannot be overemphasized. Comfort with English is almost a prerequisite for success in the world today. Some of the activities conducted in toast master are as follows: Skit, Games, Group Discussion, Presentations, Video Lessons, Movie Class, Survival English, and Line Dialogue. This activity is conducted thrice in a week.

6 Conclusion

This paper discusses curriculum design process adopted in RIT IT department. The paper has also discussed about innovative practices followed to strengthen academic ambience in the department. We have attempted to give innovative thoughts to different difficulties, weakness identified in the achievement of PEOs. As curriculum design is a continuous process, we believe that exploring new methods and ideas are integral components of any good academic system.

References

1. www.ritindia.edu
2. Exploring Curriculum Design. Transforming curriculum and assessment. Ministry of Education, January 2013.
3. Sale D. Practical steps in designing a curriculum to promote critical & creative thinking. Center of Development of Teaching and Learning. December 2001;4(6).
4. O'Grady G. Designing and planning a successful course: bridging the gap between common practices and best practices. Center of Development of Teaching and Learning. 2001;4(6):1–4.
5. http://www.aicte-india.org

Essential IT Skills to Learning Community for Industry Readiness

Bhupesh Deka, K.N. Vani, Srinagesh Chatarajupalli, Manisha, Anoop Singh, and K.V.R.S. Sarma

Abstract Indian Information Technology (IT) industry has witnessed unprecedented growth since the mid-1990s. Being a people-centric business, IT companies have continuously innovated to attract and retain talent needed to sustain the business growth. The rejection rate of on-campus recruitment performance has evidently indicated the necessity to enhance the students' technical and behavioral knowledge and skills. Inadequacy in teaching and learning infrastructure such as faculty, industry-relevant content, powerful assessment methods, and accelerated learning aids has impacted employability. The Industry Readiness Program has become essential to bridge the skill gap for making the engineering graduates ready for the IT industry. Engineering students who successfully get exposure to such an industry readiness training program can become productive from day 1 in IT industry. This paper talks about the design and implementation of such an industry readiness. This program has also contributed a part of the National Association of Software and Services Companies (NASSCOM) Fundamental Skills of Information Technology (FSIT) program along with other IT organizations for the students to augment them at par with the IT industry requirements.

Keywords Industry readiness • Technical foundation • Employability • NASSCOM • FSIT

1 Introduction

Infosys has an extensive educational infrastructure and is one of the largest in the world. The charter of the Education and Research Department, the corporate learning unit, is to enhance the client delight, support the company's vision, and meet

B. Deka (✉) • K.N. Vani • S. Chatarajupalli • Manisha • A. Singh • K.V.R.S. Sarma
Education and Research Department, Infosys Ltd, Bangalore, Karnataka, India
e-mail: Bhupesh_deka@infosys.com; Vani_kn@infosys.com; Srinagesh_C@infosys.com; Manisha@infosys.com; Anoop_Singh@infosys.com; SARMAKVRS@infosys.com

R. Natarajan (ed.), *Proceedings of the International Conference on Transformations in Engineering Education: ICTIEE 2014*, DOI 10.1007/978-81-322-1931-6_23,

Fig. 1 CC program approach

employee aspirations by developing the competency of Infoscions in technical, behavioral, domain, project management, and leadership areas. One of the strategic functions of the unit is the "industry-academia partnership program" – Campus Connect (CC) Program – which aims at implementing a large systemic change in engineering education by collaborating with colleges to create industry-ready IT workforce. The problem exists in the two dimensions: quality and quantity. The quality gap is between the competencies such as technical and soft skills, process awareness, and English language of fresh engineering graduates that are essential for the IT industry proficiency. The program was designed with a principle of Engage-Enhance-Sustain. We have successfully completed the Engage and Enhance phases and are currently in the Sustain phase.

CC was launched in May 2004, with 60 Indian colleges. The initiative was aimed at deepening industry-academia bonds and building a strong footing for the future needs of both academia and the IT industry. The global initiative focuses on crafting industry-ready IT professionals, by aligning engineering talent with the current needs of the industry. As of June 2013, CC partnered with 355 colleges in their pursuit of excellence and trained 198,267 students through technical programs (generic and domain) and 68,348 students in soft skills, totaling over 266,615 students nationwide. It has enabled over 9,245 faculty in technical and soft skills through various faculty development programs (FEP). Currently working with policy-making bodies to include industry centric electives, including behavioral skills, in autonomous engineering colleges to make students industry ready, when they graduate (Fig. 1).

Its initiative is driving the quality enhancement of the IT pool by offering about 12 different types of programs. These programs are targeted for various stakeholders such as students, faculty members, and college in a consistent, predictable, and effective way. This paper describes the initiative Fundamental Program, its proven courseware, methodologies, and education experiences that promoted developing technical skills for the students with integrated tracking capability for end-to-end deployment management.

2 Literature Survey

Software lobby NASSCOM says only 25 % of graduates coming out of academia are employable. The World Economic Forum's Global Talent Risk report (2011) cautions that developing countries like India and Brazil will also face low employability due to lack of skilled talents. There is a wide gap between the skills required in industry and those provided by the education system [1]. The report, *Global Talent Risk – Seven Responses*, analyses projected talent shortages by 2020 and 2030 in 25 countries, 13 industries, and 9 occupational clusters [2]. According to the Computing Technology Industry Association (CompTIA), more than 15 million businesses rate the aggregate skill levels of their IT staff as less than optimal, and 93 % of employers indicate that there is an overall skills gap among employees.

NASSCOM is predicting that IT industry would continue to grow at healthy growth rates over the next decade leading to a continuing demand for a skilled workforce. The IT industry and its human resource would need to build consulting capabilities (architecture, application selection, solutioning, and process engineering/reengineering) to help them tap into higher value projects which would have several downstream revenue spin-offs [3]. The challenges that today's IT company is facing are (1) incomplete competencies from the input talent pool and (2) inadequate industry orientation and inability to apply the concepts learned during undergraduate engineering education. The gap exists and hence IT industries provide intensive training to joinees for about 3–6 months before they could contribute to client project. If students in the institutes are exposed to what they get trained, then they are better organized to take on the challenges once they join the industry.

3 Background

The IT industry is mounting at a vigorous rate with businesses deeply dependent on IT applications for their operations. Today's enterprises are utilizing IT outsourcing as a strategic means to achieve business goals such as reduction of cost, business innovation, and competitive advantage. The IT services industry is, by nature, people-centric. The quality manpower demand has been rising continuously. Quality manpower is indispensable for IT industries to remain ahead of competition in the global marketplace. The necessity for industry-academia partnership has been intensely felt to enhance the industry readiness of engineering graduate workforce to address the problem of quality engineering education in India that has got around 4,753 engineering and MCA institutions. The problem is colossal and necessity is to work with a significant number of these institutions to make an impact. The offerings of this practice should be designed with scalability as one of the crucial criteria.

It is extremely important that if students in the college are prepared to become industry ready, then they are better equipped to take on the challenges once they join the industry. The two ways to deal with this are as follows:

- Designing and deploying training programs to the students – scalability and sustainability are the key issues of this approach.
- Reaching out to the students through the faculty and leveraging technology of the alliance institutions.

This approach involves enabling faculty members to develop the students with the aid of technology. This model was adopted in designing the framework.

An initial survey was conducted with target stakeholders – engineering students, faculty members, institution management, and experts from the industry. This helped in identifying their needs of this program. Based on the research, industry expectations, and the recommendations from several stakeholders, two courses were identified, namely, technical and soft skills, and the courseware is deployed on a digital platform to reach the large scale. Several external agencies/partners are involved in the execution of this project. The program has carved out logical parts of the technical and soft skills courses, leveraged IT industry best practices, and infrastructure of institutions to deploy offerings and training programs in a consistent, scalable, effective, and sustainable way. The design decision was reached based on three important criteria, namely, scalability, measurability, and sustainability.

4 Design

The core of the CC program is the Fundamental Program (FP) rollout for the students at the institutes. The program's courseware represents the intellectual property and experience of training thousands of entry-level engineers from diverse backgrounds, disciplines, and regions to deliver solutions to global customers. It provides the faculty members with unique teaching aids and consolidated training material to help them make students industry ready. This courseware has been used to train entry-level engineers. This is typically delivered as classroom sessions coupled with extensive hands-on exercises and assignments.

The Program 3.1 is the latest version offered in the year 2011 and currently offers six courses and Integrated Project along with newly incorporated ways with the help of rubrics methodologies for assessing students' performance. Integrated Project is the crucial component of it. The duration of this program rollout is 55–65 classroom contact hours along with dedicated 25–32 working hours at the laboratory for Integrated Project Development. The enhancements on these program offerings and processes are based on the amendments in the Generic Program at the company's Global Education Center (GEC) and the inputs and feedback received from the partner colleges.

The entire modules are followed in a sequential manner. The rationale behind is the competencies needed to be developed for becoming a software professional. The students need to comprehend the computer and its components as they would be working on it to develop solutions. To identify and solve a problem, knowledge of the problem is very essential. Any software applications developed require an operating system – development and deployment. The knowledge of programming and testing is required to build and test an application. There has been a paradigm shift in the development methodologies and hence the object-oriented concepts.

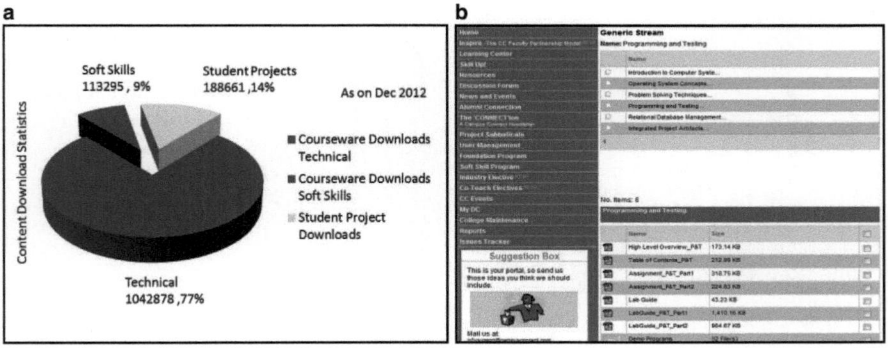

Fig. 2 (**a**) Courseware download statistic. (**b**) Courseware in the portal

Java has been the popular language and hence taken as a vehicle. Lastly, an application interacts with data; hence, the knowledge of data management is required. Project development is associated with it, keeping in mind the hands-on experience and understanding the courses cutting across. In addition, there are few tools included in order to make the student familiar working with it. Tools provide an effective means to implement a repeatable and predictable process. The advantages of using tools are as follows:

- Reduced development effort, time, and cost
- Increased productivity and quality
- Get products to the market faster

Raptor, PMD, and JUnit are open-source tools recommended for the PST and OOC with Java Module. Moodle, an open-source course tool, is recommended to conduct the program assessments. The courseware is made available digitally in the CC portal (http://campusconnect.infosys.com). The accessibility has been made easier with the better representation of the different version of courseware at the same page for easy readability (See Fig. 2a and b).

Faculty members are responsible for rolling out the program at the institute. This facilitates the scalability and sustainability of the program. FEP are conducted at institution to enrich the knowledge of faculty members with technology application examples and experiences, usage of latest tools, case studies, generation of new case studies, project execution, and assessment methods to enhance courseware delivery effectiveness and leverage the same to the students.

5 Methodology

The process of rollout is divided into three phases – pre-deployment, deployment, and post-deployment – for the ease of understanding. Pre-deployment involves planning and preparation for the program rollout. Deployment involves the delivery

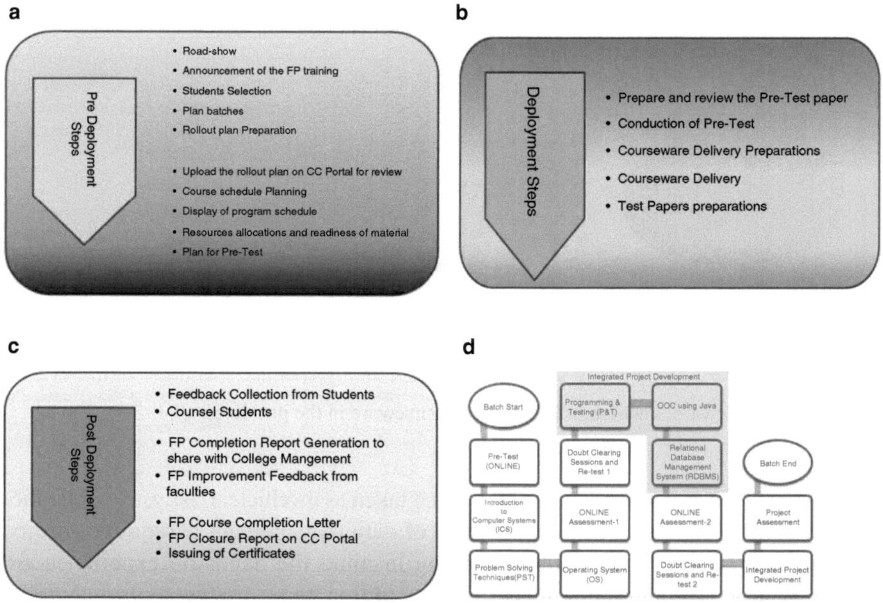

Fig. 3 (**a**) Pre-deployment steps. (**b**) Deployment steps. (**c**) Post-deployment steps. (**d**) Program flow diagram

of courses and assessing student performance. Post-deployment involves student counseling, reporting, and feedback. Figure 3a, b, and c depict the entire process. Figure 3d depicts program execution.

The batch data is fully managed using the portal. The College Single Point of Contact (SPoC) uses their login credentials and selects the option Fundamental Program → Inbox → "Plan a batch" screen. College SPoC provides the details like batch anchor, faculty members, course dates, and student details while uploading the batch plan. Every new batch submitted on the portal goes through an approval process. Development Center (DC) SPoC approves/resends the batch after careful evaluation of the batch plan. A unique batch identifier gets generated for every approved batch. This batch identifier is used for the purpose of tracking and closure of the batches. After the Fundamental Program batch plan is created and approved for execution and the student selection is completed, the college is all set to start the first component of the Fundamental Program, i.e., pretest. It is to be conducted before the delivery of the first module of the program. The expected mode of deploying pretest is online. The objectives of this pretest are as follows:

1. To gain insight into the practicality of this program
2. To provide understandings of relevance and importance of the each module
3. To provide a feedback to students

After the completion of pretest, the session delivery for the first three courses – Introduction to Computer Systems (ICS), Problem Solving Techniques (PST), and Operating Systems (OS) – would start. These three courses set the necessary groundwork required for the next level courses. The first online assessment 1 is to be conducted for the batch after completing the courses – ICS, PST, and OS. The case study or assignments and the Conceptual (C) Analytical (A) Memory (M) and Practice (P) (CAMP) Model-based question papers should be sent to respective DC SPoC for review before conducting the assessments. The retest option is available. A doubt clearing session is to be arranged before Course Delivery 2. The objective of conducting the doubt clearing session is to provide a platform to students to clarify doubts about the course focus areas if any. Programming and testing (P&T), Object-Oriented Concepts (OOC) using Java, and Relational Database Management Systems (RDBMS) are the courses to be delivered in the second set of course delivery. During the course delivery of these, few classroom hours are to be utilized to provide the understanding on the Integrated Project Development. This includes giving exposure to the system requirements, deliverables, standards to be followed, and execution map and project assessment framework. Online assessment 2 test is based on the courses such as P&T, OOC using Java, and RDBMS. The case study or assignments and the CAMP-based question papers should be sent to respective DC SPoC for review before conducting the assessments. Like online assessment 1, a retest 2 and a doubt clearing session are to be completed. The Fundamental Program test question papers are set based on the CAMP model of assessment. The distribution of questions recommended over the different aspects of CAMP model should typically be C + M = 20–30 % and A + P = 70–80 %. The context setting for Integrated Project starts during Course Delivery 2. The project briefing should start along with P&T course for better understanding of it. Faculty can spend approx. 8 h for setting context for the Integrated Project Development and exposure to the system requirements and deliverables. Every student fulfilling the following certification criteria for successful completion of the program is eligible to get Fundamental Program Certificate:

- Attendance with minimum attendance required for classroom sessions is 75 %.
- All the tests, case studies, and assignment are completed.
- Integrated Project Development.
- Total score in the assessments equal or more than 60 % (i.e., total number of marks earned equal to or more than 150 out of 250).

On completion of rollout, the College SpoC uploads the closure report on the portal and informs the DC SPoC so that he/she could add further information to the file for internal processing and analysis. On receiving the closure report from the portal, the team verifies the data and processes the issuing of the certificate. The number of students who should receive certificates from the data provided in the closure report is identified. The College SpoC is provided with a complete workflow of the Fundamental Program process document. This is also uploaded in the portal for reference.

6 Result and Discussion

The data is shown from the year 2009 as it is automated in digital infrastructure (see Fig. 4a and b). Figure 4c and d highlight the distribution of the program in terms of batches and students.

a

	FY2009-10			FY2010-11		
Hub DC	# Batches Completed	# Students Completed	% Students Eligible for Certificates	# Batches Completed	# Students Completed	% Students Eligible for Certificates
Bangalore	44	2584	92.18%	54	3147	92.02%
Bhubaneswar	64	4071	94.45%	35	2092	91.83%
Chandigarh	60	3946	93.08%	57	3414	92.59%
Chennai	46	3306	66.36%	53	3386	80.36%
Hyderabad	9	606	88.78%	17	1177	88.11%
JKC	86	5645	94.90%	44	2949	82.64%
Mangalore	10	585	90.26%	12	689	85.63%
Mysore	22	1389	76.96%	14	1033	88.38%
Pune	18	1270	77.80%	11	756	74.34%
Trivandrum	14	931	95.06%	10	908	84.91%
Total	**373**	**24333**	**88.19%**	**307**	**19551**	**87.00%**

b

	FY2011-12			FY2012-13		
Hub DC	# Batches Completed	# Students Completed	% Students Eligible for Certificates	# Batches Completed	# Students Completed	% Students Eligible for Certificates
Bangalore	86	5116	94.38%	63	3510	94.79%
Bhubaneswar	27	1778	81.77%	78	4709	94.10%
Chandigarh	67	3998	97.39%	29	1717	98.00%
Chennai	49	2881	90.49%	49	2864	92.30%
Hyderabad	15	1002	88.82%	17	1004	84.87%
JKC	24	1737	78.58%	71	3894	76.90%
Mangalore	18	1161	86.56%	24	1348	87.76%
Mysore	27	1902	97.06%	29	1890	92.29%
Pune	10	632	86.39%	7	386	69.93%
Trivandrum	16	1458	80.18%	10	840	90.71%
Total	**339**	**21665**	**90.48%**	**377**	**22162**	**89.11%**

Fig. 4 (**a**) Program status FY 2010 and 2011. (**b**) Program status FY 2012 and 2013. (**c**) Program batches from FY 2010 to 2013. (**d**) Students from FY 2010 to 2013

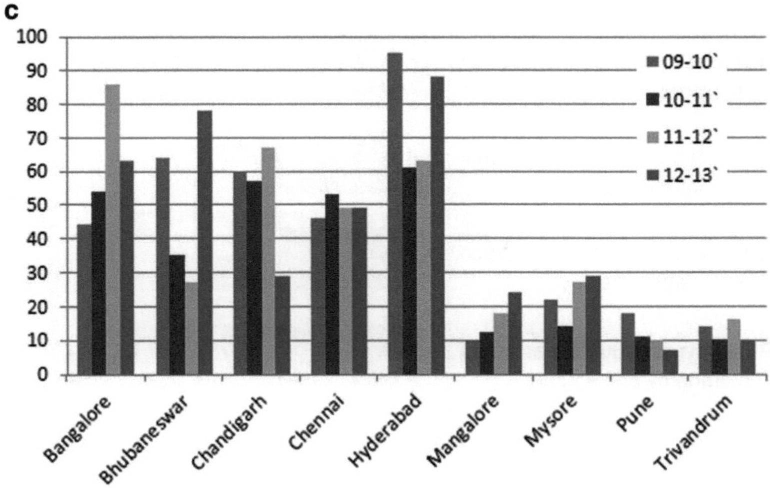

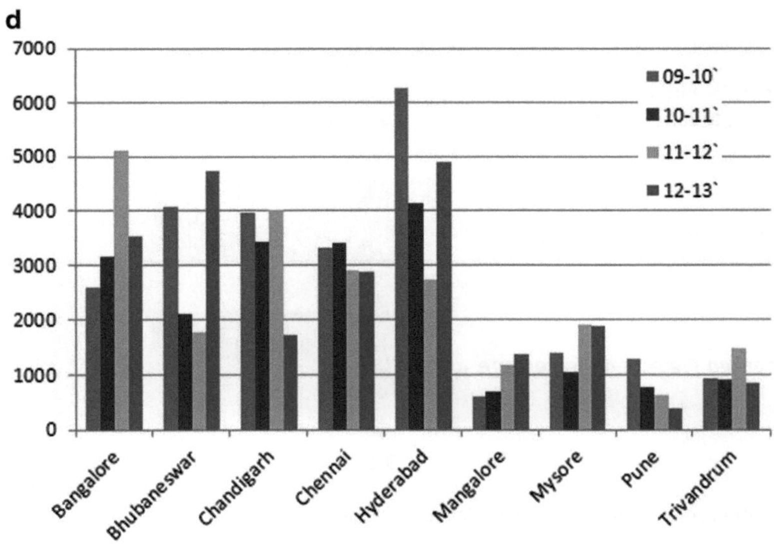

Fig. 4 (continued)

7 Impact

Feedback had been gathered for the better alignment of the offerings to the need of the stakeholders and business. Survey has been conducted to understand the faculty/college management perception on the usefulness and relevance and also to measure the impact of it on employability/academic performance of the students to gather inputs on modifying its offerings. The portal had been used to upload the questionnaires to reach out to the stakeholders – faculty and management (Fig. 5).

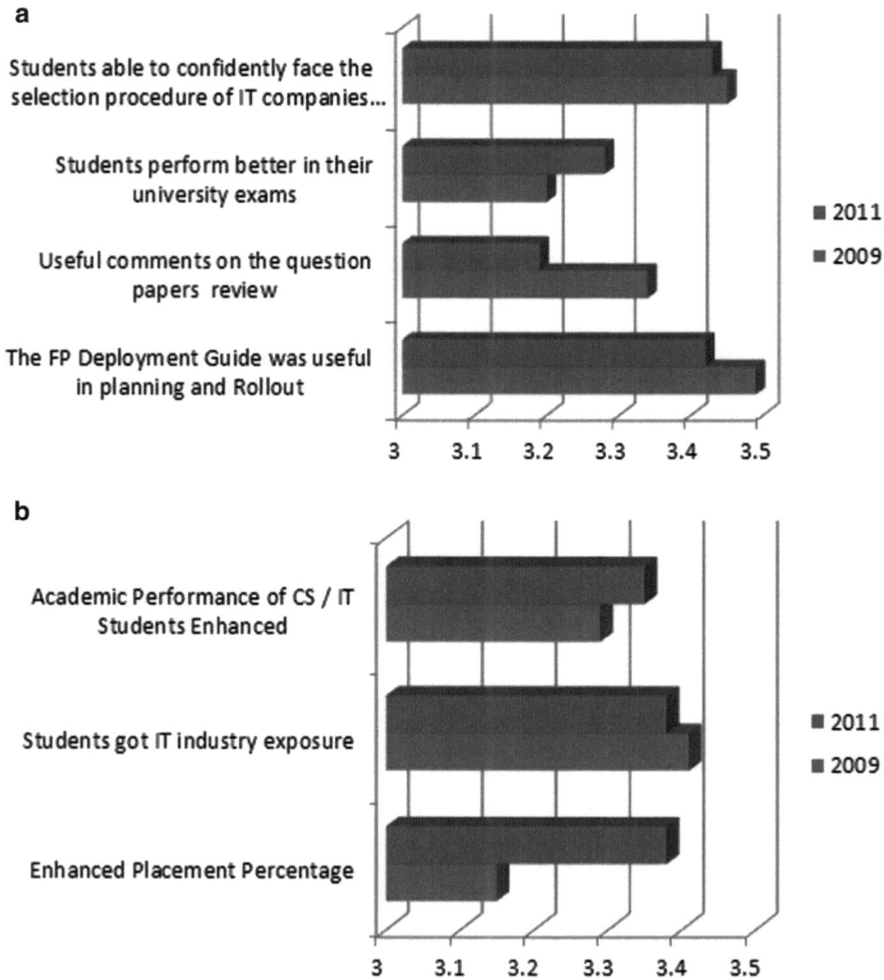

Fig. 5 (**a**) Faculty survey results. (**b**) Management survey result

The ratings are on the scale of 0–4. From the survey of the faculty and management, it is observed that Fundamental Program has been able to provide the building blocks for the students in terms of their academic results and enhancing the employability to large extent.

8 Conclusion

- Fundamental Program is being delivered to students of engineering colleges as one of the levers to augment the technical foundation and enhance employability by creating awareness for problem-solving skills without any business interest.

- This has demonstrated that if the student is motivated and provided with industry best practices and requirements for learning, he/she can be made industry ready.
- Currently, thought process is on to move the program to blending learning methodology with most of the content available online along with limited intervention of faculty members for doubt clearing sessions, project guidance, and proctoring the online assessment.
- As per the learning needs of current generation, the plan is also to convert the session presentations into e-learning contents and spoken tutorials. In order to check the learning goals of the learner, various online quizzes and assignments are getting design for formative assessment.

References

1. Bridging the skill gap: opportunities for skills training in India's organized retail sector. http://iimksydneyconf.iimk.ac.in/
2. Human resource and skill requirements in IT and ITES industry sector 2022 a report. http://www.nsdcindia.org/pdf/it-ites-industry.pdf
3. Infosys Education and Research Departments Campus Connect. http://campusconnect.infosys.com

Benefits of Cloud Computing in Education During Disaster

Kiran Bala Nayar and Vikas Kumar

Abstract The application of cloud computing in education not only relieves the educational institutions from the burden of handling the complex IT infrastructure management but also leads to huge cost savings. The motive of this paper is to encourage the usage of cloud computing in the education sector, especially during disaster for the smooth running of the system. Traditional techniques used for disaster recovery were very costly. The education sector could not afford this because of limited funds. But with the advent of cloud DR, it is now possible for education system to go for disaster recovery techniques for securing of data during disaster. In this paper, the various cloud computing-based disaster recovery techniques and their benefit to the schools and university systems have been discussed. The traditional education management systems have also been discussed and a model has been proposed for their implementation using the cloud computing platforms.

Keywords DR metrics • Challenges • Benefits of cloud computing • Proposed model

K.B. Nayar (✉)
Research scholar, Computer Science Department,
JJT University, Jhunjhunu 333001, Rajasthan, India
e-mail: mail2kiran7@gmail.com

V. Kumar
Associate Professor, Information Technology Asia Pacific
Institute of Management, Jasola, New Delhi 110025, India
e-mail: prof.vikaskumar@gmail.com

R. Natarajan (ed.), *Proceedings of the International Conference on Transformations in Engineering Education: ICTIEE 2014*, DOI 10.1007/978-81-322-1931-6_24,

1 Introduction

1.1 Disaster in Education

A disaster is a "natural" or "man-made [or technological]" hazard resulting in an event of substantial extent causing significant physical damage/destruction, loss of life, or drastic change to the environment. When it comes to the education sector, the repercussions are much more intense. In the education sector, the important features are repository, examination, result, fees, and all important circulars related to students and faculty and future endeavors of an institution. A disaster event that causes significant loss of university's records has the potential to cause major disruption to the university's ability to operate effectively. Further, this can result in monetary loss and loss of creditability and goodwill of the university. Traditionally, for recovery of documents, "tape backup" was generally used by various organizations, but there were more chances of failure of tape backup during "disaster." Also tape backup solution requires retrieval and loading of the backup through a mechanical solution which is further a time and resource issue. Disaster had very strong impact on the work of universities and colleges in the past and present too. One of the examples is Takshila High School in Agastyamuni, Uttarakhand, which was washed away in the floods. In fact, there are so many colleges and schools of Uttarakhand submerged and damaged in the recent floods. Since all schools and colleges were using the traditional ways of teaching and document recovery methods, now it will take several months for the normal work of damaged schools and colleges to resume.

To overcome similar instances of disasters and the smooth running of educational institutions, "cloud computing" is the best way to survive. It is a recent concept that is still evolving across the information technology industry and academia. Lots of definitions have been given for cloud computing. The National Institute of Standards and Technology [1] defines cloud computing as "a model for enabling convenient, on-demand network access to a shared pool of configurable computing resources [e.g., networks, servers, storage, applications, and services] that can be rapidly provisioned and released with minimal management effort or service provider interaction". The advantage of using cloud computing in the education sector is that the data can be retrieved and restored almost immediately and also it offers mobility to ensure the continuity of the education system. It also ensures the data integrity by replication and mirroring of the existing systems along with the backups, resulting in various solutions during disaster. This can be done by setting automatically upload data from your servers at predetermined times and dates. All data being uploaded and/or downloaded is encrypted so it is safe and secure to use cloud. Moreover, due to its cost-effectiveness and scalability and flexibility features, cloud computing is gaining success in the education field. Cloud computing can be used to support cooperative learning and e-learning based on the environment of cloud computing and reform the computer fundamental curriculum in universities for handling the disaster effects on the education sector; various players like IBM, Amazon, etc., came forward to help the universities and colleges. For this, IBM constructed an education platform based on cloud computing for teachers and students with high performance and flexible scalability at Carnegie Mellon University, Massachusetts Institute of Technology (MIT), Stanford University, and the University of California, Berkeley, in 2007. The North Carolina Graham primary school and SIMtone Corporation launched "universal cloud computing services" for the school teachers and students through virtual computer desktop. Many colleges and universities had collaborated on the formation of Virginia Virtual Computing Lab aiming at providing online education materials, including the use of application software, computing, and storage service [2].

2 Traditional Disaster Recovery Versus Disaster Recovery Through Cloud

2.1 Traditional Disaster Recovery [DR] in Education

Traditionally, the education system relied on the level of service required either in terms of cost or speed during disaster. Basically in education we are sensitive about the cost as most of the organizations work on "no profit, no loss" scheme. Traditionally, disaster recovery [DR] in education sector focus more on investments in data server and connectivity; which itself is a big burden for colleges and universities. A lot of time is consumed to make DR site live that will lead to huge data loss. Sometimes manual operations may be required to start the site at the time of nonavailability of connectivity. But it can take days or weeks to recover the data, and because of the cost constraints, the education system cannot go for high recovery models; adopting those models can increase the cost, so traditional disaster recovery solutions, including tape backup, totally failed to deliver the desired RTO and RPO within the budget.

2.2 Cloud-Based Disaster Recovery: A New Approach

Cloud-based disaster recovery is becoming more popular in the education sector because of its number of advantages. As the DR exists within the cloud, it reduces the up-front capital expenditure. Availing DR as a service [DRaaS] confers many benefits, out of which cost plays an important role. "Compared to traditional DR, there is a significant cost saving in the case of Cloud-based DR, and also, CAPEX is near zero," as per Wasim of Sapient [3].

Forrester defines cloud-based DR solutions as falling into one of the three main categories: do it yourself [DIY], DR as a service [DRaaS], and cloud-to-cloud disaster recovery [C2C DR]. Each model has a distinct set of benefits. The cloud also offers better flexibility and tracking of RTO and RPO. Some cloud providers including Microsoft with Azure provide inbuilt DR for data services such as storing data copies on multiple data centers across geographies as part of the storage service itself. According to Cisco's D'Souza, a key advantage of cloud-based DR over the manual, runbook style of DR process execution was nothing less than the fact that it minimized downtime and offered the lowest RTO. There are various metrics which are used to calculate the disaster recovery services with respect to different factors, and these are as follows.

2.2.1 Metrics Used for Good DR Services According to Cost

- Recovery Time Objective [RTO]: It can be calculated by the amount of time between an outage and the restoration of operations. RTO is also dependent on the various techniques used for the backup and type of application; this may involve additional manual steps such as verifying the integrity of state or performing application, restoring of data operations, and scheduling of recovery tasks to be done efficiently [4].
- Recovery Point Objective [RPO]: It can be calculated by the point in time where data is restored and reflects the amount of data that will be ultimately lost during the recovery process. The necessary RPO is generally a business decision – for some applications, absolutely no data can be lost [RPO = 0], requiring continuous synchronous replication to be used, whereas in other applications, the acceptable data loss could range from a few seconds to hours or even days [4].
- Performance: It can be calculated by the impact of normal operations of system after the recovery of documents.
- Geographic Separation: It is important that the primary and backup sites are geographically separated in order to ensure that a single disaster will not impact both sites. Increasing geographic distance can lead to high cost and greater network latency.

2.2.2 Metrics for Good DR Services According to Backup

- Hot Backup Site: Hot backup site is the most efficient and expensive site. This site provides the access to the database even after the disaster occurs with minimal RTO and RPO. It can also have the largest impact on normal application performance since network latency between the two sites increases response times. Recovery through a hot backup site can take place within a few hours due to the fact that the hot backup contains a replica of the current data in the data center.
- Warm Backup Site: A warm backup site is a site that is already equipped with hardware that contains a backup of the information that is contained in the data center. Warm backup uses the standby servers to run the application after failure but is kept only in a "warm" state where it may take minutes to display them

online. Warm backup slows down the recovery process but also reduces cost. So in warm backup, the server resources to run the application need to be available at all times, but current cost such as network bandwidth and electricity is lower during normal operation.

- Cold Backup Site: It is also known as offline backup. In this site, data is often only replicated on a periodic basis, leading to an RPO of hours or days. It can be difficult to support business continuity with cold backup sites, but these are cheapest options for applications that do not require strong protection or availability guarantees. With a cold backup site, everything that is required to restore service must be delivered to the site making restoration efforts from the cold backup site to full operation tedious and time consuming. It is the best suited metric for the education department.

3 Cloud to Education During Disaster

3.1 Challenges in Education Before Cloud

Schools and colleges remain grounded and governed by a highly rigid world that depends on the certain roles, strict powerful rules and regulation of the system, a concrete place, a strict set of content, syllabi, and a bureaucratic environment where students are considered inferior to the faculty in each and every aspect of the learning process. One of the main challenges of the education system before cloud was the forceful closing of the classroom teaching because of an unexpected natural disaster like cyclone, fire, etc. The second challenge was the non-recovery of exact documents due to server failure or fire in the server room or any man-made disaster like intentional deletion of files by someone. As we know, the education system does not have so many funds to spend on DR techniques. Traditionally, the education sector provides their own disaster recovery procedures in isolation of other organizations and is often disparate in nature. This leads to different approaches used to ensure data recovery, and these approaches involve storage area network (SAN), tape backup, and third-party contractors offering an off-site solution.

3.2 Solution

Cloud computing is a flexible delivery model for information and communication technology (ICT) services at the time of disaster that uses powerful systems and networks with high transfer rates. It leverages distributed hardware and software resources and shared redundant, multi-tenant platforms that deliver a high degree of scalability. Utilizing a cloud solution to deliver disaster recovery [DR] and business continuity planning [BCP] is an emerging technology, although not supporting directly the pedagogy of teaching and learning as such. The DR within the education system is to protect data and run the system smoothly during disaster. There should be proper contingency procedures and plans for accommodating the changing needs of the organization and ensuring data integrity during disaster. Cloud computing offers schools, colleges, universities, and others a low-cost option for using high concept computing systems. Only an Internet connection is needed by universities or college which is further available at a low cost. The advantage of cloud is that the data can be retrieved within a few seconds, whereas a current tape backup solution will require the retrieval and loading of the backup tapes into a mechanical solution which is further a time and resource issue, and also since data recovery is paramount to the institution, recovering the data is time sensitive to the system especially when universities have to declare the result and admission [5]. The cloud solution offers the mobility to ensure the continuity of the education system and provide data integrity during disaster.

There are so many cloud providers ready for providing scalable cost-efficient solutions for the education community while delivering industry-shaping technology and high-performance computing necessary to facilitate the most demanding research projects and course objectives at public and private universities, community colleges, and vocational schools. For example, AWS is one of the cloud providers giving these services to the education sector. To assist educators around the world in providing cloud computing instruction, AWS offers Teaching Grants supporting free usage of AWS for students in eligible courses. The grants will provide educators up to $100USD in free usage for each student enrolled in courses with Amazon Web Services as part of the curriculum. AWS services supported in the grants include Amazon EC2, Amazon S3, Amazon SimpleDB, Amazon RDS, Amazon SQS, Amazon CloudFront, and Amazon Elastic MapReduce [6]. All India Council for Technical Education (AICTE) is well known for its technical and quality information. AICTE recently signed MOU with Microsoft and Autodesk to raise the level of technical learning and education in India. Students and faculty across the entire nation in various PG and UG Colleges, in both rural and urban areas, will be able to quickly and easily access world-class software. Now no school and colleges take tensions for loss of documents at the time of disaster as DR is taken care of by cloud itself. According to Dr. S.S. Mantha [Chairman of AICTE], "Microsoft's cloud platform will make for a truly progressive ecosystem and contribute to the country's technical education by providing a better communication and collaboration platform for institutes and students" [7]. Cisco is also playing a good role for the growth of the education sector even during disaster; they are giving services to universities or colleges at very reasonable cost. Berlin's University of Technology [TU Berlin] is one of the customers of Cisco who has virtualized much of its IT infrastructure using the Cisco Unified Computing System™ [Cisco UCS™]. As a result, TU realizes high scalability through maximum utilization of resources while simplifying management and gaining a clear edge over international competitors. Cost savings have been particularly impressive. For example, TU now requires only eight cables per blade chassis, achieving a 90 percent reduction in cabling-related costs [8].

3.3 Benefits of Cloud to the Education Sector

Higher education was acknowledged in time as one of the pillars of society development. Through the partnerships between universities, governments, and industries, researchers and students have proven their contribution to the transformation of society and the entire world economy [9]. Cloud computing offers to universities the possibility of concentrating more on teaching and research activities rather than on complex IT configuration and software systems [10], through a fast IT implementation. So cloud computing offers many benefits to e-learning solutions by providing the infrastructure, platform, and educational services directly through cloud providers and by using virtualization, centralized data storage, and facilities for data access monitoring [11, 12]. The benefits of cloud to the education sector are the following:

- Real-time learning: The main advantage of the cloud is the ability to access real-time information from anywhere in the world in seconds, which further encourages an open learning environment for the youngster to share their work, ideas, and information.
- Energy saving: In cloud computing platforms, documents, software, applications, etc. are stored on a remote server instead of a local network. Keeping them on remote server saves more energy than keeping them on the personal system. So energy saving offers great benefits to the education sector.
- Cost saving: Adopting cloud computing technologies, such as remote desktops and cloud-based IT support, for example, helps schools and colleges to cut down the costs and equipment. Running these services through the cloud reduces the need for on-site technical support and maintenance services, while upgrades can be carried out automatically, for all end users, at a much lower cost.
- Content growth: Earlier contents were part of limited channels like textbooks, encyclopedia, newspaper, and television. Most content now comes from relatively "unknown" sources through the web. But the web is not simply a less trustworthy encyclopedia – it is also a place to publish and interact with content. Another drawback of the traditional system was that students were traditionally limited to a class, a school, and possibly a small community. The main benefit of cloud is that anyone can create a content which is available to the world instantly and can last for many years – possibly "forever" [13]. Content can be constantly evolving through collaboration and interaction and updates. People do not just refer to information, or just copy it, but they interact with it. They modify it, and they add to it – and this is to be encouraged. Tools for publishing, creating, and interacting with content are changing every year.
- Collaboration: Technology is rapidly improving the ability to communicate and collaborate with others. Traditionally, we are connecting with people in person or by telephone; teams are formed and work face-to-face. One of the major drawbacks is limited social circle, but cloud computing overcomes these drawbacks and connects us with so many social tools like Facebook, Twitter, etc. Earlier it was started as personal tools but now is more and more rapidly moving from the

personal to the professional world. It is becoming easier to find and connect with anyone in many new and expanding ways like mobile phones, e-mail, instant messaging, social and collaborative software, and blogs. Collaboration has been a one-time, relatively static, and sequential process. Cloud computing makes interactive collaboration possible on the web, between students in the same class, or around the world.

- Storage: As confidential and critical data can be stored centrally in the cloud, there is less exposure to threats such as the loss or theft of laptops or USB flash drives.
- Convenience: Because of the online nature of cloud, it is very convenient for the users to adopt this technology. We just need an Internet connection. No matter where we are in the world, we can access our data any time and can do modifications.
- Easier integration: Applications running in private clouds are easier to integrate with other in-house applications, such as identity management systems.

4 Proposed Cloud-Based Model for Education During Disaster

In the education system, the important entities are students, faculty, and staff. The important data related to all these entities are student's result, fee structure and student's assignment, teaching materials, curriculum, etc. The loss of any documents related to any entity is a big issue at the time of need. Traditionally, documents were there on the local server, and loss of any documents during disaster leads to stopping the entire system. Earlier, faculties were using USB for the backup of their documents. The intension of the faculty for using USB can be to take the backup of student's results, notes, and lab manuals at home for the smooth running of classes in case of emergency. But when the faculty plugged the USB in a laptop at home, found that the USB had an undetected and malicious virus. Without warning the virus multiplies and within a matter of hours laptop crashed; most of the data thereon has either been corrupted or is completely inaccessible. Cloud is helpful for all categories of users to access stored files, e-mail, database, and other applications from anywhere, which leads to a more efficient use of information [14]. The objective is to identify the new technologies fulfilling the requirements of all the important members like students, faculty, and staff with a low cost. It also provides the security and availability of documents at a very low cost [15].

From Fig. 1 we can see the students are the main entity and already using many of the cloud platforms like Software as a Service (SaaS) and Platform as a Service (PaaS) for obtaining the advantages of the ability of working and communicating in the educational sector. The faculty can also benefit from the cloud platform for teaching and research purposes. Now, they will not worry for loss of lecture material, not need to carry study materials for security in pen drives or disks, etc. Third, the entity staff, specially developers, can benefit from Infrastructure as a Service (IaaS)

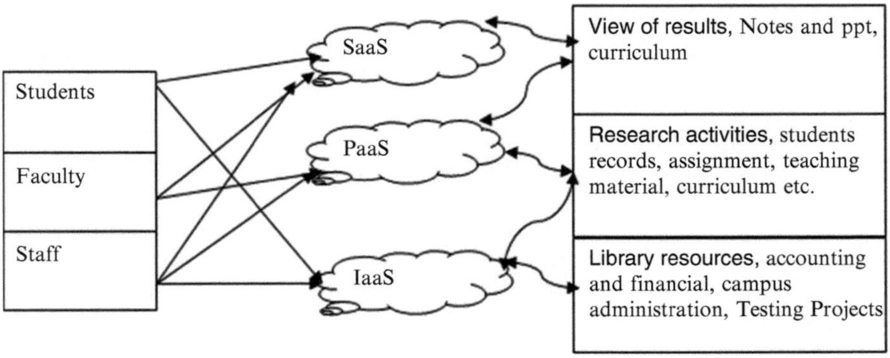

Fig. 1 Cloud users on different platforms of cloud

for designing, building, and testing the applications that are executed on the infrastructure of the cloud provider and delivering those applications directly from the servers of the provider to the final users. In the traditional system, developers were restricted to the local networks where they have limited resources. So the model will be helpful to developers.

5 Implementation

From Fig. 2 we can see that the cloud is providing a platform to those schools; universities that wish to run online platform for the students and teachers can use infrastructures such as servers, storage, communication, and e-learning software required for hosting and running an online school. The school will facilitate teacher-student interaction for every learner to utilize services of the best teachers to get quality education from anywhere at affordable cost while encouraging a competitive environment for teachers to innovate new techniques in teaching. The model facilities will include audio/video sessions or online chats that would enable a convenient and effective environment for learning. In addition, there will be a facility to use educational content enrichment tools for publishing, reviewing, and authoring to promote creation of new content.

6 Conclusions

With the use of cloud computing, the burden of taking daily backup of documents can be greatly reduced (for the security at the time of disaster). By doing so, we can reduce the cost of DR solution, as it is within the cloud. Cloud computing also offers better storage, enhances the productivity, and reduces the IT management work.

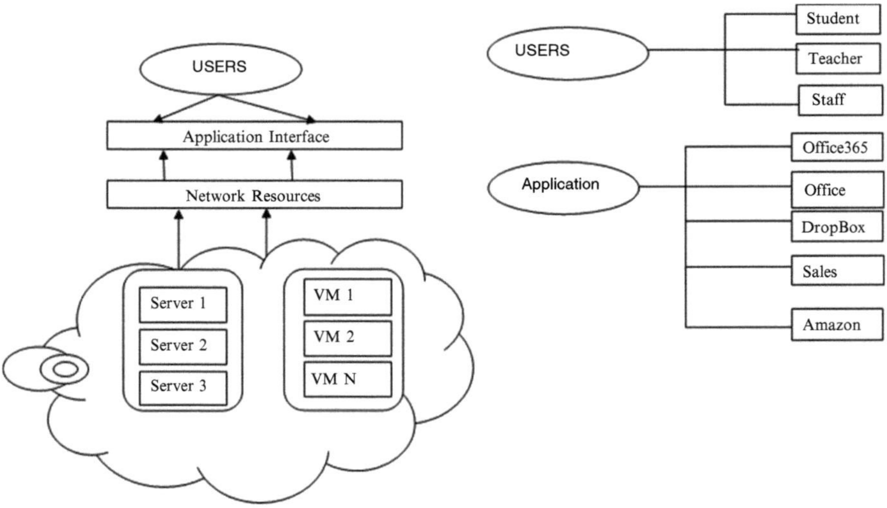

Fig. 2 Cloud architecture in education

It also increases the speed of work with very reasonable cost. Thus, every organization wants to update its IT infrastructure to keep updated with the technology. Present economic situation will force more and more organizations at least to consider adopting a cloud solution. Universities have begun to adhere to this initiative and there are proofs that indicate significant decreasing of expenses due to the implementation of cloud solutions.

References

1. Mell P, Grance T. The NIST definition of cloud computing. Gaithersburg: National Institute of Standards and Technology; 2009.
2. Lin Yu-hua. Practice and innovation of an experimental teaching mode in universities under environment of cloud computing. Res Explor Lab. 2011;30:271–4.
3. Express Computer. June 2012. http://computer.financialexpress.com/features/711-dr-in-the-cloud. Accessed October 2013.
4. Wood T, Cecchet E, Ramakrishnan KK, Shenoy P, van der Merwe J, Venkataramani A. Disaster recovery as a cloud service: economic benefits & deployment challenges. In: Proceedings of the 2nd USENIX conference on hot topics in cloud computing [HotCloud'10], Berkeley, 2010; pp. 8–8.
5. Alabama Department of Education improves recovery time from 24 hours to 2 minutes with DellAppAssure. http://www.appassure.com/wpcontent/uploads/2013/01/Dell_AppAssure_Case_Study_AL_State_Dept_of_Education.pdf
6. Amazon Web Services Education Grants, Purdue Libraries News, 20 Sept 2011, CFP. http://aws.amazon.com/education/#educators. Accessed on October 2013.

7. Microsoft. Press release on AICTE adopts cloud to improve technical education and prepare students for the workforce of tomorrow; April 2012. http://www.microsoft.com/india/msindia/pressreleases/aicte-adopts-cloud-to-improve-technical-education-and-prepare-students-for-the-workforce-of-tomorrow/355/
8. Cisco. White paper on Cloud 101: developing a cloud-computing strategy for higher education, 2012. http://www.cisco.com/en/US/services/collateral/ps10658/ps11785/cloud_101_higher_education_wp.pdf
9. Lazowska E, Lee P, Elliott C, Smarr L. Infrastructure for escience and elearning in higher education. Computing Community Consortium (Online); 2008. Retrieved 5 Oct 2010. http://www.cra.org/ccc/docs/init/Infrastructure.pdf
10. McCrea B. IT on demand: the pros and cons of cloud computing in higher education. Campus Technology (Online); 2009. Retrieved 5 Oct 2010. http://campustechnology.com/Articles/2009/08/20/IT-on-Demand-The-Pros-and-Consof-Cloud-Computing-in-Higher-Education.aspx?Page=1
11. Pocatilu P, Alecu F, Vetrici M. Using cloud computing for e-learning systems. Recent advances on data networks, communications, computers; 2009. ISBN: 978-960-474-134-2 (Online). Retrieved 5 Oct 2010. http://www.wseas.us/elibrary/conferences/2009/baltimore/DNCOCO/DNCOCO-06.pdf
12. Armbrust M, Fox A, Griffith R, et al. A view of cloud computing. Commun ACM. 2010;53:50–8.
13. Bittman T. Cloud computing and K-12 education, 26 Nov 2008. http://blogs.gartner.com/thomas_bittman/2008/11/26/cloud-computing-and-k-12-education/. Accessed October 2013.
14. Nicholson J. Cloud computing's top issues for higher education. University Business Solution for Higher Education Management (Online); 2009. Retrieved 5 Oct 2010. http://www.universitybusiness.com/viewarticle.aspx?articleid=1342&p=4#0
15. Mircea M, Andreescu AI. Using cloud computing in higher education: a strategy to improve agility in the current financial crisis with vol. 2011, Article ID 875547, 15 p.; 2011. doi: 10.5171/2011.875547. Communications of the IBIMA. http://www.ibimapublishing.com/journals/CIBIMA/cibima.html

Technology-Enhanced Learning Through ICT Tools Using Aakash Tablet

Vaman Ravi Prasad Dasu and Balram Gujjari

Abstract Today, with advancement in technology, teachers have been conscious about the quality of teaching. Teaching can be enhanced through information communication technology (ICT) tools in the education system for better understanding of the subject. Information communication technology can be utilized for the education sector. ICT can play a great role in formal and nonformal forms of education. Lots of companies are coming up with digital materials for poor students. Many are making specialized hardware to meet the requirement of the elementary education system.

Keywords ICT • Gadgets • Technology • Traditional • Education

1 Introduction

> Students cannot be effective in tomorrow's world, if they are trained in yesterday's skills – Tony Blair.

Education is a process by which a person's body, mind, and character are formed and strengthened, which enables a person's holistic development of personality through knowledge. This knowledge can be improved by the students through their learning environments. Nowadays, information communication technology (ICT) tools are widely used to interact in digital environments [1]. ICT brings more rich materials in the classrooms for the teachers and students. With the help of ICT, students can easily browse via sample examination papers, e-books, assignments,

V.R.P. Dasu (✉) • B. Gujjari
Anurag Group of Institutions, Ghatkesar, Ranga Reddy Dist, Hyderabad, Andhra Pradesh 500088, India
e-mail: dravi_09@yahoo.co.in; gujjaribalram@gmail.com

R. Natarajan (ed.), *Proceedings of the International Conference on Transformations in Engineering Education: ICTIEE 2014*, DOI 10.1007/978-81-322-1931-6_25,

videos, etc., and can also have an easy access to mentors, experts, resource persons, researchers, and professionals all over the world.

IT has become a buzzword while talking about technology and its applications. IT is used in various business and management functions but not in improving the quality of education. The quality of education has been an issue of concern in the absence of standard parameters to measure the quality. The hardware, software, methods, and know-how required or used in acquiring, storing, processing, and displaying data and information are collectively known as information technology (IT). Also on the other hand, many developments and achievements took place in the communication technology sector after the Second World War. Hardware, know-how, programs, and the methods used in ensuring that message is transmitted correctly, efficiently, and cost-effectively are collectively known as communication technology (CT). Both of these technologies became complementary to each other which means progress in one alone is not much beneficial. Hence, IT and CT started moving together, and a new term was coined named as information communication technology (ICT) [2].

The convergence of these two technologies gave birth to ICT. The education system includes formal and nonformal forms of education at various levels of education. Teaching is imparting knowledge or skill, whereas learning is skill acquisition and increased fluency. The usage of ICT is one of the ways by which India's large population base can be effectively reached. Moreover, in enhancing the quality and delivery of services through ICT especially in the case of developing relations with citizens, the government will be better positioned. Passive learning occurs when students use their senses to take in information from a lecture, reading assignment, or audiovisual. Traditional lecture is not an effective learning environment for many of our students because so many students do not participate actively during a traditional lecture [3]. This is the mode of learning most commonly present in classrooms, whereas active learning involves the student through participation and investment of energy in all three phases of the learning process [4]. This type of learning is more apt to stimulate higher cognitive processes and critical thinking. In the past few years, there has been a paradigm shift in the curriculum where the teacher acts as a facilitator in a student-centred learning. In student-centred learning, the focus is on the student's needs, abilities, interests, and learning styles with the teacher as a facilitator of learning. Here students have to be active, responsible participants in the learning process. The teacher has a key role in the whole process, whereas in the case of ICT-based education [5], various ICT tools are supplemented to make the teaching–learning process effective. With the help of blended learning, the total time devoted to teaching can be decreased. A survey says that there was a sense of pride created and interest generated among the teachers and students for gaining ICT and its privileges. ICT has the potential to remove the barriers that are causing the problems of low rate of education in any country. ICT as a tool can overcome the issues of cost, less number of teachers, and poor quality of education as well as overcome time and distance barriers. In this paper, how learning through ICT can be made effective and easier for improving the quality of both formal and nonformal forms of education is discussed.

2 Traditional Learning

2.1 Advantages of Traditional Learning

Teaching in a traditional schoolroom is vastly different from facilitating a virtual classroom. Classroom environments promote and stimulate a dynamic known as collaborative learning. Collaborative learning translates into a type of learning in which the pairing or grouping of students is required to complete a task or to come to a specific outcome. Teaching in a classroom gives students the opportunity to engage in live discussions in which they are forced to use their critical thinking skills to formulate opinions or arguments. When students are placed in a live classroom, they experience social interactions with peers and establish rapport with teachers. Helping children develop socially is an important aspect within the realm of their academic education. Classroom teaching environments help students figure out how to resolve conflicts, work in teams, get along with those from different cultural backgrounds, and give presentations in front of peers. Such experiences are valuable in shaping students' communication and listening skills, as well as growing and maturing emotionally.

Teaching in a classroom environment opens up opportunities for teachers to do more with their lesson plans [6]. Traditional classroom settings teach students how to develop organizational skills, beginning with the basics, such as arriving to school on time. In a live classroom, students are held accountable for being prepared to do school work, which includes having done their homework the night before, being ready for pop quizzes, turning in assignments by their due date, and being prepared for in-class discussions.

2.2 Disadvantages of Traditional Classroom Teaching

A drawback of traditional training is that it inherently places the most value on standards, curriculum, and passing tests as opposed to student-focused learning. Student-focused learning places value on the student and builds the curriculum around the questions young people need answered in order to understand the material. Constructivist learning builds on the knowledge students already have, allowing them to form concrete associations to new information, which improves retention. Traditional learning is based on the repetition and memorization of facts that students care less about and retain at lower rates after testing. Traditional classroom training does not encourage critical thinking skills and the ability to actively apply information gained through experience and reasoning. Instead, traditional training emphasizes the role of teachers as knowledge dispensers and students as repositories. This style of learning does not allow students' deeper levels of understanding required for complex concepts and lifelong learning. Traditional training emphasizes passing tests, whether or not students under testing material. The learning process is thus devalued, and students are not encouraged to understand the methods, techniques, and skills required to find answers. Traditional training emphasizes individual student work and projects and is a poor preparation for a student's future endeavours, which are likely to include working on teams and collaborating with colleagues. Under this training model, students receive few opportunities to practice group dynamics and teamwork [5].

3 ICT Application for Quality Improvement in Formal and Nonformal Education

ICT applications are becoming indispensable parts of contemporary culture, spreading across the globe through traditional and vocational education. In the Indian scenario, mainly the education system has three tiers: primary level, high school or secondary level, and college or higher level. In all these levels of education, ICT can be utilized for a better teaching–learning process and improving the quality of education. Using multimedia in education results in the increasing productivity and retention rates, because people remember 20 % of what they see, 40 % of what they see and hear, but about 75 % of what they see and hear and do simultaneously. An interactive whiteboard helps teachers to structure their lessons, supports collaborative learning, can help to develop student's cognitive skills, and enables ICT use to be more integrated into the classroom. The government of India has announced 2010–2020 as a decade of innovation [2]. Reasoning and critical thinking skills are necessary for innovation. The foundation of these skills can be laid only at the primary level of education. Students who enter school are very curious, creative, and capable of learning many things. At this level, a statement picture is worth more than a thousand words is very much true in the case of the teaching–learning process. Befriending ICT in the initial stages of education will help young people come to terms with what lies ahead. Students studying at this level take much interest in cartoons. They understand more through animated pictures. Hence, if the same environment is created in schools by using ICT for teaching kids at the primary level may bring drastic changes in the education scenario. Nursery students can be taught by showing pictures, animals, fruits, etc. With the help of ICT tools, students at this level are able to grasp a lot by hearing voices or sounds and animated motion of various animals. Language learning is also taught at this level [7]. Learning a new language at this age is easier as compared to other levels. A multimedia projector and a computer can be used to teach phonetics and pronunciation. Lessons, poems, and lectures by eminent scholars stored in computers or other ICT tools can easily be shown to the students time and again anywhere. Such type of teaching and learning enables for long-time retention in the minds of the children. At the high school level, subjects like history, geography, political science, physics, chemistry, biology, physical education, etc., are taught. Lessons in these subjects can easily be taught by showing a short movie related with the subject to create interest among the students. Such type of movies and related multimedia material is easily available at academic repositories and from various related sites with the help of the Internet. The Internet is a basic tool which can be utilized by teachers and students to find any information on any topic. This type of teaching–learning makes the environment very interactive and is liked by students. Educational and practical CDs available in the market make this task easier to implement. At the college level, various facilities like computers, electronic board, Edusat facility initiated by various state governments, MM projector, and other peripheral devices related with the teaching–learning process are easily available. Easy availability of Aakash tablet will help in providing and getting more education for both teachers and students. Repositories are libraries where these digital resources are stored and provide teachers, students, and parents with information that is structured and organized to facilitate the finding and use of learning materials regardless of their source location. Various programs running on Edusat are also very helpful for the students. Soft skill program can help students in getting their placements in reputed multinational companies (MNCs). A state-level quiz and seminar can also be conducted with the help of the Edusat infrastructure and can be transmitted throughout all institutes. Edusat can be used for providing training to teachers on the latest subjects and technologies and can save a lot of time and money of governments. In Haryana, the Edusat project is being implemented at school and college levels and is being used for transmitting lectures according to syllabi. In nonformal learning, learners can access information and learning materials from anywhere and at any time. It includes distance education and other open learning systems. There are various functions to be performed with the enrolment of students in any course of distance education in any university or institute [7]. Functions include allotment of unique number, providing books, and providing information related with instalment of fees and details thereof to name a few. Out of all these activities, some of these may be performed well with the help of ICT tools. In distance education, ICT can be used for better management of records by making a complete database of all the students in various courses. Once the students are enrolled, a unique number is generated called reference number, and it is provided to the particular students. Short message service (SMS) of mobile phones may be utilized for this purpose. The mobile phone is one major ICT tool and can be used for the purpose. Other information related to the PCP such as exam dates can easily be sent to the students through SMS by universities/institutes concerned. Moreover, the enrolled students can be given a username and password for using various online services and resources in the form of academic repositories maintained by institutes. All such instructional materials may be uploaded at the university portal, and CDs of those lectures may be provided to the students instead of printed or hard copy material. An online fee payment system can also be made on the portal of the concerned university or institute. Students will be saved from a lot of hardships they face in depositing fees, attending PCPs, taking exams, and many more. Exam results in such cases may be provided online on the same day as is happening in the case of online exams and entrance tests. This would help to sort out the problem of the delay in the declaration of results of various exams by various universities. But all this must be the case for the nonformal education system. Advantages of utilizing such tools include saving of lot of paper [5].

3.1 IT-Enabled Classrooms

Keeping in view the abovementioned advantages and disadvantages of traditional classroom teaching, where usually a lecture method is followed, in the case of higher education, we can add some IT gadgets to the classroom [8]. Projectors are already being used in classrooms for teaching. No doubt that this (projector) is an excellent IT gadget used in the classroom, but our teachers as well as students failed to make optimum use of it, and it is being used as a presentation tool only. With the passage of time, many new gadgets are coming to the classroom which include laptops, netbooks, smartphones, and tablets. All these devices can be used in the classroom for teaching purposes if planned properly. We had to set up the following infrastructure to set up an IT-enabled classroom to teach some IT-related subjects. A classroom with the capacity of 30 students was selected with the following IT gadgets:

- One standard computer with speakers
- One multimedia projector
- One Wi-Fi router

The following are some more electronic gadgets proposed to be used in class:

3.1.1 Laptops

The laptop is the need of any engineering student. This is used by the student to keep the records of all his notes, and as the Internet is available almost everywhere in the campus, students use the Internet to complete their assignments. There are many eBooks available free of cost; students download such eBooks and study using laptops.

3.1.2 Netbooks

Netbooks are smaller laptops, usually with 10″ to 12″ display with reasonable processor, memory, and disc space. This is the second type of electronic device preferred by students due to its cost, size, and battery life.

3.1.3 Mobile Phones

Mobile phones are usually not recommended in the classroom, and in most of the classrooms, mobile phones are banned, but if monitored properly, smartphones can be used a learning tool.

Fig. 1 Aakash 2.0 tablet

3.1.4 Tablet

Tablets are gaining popularity, and with the regular advancement in technology, the prices of such devices are coming down. Tablets are available in the price range of Rs 4,000. The government has taken an initiative by introducing Aakash tablet that perfectly suits the pocket of undergraduate students. One common feature that is required is Wi-Fi, and almost all the tablets have inbuilt Wi-Fi adapter so that they can communicate over local area network [9] (Fig. 1).

About Aakash 2.0 Tablet

Aakash 2.0 is a complete educational purpose tablet and not designed with hi-fi configurations. Aakash 2.0 includes 3D-modelling, C++ programming, remote and collaborative training applications, robotic control, and live assessment tools. IIT Bombay has partnered with the Center for Development of Advanced Computing (C-DAC), to assist with the hardware testing and logistics.

Aakash Applications

The tablet comes with a variety of applications. A sample set of applications is shown below (Fig. 2).

The tablet also includes Mango learning solutions for game-based educational modules, interactive smart books, and the full CBSE curriculum and assessment

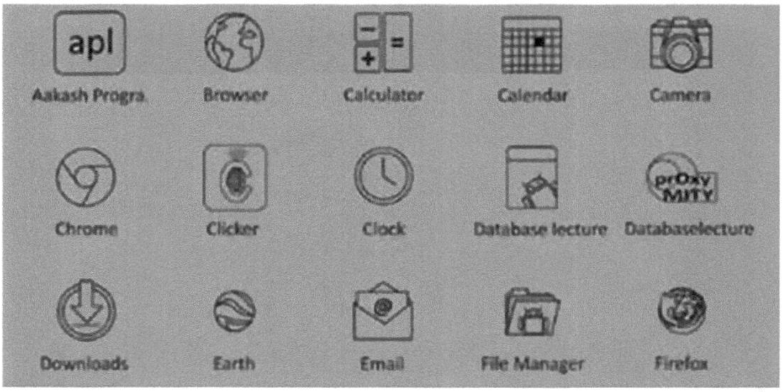

Fig. 2 Sample page of applications

tools. Intelligia finger tracing apps have also been included to teach letter writing to toddlers. Some of the salient applications are highlighted below:

(a) *Blender:* Blender is a free and open-source software product, used for creating animations, rendering, video editing, etc.
(b) *Clicker:* Clicker devices are used to collect instant feedback from a large number of students, either as a response to a question or to a quiz. A quiz question now gets fully downloaded on an individual student's Aakash tablet, through Wi-Fi. A multiple-question test can now be conducted. All the questions of such a test are downloaded on the student tablet. Time control is maintained by Aakash. At the end of the test time, all answers are automatically collected, and individual scores get recorded in the back-end system.
(c) *Content Distribution:* Apart from displaying text files in various formats, Aakash can play video and audio content. A lot of content is available through projects such as NPTEL, Spoken Tutorials, Teachers' Training Workshops, etc. Additionally, Wikipedia, Gutenberg project, Connexions project at Rice University, and many such global efforts have generated a large pool of knowledge content in open source. The contents can be kept in the tablet or they can be accessed.
(d) *Proximity:* proxyMITY enables the creation of interactive lessons, by importing lecture video and presentation slides. The name stands for Proxy Multimedia Integration Tool for You. The entire lesson can be published in the form of either a desktop stand-alone application or as html content to be viewed within a web browser.
(e) *Robot:* The project aims at the design and deployment of robots, for enhanced teaching of subjects in engineering colleges and to create open courseware for embedded systems for engineering students, based on robots. A robot-control application runs on Aakash. Students can control the movement of the robot using a simulated touch-controlled joy stick provided on the tablet. Simultaneously, the video stream captured by the camera mounted on the robot is transmitted to Aakash, which can be viewed in a window on the tablet.

Tools and Languages

Aakash tablets support C, C++, and Python for programming activities. Also, SciLab is available for numerical computations and for research activities.

Purpose and Usage

IIT Bombay has planned to deploy the Aakash tablets in engineering institutions in India, with two objectives:

- The first is to test and enhance the effectiveness of these tablets for use in classrooms.
- The second is the development of new educational applications and contents on Aakash, largely through final year research and development projects done by BE/ME students at institutes.

eBook Readers

eBook readers are small tablet-like devices, which can display the content of an eBook and can also be used to keep notes. Kindle from Amazon is the most popular eBook reader. eBook readers also come with Wi-Fi connectivity.

3.2 Setting Up an IT-Enabled Classroom

Our college has set up the following infrastructure to set up an IT-enabled classroom to teach some IT-related subjects. A classroom with the capacity of 30 students was selected with the following IT gadgets [8]:

- One standard computer with speakers
- One multimedia projector
- One Wi-Fi router that can support up to 30 wireless connections
- We installed Linux Mint 13 (Cinnamon 64 bit) and configured with the following softwares/services:
- Apache2 (Web server)
- Squid proxy server
- Samba (file server)
- MySQL (database server)
- phpmyadmin
- LibreOffice
- Moodle (learning management system)
- Aptana3 Studio (IDE for programming)
- JDK
- Java Netbeans

This computer system was connected to a Wi-Fi router and projector. And we were ready with our IT-enabled classroom, where the teacher can teach with the help of presentations and can also demonstrate examples by running them on this computer. We selected those 30 students for our class having either of such electronic gadgets. Students were asked to bring their devices to the classroom, with IT technicians and lab attendants helping on the first day. All devices are configured to use classroom Wi-Fi so that they can access the resources stored on our classroom server. Though there were more than 30 students having such devices, we had to select only 30 so that we can conduct the trial; we decided to select those students who were considered to be weak in class.

4 Proposed System Architecture

4.1 Proposed Features of the System

As stated in the earlier chapters, the project aims at producing a cross-platform in-class teaching and classroom management system that can serve both as a whiteboard and a notebook. According to it, some of the important features of the proposed system are described below.

4.1.1 Registration and Session Management for Teachers and Students

The system should keep the record of all the teachers, students, and courses. Hence, it should allow registration of teachers and students and enrolment of students in different courses. In running state, the system should keep track of the present students at any point in time. For this, the system needs to maintain up-to-date session information for both teachers and students. Also, the system should automatically record the attendance of any student who logs in and enters a particular course so as to reduce the teacher's burden of taking attendance. The system should make sure that no student registers himself more than once in a single lecture. The system should enable any student to check his attendance in any course. It should allow teachers to check the attendance of all the students enrolled in any particular course, so that he can inform the students having less attendance than expected.

4.1.2 Teaching and Note Taking

This is the most important feature of the proposed system. It should present every teacher and student with an interface having features of basic drawing using a stylus pen on a tablet PC. The system should also enable them to change the colour and line width of drawing as they may want to highlight something. As the teacher starts teaching (solving a problem) on his tablet (instead of whiteboard), the teacher's screen should be transmitted live to all the present students. Now all the students should be able to see the problem being solved on their own screens instead of the whiteboard. Moreover, they should be able to make their own notes on top of the live feed that they receive from the teacher. In that way, the teacher will not have to keep standing for the whole duration of a lecture, students will not have to switch their eyes from whiteboard to their notebooks, and they will have better organized class notes. After a page gets full, users should be able to save it as a file for future use, and a new blank page should appear. Users should also be able to load a previously saved page, if they want to solve or change an earlier solved problem.

4.1.3 Monitoring

In real-world classes, many times a teacher gives the students a problem to solve. Students solve it and the teacher walks across the classroom to track the progress of any student he wants. In an effective teaching system, this should be automated. The teacher should be able to switch the mode from *Teach* to *Monitor*. In the monitor mode, the teacher should be presented with a list of all the students present in the class at that point in time. After the teacher enters the user id of the student he wants to monitor, the screen of that student should be visible on the tablet of the teacher. This will reduce the teacher's efforts in tracking the progress of every single student.

4.1.4 Sharing

This is another important feature of the proposed system. In real classes, the teacher sometimes asks a student to solve a problem on the whiteboard. This should also be automated. In the proposed system, the teacher should be able to switch the mode to *Share*. After switching the mode, the teacher should again be presented with a list of students present in the class. After the teacher enters the user id of a student, the screen of that student should be visible on the tablets of all the students as well as the teacher. The difference between monitor mode and share mode is that in the former, only the teacher can view the selected student's screen, while in the latter, every present member of the class can view it.

4.2 System Structure

The system is based on the simple client–server architecture. There is a central server on which the entire program and database are kept. The teacher and each student have a tablet. There are ten different clients for the teacher and student. Clients are able to connect to the server via a wireless network set up in the classroom. The system structure is shown in Fig. 3

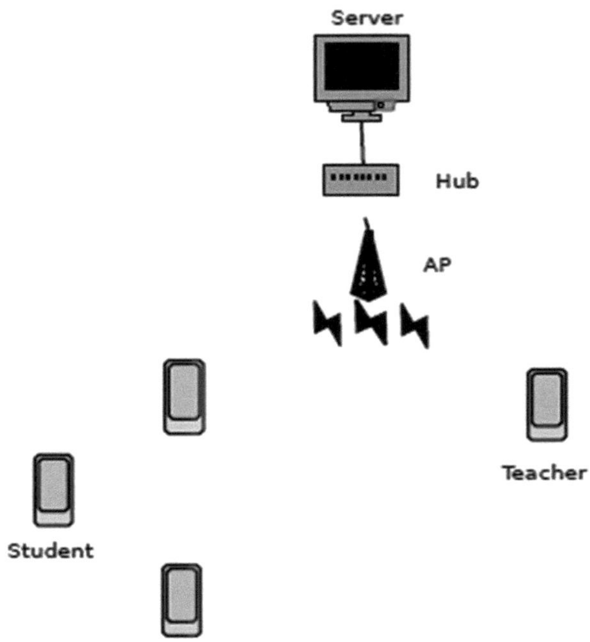

Fig. 3 System structure

4.2.1 Client–Server Communication

- There are different messages exchanged between clients and the server over the period of a lecture. These messages are as follows:
- At the beginning of a lecture, clients log in to the server. At this time, clients send their authentication information, i.e., username and password. After authenticating the clients, the server sends them their respective home pages.
- Clients send the id of the course that they want to enter to the server. The server, after verifying the course id, sends them the current page of that course.
- In *Teach* mode, the teacher continuously (e.g., once in every 500 ms) sends the teaching data to the server, which is then sent to all the students by the server.
- As soon as a student enters a course, he continuously (e.g., once in every 500 ms) sends a request to the server to know the current mode of teaching. The server queries the database and sends the current mode (i.e., *Teach*, *Monitor*, or *Share*).
- When the teacher switches to *Monitor* mode, the teacher stops sending teaching data; instead the selected student sends data (his screen) to the server and the server then sends it to the teacher.
- When the teacher switches to *Share* mode, the selected student sends data (i.e., his screen) to the server which the server then sends to all the other connected clients.
- At the end of the lecture, the entire clients log out of the server and the communication terminates.

4.2.2 Observations

During the 1 h session, the first part was usually silent, but in the self-practice period, there was lot of noise due to peer-to-peer discussions and the teacher had to announce to keep the noise under control. Secondly, in case there are common doubts, it was difficult for the teacher to go to each individual and address their doubt. A kind of indiscipline was observed in the classroom during the session, and for this reason only, we had to employ more than one teacher [1]. During some sessions, we felt that organizing such type of classes is not feasible because there was an expenditure of over 1 lakh per classroom (that includes the cost of computer, Wi-Fi router, projector, and other installations). Also there are many students who are financially weak and cannot afford expensive electronic gadgets for such classes. The initiative by IIT Bombay in this regard is worth mentioning here as they are trying to develop a low-cost Aakash tablet which can be used in this process. It was observed that there was a remarkable improvement in the performance of students as those who were literally bottom liners (used to pass with grace, sometimes) successfully passed the test, of course not with flying colours, but with reasonable scores. But this was restricted to those students who participated actively in the self-learning sessions and vigorously surfed the Internet and our local resource repository for information and studied it [4]. There were some students who were more interested in copying the content only; we could control this by motivating students to solve the assignment during class only and helping them doing so.

5 Conclusion

This paper has sought to explore the role of ICT in education as we progress into the twenty-first century. Taking the trends discussed in this paper into account, it can clearly be seen that the education system should change to adapt to modern requirements and to incorporate new technologies [5]. By incorporating these technological trends into the educational system, a higher-quality education can be provided at a cheaper cost and spread over a larger segment of the population. Children are required to apply their ICT know-how in other lessons, but little emphasis has been placed on how they are taught ICT. This paper discusses the current teaching practices and techniques used in the teaching and learning process. Today, we need to create a perfect balance between the quantity and quality of education. We cannot expect educational authorities to cut down the syllabus so that the student learns only what he can understand, and at the same time, we need to make sure that whatever is being taught must also be understood. The proposed technique of teaching with the help of ICT will surely increase the learning level of the student, but we might end up in a situation where we could cover only half of the syllabus [6]. Secondly, setting up such infrastructure involves a lot of expenditure on hardware (IT devices) as well as manpower, because we need to employ multiple people in a classroom. This approach might seem infeasible in some situation, but still we cannot deny the output. Students learned better and have shown remarkable improvement in their understanding. Leaving aside the financial aspect, we recommend that educational institutes must set up a few IT-enabled classrooms if they really want their students to be ready for the competitive world ahead. Increased student satisfaction can lead to both increased faculty satisfaction and higher student retention [3].

References

1. Backer PR. Using multimedia to teach a class on technology and society. J Technol Stud. 2004;30(2):70–9.
2. Mukhopadhyay M. Universal quality school education and role of ICT. Available at www.ciet.nic.in
3. Balaveera Reddy K. A new paradigm in teaching-learning and evaluation process.
4. White paper: Malinalco, Mexico: using innovative technologies to improve learning in a rural community.
5. Oliver R. The role of ICT in higher education for the 21st century: ICT as a change agent for education; 2002.
6. Kozma R. National polices that connect ICT-based education reforms to economic and social development. Hum Technol. 2005;1(2):117–56.
7. Govt. of India, National policy on Information and Communication Technology (ICT) in school education, Draft version 1.5. Department of School Education and Literacy, Ministry of Human Resource Development. 2009.
8. Creating classrooms for everyone how interactive whiteboards support universal design for learning. White paper published by SMART on smarttech.com.
9. Learning, innovation and ICT lessons learned by the ICT cluster education & training 2010 programme.

Impact of IUCEE on Institutional Performance: A Case Study of Rajarambapu Institute of Technology

Prakash M. Jadhav, Sushma S. Kulkarni, and Martand T. Telsang

Abstract RIT, established in 1983 as a self-financing institute, is well known for its continuous experiments, leadership in academic practices, and excellence in engineering education. In the journey so far since 2008 with the help of IUCEE, RIT has progressed in leaps and bounds right from curriculum development to teaching-learning process and assessment. The Institutional Performance is measured on the basis of competent curriculum, achieving course learning objectives in turn achieving program educational objectives and program outcomes of all the programs. Various IUCEE Faculty Leadership Institutes (FLIs), webinars, Virtual Academy courses, and visit to US universities have helped RIT in transforming itself from a well-known institute in western Maharashtra to one of the leading private institutes at the national level. In this paper, the attempt is made to present how this transformation took place. With the help of IUCEE, RIT shifted from the old paradigm of one-way lecturing to two-way joyful learning experiences with better, advanced technical knowledge and globally adopted advanced pedagogical tools. And also it explains new practices we learned and followed and its impact on students' result and overall students' satisfaction index (teaching index).

Keywords Academic audit • Academic development • Institutional performance • Outcome-based education

P.M. Jadhav (✉) • M.T. Telsang
Mechanical Department, Rajarambapu Institute of Technology,
Rajaramnagar, Urun Islampur, Maharashtra 415414, India
e-mail: prakash.jadhav@ritindia.edu; martand.telsang@ritindia.edu

S.S. Kulkarni
Director, Civil Department, Rajarambapu Institute of Technology,
Rajaramnagar, Urun Islampur, Maharashtra 415414, India
e-mail: sushma.kulkarni@ritindia.edu

R. Natarajan (ed.), *Proceedings of the International Conference on Transformations in Engineering Education: ICTIEE 2014*, DOI 10.1007/978-81-322-1931-6_26,

1 Introduction: About RIT

Rajarambapu Institute of Technology, Rajaramnagar, Sakharale (RIT), is established in 1983. It has a lush green campus spread over 42 acres. It offers 7 UG and 11 PG programs along with a Ph.D. center with intake of 666 students in engineering and management and a total of 2,500 students on campus. All its programs are accredited and reaccredited by National Board of Accreditation (NBA), New Delhi. The institute has successfully implemented World Bank-funded TEQIP Phase I and received excellent grade and has been selected for TEQIP Phase II. The institute has become autonomous in 2011.

RIT is a self-financed institute situated in the rural area of western Maharashtra, but it is forward looking and adaptable to change. RIT is rated as Grade "AA+" institute by Careers360 Survey 2013. It has received "The Best Institute In Maharashtra," "Best Principal Award," and "Best Teacher Award" in 2012. RIT is ranked among the first ten in the Survey of Industry-Linked Engineering Institutes done by AICTE. It is ranked among the top 50 private engineering institutes in India and top 10 in Maharashtra by various magazines and survey agencies. RIT is marching forward on the core values like practice what we preach, joy of learning, each one can excel, fairness and equality, cooperation, honesty, and sincerity.

2 RIT-IUCEE Collaboration

RIT-IUCEE collaboration started in 2008 with attending the FLIs at Infosys, Mysore. In 2009 IUCEE Regional Centers were established with the mission of improving the quality and global relevance of engineering education in India. Since then we have many interactions with immanent professors and institutions of good repute worldwide, which encouraged us to follow best practices like outcome-based learning (OBE), changing the classroom dynamics from traditional one-way lecturing to two-way communication process.

3 First Step Toward Improving Teaching-Learning Process

RIT faculties have attended 50 different FLIs in emerging areas since 2008, and they are proved to be very expedient for the students in this rapidly changing and competitive world. Figure 1 shows the number of faculty who attended the FLIs year wise. These numbers itself show how these FLIs are popular among them.

After successful completion of the FLIs, each one is implementing knowledge and techniques acquired on continuous basis in their teaching activities. Also many faculty got confidence in selecting those areas for their research work. RIT has designed a system wherein after completing the training program, each faculty has to fill and submit the Commitment Form to respective HOD which contains information about his implementation of the knowledge and/or technique learned in the

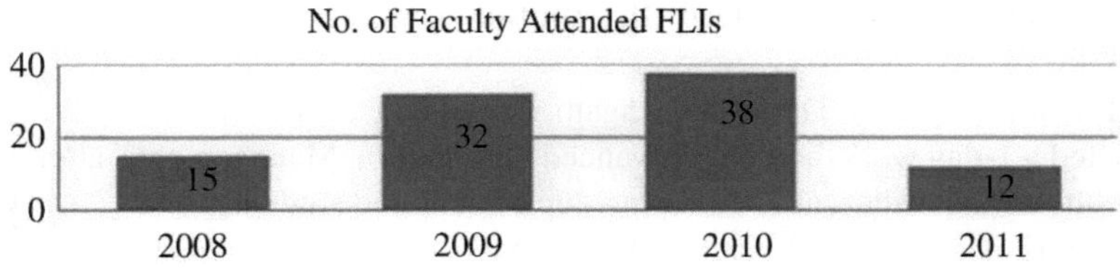

Fig. 1 Graph showing the number of faculty attending FLIs

training program. After implementing those, each one gives a presentation in front of the Director, Dean (Academics), all department heads, and all faculty on his experiences, feedback received from students, and the testimonial of benefit. This has helped a lot in gaining subject knowledge, planning and delivery of the curriculum, classroom management, etc.

RIT's Director, Dean (Academics), and other senior faculty attended workshops by Richard Felder and Lueny Morell and took a challenge to train other faculties to change the classroom dynamics for the better teaching-learning process. Because of IUCEE FLIs, we got confidence of preparing outcome-based teaching plan and execute it effectively.

4 Activities of IUCEE, Islampur Regional Centre

The regional centre is established with the objective of organizing national and international workshops in recent advances in engineering and pedagogical aspects.

4.1 Organizing Workshops with International Faculty

4.1.1 International Workshop on "Mechatronics and Product Design"

The IUCEE Islampur Regional Centre organized a 1-week workshop (FLI) for 40 participants on "Mechatronics and Product Design" from July 12 to July 16, 2010, by internationally known Dr. Devdas Shetty of the University of Hartford.

Outcomes and Impact: Seven industry personnel attended the workshop and we could interact with them for 5 days. Dr. Shetty suggested to learn a software, namely, "VisSim," for modeling and simulation of mechatronics system to the faculties which is widely used in the industry. Our two faculties have mastered it and conducted batches of size 25 students each year, which is beneficial for the students' placement. Also we could modify the syllabus as per the guidelines given by Dr. Shetty and included "mathematical modeling" in the mechatronics course syllabus for UG students. Students are encouraged to take up project work in mechatronics and robotics, and they are participating in various robotic competitions.

4.1.2 Workshop on "Advanced Distribution Management"

IUCEE resource person Dr. Manivasagam visited RIT on December 31, 2008, and conducted a 1-day workshop on "Advanced Distribution Management," attended by the faculty of the electrical engineering department and students.

Outcomes and Impact: Because of his inputs and encouragement, our electrical engineering department added the new course "Power Distribution System" in the M. Tech. (Electrical) curriculum. With the knowledge of this newly introduced course, faculties and students got very good insight and could complete four dissertations.

4.1.3 A Motivational Presence of IUCEE Executive Director in RIT

In the first visit of Dr. Krishna Vedula, Executive Director of IUCEE, IUCEE Islampur Regional Centre was established. He highlighted on IUCEE Mantra "I am teaching, are they learning. Say What You Do, Do What You Say. Prove it and improve it." After his visit, RIT proactively initiated steps to transform teaching-learning process to make learning a joyful experience for the students. This shift of focus from teaching and learning started yielding good results. Because of the new and innovative practices learned, within a short span of a year of autonomy, RIT is able to design and implement the academic systems to raise the bar of academic standards.

In his meeting with the Board of Governance of RIT, it was decided to organize various workshops on outcome-based education (OBE) at RIT and in nearby institutions. Dr. Sushma S. Kulkarni, Director of RIT; Prof. Martand T. Telsang, Dean (Academics); and Prof. S. S. Gramopadhye conducted 10 workshops for 2–5 days on various aspects of OBE. As a result, teaching-learning process has changed drastically and reflected in students' results and feedback, which is discussed in Sect. 8. Also each year RIT conducts in-house induction training program for newly recruited faculty to train them in OBE.

5 Visit to US Universities

In November 2008, Chairman, BOG, RIT, Shri. Bhagatsingh Patil along with Director Dr. Sushma S. Kulkarni and Prof. Martand T. Telsang joined the Indian delegation who visited prominent US universities. Subsequently, RIT signed MoUs with UMass, Lowell, Purdue University, and Lawrence Tech. University. The team got exposure to the working system of top US universities and could interact with eminent researchers and academicians and make righteous changes.

5.1 Outcomes and Impact

- Industry institution interaction: lab setup in collaboration with industry
- Developed research culture
- Doing Engineering Projects on Community Services (EPICS) started incubation center
- Effective use of the Internet and communication technology in teaching-learning process
- Keeping flexibility in curriculum
- Transformation and some best practices implemented

6 Webinars

In the year 2011, IUCEE introduced a new concept called "webinar." About 25 webinars are attended by RIT faculties and students on different advanced topics and pedagogy. This has given a good exposure to the faculties and students and helped in gaining knowledge of very recent courses and their teaching methodology. Also all the webinars' recordings are available on Lensoo which can be viewed at any time later on. Many faculties and students are interacting with US experts and seeking guidance through mail.

7 Virtual Academy Courses

Using IUCEE Virtual Academy, a new, innovative, and unique approach in India, faculty experts from the USA are delivering complete course directly to students through online webinar series of ten lectures. In 2011, our 72 second year electrical students and 2 faculties imbibed the subject "Electrical Machines-II" through this facility. Students showed great interest in interacting with the resource person Dr. Sarma Mulukutla of Northeastern University, Boston, USA, through mails. They gave very positive feedback and wished to learn more subjects through IUCEE Virtual Academy. In view of this impact, in 2012 we subscribed three VA courses for second year electrical, second year EnTC, and M. Tech. (CSE) students.

This year our 32 faculties are being trained through a webinar series of ten lectures on outcome-based education, which is very useful in designing program outcomes, program educational objectives, course learning outcomes, and students' outcomes and mapping the same to create engineers to face global challenges.

Faculty and students have given very positive feedback about this methodology. It has added values to their knowledge; the shared passion of US experts has created a lasting impression on the minds of the faculty and students.

8 Overall Measurable Impact

Overall impact on the institution is measured with the help of many parameters like students' results and Students' Satisfaction Index, placement, industry institution interaction, publications, research funding and patents, etc. In this paper, two highly impacted major academic parameters, namely, students' results and Students' Satisfaction Index.

8.1 *Students' Result*

Director Dr. Sushma S. Kulkarni; Prof. Martand T. Telsang, Dean (Academics); and other senior faculties trained almost 100 % of the faculty in the crucial theme of OBE. They could deliver the curriculum with ease and assess it properly. Hence, we found an increase in transition rate in the last 2 years. Transition rate is percentage number of students passing successively in each year to complete the course in 4 years. Figure 2 shows increased transition rate for all programs.

8.2 *Academic Audit*

The strategic plan of the institute focuses on faculty development wherein performance indicators are influenced by the "IUCEE Way." RIT has a system of academic audit to ascertain the degree of teaching-learning process occurring at the institute level, which is carried out based on a 10-point scale pertaining to teaching dimensions like subject knowledge, planning and delivery, class management, communication and English, and accountability and concern. The outcomes of the audit are summarized first at the faculty level, then at the department level, and finally at the institute level. Comparison for the last 3 years shows that there is continuous improvement in all teaching dimensions in which IUCEE has played very important

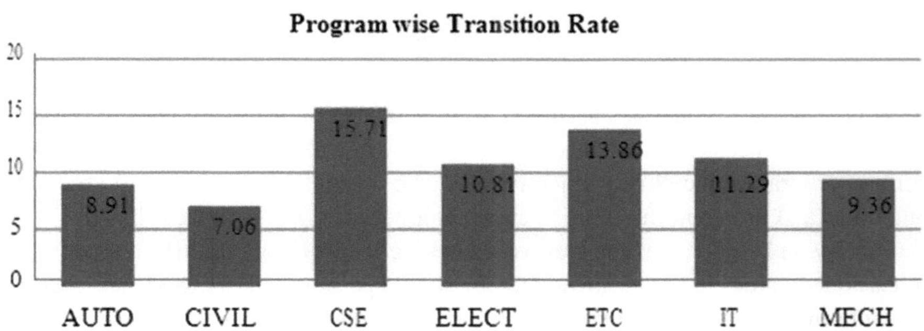

Fig. 2 Graph showing program wise increase in transition rate

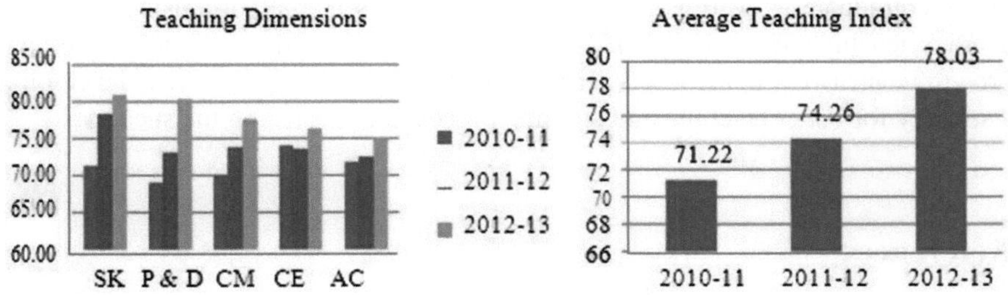

Fig. 3 Graphs showing increase in all teaching dimensions and teaching index for 3 years

role. The average teaching index (Students' Satisfaction Index) of the institute has shown growth in each year. It is increased by 3.04 % in 2011–2012 and by 3.77 % in 2012–2013 as shown in Fig. 3.

9 Transformation

Due to the IUCEE activities described above, RIT could develop its own instructional methodology based on outcome-based learning which has potential for long-term, sustained benefits. RIT has shifted from the old paradigm of one-way lecturing to two-way joyful learning experience with better, advanced technical knowledge and globally adopted advanced pedagogical tools.

It has achieved more height in the following:

- Enhanced innovation in teaching and learning
- Enhanced the teaching-learning experience promoting student engagement, proper assessment, and success in examinations
- Enhanced student technology access
- Provided enhanced training for the students and faculty for use of technology
- Provided enhanced resources for the students and faculty with special needs

RIT appreciates the importance of the student's involvement in learning and teaching process for achieving the desired student success. Thereby, exploring teaching philosophies and students' views of teaching practices yields insights on the teaching and learning process in education. Students have become more objective in the feedback of teachers. Experiments by faculty have influenced and motivated their learning which is reflected in their academic audit (feedback).

10 Conclusion

RIT, in order to transform from traditional teaching-learning process to a student-centered learning, has adopted most of the techniques from IUCEE. These techniques have helped RIT in increasing the overall result of all the students. The transition rate of undergraduate as well as postgraduate students is increased considerably. The average teaching index (Students' Satisfaction Index) of the institute has shown growth in each year in the last 3 years. Thus, IUCEE has helped RIT in developing a strategy for better teaching and learning process and achieving the success.

References

1. Card DN. Defect analysis: basic techniques for management and learning. q-labs 115 windward way Indian harbour, FL 32937 USA.
2. Felder RM, Forrest KD, Baker-Ward L, Dietz EJ, Mohr PH. A longitudinal study of engineering student performance and retention. North Carolina State University, Raleigh.

Green Electronics Design: Curriculum, Content, and Learning for Engineering

Ravi Sankar Chandu, Rakesh Kumar Singh, and Santosh Kumar Singh

Abstract Energy consumption, material application, packaging, and recyclability are the major areas to be considered for greener electronic products with reduced environmental effects and increased eco-efficiency. Design for green electronics includes lead-free solders, conductive adhesives, environmentally friendly packaging and design technologies, LCA and assessment, life cycle cost analysis, and data management. Engineering graduates should be taught sustainability, its principles, and how these principles can be incorporated into the design and manufacturing processes. Engineering students have to be exposed to design for the environment, directives, environmental law, technology, waste management, and electronic waste recycling. This paper gives introduction to green electronics, sustainability principles, curriculum, and content, teaching, and learning methods needed for green electronics. This paper analyzes the content necessary and the methods for educating the engineers for sustainability, problem solving, and the skills required. This paper enlightens the reader of the content, skills, and the design methods for sustainable engineering.

Keywords Sustainability • Green electronics • Learning

R.S. Chandu (✉)
Vignan University, Vadlamudi, AP, India
e-mail: rsc_ece@vignanuniversity.org

R.K. Singh
Uttaranchal Technical University, Dehradun, Uttaranchal, India
e-mail: rksinghkec12@rediffmail.com

S.K. Singh
Environment Engineering Department, Delhi Technological University, New Delhi, India
e-mail: sksinghdce@gmail.com

R. Natarajan (ed.), *Proceedings of the International Conference on Transformations in Engineering Education: ICTIEE 2014*, DOI 10.1007/978-81-322-1931-6_27,

1 Introduction

Sustainability aims to reduce the level of material consumption and environmental damage without affecting quality of life. It aims at poverty alleviation and redistribution of opportunity in the long run. It advocates reduction in pollution levels, excessive consumption, and action for alleviating poverty, achieving greater equity and distribution of opportunity. It requires protection of biodiversity and reduction of global pollution and poverty [1]. Sustainable development demands change in consumption patterns. It can be achieved through lower consumption rates as well as increased energy efficiency. Though the UN declared the period between 2005 and 2015 as the decade for Education for Sustainable Development (ESD), there is not much progress on education toward sustainability in many countries. Education for sustainability will not suffice with just a module on environmental principles or studies. Problem solving toward sustainability and other required skills have to be included in the curricula. Many of the engineering faculty themselves are ignorant of the sustainability-related issues, problems, and solutions. As engineers play an important part in sustainability processes, achieving sustainable development involves inclusion of sustainability and equity in design criteria. Sustainability is not just minimizing material and energy use as was felt earlier by the engineering community, but understanding stakeholders, society, and equity toward a better environment. Engineering education has to prepare engineers who accept sustainability as a basic design requirement in development of products and processes and as a policy criteria in future industrial developments. Sustainability involves many disciplines such as social sciences, environmental economics, law, environmental politics, and others.

2 Green Electronics

Hazardous materials such as lead and bromine in the electronics production sector have been advised against use through WEEE (Directive 2002/96/EC on Waste Electrical and Electronic Equipment) and RoHS directives (Directive 2002/95/EC on the Restriction of certain Hazardous Substances Directive) in the legislation of most European Union (EU) member states. The three aspects under focus are green electronics design, assembly, and materials [2]. Mainly electronic engineers are concerned with reduction in energy guzzled by the ICs, reduction of materials consumed in the process of making a component or process. Green processes and electronics avoid hazardous materials and high-energy-consuming products and processes. In order to save materials and energy, innovative ideas are used to integrate several products into one (e.g., a photocopier has several functions and features like scanning, fax machine, and clock besides photocopying). Also, when products are being made, supply chains are also made green. New materials are being used for materials that enhance greenness of the products and processes. Low-dimensional allotropes of carbon (including carbon nanotubes (CNTs), graphene, and graphene nano-ribbons (GNRs)), known as carbon nanomaterials, have extraordinary physical properties giving advantage in the manufacture of active and passive devices used in integrated circuits [2, 3]. Carbon nanomaterials hold exciting prospects in designing ultra-energy-efficient electronic devices. Hence, electronic engineering students need to be taught about nanomaterials [4].

3 Engineering, Sustainability, and Education

Engineering professionals are coming under increased pressure to practice engineering sustainably [3]. Engineers are seen to have a key role in delivering sustainability [4]. Engineers have to become knowledgeable of SD principles and be continuously trained about current sustainable development technologies applicable to their profession. Lack of knowledge and practice for sustainability in engineering education, pressure from the industry, demand from the society, and requirements from the professional bodies point to the urgent need for integrating or embedding sustainability into our engineering curricula. But the universities have so far failed to train new generation of engineers effectively in SD. The professional and academic accreditation bodies introduced sustainability as one of the major criteria for professional/ engineering program accreditation. Till now, engineering is thought as applying science to solve practical problems based on empiricism. Engineering education should not only embrace empiricism but also ethics, creativity, and social responsibility. This demands that the concepts of SD must be embedded into all stages and topics of engineering education. Broadening the boundaries of the design space is a key challenge in redefining the requirements of future engineering education [5].

Engineers do not have much exposure to sociology, ethics, politics, philosophy, and other such areas [6]. To broaden the remit of engineers, they have to be taught wider range of subjects. This may lead to additional topics taught superficially by which technical rigor of traditional subjects may be dissipated by the width. This can lead to a UG engineer who is aware of SD with breadth in masters (tools and skills) [7]. Engineering curriculum needs redesigning so that students are exposed to issues and problems and imbibe the right thinking process involving sustainability concepts and their use in engineering problem solving. It is a philosophy similar to that of quality management. It is a paradigm shift from an approach based on social and political rather than Newtonian approaches.

3.1 Environmental Legislation, Policy, or Protocols

As engineers work across borders, they should know the international and local environmental legislation, policy, or protocols, acts, directives, declarations, permits, and laws and hence should be taught. The content that should be taught is (1) Protection of the Environment Act, (2) Integrated Pollution Prevention and

Control (IPPC) License, (3) Restriction on Hazardous Substances Legislation, (4) Water Framework Directive, (5) Air Quality Framework Directive, (6) Strategic Environmental Assessment, (7) Wildlife and Habitats Legislation, (8) Kyoto Protocol, (9) Rio Declaration, (10) Tradable Permits, (11) Planning Laws, (12) Environmental Impact Statement (EIS), and (13) Energy Use in Buildings. Some of them are related to environmental engineering.

3.2 Tools, Technologies, and Approaches

Developing a technology, engineers have to differentiate between a clean and polluting technology and, if avoidable, should develop cleanup technology. Important concepts, tools, technologies, and approaches an engineer should know are (1) Clean Technology, (2) Cleanup Technology, (3) Design for Reuse, (4) Renewable Energy Technologies, (5) Dematerialization, (6) Design to Reduce Environmental Impacts, (7) Eco-labeling, (8) Life Cycle Assessment, (9) Waste Minimization, (10) Design for Energy Efficiency, and (11) Sustainability Impact Assessment.

3.3 Sustainability Principles

To implement sustainability, many important topics related to socioeconomic sphere should be taught to the engineers. These are (1) Precautionary Principle, (2) Intergenerational Equity, (3) Intragenerational Equity, (4) Earth's Carrying Capacity, (5) Resource Renewability, (6) Ecological Footprint and Environmental Space, (7) Stakeholder Participation and Management, (8) Triple Bottom Line Approach, (9) System Thinking and Approaches, (10) Principles of Sustainable Development, (11) Engineering Ethics, and (12) Engineering Community's Response to Sustainable Development.

3.4 How to Teach Sustainability to Engineering Students

Sustainability concepts make an engineer understand the interdependence of natural, social, and economic systems, the needs and rights of future generations, equity, and precaution. The engineers realize their role in promoting sustainability with a sense of social responsibility through these concepts. In engineering education, these concepts and solutions should be transformed into specification and expressed as learning outcomes (that are assessable and measurable) [6].

Table 1 Necessary skills, technologies, and tools needed for sustainability

Sl. no.	Area	Engineers should know
1.	Management skills	Management, ethics, communication skills, basic accounting skills of process, project management, time and people management, cost-benefit analysis, and team working
2.	Other skills	Critical and creative thinking, logical thinking skills, system thinking

3.5 *Skills Needed for Sustainability*

Sustainability-included curricula develop appropriate levels of awareness/attitudes, skills, and knowledge (Table 1).

3.6 *Teaching and Learning Processes*

Embedding sustainability does not mean including new contents. It is not just adding few courses to the program either. Educators have to pool their knowledge from technical and nontechnical areas and develop appropriate problem-solving approaches. If engineers are to contribute truly to SD, sustainability should affect their everyday thinking. This can be achieved if SD becomes an integral part of the engineering education program, not a mere add-on to the core parts of the curriculum [8]. Integrated and embedded approaches are advised.

Table 2 gives different approaches in teaching and learning processes for sustainability.

Teaching methods such as short intensive courses, team project work, structured seminar discussion, interactive role-plays, small group lectures, fieldwork, residential field courses, and people-planet-prosperity entrepreneurial competition can be used to give a variety of experiences to the students. These should be brought into the curriculum by learning processes that are amenable to the use of case studies, studio based, issue driven, process based, team based, and design/delivery focused [9].

Sustainability integration can start with a basic course on sustainability in the first year of an engineering program. There are several problems in incorporating sustainability in engineering education. They are (1) maturity of students, (2) knowledge of sustainability among lecturers, (3) acceptance of sustainability engineering, (4) lack of textbooks, (5) lack of examples, and (6) lack of time [10].

Table 2 Approaches for teaching sustainable development

Approach	University	Remarks
One tier	Many universities	Difficult to develop concepts, teaching models, tools, and basic skills for application during a professional life in a single module
Semesters	Carnegie Mellon	Sustainability incorporated into the design process of a specific discipline the student belongs to
Three-tier approach	University of Surrey	Dedicated lectures and tutorials on sustainable development, specific case studies and integration of sustainability into the overall curriculum, integration of different sustainability criteria into conventional design approach using life cycle thinking, industrial ecology approach, and appropriate ethical principles
Integrated approach	University of Cambridge	The ability to work complex/ill-defined problems as a team and acquire two-way communication skills and the ability to evaluate merits and demerits of the options
	Delft University of Technology	Through interconnected operations of elementary course, development of graduation in SD in each faculty and department, and intertwining of sustainable development in all regular disciplinary courses

4 Design

The design approaches toward sustainability of electronic products can be Emotionally Durable Design (reduce, reuse, recycle, emotional connection with the products), Cradle to Cradle (the term "waste" has no meaning; material is simply a biological or technical nutrient and hence be recycled), Bio-mimicry (nature produces everything without producing toxic waste and it should be to copied/emulated in the design), Product Attachment (encourage consumer's attachment to products), Behavioral Design (change the way people behave, reducing carbon footprints), Slow Consumption (people reduce their desire to acquire more and thus slow down the cycle of consumption), Products to Services (solutions useful to wider community), Dematerialization (question the need for a product in the first place), and Product Longevity (move to reusing products and parts of products).

5 Conclusion

Engineering is no more Newtonian where problems are solved through mathematics and design of single discipline. Due to the ill effects of engineering solutions on the environment, engineering education has to incorporate a system approach in problem solving. As a result, engineers have to handle society and stakeholders in such problem solving, and hence content related to the environment, society, social responsibility, stakeholder management, and politics should be included into engineering curricula. Hence, engineering students should be allowed to take subjects in the above areas during their undergraduate education. The resulting curricula will lead to general engineering programs rather than single discipline oriented. Green design is achieved through sustainability education enhanced by the curriculum. This should be at all levels and areas achieved through embedding sustainability in all modules. Engineering students should be made to understand that technology affects environment and society. Engineering students should be educated to identify these effects and be equipped with the right thinking process and skills to overcome these problems. In order to achieve sustainable development, engineering students have to understand the principles behind sustainability and hence have to be taught, as today's problems are interconnected and solving them need knowledge and skills from different disciplines. Hence, students should be taught working in diverse teams. Students have to imbibe sustainable thinking processes, and this is only possible when sustainability is embedded into all programs, all subjects, all levels, and all activities during the program. Teaching and learning processes used should give the student a variety of experiences in understanding the problems and expertise in solving them.

References

1. Perdan S, Azapagic A, Clift R. How much do engineering students know about sustainable development? The findings of international survey and possible implications for engineering curriculum. Eur J Eng Educ. 2005;30:1–19.
2. Avouris P, Chen Z, Perebeinos V. Carbon-based electronics. Nat Nanotechnol. 2007;10:605–13.
3. Banerjee K, Dadgour H, Khatami Y, Li H, Xu C. Carbon based active and passive devices for next- generation ICs. In: Proceedings of the global COE international symposium on silicon nano-devices in 2030: prospects by world's leading scientists, Tokyo. 2009.
4. Banerjee K, Li H, Xu C, Khatami Y, Ddgour HF, Sarkar D, Liu W. Prospects of carbon nanomaterials for next-generation green electronics. In: Proceedings of the 10th international conference on nanotechnology joint symposium with nano Korea, Korea. 2010.
5. Fenner RA, Ainger CM, Cruickshank HJ, Guthrie PM. Widening engineering horizons: addressing the complexity of sustainable development. In: Proceedings of the institution of civil engineers, engineering sustainability. 2007.
6. Azapagic A, Perdan S, Shallcross D. How much do engineering students now about sustainable development? The findings of an international survey and possible implications for the engineering curriculum. Int J Eng Educ. 2005;30:1–19.

7. Fenner RA, Ainger CM, Cruickshank HJ, Guthrie PM. Embedding sustainable development at Cambridge University Engineering Department. Int J Sustain High Educ. 2005;6:229–41.
8. Sterling. Higher education, sustainability and the role of system learning. In: Higher education and the challenge of sustainability curriculum. Boston: Kluwer Academic Publishers; 2004.
9. Jowitt PW. Sustainability and the formation of the civil engineer. In: Proceedings of the Institution of Civil Engineers, Engineering Sustainability, ES2; 2002. p. 79–8.
10. Boyle C. Considerations on educating engineers in sustainability. Int J Sustain High Educ. 2004;5(2):147–55.

Should We Have Compulsory "Engineering Ethics" Course for Engineers? Why Should It Be?

Satya Sundar Sethy

Abstract Human beings are ethical beings. They judge human actions as good or bad based on their mere understanding of "ethics." But this is not adequate for professionals to judge their actions as good or bad. They need knowledge of "applied ethics." For example, physicians need medical ethics; lawyers require legal ethics, etc. Similarly, engineers need "engineering ethics" for their professional tasks, whether it is about designing machines or innovating technology, etc. No doubt, they deserve appreciation for their professional tasks, which they carry out through rationality and quantitative aptitude. But mere rationality and quantitative skills could hold them accountable for societal insensitiveness and engineering disasters. They cannot shirk their responsibility for engineering disasters that are the consequences of their designed products. So, it is imperative for them to incorporate "engineering ethics" in their professional tasks. In this context, this paper attempts to answer how "engineering ethics" is different from "ethics." Why do engineers need to do a course on "engineering ethics"? Should it be made compulsory for engineering students? What to teach, how to teach, and who should teach this course? How should engineers integrate "engineering ethics" into their tasks? What type of responsibility do they have towards the society at large?

Keywords Applied ethics • Curriculum • Engineering design • Engineering ethics • Responsibility • Technological innovations

---

S.S. Sethy (✉)
Department of Humanities and Social Sciences,
Indian Institute of Technology Madras, Chennai 600036, India
e-mail: satyasuundar20012001@gmail.com

R. Natarajan (ed.), *Proceedings of the International Conference on Transformations in Engineering Education: ICTIEE 2014*, DOI 10.1007/978-81-322-1931-6_28,

1 Introduction

Human beings are born free, but subsequently they get associated with ethical norms, principles, and regulations depending on where they are born and brought up. Thus, a human being bereft of ethical life is impossible, unless one consciously avoids it. Ethics help us judge our actions as good or bad, right or wrong. So by implication, engineers possess ethical life and thereby ethical knowledge. But, is it adequate for them to deal with their professional tasks only with ethical knowledge? Will they be able to judge their professional actions as good or bad by possessing mere ethical knowledge?

Engineers need knowledge of "professional ethics" to deal with their tasks just as other professionals like physicians, lawyers, and accountants. Professional ethics are applied ethics as applicable to a specified group of people, such as business ethics is for businesspersons, medical ethics for physicians, legal ethics for lawyers, etc. In a similar vein, engineers need "engineering ethics" for their professional tasks, whether it is about designing machines or innovating technology, etc. A question arises, how is "ethics" different from "engineering ethics"?

2 Ethics Versus Engineering Ethics

Ethics is a reflective consideration of our moral beliefs and practices [1, p. 6]. According to Harris et al. [8] "ethics refers to those standards of conduct that apply to everyone rather than only to members of a special group. Ideally, these standards are ones that every rational person wants every other to follow, even if everyone else's following them would mean that s(he) had to do the same" (p. 93). Ethics teach us what we should and should not do, good and bad behavior, etc. But ethical judgments are subjective in nature because judgments rely on social context, cultural atmosphere, and religious condition of the judgment makers, among others. Some ethical norms that we learnt from our forebears and can be treated as ethical norms are don't steal, keep your promise, maintain integrity and impeccability in your life, etc. These ethical guidelines are known to engineers, but what is unknown to them or not taught to them is their "professional ethics," that is, engineering ethics. Thus, engineers are required to learn "engineering ethics."

Unlike ethics, engineering ethics sets standards for engineering practice. It guides engineers' actions as members of a profession. According to Harris et al. [8], "engineering ethics is as much a part of what engineers in particular know as factors of safety, testing procedures, or ways to design for reliability, durability, or economy" (p. 93). Schinzinger and Martin [23] define "engineering ethics … is the study of the moral values, issues, and decisions involved in engineering practice." Kroes and Meijers [16] state that "engineering ethics is the field of study that focuses on the ethical aspects of the actions and decisions of engineers, both individually and collectively" (p. 21).

Harris et al. [9] explain the concept "ethics" by splitting it under three heads: general ethics, personal ethics, and professional ethics. "General ethics" suggests ethical standards that are known to most of the persons irrespective of their religion, caste, sex, creed, and education; for example, prevent killing, cheating, stealing, etc. It advises individuals to restrain from all sorts of violations or invasions. It distinguishes evaluation of an individual's actions from evaluation of his/her intention. "An evaluation of action is based on an application of the types of moral precepts we have been considering, but an evaluation of the person himself is based on intention" [9, p. 8]. An example can clarify this notion. If a car driver kills some persons accidentally by riding his car onto the pedestrian lane, then the driver can at best be charged with manslaughters, as his/her intention was not to kill those persons. The law treats the driver differently so long as he/she was not reckless. Here, the result is the same (killing human beings), but the intent is different.

Personal ethics is defined as a set of moral beliefs that one upholds in his/her life. For example, not lying to others, maintaining integrity, practicing honesty, etc. Thus, it is stated that personal ethics can be situated parallel to the precepts of general ethics. But there are cases where the distinction between these two models of ethics is clearly noticed, at least with reference to the fact where general ethics seems to be unclear. For example, one may oppose testing of certain drugs on human beings, although general ethics is not clear on the issue. On the other hand, professional ethics sets the "code of ethics" and professional responsibilities for professionals. In addition to that, it guides professionals on how one should give priority to his/her professional ethics rather than personal ethics and general ethics.

"Engineering ethics" as professional ethics is applicable to engineers only and to their profession. So, engineers should possess the knowledge of "engineering ethics" and its application to their professional tasks. Otherwise, having knowledge of "general ethics" and "personal ethics" and taking decision on engineering tasks will inadvertently invite dangers, such as unwanted tragedies and engineering disasters. Thus, "general ethics" and "personal ethics" will not be adequate for them to make the right decision in a given situation. It is suggested that when engineers are in trouble for making a decision, depending on their subject discipline, they can refer to the code of ethics prescribed by the statutory and governing bodies. For example, one can refer to the code of ethics of NSPE, IEEE, ASME, etc., to make a decision in a complex situation. Thus, it is necessary that engineers should read and understand their professional code of ethics before practicing engineering. In this context, studying engineering ethics can help students in various ways. Among others, they can recognize ethical issues involved in engineering tasks and understand intricate concepts associated with their profession, such as public health and safety, quality retention, usefulness of technology, efficacy of a design, cost–risk–benefit analysis, truthfulness, trustworthiness, and loyalty to employer and public. Further, they can improve on their ethical judgments and accelerate ethical will power. Calhoun [3] states that studying engineering ethics can increase students' knowledge of relevant standards. Thus, it is said that a student who reads code of ethics is more likely to know what is in it and how to refer to it for a crucial decision in a complex situation than a student who does not read it. Hence, it is necessary for engineers to undergo engineering ethics course in their educational career. "Engineering ethics" course should be made compulsory for engineering students because they are professionals and have some responsibilities towards their profession and the society. Their responsibilities include the following: (a) protection of public safety, (b) technical competence, (c) timely communication of negative and positive results to management, and (d) resolve ethical dilemmas by referring to code of ethics.

Engineering artifacts are made mostly for public use, whether it is a bridge, or medical equipments, or high-rise buildings, etc. The engineering artifacts are not risk-free as they are associated with either bad or good consequences. The consequence is good when people use the artifact and benefit from it and bad when disaster occurs. It is hence an obligation of engineers to communicate the risks of the artifacts to the public if the problems are not sorted out during the design process itself. One may ask, engineers do innovations, so how are they going to know all the risks of a new technology? So how can we expect a completely safe technology from them? Most of the scholars including engineers did not agree with the supposition that a completely safe technology is not possible. Baura [1] says that even in innovation, safe technology can be engineered through suitable design. She defines "safety through design is the integration of hazard analysis and risk assessment methods early in the design and engineering stages and the taking of the actions necessary so that the risk of inquiry or damage are at an acceptable level" (p. 14). This concept embraces facilities such as hardware, equipment, products, tooling, materials, energy controls, layout, and configuration.

As engineering projects are time bound and need to accomplish the task within a stipulated time, they should have subject domain knowledge along with will power and knowledge of professional ethics. These three components define engineering competency. Thus, if engineers were having some lacunae in any of these components, then the final designed product would be either shoddy or of poor quality. Even the slightest compromise in any of these components may endanger public safety. So to enhance understanding and knowledge of these components, engineers are recommended to go through the "engineering ethics" course. The honorable Supreme Court of India also gave a judgment on whether engineering ethics course should be made compulsory for engineers. The verdict says as engineers are professionals and they are involved in many projects related to human safety and design ingenious artifacts for public use, they should qualify for the professional ethics course. Lincourt and Johnson [17] convey that to be an engineering professional, one must be competent and good. The term "good" ascribes the ethical aspect of one's understanding about his/her profession. Further, "Engineering competence does not ensure moral goodness, nor does moral goodness guarantee engineering competence. Both are necessary; neither is sufficient" (p. 353). This suggests "engineering ethics" should be integrated into the engineering education curriculum as a core course. In the ABET criterion for accrediting engineering programs, it is written that "engineering programmes must demonstrate that engineering graduates must have an understanding of professional and ethical responsibility."

With regard to timely communication, engineers should communicate both positive and negative results of their project to their higher authorities from time to time, so that rational decisions can be made in the designing process and attempts can be made to avert the bad consequences. Thus, a timely communication phenomenon is treated as one among the other responsibilities of engineers towards the profession and the society at large. Here one can recall the case studies of the space shuttle disaster, which is described in most of the scholarly books on engineering ethics. The disaster could have been avoided had there been timely communication between engineers and managers. By adapting this behavior engineers can resolve some of the ethical dilemmas in their profession, such as conflicts of interest, accountability to clients and public, fair treatment, differentiating gifts from bribe, etc. To understand all these complex phenomena, engineers should design the engineering ethics curriculum in such a manner that it should cater to the need of current engineering practices. Lincourt and Johnson [17] claim that engineering ethics course should be an imperative course for all engineering students irrespective of their disciplines. Jessica and Shengli [15] endorse this view and state that engineering ethics course is a must for engineers.

3 What to Teach in Engineering Ethics Course?

What to teach in "engineering ethics" course has not been explored adequately in the engineering ethics literature [15], although a variety of strategies and methods of teaching engineering ethics have been researched and experimented [10, 12–14]. Engineering is something that engineers do, and what they do has profound effects on others. Teaching engineering ethics is therefore part of teaching engineering. Against this backdrop, attempting to answer this question makes me a little comfortable. As I am involved in this course in my institution (IIT Madras, India), I am expressing my opinion for other's consideration, remarks, suggestions, and comments if any. To me, the motto behind teaching "engineering ethics" course should be to produce the best possible moral engineers in today's fast-changing environment. The objectives of the curriculum should be to prepare engineering graduates to analyze, decide, and resolve ethical issues that arise in their work. Further, by incorporating engineering ethics in their tasks, they should be able to achieve the core ethical values which may be treated as objective (certainly, not subjective) and impartial. This may not be possible through either personal ethics or through general ethics. Thus, it is a prerequisite for engineering students to learn about ethical implications within the context of the social, organizational, and even political environment where engineering is being practiced [2]. To attain the objectives of the course, Herkert [11] states that this course should be integrated with the field of science, technology, society, and other related field of ethics. In this course, instructor(s) therefore can teach ethical theories and their implications on engineering profession, such as egoism and consequences of one's responsibility [18], egoism in engineering and responsible action [9], utilitarianism, duty ethics, rights ethics, and virtue ethics [1]. Further, some of the case studies can be discussed and analyzed by using appropriate methodologies. Again, some current ethical issues in engineering can be discussed by highlighting the complexities involved in the engineering profession. Last but not the least, engineering code of ethics should be discussed by referring to various registered organizations, such as NSPE, IEEE, etc. [4], and the responsibility of an engineer towards the profession and the society [2, 11, 18]. Cultural sensitivity is another area that needs to be integrated into the engineering ethics curriculum to sensitize students about the social and environmental impacts of their products [26]. The engineering ethics course contents therefore should be prepared in such a manner that at the end of the course, students are expected to have the required knowledge and skills needed to face current ethical challenges.

4 How to Teach Engineering Ethics Course?

The teaching methods of a course are determined on the basis of its curriculum and the target group to whom it will be delivered. There is a unanimous view that the best way to teach engineering ethics is by using cases [20], not just the engineering disasters but the kind of cases that an engineer is more likely to encounter [8]. Depending on the case studies, one can adopt the methodologies for its analysis. For example, "drawing the line" can be used as a methodology to explain the situation between accepting bribe and gift. Further, "conflict resolution" can be considered as a methodology to discuss a case study where two pressing phenomena are presented before an engineer and what he/she should do in that situation, for example, an engineer's obligation towards employer and towards the public, whose health he/she has to protect. Besides these two methodologies, there are many methodologies available to teach engineering ethics course, such as simulation, role play, games, and traditional strategies of lecture, group discussion, and writing assignments [15]. Again, it is stated that it would be ideal to collate all the possible methodologies and design an inclusive instructional strategy to deliver this course to engineering students.

5 Who Should Teach Engineering Ethics Course?

It is widely accepted that either academic engineers, or philosophers, or engineers involved in industrial work can teach "engineering ethics" provided they have earned the expertise in this field. This suggests three options. First, philosophers can teach this course as they have earned the mastery over ethical theories and their applications to the worldly affairs. Second, engineers themselves can teach this course as they are experts in engineering subjects, and this course is about ethical issues related to the engineering profession. Third, there can be a blend of philosophers and engineers, where philosophers can handle the theoretical aspects and engineers can take care of the practical aspects of the course. However, in all these

Table 1 Merits and lacunae of the instructor(s) of "engineering ethics" course

Type	Instructor(s)	Advantages	Disadvantages
First	Philosophers	Explain all the possible and required ethical theories to the engineering students along with their applications to engineering tasks	Emphasis is placed only on ethical theories by overlooking the engineering case study analysis
Second	Engineers	Discuss contents related to engineering and science subjects only	Ethical theories are ignored. Hence any discussion of case study will not be adequate for students to comprehend the intricacies involved in engineering practice and the need to include "professional ethics" in their decision-making process
			Students may not be able to know how to resolve many ethical problems, as they are not discussed in the class except a few case studies
Third	Philosophers and engineers	Students get an opportunity to hear from both engineers and philosophers	Due to the mixture of experts from different disciplines, it would be very difficult to stick to the instructional design of the course
		Students can obtain both theoretical and practical knowledge about engineering ethics inclusion into engineering profession	Repetition of certain content is certain

cases both merits and lacunae are found, which are mentioned below in Table 1. For example, in the Indian Institutes of Technology (IITs), which are believed to be the premier educational institutions, the third option is adopted and adhered to deliver the engineering ethics course to engineering students.

6 How Should Engineers Integrate Professional Ethics into Their Tasks?

Engineering is a pragmatic discipline which operates within a culture devoted largely to problem solving – designing machines and innovating technologies. It has been observed that engineers are insensitive towards technological designs. Their decision on an engineering artifact is based on the schedule–cost–benefit analysis.

As a result, they define risk as a function of probabilities and unwanted consequence [21]. But in reality it is not true in all contexts/situations, as with a little precaution, that is, insertion of professional ethics into the engineering design, the unwanted tragedies would be averted. If it is so, then how should engineers integrate professional ethics into their design process? Certainly, pure rational decision overlooks ethical considerations about risky technologies. Hence, engineers should relate risk with their social responsibility, which is enshrined in their code of ethics. Here, Roeser [21, 22] states engineers should assess risks of the technology by referring to their "social responsibilities" as a code of ethics. Only then, they can be able to take the right decision in a complex situation. Risk evaluation is of two types: (a) high risk and less probability of success and (b) less risk and high probability of success. It is suggested that engineers should choose the second alternative for the benefit of the society at large and avoid engineering disasters. Van den Hoven [24] states that engineers should develop "value sensitive design" products. It means that engineers should include moral values and stakeholder values in an iterative process in the technologies they develop [7, 27]. As a result, they can enhance their ethical reflection on their designs and infer probable consequences of those. Hence, they can produce humane technologies and establish themselves as role model professionals for others.

7 What Sorts of Responsibility Should Engineers Have Towards the Society at Large?

Every professional has some responsibilities towards his/her profession and the society, so does an engineer. Broadly, responsibility is of two types: forward looking and backward looking [6, 22]. The prior one is characterized by sympathy, empathy, and compassion, whereas shame, guilt, blame, etc., are attributed to the latter. The differences between them are (a) the prior one comprehends the consequences of risks involved in technological designs, whereas the latter one repents for the unwanted consequences occurred in the past and learn the reasons for that and thereby is inclined to take precautionary measures for future technological designs, and (b) the prior one is a positive responsibility [5], but the latter one is a negative or failed responsibility [19]. Some commonalities are also found between these two responsibilities. These are (a) they associate with "emotion," which is a component of ethics; (b) they deal with social responsibility on the one hand and technological growth on the other hand; and (c) they suggest that professionalism does not mean to perform tasks only through rationality and quantitative skills, but also through ethics. Thus, Williams [25] says that "social responsibility" is a virtue for professionals. Roeser [22] conveys that a virtue-responsible person is aware of the different normative claims that rest on him/her and makes the right decision. Thus, these two responsibilities are suggested for engineers' consideration in their engineering practice.

8 Conclusion

The insertion of "engineering ethics" course in the engineering curriculum forms the central point of discussion for many scholars belonging to engineering and other disciplines. The discussion is in full swing because of its relevance to engineers belonging to the present technological era. To me, this course primarily aims at training engineers to look beyond demonstrable technical skills in their engineering practices and to assist them in acting ethically and rationally in their professional tasks. Acting ethically can help them resolve some of the ethical problems associated with technological designs in particular and engineering profession in general – problems like public safety and welfare, risk and informed consent, conflicts of interest, conflicts of commitment, and trade secrets. Thus, engineers are indeed in need of "engineering ethics" education to resolve ethical problems and understand their social and technical responsibilities. To develop ethical understanding, engineers therefore should go though the engineering ethics course. Hence, "engineering ethics" course should be integrated into the engineering curriculum and is a must for engineering students.

References

1. Baura GD. Engineering ethics: an industrial perspective. Boston: Elsevier Academic Press; 2006.
2. Bucciarelli LL. Ethics and engineering education. Eur J Eng Educ. 2008;33(2):141–9.
3. Calhoun D. Goals in the teaching of ethics. In: Callahan D, Bok S, editors. Ethics teaching in higher education. New York: Plenum Publication; 1980.
4. Colby A, Sullivan WM. Ethics teaching in undergraduate engineering education. J Eng Educ. 2008;97(3):327–38.
5. Eisenberg N. Emotion, regulation, and moral development. Ann Rev Psychol. 2000;51:665–97.
6. Fahlquist NJ. Moral responsibility for environmental problems – individual or institutional? J Agric Environ Ethic. 2008;22:109–24.
7. Friedman B. Value sensitive design. Encyclopaedia of human-computer interaction. Great Barrington: Berkshire Publishing Group; 2004. p. 769–74.
8. Harris CE, et al. Engineering ethics: what? why? how? and when? J Eng Educ. 1996;85(2):93–6.
9. Harris CE, Pritchard MS, Rabins MJ. Engineering ethics: concepts and cases. Belmont: Wardsworth Cengage Learning; 2009.
10. Haws DR. Engineering the just war: examination of an approach to teaching engineering ethics. Sci Eng Ethic. 2006;12(2):365–72.
11. Herkert J. Ways of thinking about and teaching ethical problem solving: microethics and macroethics in engineering. Sci Eng Ethic. 2005;11(3):373–85.
12. Herkert J, Wetmore J, Canary H, Ellison K. Integrating microethics and macroethics in graduate science and engineering education: developing instructional models. Retrieved August 12, 2013, from American Society for Engineering Education: http://www.asee.org. 2009.
13. Hollander R. Arenberg CR. Ethics education and scientific and engineering research: what's been learned? What should be done? Summary of a workshop. Retrieved July, 2013, from National Academy of Engineering: http://nap.edu/catalog.php?record_id=12695. 2009.
14. Huff CW, Frey W. Moral pedagogy and practical ethics. Sci Eng Ethic. 2005;11(3):389–408.
15. Jessica L, Shengli F. A systematic approach to engineering ethics education. Sci Eng Ethic. 2012;18:339–49.

16. Kroes P, Meijers A. Ethical issues in engineering design: safety and sustainability. Tilburg: Anke Van Group; 2005.
17. Lincourt J, Johnson R. Ethics training: a genuine dilemma for engineering educators. Sci Eng Ethic. 2004;10:353–8.
18. Martin MW, Schinzinger R. Ethics in engineering. 4th ed. New York: McGraw Hill; 2005.
19. McGraw KM. Guilt following transgression: an attribution of responsibility approach. J Personal Soc Psychol. 1987;53(2):247–56.
20. Rabins MJ, et al. Introducing ethics case studies into required undergraduate engineering course. Retrieved August 29, 2010, from Texas A&M University: http://ethics.tamu.edu/. 2009.
21. Roeser S. The role of emotions in judging the moral acceptability of risks. Saf Sci. 2006;44(8):689–700.
22. Roeser S. Emotional engineers: toward morally responsible design. Sci Eng Ethic. 2012;18:103–15.
23. Schinzinger R, Martin MW. Introduction to engineering ethics. Boston: McGraw-Hill Publication; 2000.
24. Van den Hoven J. ICT and value sensitive design. In: Goujon P, Lavelle S, Duquenoy P, Kimppa K, Laurent V, editors. The information society: innovation, legitimacy, ethics and democracy. Boston: Springer; 2007.
25. Williams G. Responsibility as a virtue. Ethic Theor Moral Pract. 2008;11(4):455–70.
26. Yutrzenka BA. Making a case for training in ethnic and cultural diversity in increasing treatment efficacy. J Consul Clin Psychol. 1995;63(2):197–206.
27. Zwart SD, et al. A network approach for distinguishing ethical issues in research and development. Sci Eng Ethic. 2006;12:663–84.

Tracking Lab Activity in Technical Education System: A Case Study at the Guru Nanak Institutions (India)

Sandeep Singh Rawat, H.S. Saini, S. Sreenatha Reddy, and R. Ranjan Dandpat

Abstract Nowadays, it can be observed that computer laboratories have encountered many challenges which prevent them to achieve their quality objectives. The management of computer laboratories is a time-consuming manual process which is often a tedious job of maintaining the records in books. In a university, the systems are maintained by the lab assistant/in-charge. To know about who is using the system, they maintain a lab record in which the user details are stored. But this can easily be modified and is not very reliable. In case of any theft, damage, and breakage of components, it is difficult to trace who the actual culprit is. This work helps the lab assistant/in-charge to trace the person who has committed theft, damage, and breakage to components by identifying the system ID.

We study the application of log tracking to educational data collected from the Department of Information Technology at Guru Nanak Institutions, Hyderabad, India. The results are automated and accurate and thoroughly help the lab assistant/in-charge in further enhancing the efficiency and effectiveness of the traditional processes. It can be used to analyze the existing work and identify existing gaps and further works. As our further work, we use our model as a process model to develop an appropriate system for any other technical institute; if found useful, this model can be adapted, so that the university will reach a higher academic standard.

Keywords Automated • Decision • Log • Technical education system

---

S.S. Rawat (✉) • H.S. Saini • S.S. Reddy • R.R. Dandpat
Guru Nanak Institutions, Hyderabad, India
e-mail: hodit.gnit@gniindia.org; md@gniindia.org;
principal.gnit@gniindia.org; ranjandandpat@gmail.com

R. Natarajan (ed.), *Proceedings of the International Conference on Transformations in Engineering Education: ICTIEE 2014*, DOI 10.1007/978-81-322-1931-6_29,

1 Introduction

Today, even if we have computers, CDs, and other forms to store data, we still rely on pen and paper. So there is a need to develop efficient system that can handle huge amount of data. In a context of labs in various institutions, the data of students are still maintained in books. This is not a safe way of maintaining records, and moreover, it is easily changeable. The integrity of data can be destroyed. So in case of any damage to the system, there is no way to find who the actual culprit is. Even if we maintain a logbook to make a note of who is sitting on which system, it certainly does not provide the accurate information because students may change the system and never update the log record. So there is a need to find an alternative to this problem. This is where the lab tracking system can be used.

With the rapid development of laboratory in a university, the traditional management system of equipments in teaching and experiment faced lots of problems with management by manual work. To solve these existing problems and analyze the actual demand, this study integrates RFID, Java, and MySQL database technologies to realize a new laboratory equipment management system (LEMS) in both teaching and experiment, which positioning and monitoring real-time tracking of each device, identities teacher identification automatic. This management system improves the management of equipment's efficiency, reaches each management goals, and lowers the cost in deployment with the Internet of Things technology [1].

The rest of the paper is organized as follows. Section 2 discusses the related work. Section 3 formulates the proposed systems. Section 4 explains the experimental results and presents simulations that demonstrate the tracing of students and give some intuitions about the observed results. We conclude in section 5 with mention of the future likely enhancements of the system.

2 Literature Survey

Up until the late 1970s, the management of laboratory samples and the associated analysis and reporting were time-consuming manual processes often riddled with transcription errors. In 1982, the first generation of LIMS (Laboratory Information Management Systems) [2] was introduced in the form of a single centralized minicomputer, which offered laboratories the first opportunity to utilize automated reporting tools. By 1988, the second-generation commercial offerings were tapping into relational databases to expand LIMS into more application-specific territory, and International LIMS Conferences were in full swing. As personal computers became more powerful and prominent, a third generation of LIMS emerged in the early 1990s. These new LIMS took advantage of the developing client/server architecture, allowing laboratories to implement better data processing and exchanges.

By 1995, the client/server tools had developed to the point of allowing processing of data anywhere on the network. Web-enabled LIMS were introduced the following year, enabling researchers to extend operations outside the confines of the laboratory. From 1996 to 2002 additional functionality was included in LIMS, from wireless networking capabilities and georeferencing of samples to the adoption of XML standards and the development of Internet purchasing.

According to Tang Youchun [3], computer lab management is part of the Information Technology Department at UC Merced and is responsible for overseeing instructional use and open-access computer labs. Access to computer labs is granted to currently enrolled students, active staff, and faculty. To log on to a computer, you will need to have an active user ID and password.

According to Liu Yajing [4], currently, the problems which the computer laboratories in higher learning institutions are confronted with are as follows: lacking consciousness for service and standardized management, lacking means for executing and maintaining, no standardized process, no record for software and hardware configuration and change as well, and also changing configuration randomly. All these lie in that management concept of these laboratories stay in elementary mode of old mechanism, with the executing mechanism lacking the consciousness of service and practice. The system combines with the current management situation in higher learning institution and makes some progress in the executing and maintaining management in a computer center.

3 Proposed Approach

This work provides a better way to maintain and store log records. The lab assistant/in-charge does not have to do much of the work, thus reducing the burden on the administrator. Once the students enter the laboratory and switch on their system, the first thing they need to do is to give their login credentials. These log details are securely stored on the server and are not accessible to students, thus preventing the manipulation of data by students. So even if the student leaves the current system and tries to access the other system, he will not be allowed. So in order to shift from one system to another, the person has to first log out of the current system. The log record is automatically updated when he logs into another system. This saves time of making changes to the book which is generally maintained. The admin can anytime check the number of students logged in currently and also between some days.

Figure 1 shows the different classes used to represent the system. Basically, there are only five classes. The classes are associated with each other, and the student and admin class are generalized classes of users. A lab may have one or more PCs, and each PC in the lab is connected to one server. One user may use one or more PCs.

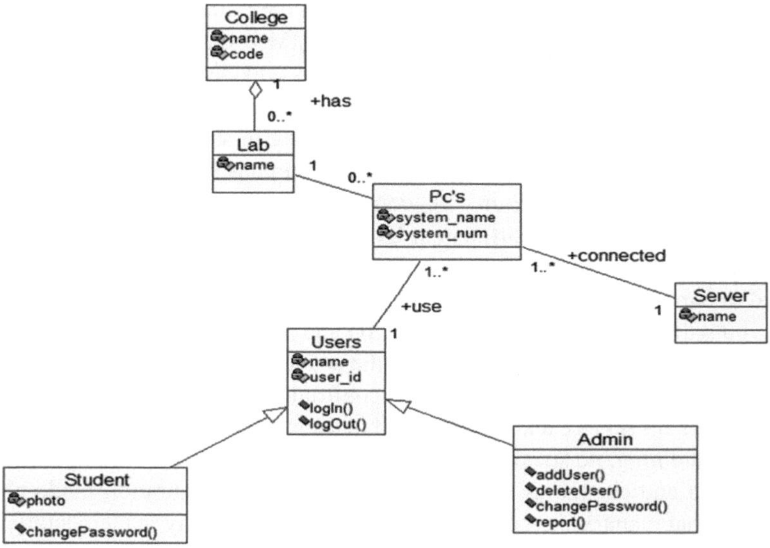

Fig. 1 Class diagram of the proposed system

4 Experimental Results

We have developed simulated examples to trace the person who has committed theft, damage, and breakage to components in distributed environment. In our experimental testing, we have used Windows 8 with SQL Server and 64 homogeneous machines at the Department of Information Technology, Guru Nanak Institute of Technology.

Figure 2a shows the currently logged in users. This is visible only to the admin. So only when the admin logs into the system with his credentials does he then have the permissions to see the list of currently logged in users. It contains details about the student ID, their names, branch, system number, login time, and date of use. Figure 2b shows the report that is generated date wise. This report is again generated or available to the admin. The admin is also given with an option to print the generated report. This would help him to analyze the report more closely. All the reports can now be maintained easily.

5 Conclusion and Future Scope

This paper has been an effort in providing the result toward advancing the technical educational system. The main idea is organized into a model to represent how technology can be used in educational system to improve the efficiency and effectiveness of the traditional processes. The model is presented as a guideline for technical

a

stu_id	uni_id	stu_name	branch	system_no	login_time	logout_time	date
10831A1204	57037	Harish	CSE	RANJAN	18:38:35	NULL	2013-06-21
10831A1204	15055	Harish	CSE	RANJAN	18:39:51	NULL	2013-06-21
10831A1204	119681	Harish	CSE	RANJAN	20:05:56	NULL	2013-06-24
10831A1204	424006	Harish	CSE	RANJAN	20:17:36	NULL	2013-06-24
10831A1204	325984	Harish	CSE	RANJAN	20:26:45	20:26:52	2013-06-24
10831A1201	384359	laskd	lksal	RANJAN	21:21:02	NULL	2013-07-02
10831A1201	332404	laskd	lksal	RANJAN	21:25:32	NULL	2013-07-02
10831A1201	163964	laskd	lksal	RANJAN	21:33:16	NULL	2013-07-02
10831A1201	478814	laskd	lksal	RANJAN	22:22:37	NULL	2013-07-03
10831A1201	122616	laskd	lksal	RANJAN	22:23:31	NULL	2013-07-03
10831A1201	441241	laskd	lksal	RANJAN	22:27:50	NULL	2013-07-03
10831A1201	359824	laskd	lksal	RANJAN	22:33:47	NULL	2013-07-03
10831A1201	468136	laskd	lksal	RANJAN	22:42:15	NULL	2013-07-03
10831A1201	349372	laskd	lksal	RANJAN	14:57:37	14:57:57	2013-07-06
10831A1201	131428	laskd	lksal	RANJAN	16:02:48	16:03:08	2013-07-06
10831A1201	396789	laskd	lksal	RANJAN	16:04:56	16:05:17	2013-07-06
10831A1201	153436	laskd	lksal	RANJAN	14:23:37	NULL	2013-07-07
10831A1201	381786	laskd	lksal	RANJAN	15:02:46	NULL	2013-07-07
10831A1201	367003	laskd	lksal	RANJAN	19:16:04	NULL	2013-07-07
10831A1201	342885	laskd	lksal	RANJAN	12:23:15	12:23:43	2013-07-18

1 of 200

b

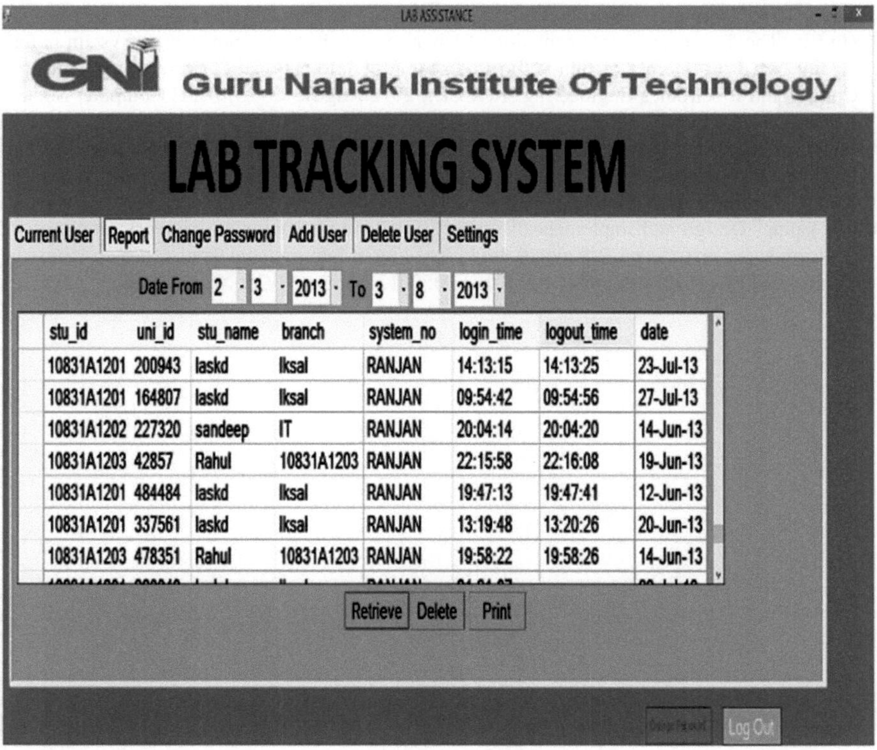

Fig. 2 Report generation. (**a**) Currently logged in users and (**b**) report generated date wise

educational system to improve their decision-making processes. It can also act as a guideline or roadmap for lab assistant/instructor to identify which system is not working or has been stolen/damaged/broken as part of their educational processes and how they can improve their traditional processes by getting advantages of it.

Many directions for future enhancements are open. Among them, we can mention:

- The testing of the proposed work on a real distributed platform
- The use of other techniques like data mining and artificial intelligence
- The use of our model as a process model to develop an appropriate system for any other technical institute
- The use of the grid computing paradigm to solve even more challenging data mining problems

References

1. Tang Youchun. Design of management system for computer laboratory based on ITIL. In: Proceeding of 4th International Conference on Computer Science & Education ICCSE; 2009 July 25–28; Nanning, China.
2. Hu Bin. The design and implementation of laboratory equipments management system in university based on internet of things. In: Proceeding of International Conference on Industrial Control and Electronics Engineering (ICICEE); 2012 August 23–25; IEEE Computer Society Washington, DC, USA.
3. Piho G. Towards LIMS (Laboratory Information Management Systems) software in global context". In: Proceedings of the 35th International Convention, MIPRO; 2012; IEEE Computer Society Washington, DC, USA.
4. Liu Yajing. Design and implement of university laboratory equipment network management platform based on C/S and B/S model. In: Proceeding of International Conference on Industrial Control and Electronics Engineering (ICICEE); 2012 August 23–25; Xian, China.

Curriculum Redesign in Higher Education Using QFD: A Case Study

Prabhushankar G.V., B. Latha Shankar, and T.R. Veena

Abstract In the twenty-first century, higher education institutions are under pressure to attract meritorious students and develop them as graduates with superior skills, pertinent capabilities, and right attitude. Achieving this requires adapting to flexible curricula – relevant to stake holders, creative delivery, and outcome-based learning. In this research, an effort has been made to redesign the curriculum of Industrial Engineering and Management (IEM) program at Siddaganga Institute of Technology (SIT), an autonomous institute, in Karnataka, India, using quality function deployment (QFD). The methodology of research involved design of questionnaire for capturing the voice of customer (VOC) in the context of changing curriculum requirements that are synergistic with real time. Then data was collected using the questionnaire, from 148 stakeholders who included faculty, alumni, employers, and students. Further QFD was used to transform customer needs (VOC) into design requirements of curriculum. While analyzing the data, analytical hierarchical process (AHP) was employed as a prioritizing method. The results of the study indicated that to improve the IEM education in general, focus should be given on specific courses in the curriculum that are highly desirable to meet present-day challenges in the industries. Also the curriculum should adopt teaching–learning practices that produce competent engineers and managers with a flair for lifelong learning.

Keywords Higher education • Curriculum design • AHP • House of quality • QFD • Voice of the customer

P. G.V. (✉) • B.L. Shankar • T.R. Veena
Department of IEM, Siddaganga Institute of Technology,
B.H. Road, Tumkur 572103, Karnataka, India
e-mail: prabhu.gvp@gmail.com; blathashankar@gmail.com; trveena@yahoo.com

R. Natarajan (ed.), *Proceedings of the International Conference on Transformations in Engineering Education: ICTIEE 2014*, DOI 10.1007/978-81-322-1931-6_30,

1 Introduction

Traditional on - campus higher education assumes that all students in a course are alike in their learning needs, interests, and abilities and allows individual faculty members to take the whole responsibility of course content development and delivery. But globalization and changing needs of industries in the twenty-first century treat alumni, employers, students, and parents as stake holders along with faculty and call for flexible course designs and creative teaching–learning practices for effective learning for all. According to Boonyanuwat, Suthummanon et al., an effective curriculum is one of the most important factors in order to have an attractive and desirable program [1]. Further proposals for curriculum enhancements are scrutinized to ensure that they are well suited to deliver the course content needed by stakeholders. Quality function deployment (QFD), a planning and design tool traditionally employed to facilitate integrated product development, can be modified to provide a flexible, integrated planning framework for curriculum planning [2]. In this direction, an effort has been made to redesign the curriculum of Industrial Engineering and Management (IEM) program at Siddaganga Institute of Technology (SIT), an autonomous institute, in Karnataka, India, using quality function deployment (QFD) tool.

2 The Tools Used

2.1 House of Quality

House of quality (HOQ) is the central component in constructing QFD. HOQ is one of the QFD products, which correlates customer requirements (what's) to a large variety of means (how's) by which customer desires can be satisfied. It is utilized by a multidisciplinary team to translate a set of customer requirements, drawing upon market research and benchmarking data, into an appropriate number of prioritized engineering targets to be met by a new product design. The HOQ matrix arranges important data in such a way that establishes criteria for a successful customer satisfaction. Typical HOQ is shown in Fig. 1.

The general format of the "House of Quality" is made up of six major components which are completed in the course of a QFD project. The six major components are:

- Customer needs (what's): a list of stakeholder's requirements.
- Service elements (how's): a list of ways of achieving what's, which is also called as technical requirements.
- Importance ratings: the relative importance scores of each need which is derived using the AHP method.

Fig. 1 House of quality

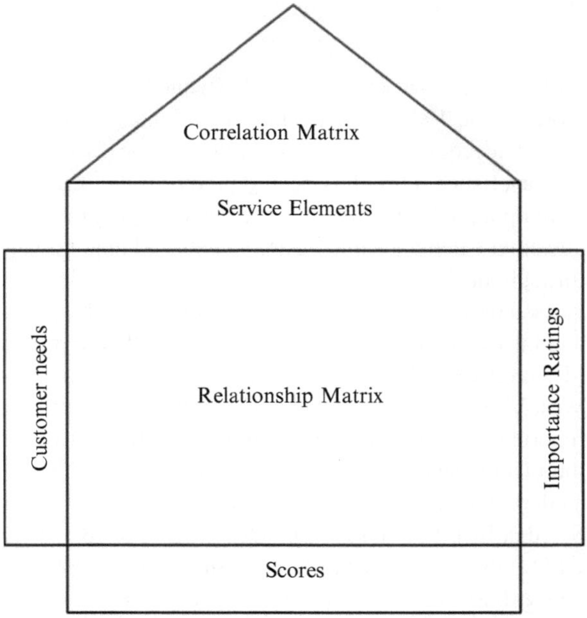

- Relationship matrix: a matrix that shows the relationship level between what's and how's. This portion is also called as technical portion.
- Correlation matrix: a matrix that shows the relationship between the design requirements (how's).
- Scores: a series of computed numbers for indicating the importance of each how, also called as absolute weight.

2.2 *Analytical Hierarchy Process*

The analytic hierarchy process (AHP) was developed by Thomas Saaty (1980) and is often referred as the Saaty method. It is one of the ways to structure decision problem which is used to prioritize alternatives, building an additive value function. AHP attempts to mirror human decision process. It uses pair-wise comparison matrix method to assess the relative value of the stakeholder requirements. The pair-wise comparison method typically uses a 1–9 point scale which is called as Saaty's rating scale. As the comparisons are done through personal judgments, some degree of inconsistency may arise in the matrix. Consistency ratio (CR) is being computed to guarantee the judgment of the matrix. If the CR is less than 0.1 (the limit), then the pair-wise matrix constructed is said to be trustworthy, else the decision makers should review and revise the comparisons.

3 Literature Review

QFD originated in Japan in the 1970s and became increasingly popular in the Western world in the 1980s. It helps multifunctional teams in identifying and prioritizing the customer requirements and relating these needs to corresponding product or service characteristics. QFD has been defined by its originator Akao [3] as "a method for developing a design quality aimed at satisfying the customer and then translating the customer's demands into design targets and major quality assurance points to be used throughout the production phase." The application of QFD in higher education is classed into three broad categories, namely, teaching effectiveness, curriculum design, and instructional resources. Clayton used QFD to provide productive quality learning [4]. Jaraiedi and Ritz applied QFD to improve advising and teaching processes at West Virginia University [5]. QFD application in research and strategic planning was reported by 6. The design requirements which satisfy each customer by considering the faculty, students, and industry as clients were shown by 7. The improvement of student satisfaction and learning was studied at the Lulea University of Technology in Sweden in 1999, and for this QFD was used [8]. Hwarng and Cynthia in 9 applied QFD at the National University of Singapore. In 2003, QFD was applied to design and deliver basic statistics at the King Fahd University of Petroleum and Minerals in Saudi Arabia [10]. Further Shaney, Banwet, and Kaurnes used QFD to improve the quality of education and student satisfaction at the Indian educational institutions [11]. Denton, Kleist, and Surendra in 2005 proposed QFD to aid the design of curriculum and course in the academic domain of management information system [12]. Later Thakkar, Deshmukh, and Shastree successfully used QFD model with four HOQs to improve the quality of education in the self-financed technical institutions [13]. Sohn and Kim applied QFD for engineering curriculum redesign at Korea [14].

4 Methodology

The objective of this study is to redesign the curriculum for IEM at SIT, based on the importance of stakeholder's requirements by the application of QFD.

The methodology is based on the adaptation of QFD. The proposed methodology consists of the following steps:

Step 1: Identification of Stakeholders

The stakeholders were identified and classified into employers, alumni, faculties, and students. The employers, alumni, and few faculties of other colleges act as external stakeholders, while the faculties and students act as internal evaluators.

Step 2: Questionnaire Design

The main aim of the questionnaire design is to capture the voice of customer (VOC) in redesigning the curriculum. The questionnaires were designed and modified on expert's suggestion.

The objective of this study as well as the questionnaire was included in the cover page of the questionnaire to guide the stakeholders in providing the feedback.

The questionnaire set had 4 multiple choice questions and 4 open-end questions. Multiple choice questions designed for the study were of the form:

1. *Which type of subjects would you suggest as open electives for IEM?*
   - Software related
   - Hardware related
   - Management related
   - Production related

Open-end questions were designed, so that the customer's true voice can be captured, and were of the form:

2. *Mention any course content relevant to the present industrial needs/requirements.*

Before circulating the questionnaire to the study group, it was first given to a selected small group of 6 stakeholders to find flaws that can be corrected and any questions that are superfluous. The criteria for the selection were based on their experiences and voluntaries. Finally the questionnaire was modified for each customer group respectively.

Step 3: Sample Size.

The study group consisted of employers, alumni (passed out students), faculties, and students. 99 students, 22 faculties, and 27 alumni and employers were selected to participate in the questionnaire survey. Of 148 stakeholders, 57 responded (response rate = 38.51 %). Among the respondents, 36 (36.36 %) were students, 7 (25.92 %) were alumni and employers, and 14 (63.63 %) were faculties.

Step 4: Collecting the VOC.

The VOC was collected through survey, direct interview, and mails. To arrive with a list of requirements, in-depth discussion was made with 13 stakeholders in order to find any redundant requirements. Also we reviewed and reanalyzed the stakeholder's requirements, after meeting the customers, to ensure that they were correct, complete, and clear.

Step 5: Analyzing the Customer's Feedback.

After collecting the VOC, an affinity diagram was drawn to analyze the information obtained. Affinity diagrams are used to surface the "deep structure" in voiced customer requirements. The affinity diagram, also known as the KJ method, is used to unlock creative thinking. This method looks for a different way of describing some relationship. The said diagram is shown in Fig. 2.

Step 6: Ranking the Stakeholder's Requirements.

We decided to use analytical hierarchical process (AHP) as a prioritizing method. 11 students and a faculty were involved in this prioritization process. For which a discussion was made with the stakeholders to give instructions about the process.

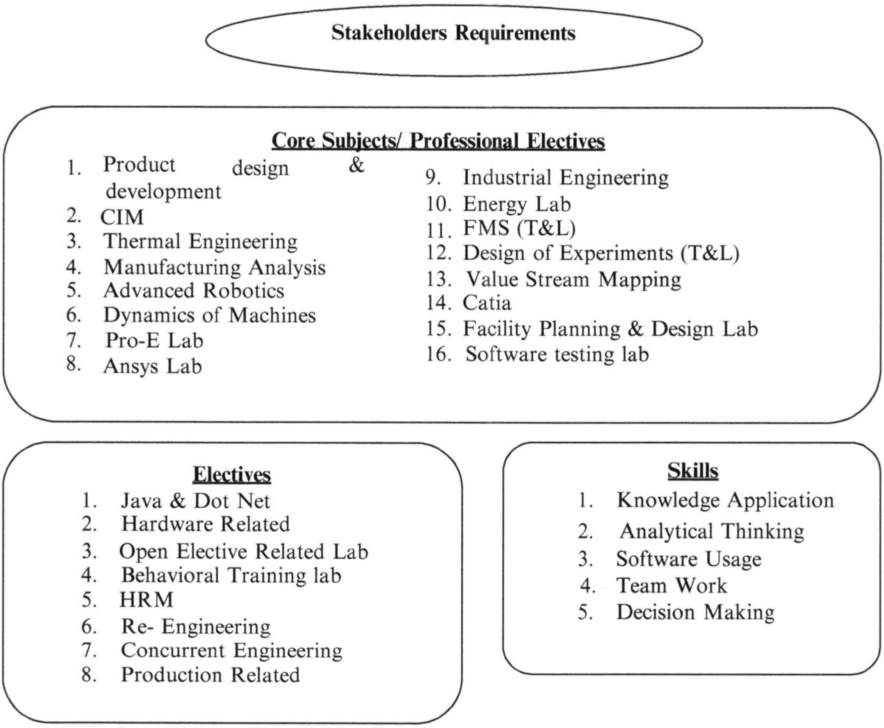

Fig. 2 Affinity diagram

The stakeholders were involved in generating the pair-wise comparison matrix which was normalized further (a) to express each requirement's importance weights and (b) to find the internal consistency of the matrix. According to the results provided by the questionnaires, each representative's importance rating is expressed in an AHP matrix. Subsequently, the AHP matrices are aggregated within each stakeholders group by calculating the geometric mean of the stakeholder's importance weights. In order to complete this task, the stakeholder importance weights were multiplied by stakeholder requirement values, which result in the stakeholder requirement importance weight.

Step 7: Building the HOQ.

A HOQ matrix is constructed to analyze the relationship between stakeholder's requirements (what's) and service elements (how's); the matrix is shown in Fig. 3. The relationship values between stakeholder requirements and service elements are entered on expert's suggestion. The relationship value differs from the type of relationship existing between what's and how's, i.e., 9, strong relationship; 3, moderate relationship; and 1, possible relationship. Correlations between each pair of service elements are indicated at the roof of the HOQ. These indications show that the service elements affect the performance of each others. The importance values for each curriculum design feature are found by multiplying the relationship values with the

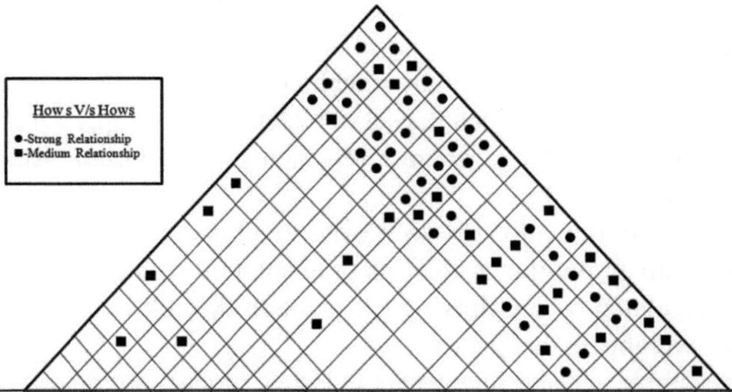

What s V/s How s

9- Strong relationship
3- Moderate Relationship
1- Possible relationship

	Stakeholder's Requirements (What s)	Design Requirements (How s)	Product design & development (T & P)	CIM (T& P)	Thermal Engineering	Manufacturing Analysis	Advanced Robotics	Dynamics of Machines	Industrial Engineering	Rapid Prototyping	Java &Dot Net/ Software testing (T&P)	Computer Hardware	Concurrent engineering / Business Process Re-engineering / Risk Management	Design of Experiments (T&P)	Facility Planning & Design(T&P)	Industrial Psychology(T&P)	Principles of Lean Manufacturing	Project Work	Industrial Training	Industrial Visits	Case Study Approach	Conducting Knowledge activities	Customer Requirements Importance Weight	Importance Ranking
			Course Contents															Enablement						
Core Subjects/ Professional Electives	Product design & development		9	3		3	3	3	3	9		3	1	3			1	3	1	1	1	1	0.01928	5
	CIM		3	9		3	3	3	1	3		1	1	1			3	3	1	1		1	0.03077	1
	Thermal Engineering				9													1	1	1	1	1	0.01478	13
	Manufacturing Analysis		3	3		9	1		1	3		1	1	1	1	1		3	3		3	1	0.02192	3
	Advanced Robotics		3	3		1	9	1	1	3		1	1	1			3	3	1	1		1	0.00138	28
	Dynamics of Machines		3	3			1	9	1	3		1		3			1	3	1	1		1	0.00178	27
	Pro-E Lab		9	3			3		3	9			1					3	1	1	1	1	0.00201	25
	Ansys Lab		9	3			3		3	9			1					3	1	1	1	1	0.00197	26
	Industrial Engineering		3	1		1	1	1	9	1		1	1	3	3	3	1	9	3	3	9	3	0.01483	12
	Energy Lab				9													1	1	1	1	1	0.00084	29
	FMS (T&P)		3	9			1		1	3			1	1			3	1	3	1		1	0.02153	4
	Design of Experiments (T&P)		3	1		1	1	3	3	1				9			1	9	1		3	3	0.02233	2
	Value Stream Mapping		1	3					1					1			9	3	3	1	3	3	0.00594	22
	Catia		9	3					3	9				3			1	3	1	1	1	1	0.00307	24
	Facility Planning & Design Lab					1			3						9			3	3	1	3		0.0079	18
	Software testing lab										9							3	3		3	3	0.00777	19
Electives	Java & Dot Net										9	3						3	3		3	3	0.01864	6
	Hardware Related		3	1		1			1		3	9						3	9		3	3	0.00563	23
	Open Elective Related Lab										9	3				9		3	3		3	3	0.00759	20
	Behavioral Training lab					1										9		1	3		1	3	0.0167	7
	HRM					1							3			9		1	3		3	3	0.0167	8
	Re- Engineering		1	1		1			1				9			3		3	3	1	9	3	0.01233	16
	Concurrent Engineering		1	1		1			1				9			3		3	3	1	9	3	0.01233	16
	Production Related		9	3		3	3	3	1	9				1	1		1	9	9	1	3	9	0.01648	9
Skills	Knowledge Application		3	3	1	3	3	3	3	3	3	3	3	3	9	3	3	9	3	3	9	9	0.01545	10
	Analytical Thinking		3	3	1	3	3	3	3	3	3	3	9	3	9	3	9	9	3	1	3	1	0.01515	11
	Software Usage		3	3	1	1	3	1	1	3	9	3	1	3		1	1	3	3	1	1	3	0.01306	15
	Team Work																	9	3	1	3	3	0.01449	14
	Decision Making		3	3	3	9	3	3	3	1	1	1	3	3	3	3	9	9	1		1	3	0.0074	21
	Relative Importance weights of Design Requirements		0.929	0.91	0.21	0.68	0.46	0.43	0.56	0.79	0.5396	0.4	0.603664983	0.569	0.4516	0.6364	0.5546	1.546	0.97	0.286	1.06	0.949		
	Importance Rankings		1	2	15	4	11	13	8	3	10	14	6	7	12	5	9	1	3	5	2	4		

Fig. 3 House of quality matrix

corresponding stakeholder requirement importance weights and adding all these values column wise. Then, these proposed changes should be implemented to fulfill the stakeholders' needs. The steps in building the HOQ are as follows:

1. The stakeholder's requirements, identified through the survey and discussion with several stakeholders, are listed on the left side of the house.
2. The stakeholder requirements importance weights, obtained from AHP, and their ranks are listed on the right side of HOQ.

3. The necessary service elements (how's) are listed horizontally below the roof. The service elements are grouped into course contents and enablement, based on the requirements listed by the stakeholders.
4. The correlations between the service elements are represented using symbols in the roof of the HOQ, where • represents strong relationship and • represents medium relationship.
5. The relationship value between what's and how's is entered.
6. The relative importance weights or absolute weights of the design requirements are calculated by multiplying the value in the cell by the corresponding importance weights in the last row.
7. For example, the absolute weight in Fig. 3, product design and development (T&P) =(9*0.01928)+(3*0.03077)+(3*0.02192)+(3*0.00138)+(3*0.00178)+(9*0.0020 1)+(9*0.00197)+(3*0.01483)+(3*0.02153)+(3*0.02233)+(1*0.00594)+(9*0.00 307)+(3*0.00563)+(1*0.01233)+(1*0.01233)+(9*0.01648)+(3*0.01545)+(3*0. 01515)+(3*0.01306)+(3*0.0074)=0.929.
8. Absolute weights are listed horizontally at the bottom portion of HOQ.
9. The absolute values are ranked in the descending order. Larger absolute weight indicates fulfillment of most of the requirements of the stakeholders.

5 Results and Discussions

In all, 57 stakeholders (of 148, 38.51 % response rate) participated in responding to the questionnaire. The questionnaire consisted of 5 course contents, 4 open electives, and 3 practical subjects and a skill set (consists of 5 skills), to be opted from the list. The use of the AHP process enabled a better handling in the prioritization of the stakeholder's requirements. This resulted in the generation of weights as a degree of merit for the requirements of the stakeholders. The so-generated weights were further used in developing the HOQ.

The most important requirements of the stakeholders are the subjects like CIM, DOE (theory and practical), manufacturing analysis, flexible manufacturing system (FMS), product design and development, java and .NET, behavioral training lab, human resource management, production-related elective, industrial engineering, thermal engineering, reengineering, concurrent engineering, facility planning and design lab, and software testing lab. From the point of necessary learning enablement, all the skills listed in the questionnaire are found to be important from the stakeholder's perspective.

This work resulted in classification of service elements into two categories, viz, course contents and enablement. Product design and development (theory and practical), CIM (theory and practical), rapid prototyping, manufacturing analysis, industrial psychology (theory and practical), concurrent engineering, business process reengineering, risk management, DOE (theory and practical), industrial engineering, principles of lean manufacturing, and java and .NET/software testing (theory and practical) are ranked as the ten most important course contents, for fulfilling the stakeholder's requirements.

6 Conclusions

- QFD is a simple, very effective, and efficient tool, yet powerful of discovering key characteristics of a successful curriculum. Using the QFD principles helps the faculty to design the curriculum with ease. Augmenting QFD in early stages of planning a curriculum will maximize the learning process; this leads to increased knowledge and experience for both the students and faculty members. This study considers the IEM curriculum for its quality enhancement.
- The results clearly suggest that in order to improve the IEM education, a thorough scrutiny of the curriculum pertaining to the following course contents is required: product design and development (theory and practical), CIM (theory and practical), rapid prototyping, manufacturing analysis, industrial psychology (theory and practical), concurrent engineering, business process reengineering, risk management, DOE (theory and practical), industrial engineering, principles of lean manufacturing, and java and .NET/software testing (theory and practical). From a practical view point, industrial training, industrial visits, conducting knowledge activities, and case study approaches are to be given appropriate consideration.
- Although QFD is more effective than other techniques in determining customer needs and translating them to the service element, the main problem of using QFD is in managing and analyzing large relationship matrices. This is owing to a large number of requirements and service elements. To bring down the dimension of relationship matrices, it would be appropriate to make a project into different modules, each generating its own sub-matrix.

References

1. Boonyanuwat N, Suthummanon S, Memongkol N, Chaiprapat S. Application of quality function deployment for designing and developing a curriculum for Industrial Engineering at Prince of Songkla University. Songklanakarin J Sci Technol. 2008;30:349–53.
2. Bossert JL. Quality function deployment: a practitioner's approach. Milwaukee: ASQC Quality Press; 1991.
3. Akao Y. QFD: integrating customer requirements into product design. Cambridge: Productivity Press; 1992.
4. Clayton M. Treading the quality path: a progress report from Aston University. Springer: Total Quality Management; 1995.
5. Jaraiedi M, Ritz D. Total quality management applied to engineering education. Qual Assur Educ. 1994;2:32–40.
6. Chen CL, Bullington SF. Development of a strategic plan for an academic department through the use of quality function deployment. Comput Ind Eng. 1993;25:49–52.
7. Ermer DS. Using QFD becomes an educational experience for students and faculty. Qual Prog. 1995;28:131–6.
8. Wiklund PS, Wilkund H. Student focused design and improvement of university courses. Manag Serv Qual. 1999;9:434–43.
9. Hwarng HB, Cynthia T. Translating customers' voices into operations requirements A QFD application in higher education. Int J Qual Reliab Manag. 2001;18:195–225.

10. Duffuaa SO, Al-Turki UM, Hawsawi FM. Quality function deployment for designing a basic statistics course. Int J Qual Reliab Manag. 2003;20:740–50.
11. Sahney S, Banwet DK, Kaurnes S. Enhancing quality in education: application of quality function deployment – an industry perspective. Work Study. 2003;52:297–309.
12. Denton JW, Kleist VF, Surendra N. Curriculum and course design: a new approach using quality function deployment. J Educ Bus. 2005;81:111–17.
13. Thakkar J, Deshmukh SG, Shastree A. Total quality management (TQM) in self-financed technical institutions: a quality function deployment (QFD) and force field analysis approach. Qual Assur Educ. 2006;14:54–74.
14. Sohn SY, Kim A. Quality function deployment for engineering curriculum redesign. In: Proceedings of the 20th annual conference for the Australasian association for engineering education. Barton: Engineers Australia; 2009. p. 343–9.

Evolution of Engineering Education in India

B.S. Nagabhushana, Rajeshwari Hegde, and P.S. Gowra

Abstract Engineering education in the Indian context has evolved over a long period of time having witnessed varied revolutions across the globe and having assumed strategic decisions at varied levels. Engineering education in India is, steadily, marching towards higher levels of maturity with sincere attempts to benchmark against renowned world standards. In order to, continuously, discover the scope for improvements, it is essential to observe the evolutionary trace on the timeline. This paper is a contribution in this direction.

Keywords CMM • P-CMM • NBA • Technical education

1 Introduction

India being a land of rich heritage has witnessed a series of revolutions in varied sectors with different levels of absorption, assimilation and sustenance. The first in the series is, probably, the 2nd Industrial Revolution during the period 1850–1890 amidst India under the British colonial rule [1]. Although visionaries from the Indian soil advocated the need of industrialization, owing to the intrinsic strength and expertise in agriculture, emphasis was laid mostly on the agricultural front for the sustenance of the country. Nevertheless, India, as a country, kept a watch on the industrial growth and solely owing to the financial constraints took the strategy of liberalization to achieve industrial growth. Implementation of this strategy led to an undocumented

B.S. Nagabhushana (✉)
Department of Electronics and Communication Engineering, BMS College of Engineering, Bull Temple Road, Bangalore, India
e-mail: nagabhushanabs.ece@bmsce.ac.in

R. Hegde • P.S. Gowra
Department of Telecommunication Engineering, BMS College of Engineering, Bull Temple Road, Bangalore, India

R. Natarajan (ed.), *Proceedings of the International Conference on Transformations in Engineering Education: ICTIEE 2014*, DOI 10.1007/978-81-322-1931-6_31,

Industrial Revolution in India. However, during the initial part of this revolution, due to the liberalization policy being not in full form, India faced setbacks, a major of which is its failure to set up semiconductor plants despite all efforts.

As time passed by, owing to the easy adaptation of the Indian community to the knowledge-based industry like information technology (IT) and software (SW) development, the liberalization policy of the Indian government getting into a full farm and affordable infrastructure requirement of the industry itself, India could establish well in the IT sector and create a global icon in this space [2]. As such, through in-house software development infrastructure and also through outsourcing business model, India started provisioning the IT intellectual capital taking its share of the global market. To sustain the global demand, India had to take widespread technical education which it could do very successfully during the late 1980s and early 1990s. However, the quality of technical education just met the skill set required for rendering nominal IT and SW services. With the growing work experience of the IT and SW sector coupled with the increasing demand for skilled manpower, a broadened and deepened skill set is warranted even from a fresh engineer just out of college. This situation awakened the technical education sector to review the quality of technical education provided and discover the prevalent lacuna. Although, initially, the industry might have remarked at the education sector for the quality of engineers produced, eventually, the industry approached the problem with a collaborative approach by having co-ordinated interaction with the education sector in terms of workshops, symposia, finishing schools, etc. [3]. The education sector also reciprocated to this approach in the affirmative and within itself started exploring the adaptation of process orientation which has, by then, been proved as an effective mechanism in the industry to build quality the first time.

On the other hand, corporates taking the leadership position in the IT and allied industry sector have discerned the benefit of process orientation in gracefully handling their intellectual capital. Eventually, many corporates have achieved remarkable success by adapting process orientation. Having witnessed this scenario and also owing to the commonality of the need to handle knowledge assets and intellectual capital, the education sector also got into adapting process orientation. This paper is focussed in this direction and is organized as follows: Section 2 presents few successful process models from the industry sector, whereas Sect. 3 presents the NBA process model which is, by far, the de facto standard for the education sector. Section 4 discusses the possible extensions to the NBA process model, and conclusions for the present work are drawn in Sect. 5.

2 Models

While the ISO process model ruled all the manufacturing industries, the Capability Maturity Model (CMM) and its variants, SPICE and ISO26262 process models, provided the IT and related SW service sectors with the process framework. Process models employed by IT and SW services assume relevance owing to their

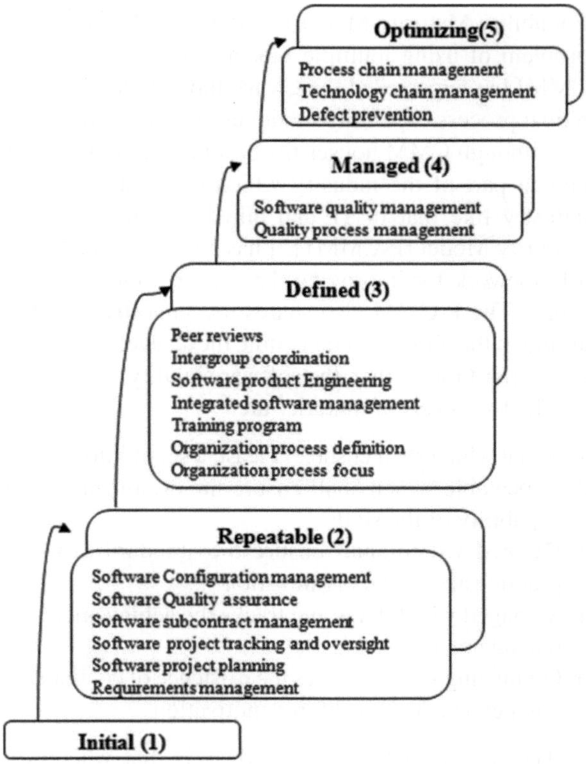

Fig. 1 CMM in software development process [4]

commonality with the education sector. Also, predominantly, the CMM and its variants were grown in an educational institution in the hands of able educationalists.

The CMM, a process model nurtured, extensively, by the Software Engineering Institute at Carnegie Mellon University, is employed by the software industry to build software in a process-oriented manner. A diagrammatic representation of the CMM is depicted in Fig. 1. The model houses five levels of maturity, listed below:

1. Initial
2. Repeatable
3. Defined
4. Managed
5. Optimizing

Each of these levels is associated with a fixed number of key process areas (KPAs) as indicated in Fig. 1. There are 18 KPAs which are distributed amongst the five levels. Obviously, stringent KPAs appear at higher levels of maturity. Each of these KPAs is associated with specific number of key practices, activities, measurements and verification mechanisms.

Applying multiple CMMs that are not integrated within and across an organization could be costly in training, appraisals and improvement activities. As such, the

Capability Maturity Model Integration (CMMI) project was formed to sort out the problem of using multiple models for software development processes; thus, the CMMI has superseded the CMM, though the CMM continues to be a general theoretical process capability model used in the public domain.

Although CMM houses the best practices of the industry, it does not address the human part of the industry which is of paramount importance in human capital industry like that of IT and allied sectors. To enable this, a People Capability Maturity Model (P-CMM) [5] has also been proposed. The P-CMM can be used as a framework for improving the way in which an organization manages its human assets. The P-CMM also shares the same five levels of maturity as that of CMM although the KPAs that constitute these levels (20 KPAs) are focussed on the human counterpart rather than the software quality.

The five levels of maturity are:

1. Initial which shall ensure, at least, a set of informal people management practices
2. Repeatable which shall ensure the establishment of policies for developing the capability of the staff
3. Defined which shall ensure the standardization of best people management practice across the organization
4. Managed which fixes and tracks the achievement of quantitative goals for people management
5. Optimizing which ensures the existence of continued focus on improving individual competence and workforce motivation

The overall summary of CMMs is that whereas the earlier version of CMM incorporated the process orientation across the industry, the P-CMM addressed the people counterpart of the industry. However, in reality, the P-CMM is not practised in the industry as closely as the traditional CMM process. On account of this, IT and especially the software industry could achieve commendable success only in rendering services that are low on the value chain amongst the spectrum of activities associated with typical service sectors. The industry for their own reasons did not allow the intellectual growth of their associates and as such there could not be any fruitful participation of theirs in executing high-valued tasks. In this direction, it is imminent that corresponding process-oriented models being employed in the education sector need improvisation as would be explained in the following sections.

3 NBA (National Board of Accreditation) Model for Technical Education

Accreditation is a process of quality assurance and improvement, whereby a programme in a technical institution is critically appraised and given credit for its academic strength and objectives of the institution. The need for accrediting technical education programmes has grown due to the unprecedented growth in the number of educational institutions and programmes. The NBA shoulders the responsibility of framing the process model and assessing the institutions for attainment of their educational objectives. The NBA has, probably, assumed an evolutionary approach for their mission of framing the process model. In this direction, technical institutions are witnessing the revised NBA process which is outcome based rather than the traditional infrastructure-based one. The present NBA model to assess the quality of technical education is a template-based one. The template provided by the NBA is manifested as the self-assessment report (SAR) and shall be used by the assessee institutions to represent the processes being followed and the outcome thereof. The NBA shall carry out the final assessment based on the validation of SAR contents and the interviews with the stakeholders of the assessee institutions. The advantages of the present NBA process include:

(a) Creativity to define the processes. In fact, the assessee institutions are at liberty to define their custom-built models as long as their outcomes generate the data required to fill the SAR which shall be validated by the NBA during their accreditation task.
(b) While the creativity is at its best as briefed above, the SAR template is also suggestive of defining a simple set of processes to meet the template requirements. This approach might still be valid for an initial run of the accreditation.
(c) Advocates, either implicitly or explicitly, the concepts of outcome-based education and graduate attributes which are vital to world standards like Washington Accord.
(d) The NBA has coined the term "programme" to represent a specific course of study (leading to a degree in specific technical area). This makes the model scalable since any institution (or even a department within an institution) could be running several programmes which can get accredited independent of one another.
(e) Customized for the present Indian scenario with coexistence of both autonomous institutions and university-affiliated colleges. The two variants of SAR shall meet this requirement.
(f) Formal introduction of process-based models which are hitherto unheard of in the educational field.

The NBA SAR expects the assessee institutions to furnish information under the following heads: (1) Vision and Mission, (2) Programme Educational Objectives (PEOs), (3) Programme Outcomes (POs), (4) Students' Profile, (5) Faculty Profile, (6) Infrastructure, (7) Programme-Specific Criteria and 8) Process Improvement.

Whereas items 4 through 7 are, predominantly, factual information, the maturity of the assessee institutions lies in the way the institutions handle items 1 through 3 and importantly item 8. In specific, item 1 shall reflect the long-term goal of the institutions, item 2 shall reflect the midterm goal of the institutions and item 3 shall reflect the just-in-time operational goal of the institutions. Remarkably, the POs closely match the graduate attributes (GAs) as defined in the Washington Accord. The maturity of the institutions get, predominantly, decided based on the following:

1. Sound delivery mechanisms
2. Assessment mechanisms
3. Conviction with which the institutions map the facts to the attainment of POs and PEOs

A typical set of processes/mechanisms that could fulfil the programme assessment requirements based on the SAR template of NBA is explained in the following paragraphs:

Assessment of the attainment of Programme Educational Objectives (PEOs) by a functional entity, formally termed the “Programme Attainment Evaluation Mechanism (PAEM)”. The PAEM forms a part of an overall Programme Execution Model which may be formalized as “Programme Execution Process Model” (PEPM). Simply stated, the PEPM is a closed-loop system that governs the execution of any programme offered by the associated department. The PAEM and other functional entities of PEPM are generic entities that manifest, differently for different programmes. This enables PEPM to be scalable across varied programmes that are, presently, being run as well as the programmes, probable, in the future.

The PAEM is a collection of mechanisms which are broadly classified as “Direct Assessment Mechanisms (DAMs)” and “Indirect Assessment Mechanisms (IAMs)”. Mechanisms typical of DAMs include Continuous Internal Evaluation (CIE)- and Semester End Examination (SEE)-based assessment mechanism. Mechanisms typical of IAMs include (1) employers’ survey, (2) alumni survey, (3) exit survey, (4) faculty survey, (5) course end survey, (6) students’ feedback, etc.

Each of these mechanisms is associated with a specific process and an associated analysis tool. The processes and the associated analysis tools of each of these mechanisms, together, on their individual merits, produce performance indices, formally, termed as “PEO Metrics”. PEO Metrics constitute the prime input for an overall analysis tool termed as “Programme Assessment Metric Tool (PAMT)”. The PAMT operates on the PEO Metrics along with other configurable and computes the overall figure of merit for each of the PEOs. These figures of merit are termed as PEO Indices. The typical approach of this type is presented in Fig. 2.

4 Motivations for Enhancements to the NBA Process

The previous sections presented the commonality of the process orientation across the manufacturing sector, knowledge-based service sectors (IT, SW and allied industries) and educational sector. This section deals with the motivations which are rather compelling on the educational sectors to incorporate enhancements to process orientations. Articulations of such motivations which are presented in the following paragraphs are reiterations of the facts, already highlighted above, with varied viewpoints:

Both the knowledge-assisted industry and more so the knowledge-based industry have achieved a good level of success having been process oriented. Especially in the Indian context with only the recently coming-up China as a close contender, the knowledge-based industry could create global icons, predominantly, owing to their process orientation. This aspect together with the fact that knowledge-based industry is based on human intellectual capital, clearly, suggests that the education sector

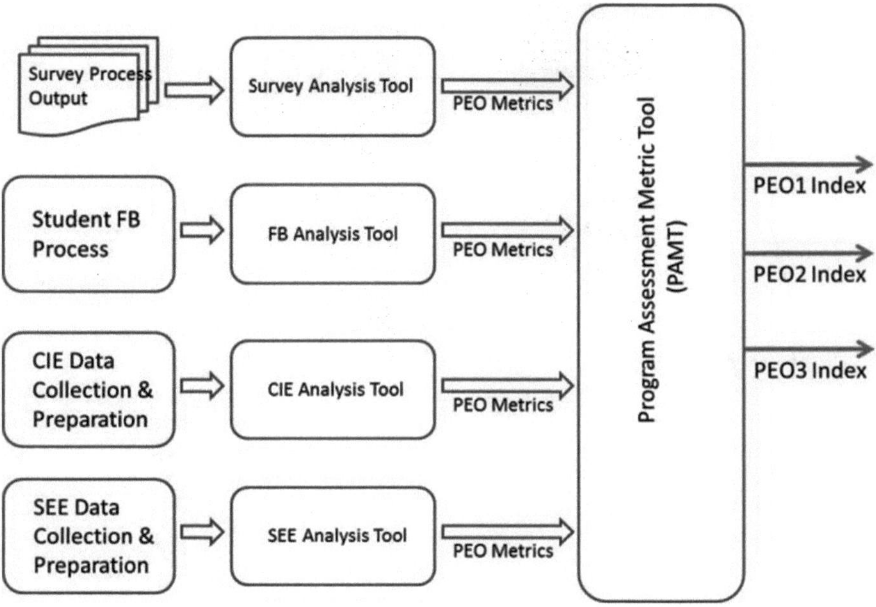

Fig. 2 Programme assessment framework

which, indeed, fulfils the need for human intellectual capital needs to have much more than the best of the process framework.

An evolving industry requires people with technical expertise to grow the industry, whereas corporates associated with most of the industry segments attach only a commercial tag and remain satisfied with the, continuously, enhancing wealth without realizing the fact that, eventually, the need or the lack of technical expertise would curtail the growth. The service sector as a part of the IT industry is a classic example of industry with this short-sightedness wherein senior technical people are not observed to grow with the industry unless they transform themselves into a management ladder which, only at lower levels, is the counterpart of the technical ladder.

One other motivation for the need of a process framework and beyond for the education sector can be drawn from the success of the IT, ITES and SW industry. The progress of these industry segments is evolutionary and a twofold one. Firstly, the constituent corporates evolved into mature multinational companies. Secondly, nature of activities that constituted the services, progressively, added more and more direct values to the end product rather than being a part of the supportive ecosystem for the product or the subsystem. This success is achieved over a period of 2–3 decades. This fact proves the sustainability and assured unprecedented growth capability of the process orientation.

Having established the need for enhancements to the process framework for technical education, as explained in the previous paragraphs, a mature model for

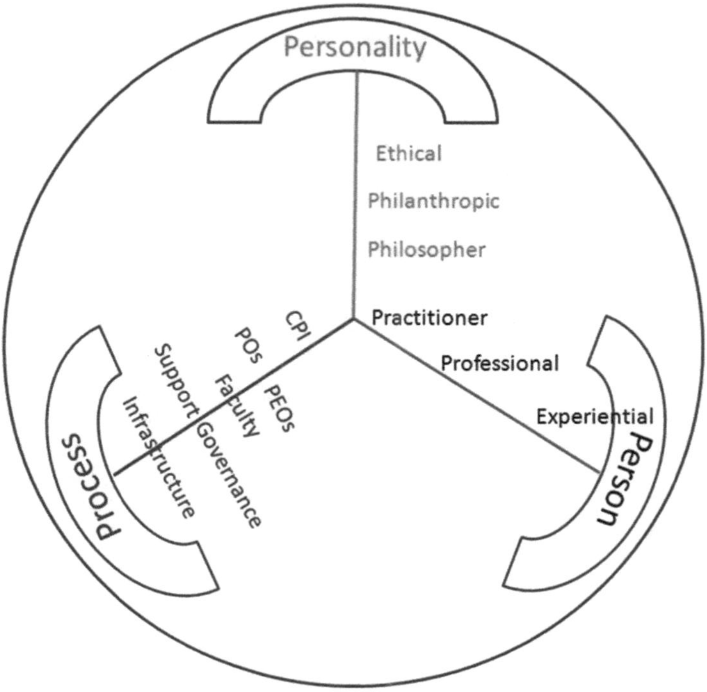

Fig. 3 Augmented process framework for technical education (3Ps)

technical education could, comfortably, rest on the 3Ps namely (1) process, (2) person and (3) personality. To the best knowledge of the authors of this paper, there is not a three-pronged model for technical education that houses the role of process, person and personality in building a mature model for technical education. These entities of the enhanced model are explained in the following sections (Fig. 3):

4.1 Process

The success of varied industry segments owing to the process orientation as explained in the previous sections, probably, has profound influence on the education sector and is manifested to a great extent in the present NBA assessment process, itself. The modified NBA process with its attempt to benchmark against the Washington Accord advocates 8 KPAs as listed in Sect. 3. Despite the, apparently, complete NBA process model, taking the example of specific industry segment, the question to be answered is whether the education sector should just follow the software industry and adapt process orientation. The answer turns out to be not in the affirmative owing to the fact that the education sector enjoys the privilege of providing the intellectual capital to the SW industry and hence needs to have attributes in addition to an assortment of processes. It is with this motivation that additional Ps are introduced which are explained in the following sections further.

4.2 Person

Section 3 indicated delivery mechanisms, assessment mechanisms and the ingenuity associated with PO-PEO mapping as the drivers for nurturing the maturity of any institution. However, Sect. 3 detailed only the assessment mechanisms but left out the other two aspects. These are the two aspects which could, effectively, be addressed by the human aspects rather than the process, all alone. In the present context, "person" refers to the faculty. Although certain attributes of the faculty are captured, directly or indirectly, in the present NBA process, there could be additional attributes which are not amenable to the nominal process framework. These additional attributes could be grouped into three vital attribute groups which a faculty gains as he/she progresses with the teaching career. These vital attribute groups are identified as (1) experiential, (2) professional and (3) practitioner attributes. The attributes which can be branded into one or more of the above attribute groups become important in causing vital contribution to technical education. Specifically, the experiential and professional attribute of the faculty helps him/her in augmenting his/her delivery mechanisms, whereas the practitioner attribute of a faculty helps him/her assuming higher roles in the institution like the "Dean of Studies", "Controller of Examination", "Registrar", etc. Although such of these aspects are, implicitly, covered in the present NBA process, a greater depth of assessment with respect to these attributes could only be carried out only if there exists a focussed formalism that would objectively describe these attributes and a specific focus is laid during the accreditation assessment process.

4.3 Personality

Whereas human attributes associated with the identified attribute groups listed above can be achieved with a reasonable stint with the teaching career, there are certain attributes which could only be achieved over a longer stretch of overall experience. These attributes albeit difficult to specifically define, collectively, dictate and make the institution demonstrate the varied manifestations of a philosophical approach to providing technical education. At a broader level, the attributes that dictate the personality can be, hierarchically, listed as (1) ethical, (2) philanthrophic and (3) philosophical. Assessment with respect to the personality attributes could only be carried out if there exists a formalism that would objectively describe these attributes and a specific focus is laid during the assessment process. However, these aspects, typically, are integral with the humans, but their manifestations are only witnessed when demonstrated. To get assessed into this level, it also becomes essential that the institutions encourage and appreciate such personality.

5 Conclusion

In this paper, evolution of technical education in India has been explained. The importance of process-based model in the education sector is explained as an extension to the prevalent process orientation in the knowledge-based industry sectors. More importantly, as a significant contribution of this paper, the importance of personal attributes and personality attributes, predominantly, of the teaching faculty, which could, additionally, elevate the maturity of technical institutions to higher levels, is also discussed.

References

1. Ahmed T, Satija MP. Growth of technical education at degree level in Punjab: a historical perspective. Indian J Tech Educ. 2005;28(3):81–91.
2. Kral J, Zemlicka M. Engineering education – a great challenge to software engineering. Computer and Information Science. ICIS 08. Seventh IEEE/ACIS international conference, 2008, p. 488–95.
3. Biswas G et al. Profile of engineering education in India. New Delhi: Narosa Publishing; 2010.
4. Hunter R, Thayer R, Paulk M. The capability maturity model for software, software process improvement. Hoboken: Wiley-IEEE Press; 2001. p. 49–88.
5. The People Capability Maturity Model. Available at: http://ifs.host.cs.st-andrews.ac.uk/Books/SE9/Web/Management/P-CMM.html

"Millimeter Techniques for Kilometer Benefits" Creating Excitement in CAED Classroom

Gurupadayya M. Hiremath and Sanjeev M. Kavale

Abstract Manual drafting improves spatial skills and helps to develop mental conversions into 3D reality, whereas computer-aided drafting is too successful because of being easy to draw complex drawings, time saving, easy editing, mass storage of drawings, and ample tools to draw complex shapes. The integration of manual drafting and the use of CAD software to draw engineering drawings bridge the gap between the spatial skills and mental conversion of 2D to 3D drawings with the advantages of computer-aided drafting as mentioned. Traditional methods of classroom teaching have proved out to be boring and inactive. Experiential learning through new techniques in Computer-Aided Engineering Drawing (CAED) classroom is a must to create excitation in the classroom and attain teaching-learning experience to the highest efficiency. This paper describes how the integration of manual and computer-aided drawing can be done indicating the Course Learning Objectives (CLOs), mapping of CLOs with Abet 3a to 3 k criteria, and chapter-wise evaluation scheme and also suggests different techniques of experiential learning and their methodology implemented in this course which enhance the manual and computer-aided drawing integration. This paper also clearly indicates the outcome of the techniques, with a significant improvement in the academic excellence of the students.

Keywords Experiential learning • Computer-aided drafting • Manual drafting • Motivation

G.M. Hiremath (✉) • S.M. Kavale
Department of Mechanical Engineering, BVB College of Engineering and Technology, Vidya Nagar, Hubli, Karnataka 580031, India
e-mail: gurupadayya@bvb.edu; sanjeev_kavale@bvb.edu

R. Natarajan (ed.), *Proceedings of the International Conference on Transformations in Engineering Education: ICTIEE 2014*, DOI 10.1007/978-81-322-1931-6_32,

1 Introduction

Work done by Sorby and Gorska [1] concludes that hands-on sketching and drawing tend to improve spatial skills more than courses that stress computer-aided drawing methods. They suggest that the physical nature of the drafting experience develops a deeper understanding of the meaning of lines and symbols on a page and helps to develop the ability to make mental conversions into 3D realities. McCardle [2] concludes that the value of manual technical drawing lies in the development of visualization and manipulation of views 2D and 3D. Technology education curriculum reviews have resulted in an increased requirement for the teaching of CAD (computer-aided drafting). CAD systems are too successful due to being too easy to draw complex drawings, time saving, easy editing, mass storage of drawings, and ample tools to draw complex shapes. Some technology education teachers believe that in a world of CAD, there is no place for drawing boards and set squares.

Considering the fact advantages of manual drafting and computer-aided drafting, the course content is so designed to accommodate both manual and computer-aided drawing. Below is the mapping of CLOs with Abet 3a to 3 k criteria [3] of the integrated course and the scheme of evaluation. Care is taken so that equal importance is given to learn manual drafting and computer-aided drafting as well.

Following are the Course Learning Objectives (CLOs) mapped with Abet 3a to 3 k criterion (program outcomes) with degree of compliance as L (low), M (medium), and H (high).

1. Use BIS conventions in all his drawings and interpret them from a drawing – 3 g (M), 3i (L), 3j (L), 3 k (M).
2. Use modern drafting tool for all his drawing needs – 3 k (M).
3. Draw orthographic projections of points, lines, solids, and sectional views of simple solids – 3a (M), 3 g (M), 3i (L), 3 k (M).
4. Draw the development of the lateral surface of solids, truncated solids, frustums, and transition pieces – 3a (M), 3 g (M), 3i (L), 3 k (M).
5. Convert pictorial views into orthographic projection with or without section and finally transform his vision into orthographic views – 3a (M), 3 g (M), 3i (L), 3 k (M).
6. Able to read orthographic projections of simple solids and their combinations and convert them into isometric view/projection – 3a (M), 3 g (M), 3i (L), 3 k (M) (Table 1).

The learning of this integrated course can be enhanced to a great extent by some experiential learning techniques. Some of them were practiced during the implementation of the course. They are mentioned in upcoming paragraphs.

2 Experiential Learning Techniques

Kolb [4] has to be given credit about experiential learning. He developed this theory as a way to evolve beyond traditional classroom teaching techniques that favor detached learning of abstract concepts that might have disconnected from direct experience.

Table 1 Chapter-wise evaluation scheme

Unit	Chapter		Manual	Computer	Total
I	1.	Introduction	–	–	–
	2.	Projection of lines	15 marks	15 marks	30
	3.	Projection of solids			
	4.	Sections of solids			
II	5.	Development of lateral surfaces	15 marks	15 marks	30
	6.	Conversion of isometric to orthographic views			
III	7.	Isometric projection/view	20 marks	20 marks	40
Total marks			50	50	100

According to Kolb, "experience is seen as an interaction or more as a transaction between a person and his environment. Abstract thinking is a product of concrete experience rather than knowledge that can be learned from books and lectures. The four modes usually assembled in is called the experiential learning cycle. They are concrete experience (CE), reflective observation (RO), abstract conceptualization (AC), and active experimentation (AE). Learners must be able to involve themselves fully, openly, and without bias in new experiences (CE). They must be able to reflect on and observe their experiences from many perspectives (RO). They must be able to create concepts that integrate their observations into logically sound theories (AC), and they must be able to use these theories to make decisions and solve problems (AE)."

In today's scenario, creativity in teaching methods is coming into light recently. Reasons for this are many. The major being lack of interest of the students in the class, technological developments, expectations of students, and practical implementation of the subject. Innovation in education is important because we need to prepare young people for an exciting but uncertain future. In a learning and teaching environment, this means creating opportunities for students to learn in new ways to meet their individual needs.

To address all these, seven experiential learning techniques are practiced to overcome conventional "chalk and talk" method. Before practicing these techniques, analysis was done on student's result, knowledge of converting pictorial view into orthographic projection and vice versa, and practical implementation of the knowledge of drawing which is more important from the industry point of view. The techniques implemented create interest in the subject and presence of mind. The expected outcomes are more familiarization of the CAD software, conversion of pictorial view into orthographic projection and vice versa easily, practical implementation of the subject, and improved imagination skills.

3 Implementation of Experiential Learning Techniques

3.1 Reverse Engineering

In industry, one of the roles of the design engineer is to generate drawings for broken or failed components. The designer is expected to go to the site, measure the dimensions of broken component, and produce the drawings for manufacturing.

Fig. 1 Students discussing dimensions of a product in reverse engineering activity

CAD users who have had hands-on problem solving through descriptive geometry and graphics taught using 3D physical models have improved spatial abilities. They argue that such experience has a positive impact on their understanding of CAD compared to those who have learned CAD systems only [1].

Cooperative learning allows students to work in teams to share ideas and experiences [5, 6]. Keeping this in mind, simple wooden components are given for a group of four students. Each group should discuss, measure the dimensions of the component, and draw different views. Each group submits only one solution after discussion. By doing so, students learn how to draw 2D drawings for 3D object and learn how to communicate with team members. Figure 1 shows students engaged in this activity.

3.2 *Model Making*

Practical exposure is essential to the learning of concepts, and the experience should be as real as possible [7]. Generally, students come across situations wherein they need to prepare some of the components by themselves, for example, a tray to house circuit board and body for small robots which they create for competitions. Though

Fig. 2 Card sheet models made by students during model-making activity

students learn theoretical part of the development of lateral surfaces, they face difficulty in preparing the actual model. To overcome this practical difficulty, the model-making task was given to each student.

To make them realize the importance of the development of lateral surface, students should prepare a 3D model using card sheets. The models prepared by students not only include truncated prisms and pyramids but also the combination of curved surfaces and flat surfaces as shown in Fig. 2. Demonstration on how to make such models was given in a class so that students learn cost-effectiveness and practical difficulties. This process is more helpful for sheet metal operations, especially for mechanical engineering sciences students.

3.3 Visual Aid

According to Kolb [4], assimilator students grasp videos or images quicker than the conventional chalk and talk. Based on experience, majority of the students find it difficult to imagine a 3D part and convert that into 2D in the mind. To make imagining in a systematic way and conveniently, videos on isometric to orthographic projection of engineering components were shown. This creates a platform to imagine in a systematic way and draw the views with less difficulty.

Fig. 3 Tetrahedron made by students during practical implementation activity

3.4 Practical Implementation

Students have a desire for practical application of theoretical knowledge to the workplace [8]. After learning 80 % of the subject, a group of students are given a problem definition in which they have to apply their knowledge on the subject, identify suitable resources, and complete the task.

For example:

- A group of students were assigned to make sheet metal box to accommodate circuit boards.
- To demonstrate solids of revolution, a group of students were assigned to prepare a working model in which the plane surfaces when rotated about an axis, which appears as solid.
- A group of students were assigned to build sheet metal prisms, pyramids, and tetrahedron (as shown in Fig. 3) by taking the dimensions from existing wooden models which were used for class demonstration purposes.

3.5 Drawing Competition

To make students more familiar with the software commands and to draw the drawings quickly and accurately, drawing competition is being conducted. Unlike the conventional drawings, these drawings include almost all the options available in the CAD software to be used. Using hot keys and faster methods to tackle such

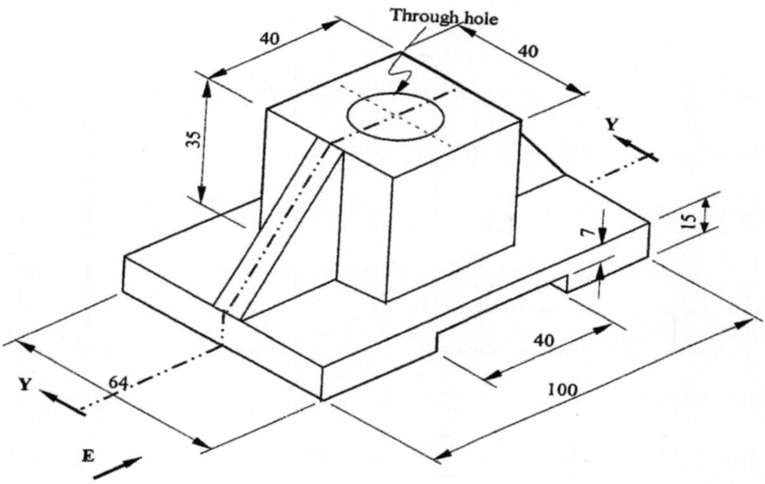

Fig. 4 Example of isometric view of a part for imagination skill activity

drawings were explained for the participants. Evaluation of the competition was made on the basis of the shortest time to finish the drawing, accuracy of the drawing, and adoption of BIS conventions for drawing. To motivate students [9], the top two students are given either motivational book or pen drive as a prize.

3.6 Imagination Skill

The interpretation necessary between 2D and 3D and vice versa involves more than the ability to follow techniques and procedures to "solve the problem." In industry, modeling engineer's responsibility is to generate 3D component for a given 2D drawings and vice versa.

At the beginning of the class, the isometric view (Fig. 4) of the component was displayed, and students were expected to draw orthographic projections. Later the solutions are displayed and students will cross evaluate themselves. The complexity of 3D component goes on increasing from class to class. After 50 % of the course completion, orthographic projections (Fig. 5) were displayed, and students are expected to draw isometric view. By the end of the semester, students were able to draw orthographic projections for the given 3D objects and vice versa.

4 Observation and Conclusion

The integrated approach towards engineering drawing will be very helpful as the student will learn both manual and computer-aided drawing and, with the certain experiential learning methods or techniques like mentioned above, will help student

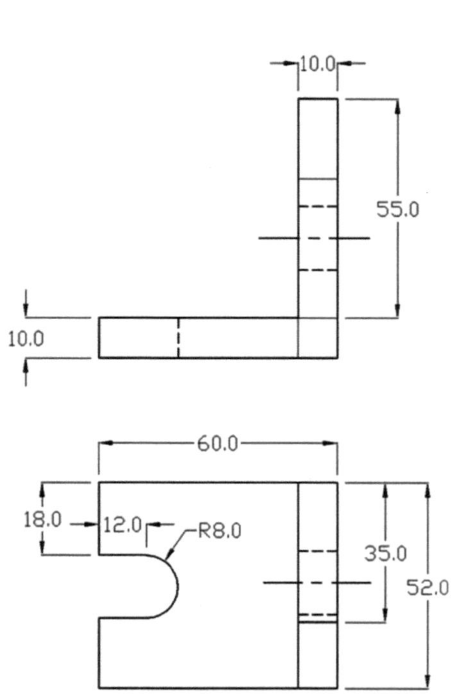

Fig. 5 Example for orthographic views of a part for imagination skill activity

to grasp the subject more easily. Very interesting things were observed during the implementation and post implementation of the abovementioned techniques.

Reverse engineering is a wonderful activity which can be conducted especially for mechanical engineering science students. Students will enjoy this session very much. It was observed that the model-measuring skills were also improved, since this activity is more practical and group involved than that of the traditional sketching.

Model making is helpful, especially for mechanical engineering sciences students. While participating in some of the robot competitions, students used this model-making skill with sheet metal and card boards, etc. for their robots. This means that students are able to implement whatever is taught in the classroom into their day to day activity.

Drawing competitions are unique, and students did show maximum enthusiasm to learn the curriculum and the CAD software. Such kinds of competitions are to be conducted wherever computer tools are used.

Imagination skills activity is also interesting because it helped the students to create orthographic views of points and lines easily.

Table 2 reflects some statistics of the academic improvement in Continuous Internal Evaluation (CIE) of the 2012 batch students who have experienced the abovementioned experiential learning techniques. The comparison is done on 2

Table 2 Statistics showing improvement in performance before and after implementation of experiential learning techniques

	2011 batch	2012 batch	% increase of marks in 2012 batch
Exam 1	12.26 (average)	15.18 (average)	23.81 %
Exam 2	12.32 (average)	17.24 (average)	39.93 %

consecutive academic years 2011 (who have not experienced these techniques) and 2012. During sampling, almost same caliber students are taken for comparison.

Acknowledgment Our deepest gratitude is towards Prof. C. M. Koti for his benevolent guidance and kind support. Our heartfelt thanks and gratitude to Dr. P. G. Tewari, Professor and HOD, Department of Mechanical Engineering, BVBCET, for his everlasting support and encouragement.

References

1. Sorby S, Gorska R. The effect of various courses and teaching methods on the improvement of spatial ability. In: Proceedings of the 8th ICEDGDE. Austin, Texas; 1998. p. 252–256.
2. McCardle J. Back to the drawing board. J Des Technol Educ. 2002;7(2):123–32.
3. Felder RM, Brent R. Designing and teaching courses to satisfy the ABET engineering criteria. J Eng Educ. 2003;92(1):7–25.
4. Kolb DA. Experiential learning: experience as the source of learning and development. Englewood Cliffs: Prentice Hall; 1984.
5. Pundak D, Rozner S. Empowering engineering college staff to adopt active learning methods. J Sci Educ Technol. 2008;17(2):152–63.
6. Johnson DW, Johnson RT, Johnson Holubec E. Cooperation in the classroom. Edina: Interaction Book Co.; 1991.
7. Christiaans H, Venselaar K. Creativity in design engineering and the role of knowledge: modelling the expert. Delft: Faculty of Industrial Design Engineering, Delft University of Technology; 2007.
8. Benjamin B, O'Reilly C. Becoming a leader: early career challenges, faced by MBA graduates. Acad Manag Learn Educ. 2011;10(3):452–72.
9. Mohammdi A. How to encourage students to learn English? Department of English Language Teaching, Kermanshah Branch, Islamic Azad University, 6718975593, Kermanshah, Iran.

Program Outcome Attainment Through Course Outcomes: A Comprehensive Approach

H.R. Bhagyalakshmi, D. Seshachalam, and S. Lalitha

Abstract The outcome-based education (OBE) system is one of the emerging trends of education which helps students to progress in their professional career and makes teachers more creative about their teaching methodologies allowing them to improve the quality of teaching by adopting suitable teaching tools. The measurement of attainment of program outcomes is an important tool which provides a yardstick to visualize how far an institution has succeeded in delivering what it visualized. This paper provides a method by which the attainment of program outcomes can be quantified by using various tools like internal evaluation and various other surveys. This method is evolved and practiced in the department of Electronics and Communication Engineering, BMSCE since practicing OBE concepts. This method can be aptly applied by all colleges which are in the line of accrediting their program to the NBA.

Keywords Outcome-based education • Graduate attributes • PO attainment • Course outcomes • Continuous improvement

1 Introduction

Outcome-based education in engineering colleges across the world is a major change that various academic institutions and other engineering colleges are willing to adopt in their academic system. In Karnataka, India, top engineering colleges are already working towards the outcome-based education system, to be accredited by

H.R. Bhagyalakshmi (✉) • D. Seshachalam • S. Lalitha
Department of Electronics and Communication, BMS College of Engineering, Basavanagudi, Bangalore, India
e-mail: hrbhagya.ece@bmsce.ac.in; dschalam.ece@bmsce.ac.in; lalithas.ece@bmsce.ac.in

R. Natarajan (ed.), *Proceedings of the International Conference on Transformations in Engineering Education: ICTIEE 2014*, DOI 10.1007/978-81-322-1931-6_33,

the National Board of Accreditation, New Delhi. The NBA has introduced a new accrediting process following the rules of Washington Accord, an international agreement bodies responsible for accrediting engineering degree programs [1]. According to OBE system, the NBA's parameters called Graduate Attributes [GAs] must reflect on a graduate of an engineering program.

OBE is an educational process that involves the restructuring of curriculum, assessment methods, and reporting practices in education to reflect a better learning and improving the skills with additional supporting qualities of an engineering graduate instead of accumulating the credits. This approach definitely demands a positive change in the teaching and assessment methods of teaching faculty of academic institutions.

In this paper, a method to assess the performance of students is discussed which is then used to show the attainment of the program outcomes. The following sections give a brief description of the requirements and preparations of the program to show continuous improvement. Section 2 presents a CO-PO model and methodology used in the program. Section 3 explains the assessment tools with PO attainment components along with the proposed process of PO attainment. Section 4 discusses the scope for further improvement with conclusion.

2 CO-PO Model

The 12 graduate attributes (GAs) prescribed by the NBA are as listed in Fig. 1. Each program has its own program outcomes (POs) defined using suitable process and must be aligned with the graduate attributes.

Authors take the example of their own program which has integrated courses as well as nonintegrated courses in the curriculum. In the program, one-to-one mapping between GAs and POs is adopted as shown in Fig. 2.

Each course of the program is designed with a set of course outcomes (COs). Each of these course outcomes maps few of the program outcomes based on its content as shown in Fig. 3, and all the courses together map all the program outcomes. For example, a course like basic engineering mathematics normally maps to PO1 and PO2. However, with suitable changes in the content and the assessment tools, mathematics can also be mapped to higher POs, say PO3 or PO4. Similarly, an integrated course like "Digital Electronics" content of the course results in a mapping with PO1, PO2, and PO4. However, by changing its content by introducing a modern tool for simulation of the circuits, one can map this course to PO5 which is related to modern tool usage. The main aim of "outcome-based education" is to make teachers use their creativity to bring transformation in engineering teaching to produce quality engineers for a better world.

Faculty handling various courses identifies the required course delivery methods and assessment tools to be used at the beginning of the course. At the end of the course, faculty check the attainment of course outcomes using various assessment tools. The resulting course outcomes are then used to find attainment of the related program outcomes. This is done by all the course instructors and for all the program

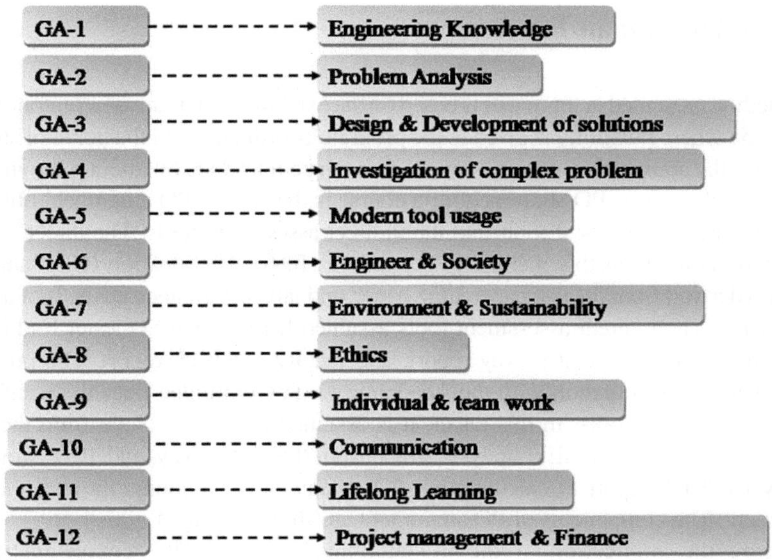

Fig. 1 Graduate attributes

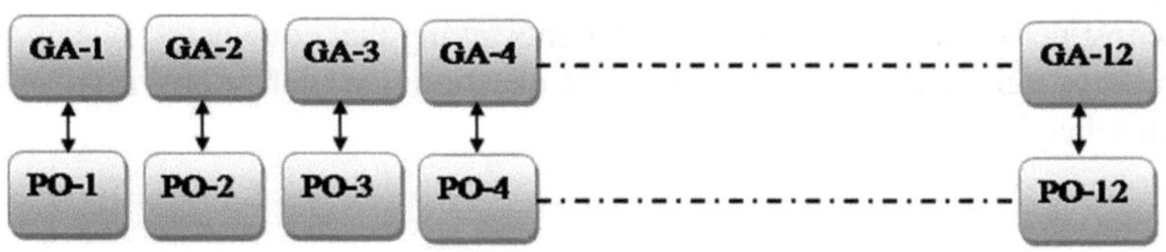

Fig. 2 One-to-one mapping between graduate attributes and program outcomes

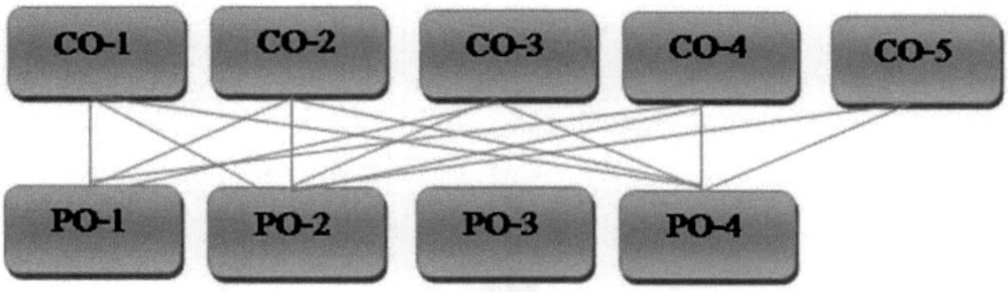

Fig. 3 Course outcomes mapping to few program outcomes

courses. The individual program outcomes are then used to find the overall attainment of the program outcomes by the program coordinator. The POs so obtained are analyzed to find the shortcomings of the existing curriculum structure, assessment method, and delivery mechanisms and revisit the set goals. Corrective measures are then recommended to improve the attainment level for a new performance target to achieve improvement in the teaching–learning process [2, 3]. So this becomes a continuous process to improve the quality of the graduates moving out of the program.

3 PO Attainment Process

The method presented in this paper is based on the existing curriculum of an autonomous institution where flexibility is given to the program coordinator and the board of studies to modify the course content and a new course can be introduced without removing the fundamental courses. PO attainment of a course is the sum of PO attainment obtained from the direct assessment tools and the indirect assessment tools. The direct assessment tool assumed in this method is Continuous Internal Evaluation (CIE) analysis which is derived from the test paper, quiz paper, and lab performance analysis of a class of students. The indirect assessment tools assumed here are the Semester End Exam (SEE) result analysis, exit survey report, and faculty feedback report. The program coordinator in consultation with the Academic Audit Committee fixes the weightage for the assessment tools. In this paper, it is assumed that 70 % come from the CIE analysis, 10 % from the SEE analysis, 10 % from the exit survey, and 10 % from the faculty feedback report to calculate the PO attainment of a course through the CO attainment. The components of PO attainment are shown in Fig. 4.

Before the application of the PO attainment process, the course instructor should have:

1. Test question papers with each question mapped to course outcome which in turn maps to program outcome
2. The test performance details of the students taking that course
3. Similarly quiz and quiz scores, lab test questions and lab score, mini projects with related scores, etc.

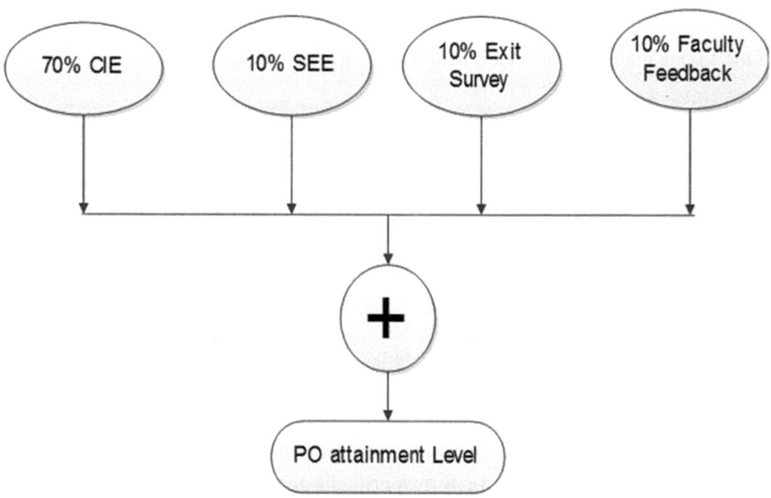

Fig. 4 Components of PO attainment

The process of program outcome attainment consists of three steps.

Step 1

1. The details of the course and students are entered in an excel sheet.
2. The test paper details – question numbers with related course outcomes mapped to program outcomes and marks allotted – are entered.
3. The test performance details are entered in the attainment spread sheet.
4. Total marks for each question (course outcome) of the course are calculated.
5. Total marks divided by the test strength gives the normalized marks.
6. Average CO attainment values are obtained from normalized marks and are mapped to find the PO attainment values of the sample course.
7. Repeat from steps 1 to 6 for other assessment tools like quiz, lab, and mini project.

Step 2

1. The performance target for each program outcome is defined. The direct and indirect assessment tools are identified.
2. The courses contributing to the same program outcome are identified.
3. The attainment values obtained for each course from the direct assessment tool (CIE analysis) are listed, and the average attainment is calculated.
4. The reports obtained from indirect assessment tools are used to provide additional information regarding the selected program outcome.
5. The weightage for the CIE analysis and for the indirect assessment method is fixed by the Academic Audit Committee in consultation with the program coordinator.

Step 3

1. The program coordinator, the Academic Audit Committee, and the Department Academic Committee analyze the attainment values and determine whether there is a need for changes in course delivery methods, curriculum content, evaluation, and assessment methods.
2. The program coordinator, the Academic Audit Committee, and the Department Academic Committee recommend the corrective measures for improvement of POs.
3. The recommended measures are followed by course instructors to improve the attainment level.
4. Repeat step 1 through step 3 by setting a new, higher target.

The total attainment value is the sum of attainment values obtained from both direct and indirect assessment tools.

Figure 5 shows the process of PO attainment calculation as explained in steps 1 and 2.

Figure 6 shows the process of PO attainment analysis as explained in step 3.

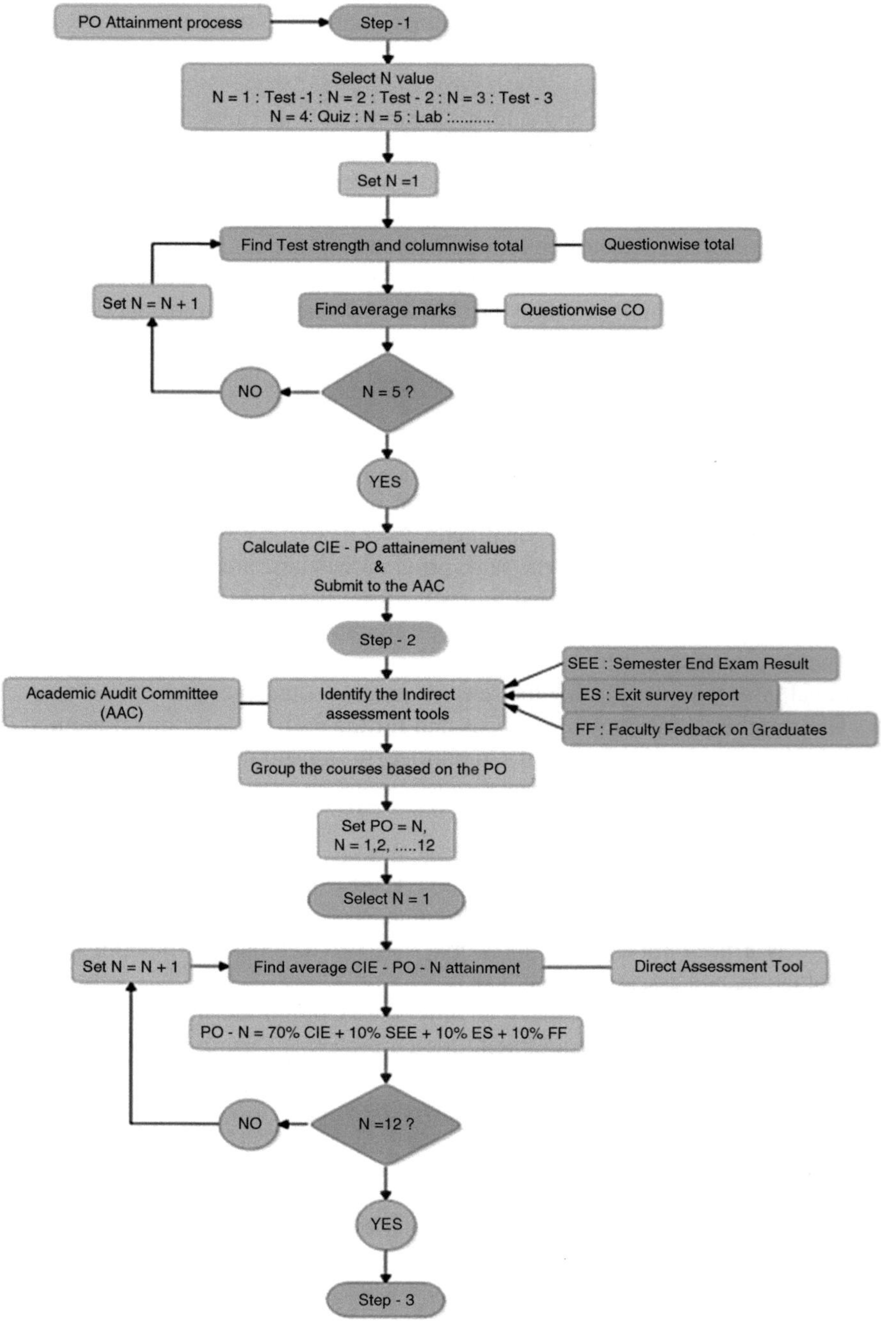

Fig. 5 Determination of PO attainment values

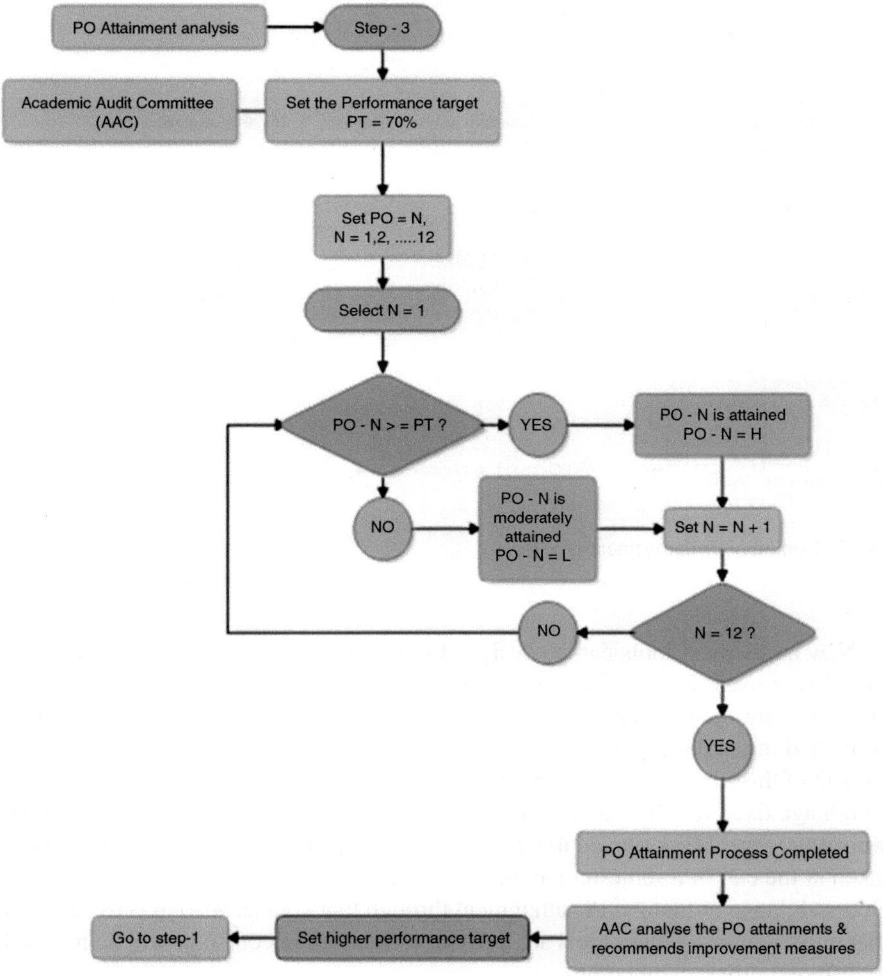

Fig. 6 PO attainment analysis process

4 Conclusions

This paper details the procedure for calculating the attainment of program outcomes through course outcomes. The paper discusses the role of various tools used such as CIE, exit survey, faculty feedback on graduates, and SEE. This methodology can be aptly applied to all the courses that have a defined course outcome, each of which is addressing an established program outcome.

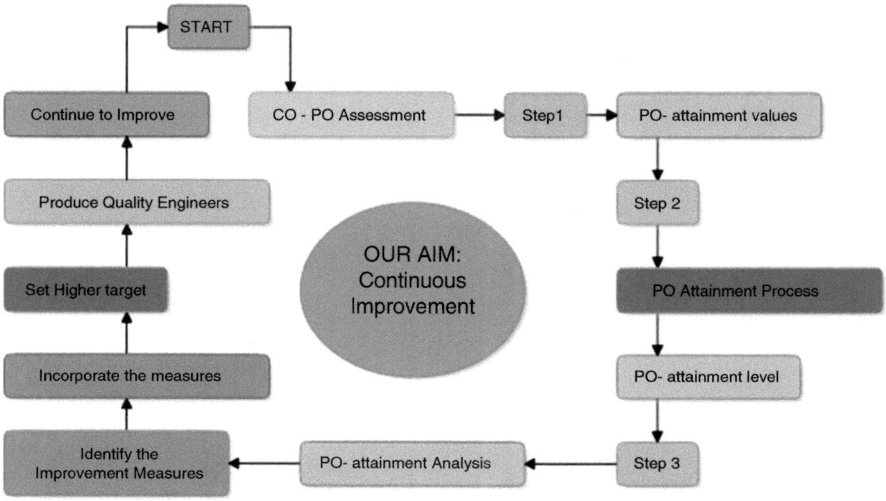

Fig. 7 Continuous improvement process

New assessment tools can be designed to find the PO attainment. For example, the SEE is used as an indirect assessment tool in this paper with a weightage of 10 % as only the final end exam result analysis is considered. However, it can be converted into a direct assessment tool once the SEE score sheets are available. As a result of this, the weightage of SEE can be increased from 10 % to 40 %, while the weightage fixed for CIE can be reduced from 70 % to 40 %, maintaining same the weightage for exit survey and the faculty feedback. Similarly, the course end survey taken at the end of a semester can be used as an indirect assessment tool with suitable weightage to find the PO attainment through that course. Also, it is to be noted that once new assessment tools are introduced to assess a course, it will be mapped to different program outcomes.

Outcome-based education has made faculty to develop creative ways of assessing student performance and good teaching practices for continuous improvement. Figure 7 shows the continuous improvement process adopted towards the PO attainment through course outcomes.

The faculty are working together to develop new innovative ways to make teaching–learning very interesting.

Acknowledgement The authors wish to express their gratitude to the Management, Principal, Vice Principal, and Dean (Academic) of BMS College of Engineering, Bangalore, for the encouragement and valuable suggestions extended towards the proposed method of finding the program outcome attainment and in bringing out this paper.

References

1. PPT – Outcome Based education downloaded from www.nbaind.org
2. Spracklin-Reid D, Fisher A. Course-based learning outcomes as the foundation for assessment of graduate attributes. In: Proceedings of 2012, Canadian Engineering Education Association (CEEA12) conference.
3. Gentemann K, Zhou Y (2005) Course embedded assessment. Virginia Assessment Group, George Mason University, 18 Mar 2005.

The Education DesignShop: A Case Study on Education Reform Through Design Thinking

Jessica A. Artiles and David R. Wallace

Abstract The innovation curve has become saturated; the low-hanging fruit has been exhausted by traditional problem-solving approaches. Any advancement in the education sector from here forward requires a new thinking paradigm: design thinking. This paper documents the process of infusing design thinking into the minds of the education system's greatest problem-solvers: policymakers, engineers/designers, educators, and students of high school, undergraduate, and graduate schools. Using the formation of the Education DesignShop as a case study, we analyze the benefits and points of contention when using a design thinking approach, typical of tangible product designs, in a large-scale application, the systemic reform of education.

Keywords Design thinking • Innovation • Social entrepreneurship • System

1 An Introduction to Design Thinking

Design thinking first appeared in Herbert A. Simon's 1969 book *The Sciences of the Artificial*. Since then, more than four decades of scholars have attempted to define "design thinking" in terms that make most sense to their contextual application (product design, architecture, healthcare, etc.). While most problems in the education system are solved from a research-based, crisis management-based, or linear, milestone-based approach, our design-based approach promises to yield more unique, creative, and effective solutions for the education system.

J.A. Artiles (✉)
Department of Mechanical Engineering, Technology and Policy Program,
Massachusetts Institute of Technology, Cambridge, MA, USA
e-mail: jartiles@mit.edu

D.R. Wallace
Department of Mechanical Engineering,
Massachusetts Institute of Technology, Cambridge, MA, USA
e-mail: drwallac@mit.edu

R. Natarajan (ed.), *Proceedings of the International Conference on Transformations in Engineering Education: ICTIEE 2014*, DOI 10.1007/978-81-322-1931-6_34,

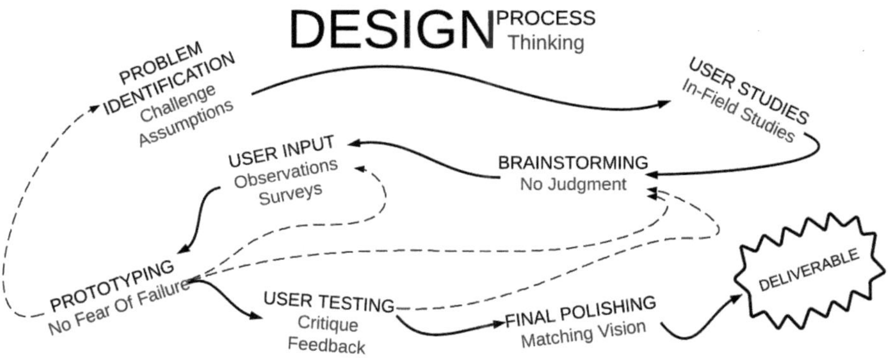

Fig. 1 A roadmap of design process steps (*black*) and their inherent elements of design thinking (*gray*), beginning with problem identification

Often called the twenty-first-century skills, design thinking offers problem-solvers a new departure from traditional standstills in problem-solving. The problem identification and brainstorming stages of problem-solving underline the starkest differences between what is achieved in previous approaches and what is achieved in design thinking practices. Non-designers tend to the dynamic of judgmental interference wherein they cut ideas short of their creative potential and do not give it the chance to grow, develop, or even transform into something that might be a suitable, if not great, solution.

This judgmental interference is unnecessary and ill-suited at the brainstorming stage as it truncates ideas while it is too early to know whether or not they could have become a viable possibility. This truncation can come from endless predispositions that are inherent to the brainstormer's psyche. Among these, the brainstormer can suffer from too much attention to pre-constructed limits (boundary conditions, assumed lack of access or zone of influence, lists of requirements) and/or from pathologically imposed constraints (complacency with the status quo, inherent resistance to change, unyielding preferences, fear of failure, judgment, and unexpected consequences, lacking the impetus for change).

A design thinking mentality, then, is exactly the antithesis of judgment interference. Elements common to brainstorming under a positive design thinking mentality include a vision-based set of goals, an experience-based specification sheet (as opposed to a constraint-based specification sheet), comfort with being risk averse, no judgment zones, and the perspiration of not giving up. Figure 1 describes other elements of design thinking seen through the steps of the design process.

It is important to note that many teams may go through the steps of the design process and still not practice design thinking. Design thinking is more than just a series of prescribed motions; it is an inherent re-wiring of thought that requires complete mental reform from the obstructions we have created in our brains in order to see pathways we would have otherwise not discovered. This paper documents and analyzes recent attempts at fixing the education system through the different mental influence of design thinking.

2 Design Thinking for the Education System

At the product level, the Education Designathon in Spring 2013 [1] demonstrated that there was space for creativity and innovation in educational tools if certain education challenges were approached with a design thinking framework. Examples of these tangible products included, among others, DynamicTable, a rotating high top table that connects to a computer's monitor such that restless kids would have to walk around the table clockwise to scroll the page up and counterclockwise to scroll the page down, as well as the Little Book of Circuits, a children's book with integrated circuits that allow for a parent and child reading pair to interact and learn directly from the book.

While design-based approaches are common in tangible engineering and design products like DynamicTable and the Little Book of Circuits, they are only now beginning to prove themselves in their application to systemic challenges, like education. This next step is imminent. The challenge of using design thinking in education rests in that policymakers—currently the bigger power holders, or stakeholders, in the education space—are trained otherwise. As IDEO explains, "the natural tendency of most organizations is to restrict choices in favor of the obvious and the incremental. Although this tendency may be more efficient in the short run, it tends to make an organization conservative and inflexible in the long run. Divergent thinking is the route, not the obstacle, to innovation" [2].

3 The Education DesignShop: A Case Study in Using Design Thinking

Design thinking and maker thinking are on their way to being infused into schools with the recent developments in nontraditional educational spaces such as the Art and Science Prize, NuVu High Schools, and "Innovation Schools" [3] like the Up Academy and the Boston Green Academy. To drive home change in the education system, however, there needs to be more.

In order to fix the education system, design thinking needs to be established as an embedded problem-solving paradigm. This paper documents the design thinking elements behind creating the first ever Education DesignShop (to occur at MIT in Spring 2014). The event began as a final class project for the then cross-registered class at the Harvard Graduate School of Education (HGSE), T-550 Design for Learning by Creating. The event experiments with bringing interdisciplinary students and professionals together in a unique mentoring and building environment to solve problems in education, all the while practicing design thinking. As follows, each section highlights the major step of the design process and the key action items within that process that, altogether, yielded the proposal for the Education DesignShop.

3.1 Problem Identification

Beginning with the initial stipulation that education still suffers from various systemic challenges that need fixing (some are centuries old), we reasoned that what it really needs is a new method of problem-solving. Notice that design thinking strategies common in product design were not present when approaching education reform. Observations from frustrated conversations across various circles identified that key stakeholders (e.g., educators and policymakers) were not communicating to find solutions that pleased (or were even informed by) both parties.

I asked myself if the root of the problem was a lack of creativity and innovation in the education space. Thinking back to products born at the Education DesignShop [1], we realized that while there were creative solutions at the bottom level from practitioners in the field (especially current students closest to the victimizing side effects), these rarely if ever trickled up to the high level change agents that could have the power of implementation at various levels. Quick investigations into the backgrounds of policymakers confirmed that these key stakeholders were lacking creativity and innovation in their practice because they had not been trained with a background of assimilating these elements into their work's problem-solving. Pressing a little harder on that hypothesis, however, revealed that regardless of the approach that key problem-solver takes, another root of the problem remained that the policymaker is too far removed from the real problems happening in the classrooms, for example, to understand enough of the problem he is trying to solve.

If, hypothetically speaking, a policymaker had been trained with design thinking elements and was somehow closely immersed into the everyday education routine, would our solutions look different? This revealed another element of the triad missing: the tools to fix the education system are still very limited and limiting to any new innovation. Space had to be accommodated for makers, such as engineers and designers, to have a chance to adapt their creative talents to the education space, too.

3.2 Brainstorming and User Studies

First steps included outlining what the event would look like if we could have limitless resources, ideal participants, and achieved goals. This envisioning is key to later specification checks to make sure that a mission-vision-goals statement has been fulfilled to some extent that is true to its purpose. Brainstorming drew inspiration from previous attempts to attack similar problems, such as the Education DesignShop [1]. From the feedback survey that followed, we learned that users wanted a more narrow scope of topics. Tools to envision the final product, like a draft of the sessions that would occur, were useful in helping my mentors visualize my prospects for the final event. Visionary tables like these are a good moment to remember the key design axiom that "Real Data is Truth" [4] and that efforts to include real or simulated data from the beginning will help frame a much more accurate picture that your brain is trying to materialize.

3.3 Requesting User Input

First rounds of presenting the brainstormed versions of the Education Designathon were necessary as a way of getting user input from, in this case, the professor of the class and the advisor that would approve and oversee the end product. A promising visual tool was creating a tree hierarchy of all the possible options at each node of a decision to be made. This gives insight into the direction of the next steps that must be taken and researched further.

To build resources and a knowledge bank of ideas and persons that could potentially be linked to the event, we attended the "How to Design a Course Workshop" at Harvard's Graduate School of Education. There, we met another potential user. Meetings with coaching figures were helpful in prying answers to different questions like the goals for a successful event and the balance between a pedagogical and competitive event.

3.4 Prototyping: A Refinement of Brainstorming

In the case of this intangible product formation, a proposal for an event, prototyping often meant writing out the format and content of the event in various forms. Each set of questions required thinking of a different set of details, thus revealing new connections and interrelations. Perhaps most helpful to the exercise of answering different questionnaires is the iteration portion of answering each question as if it were the first time considered. Without copy-pasting from previous answer sets, a trend in the development and growth of the product can be tracked over time. Other useful tools while prototyping iterations were making timeline goals of the upcoming steps in the development of the product. This exercise is most helpful in identifying what key decisions and actions need to be made from your status to the envisioned final product, as well as compiling a list of the resources available along the journey.

3.5 Requesting User Input

Almost every time design decisions are made, the user should be brought in for some feedback. At this stage, we created a survey that would be sent out two types of people: potential users and potential expert mentors that spend a lot of time thinking about these issues already. While there are many ways to gather widescale feedback, in this case, open-ended questions with many options were a better approach over a binary- or multiple-choice questionnaire.

3.6 *Iterative Prototyping with Key Risks*

I used these survey results to make some formatting decisions and to inform myself of what topics interested potential users. We took their advice of making a list of specific roles and duties for the persons that would have to be involved in order to pull off this kind of event. We used the feedback to flag new key risks that would determine the success of the event, including how to really engage the underrepresented and most critical group, policymakers, in a techy event.

3.7 *User Feedback*

For the next iteration of the feedback, we created a refined event proposal that would be shown to different parties for their feedback (sponsors, student participants, and mentors). We attended the Students for Education Reform (SFER) State Summit and led a session where these potential users were allowed to share their impressions and critiques of the proposal. The key to user feedback is that the more involved in the development of your product a person is, the more likely they are to jump on board and attend or advocate for the product. The summit also opened pathways to future meetings with Massachusetts Representative Jeffrey Sanchez and Massachusetts Secretary of Education Matthew Malone to leverage potential resources between their office and my event. These comments from user feedback sessions were later used to create a tighter value proposition for sponsor recruitment.

4 Conclusive Thoughts on Design Thinking Challenges

There is vast opportunity to applying design thinking as a driving approach to fixing the education system. Elements of this framework will inherently excite end users since they have been consulted from the beginning; iteration of prototypes will weed out the bad ideas early on. Most importantly, thorough problem identification and brainstorming will open new pathways and solutions that will, theoretically, take more accurate stabs at the root of a problem.

References

1. Artiles J. Education designathon. [online] 2013. www.edudesignathon.com. Accessed 10 Oct 2013.
2. Brown T, Wyatt J. Design thinking for social innovation. Stanf Soc Innov Rev. 2010;8(Winter):30–5.

3. Massachusetts Executive Office of Education. Innovation schools approval process. [online] 2010. http://www.mass.gov/edu/government/special-initiatives/education-for-the-21st-century/commonwealth-readiness-project/readiness-goal-4-innovation/innovation-schools/innovation-schools-approval-process.html. Accessed 10 Nov 2013.
4. Herasimchuk A, Wroblewski L, Sonin J, Knemeyer D. Design axioms: real data is truth. [online] 2013. http://designaxioms.com/quotes.php?view=single#realData. Accessed 19 Nov 2013.

Inquiry-Based Guided Learning to Enhance Interest and Higher-Order Thinking in Engineering Graduates: A Computing Education Perspective

B.S. Yalamanchili, R.S. Vaddi, and K.R. Anne

Abstract Engineering graduates should have good critical and logical thinking. Apart from having book knowledge, the student should be able to solve present engineering problems keeping in mind the impact of it towards global, economic, and environment aspects. The traditional teaching methodology, being teacher centered with passive students, using blackboard and audiovisual presentations will not impart the skills required for a professional student. To impart self-learning and inquiry-based learning abilities which help students in lifelong learning process, changing the teaching methodologies and evaluating system is required. This paper presents the modern teaching-learning methodology based on activity-based inquiry-guided learning (IGL). It also elevates the results of its implementation as a case study. IGL is more student centered and gives more focus on learning rather than teaching. Along with the steps involved in designing activities, the assessment methods used are covered here. The IGL methodology which is student centered when integrated into curriculum promotes process skills, peer communication, critical thinking, student-teacher relationships, lifelong learning, and responsibility for one's own learning.

Keywords Inquiry-guided learning (IGL) • Active learning • Student-centered learning

B.S. Yalamanchili (✉) • R.S. Vaddi • K.R. Anne
Department of Information Technology, VR Siddhartha Engineering College, Vijayawada, Andhra Pradesh, India
e-mail: bhanusree6912@gmail.com; syam.radhe@gmail.com; raoanne@gmail.com

R. Natarajan (ed.), *Proceedings of the International Conference on Transformations in Engineering Education: ICTIEE 2014*, DOI 10.1007/978-81-322-1931-6_35,

1 Introduction

The conventional teaching and learning process is teacher centered, and students are more passive. In modern education where more and better resources are available, the student-centered teaching and learning become more appealing [1, 2, 9]. This paper describes how inquiry-guided learning (IGL) can be used to overcome the pitfalls of conventional learning. IGL promotes various diverse learning styles and introduces different sources of learning to students. Further IGL helps students to progress through the recall of information to higher order of thinking such as understanding, applying, analyzing, evaluating, and creating.

The outcomes and assessment from the project "Inquiry-guided learning (IGL) in Computer Science (CS)," especially in ITCS 1002 Data Structures using C++, a course focused on object-oriented programming and basic data structures at the Department of Information Technology at V.R. Siddhartha Engineering College, India, are used as input for this paper. The maximum class strength being 36, this course has access to students with different knowledge levels in different disciplines such as vocational diploma and bachelor's/master's degree in mathematics, physics, and electronics.

Thus, IGL and related approaches have strong potential in CS, since they encourage students to interact with and understand the content rather than just try to memorize everything. IGL helps students to develop problem-solving and teamwork skills and also encourages students to collaborate and learn from each other rather than focus on an instructor [3]. IGL improves student attitudes and develops thinking and writing skills; that is why it is recommended as a teaching method to improve learning. However, few effective IGL activities exist for CS; thus, the faculty need to invest time and effort to develop or adapt them.

The procedures followed in implementing this project are explained in Sect. 2. Section 3 gives a detailed view on how to develop an activity. The outcome assessment and the student feedback are given in Sect. 4. The conclusions and future directions are given in Sect. 5.

2 Procedures and Material

The project set out to develop a set of IGL activities and evaluate their effectiveness compared to traditional approaches to teaching and learning.

The logical flow of the procedures can be shown as a swim lane diagram. It consists of 3 major sets of tasks which are differentiated as tasks before the activity, during the activity, and after the activity. Each set consists of 3 major tasks as shown in Fig. 1.

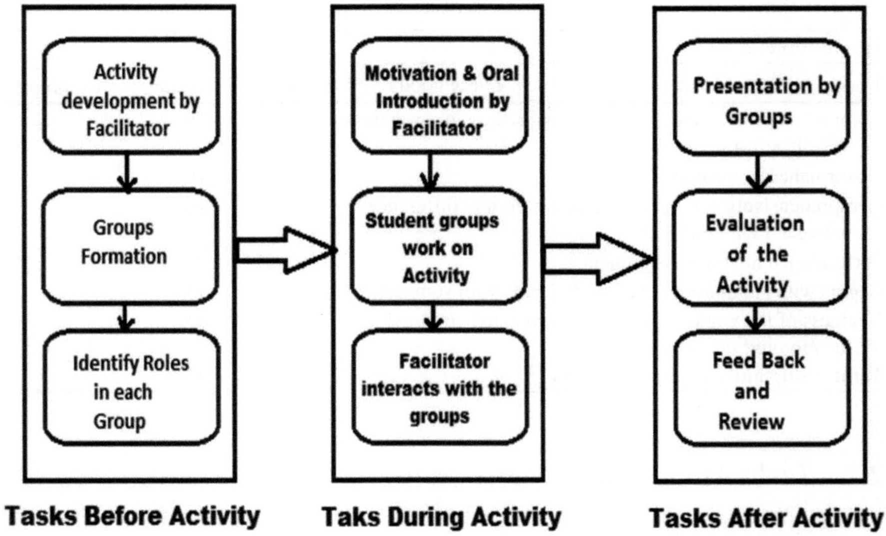

Fig. 1 Procedure in developing the IGL activity

2.1 *Tasks Before and During Activity*

The very first step in the entire IGL activity is to initiate the process, that is, to develop the tasks to be done. The facilitator has a key role in developing the activities. The team of authors cultivated a great and opening experience with IGL. After the tasks are prepared, the next step is the formation of groups to apply IGL activity. It has a greater impact in the performance outcome. The main criterion followed is that not all the students within a group are good in skills or average in skills or weak in skills. Instead, we have taken a combination of the students with difference in their skills. Each student in the group is given a specific role such as recorder, speaker, manager, and presenter. Tasks During Activity

Before starting the activity, the facilitator will give an oral introduction about the topic chosen. In the project, we have used focus question-based concept to give the overview of the topic. Every concept map responds to a focus question, and a good focus question can lead to a much richer concept map [4, 5]. The student is motivated by the focus question and has to complete the task in the class time of 120 min.

Table 1 Rubric for evaluating IGL: an assessment tool

Specific criteria 1 poor 2 satisfactory 3 good 4 very good 5 exemplary	Rating	Comments
Comprehensively synthesizes process-oriented programming and object-oriented programming	/5	
Performance in matching activity	/5	
Comprehensively synthesizes applications and difference between POP and OOP	/5	
Performance in home assignment	/5	
Begins with summary of the activity, ends with summary of questions	/5	
Neatness of the work, organization, and clarity in providing the required information	/5	
Total rating and overall comments	/30	

2.2 *Tasks After Activity*

After completion of the task, each group needs to present the work done. Finally, every task is concluded by the evaluation [6, 7]. Once all the questions have been presented and defended, students should use the IGL activity as notes for the course. We emphasize to students that IGLs are excellent study materials to use in preparing for end-unit assessments [8]. Such IGL activities are conducted for every key concept of the course, and for the present course, 9 IGL activities are prepared and conducted. The formative assessment of IGL activity in this project was done in two ways: using rubrics and quizzes.

A rubric (Table 1) is an authentic assessment tool used to measure students' work. It is a scoring guide that seeks to evaluate a student's performance based on the sum of a full range of criteria rather than a single numerical score [6, 7].

3 Activity Development

In this section, we briefly present the steps necessary to build the activity for a single concept. A sample skeleton of the activity and list of activities developed and evaluated for the ITCS 1002 course along with the evaluation rubric used is presented in this section.

In order to develop the activity, we identify and select the appropriate model for the concept. Model is an important tool in teaching-learning process to create an environment for interactive student engagement and learn quantitative skills such as graphical analysis and visualization [4, 5, 8]. In the present IGL approach, developing models to understand the concepts is fundamental, and these models are used to develop subtask section in the skeleton of the activity [9]. One such model used for searching concept is shown in Table 2. The model used can be a game, figure, problem, etc. The skeleton of the activity sheet is shown in Table 3.

Table 2 Sample model for searching concept IGL

Hi-Lo is a number guessing game with simple rules
(a) There are two players – A and B
(b) Player A thinks of a number from 1 to 100
(c) Player B guesses a number
(d) Player A responds with "too high," "too low," or "you win"
(e) Players B and A continue to guess and respond until B wins (or gives up)
Describe 4–5 different strategies that Player B could use to guess numbers (try to have a mixture of simple and clever strategies)

Table 3 Skelton of activity sheet

Contents	Page number
Facilitator information	
Learning objectives	
Activity notes	
Things to do	
Activity history	
Introduction/model	
Subtask 1 (X min)	
Subtask n(X min)	
Resources	
Home assignment	
Evaluation of the task: rubric	

4 Outcome Assessment

The outcome assessment is done in direct and indirect methods. In direct method, the student's activity sheets, presentation, and the end examination are considered. In the indirect method, students were posed with the survey questions and given an option to drop their comments in a box on their experiences with IGL. Indirect assessment method is conducted at the middle of the course and at the end of the course. Student participation in the surveys is observed to be more than 80 %. The schematic representation of entire outcome assessment is shown in Fig. 2.

4.1 Indirect Assessment

4.1.1 Student Feedback

As a part of indirect assessment, we have taken feedback from students. The responses included several recurring themes and recommendations, summarized in Table 4.

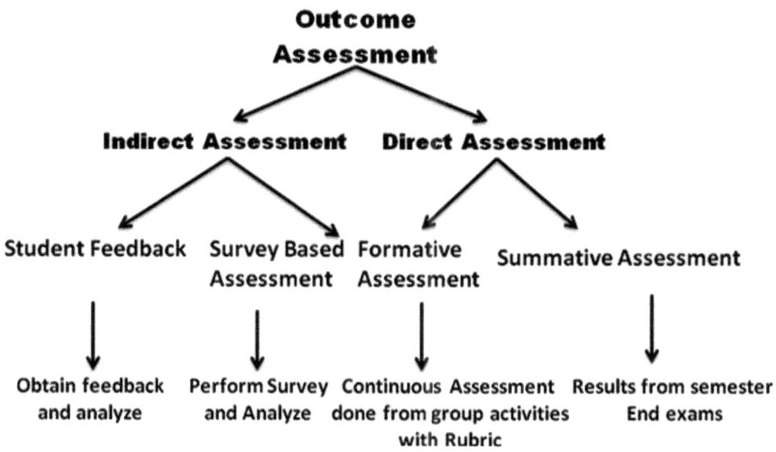

Fig. 2 Schematic representation of the entire outcome assessment

Table 4 Student feedback

Activities increase student interest and confidence	*Teammates can improve each other's understanding*
By these discussions we got more interest	Clubbing their ideas gives more views on the concept
We got the confidence of solving complex problems	Different views and different pros and cons are touched by the group discussions
I used to have an inhibition before doing something like "what if I can't do this"	
I can feel that I can learn and I can do things	
Activities increase communication skills	*Activities increase knowledge and understanding*
Soft skills and communication skills are going to be increased	Discussion of concepts is intensive so we gain more knowledge
Also improving our interactive skills	It improves the thinking levels beyond the textbooks
Improve communication skills	We got some knowledge when compared to previous learning
Ability to speak freely to professor or student	

4.1.2 Survey-Based Assessment

Apart from the comment-based feedback, we have conducted a survey at the end of the course to analyze the impact of all IGL activities; students had three choices (agree, neutral, disagree) to respond to four questions. Figure 3 shows the response distributions.

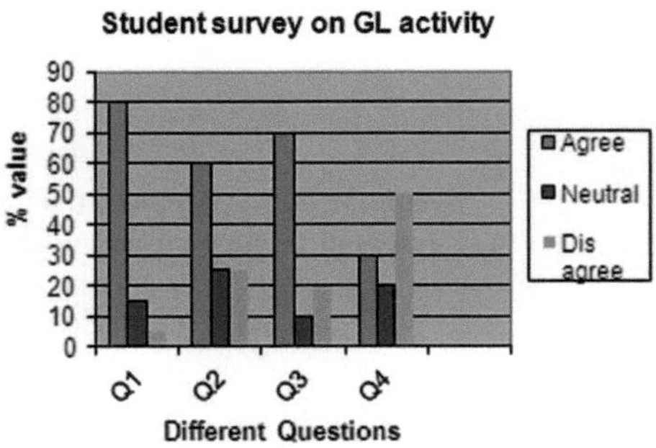

Q1. IGL activities promoted interest towards the course

Q2. IGL activities provided methods to identify a real time problem and solve it.

Q3. IGL activates improved the self-learning interest and skills

Q4. IGL activities are not giving scope for individual merit

Fig. 3 Student survey results

4.2 *Direct Assessment*

The direct assessment of the student's performance is measured continuously through their activities and with a semester-end examination.

4.2.1 Formative/Continuous Assessment

The continuous assessment is done mainly based on the activity end evaluation rubric [10]. Rubrics are developed for the present IGL activity to assess student performance by clearly indicating the two criteria's like specific to the concept and general skills with their ratings for total 20 marks, as shown in Table 1.

4.2.2 Summative/Semester-End Assessment

The semester-end assessment is prepared to check weather students have achieved the course learning outcome. The results obtained from the evaluation for 4 consecutive years and the respective average result in the form of pie diagram is shown in Fig. 4. The results obtained by the continuous assessment and semester-end assessment appeared almost similar.

5 Conclusion and Future Directions

Inquiry-guided learning (IGL) is based on learning science and has a proven track record in other disciplines. By this different teaching idea, the students as well as facilitators are well benefited. IGL classrooms are very different from lecture-based classrooms, but given the effectiveness of IGL and other forms of active learning,

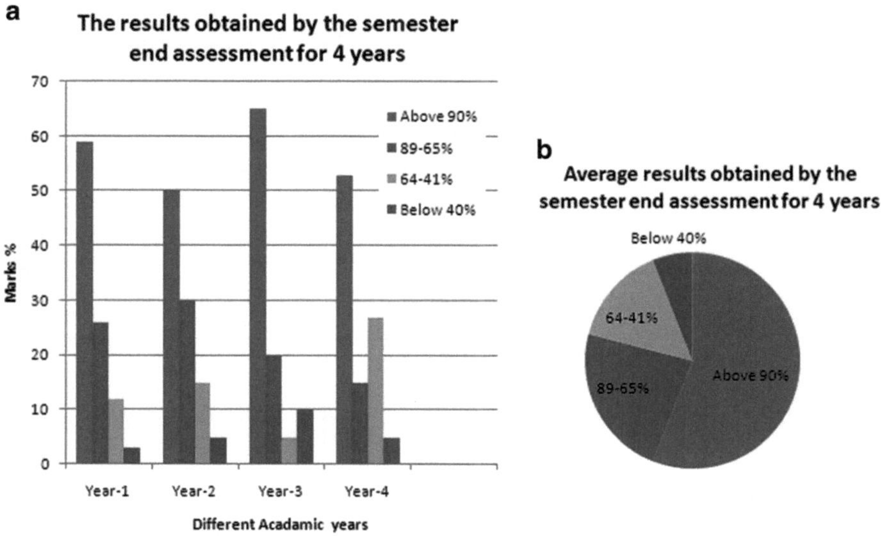

Fig. 4 (**a**) Result analysis of 4 continuous semester-end exams. (**b**) Average results of 4 years

the faculty should become familiar with such approaches. Student-centered learning integrated into curriculum promotes peer communication, builds student-teacher relationships, strengthens student motivation, and promotes responsibility for one's own learning.

Our plans for the future include replication of the experiment with other group of students and other courses. Developing course content for computing subjects like data structures, C language, and JAVA in the IGL model is one of our major goals. Finally, this project aims to develop and revise more IGL activities and extend and adapt them for courses and faculty at other institutions.

References

1. Lewis CM, Titterton N, Clancy M. Developing students' self-assessment skills using lab-centered instruction. J Comput Sci Coll. 2011;26(4):173–80.
2. Regueras LM, Verdú E, et al. Effects of competitive e-learning tools on higher education students: a case study. IEEE Trans Educ. 2009;52(2):279–85.
3. Bonwell CC, Eison JA. Active learning: creating excitement in the classroom. George Washington University 1991, SHEERIC High.Educ.
4. Schneider RF. Why concept maps? Available: http://www.ic.sunysb.edu/Class/che133/conceptmaps/Example/MapOverview.html. March 7, 2012.
5. Cañas AJ, Novak JD. Constructing your first concept map. Available: http://cmap.ihmc.us/docs/ConstructingAConceptMap.html. March 7, 2012.
6. Amador JA, Miles L, Peters CB. The practice of problem-based learning: a guide to implementing PBL in the college classroom. Bolton: Anker Publishing Company; 2006.

7. (2008, 30 July 2012). Formative Assessment Strategies. Available: http://daretodifferentiate.wikispaces.com/file/view/03+-+Formative+Assessment+Strategies.pdf
8. Hanson DM. Designing process-oriented guided-inquiry activities. In: Faculty guidebook-a comprehensive tool for improving faculty performance. 2nd ed. Lisle: Pacific Crest; 2005.
9. Giaohui wu. Discussion about active learning of undergraduates. In: Proceedings of the 2nd international conference on education technology and computer (ICECT) 2010.
10. (2009, 24 July 2012). Formative Assessment Tools & Strategies for 21st Learning. Available: http://wths-cff.wikispaces.com/file/view/Formative+Assessment+Tools.pdf

Measuring the Impact of Design Fixation on Indian Engineering Students

V. Mahesh, M. Kantha Reddy, P.V. Raja Shekar, and P. Pramod Kumar

Abstract A major concern in engineering education pertains to design fixation, which hinders the conception of novel ideas. The term design fixation refers to the designer's reluctance (or inability, in some cases) to consider multiple strategies to formulate and solve a design need. The design fixation phenomenon severely limits creativity and results in pedestrian design solutions. The primary objective of this paper is to correlate the effect of the Indian engineering education with the extent of design fixation in the engineering students of different disciplines. This work provides a deeper understanding of the design fixation phenomenon by establishing a conceptual framework for the design process based on the theories of knowledge representation in cognitive science. It can provide crucial insights that will help students to overcome the adverse effects of fixation.

Keywords Design fixation • Engineering education • Cognitive science • Conceptual design

V. Mahesh (✉)
Mechanical Engineering, S. R. Engineering College, Warangal, Andhra Pradesh, India
e-mail: v.mahesh2@gmail.com

M.K. Reddy
Computer Science and Engineering, IUCEE, Hyderabad, India
e-mail: Kanthareddy_m@yahoo.com

P.V.R. Shekar
Humanities and Sciences, S. R. Engineering College, Warangal, Andhra Pradesh, India
e-mail: pvrsleo@gmail.com

P.P. Kumar
Computer Science and Engineering, S. R. Engineering College, Warangal, Andhra Pradesh, India
e-mail: pramod1217@yahoo.com

R. Natarajan (ed.), *Proceedings of the International Conference on Transformations in Engineering Education: ICTIEE 2014*, DOI 10.1007/978-81-322-1931-6_36,

1 Introduction

According to Hershberger [1], “Everything that is absorbed and registered in your mind adds to the collection of ideas stored in the memory: a sort of library that you can consult whenever a problem arises. So, essentially the more you have seen, experienced, and absorbed, the more points of reference you will have.” This quote is true in the field of engineering design, where the field has changed from a design-scratch to a design-through-synthesis mindset, where designers transform, combine, or adapt elements of previous or existing designs to synthesize novel concepts [2, 3]. However, the use of previous designs can adversely impact the design process in the form of design fixation [4], a potentially limiting adherence to existing designs. The information that designers “absorb and register” in their mind has the potential to fixate them during the design process and result in lack of innovation. As the innovation is key to success in this economy, it is imperative that we learn to effectively manage design fixation. The first step in managing fixation is understanding how different activities affect design fixation during the design process. It is noted in the literature that fixation occurs across different levels of expertise [5] and contexts [6]. The understanding can lead to development of effective product development methods and teaching methods to reduce fixation effects. The analysis reported in this paper gives the effects of design fixation in Indian engineering students both seniors and freshmen. This report also provides evidences for design fixation and its effects on Indian Engineering Curriculum.

2 Literature Review

The field of design has explored many formal and heuristic approaches to design refinement, manufacturing, generation, and computation [7–9]. The descriptive methods research in design characterizes the natural thought process followed by humans in solving a design task. Therefore, it is an application of principles of cognitive psychology, to design. The descriptive research involves understanding how the designers perceive the problems and the supportive statements, find (creative) solutions, evaluate designs, and make decisions.

In this field, researchers vigorously pursued cognitive-based design fixation research and abstractly defined as barriers to solution based on real and perceived constraints [4, 10, 11].

Dahl and Moreau [12] demonstrated that subjects exposed to within-domain examples employed fewer out-of-domain analogies in generating solutions and that the originality of the solutions produced increased relatively when subjects were encouraged to use analogies extensively and given no example solutions. Marsh et al. [13] found that within-domain examples caused subjects to be biased toward generating solutions with similar features to those found in the examples. A common and often commented upon form of fixation is premature commitment to a particular problem solution. Consequently, the designers stop pursuing the search for alternative solutions. This premature commitment thus results in fewer solutions. These findings establish that any experiment with exposure to a within-domain example prior to ideation can cause fixation. The fixation is manifest in the form of fewer solutions and the appearance of features from the examples in the solutions [14].

3 Measuring Design Fixation

Design fixation is the designers' reluctance to consider multiple strategies to formulate and solve a design problem. The goal of the experiment is to evaluate the effects of fixation on senior academic engineering students and engineering freshmen. To answer this research issue, three cognitive skills, viz., fluency, flexibility, and originality, and three design parameters, viz., meeting the intent, simplicity, and user-friendly, are identified and are measured from the drawings given by the participants to the design task. The metric for various parameters is given in Table 1.

The participants were divided into two groups – a fixation group where the participants were exposed to an example design solution and a control group to whom only a problem statement is given without any example solution. Control group is used to establish the baseline.

3.1 Design Task

Participants in both groups were presented the design task of "design a toy for a blind child of age group 3 to 7 years." The design task is a real existing problem. Participants were provided a paper, pen, pencil, and eraser to sketch their designs. They worked on the design task individually, not in teams. Participants had 45 min to generate as many ideas as possible. After the completion of the task, participants were given 30 min to answer a questionnaire to help analyze their perceptions.

The toy for blind child problem puts most participants in the familiar domain of toys and unfamiliar domain of blindness. While the task requires minimal technical knowledge (application of scientific and mathematical principles), it uses the design

Table 1 Metrics for measuring design fixation skills

Cognitive skill	Definition	Metric
Fluency	Ability to generate many solutions consistently	Quantity of ideas generated
Flexibility	Ability to explore design space in many directions	Variety of ideas generated
Originality	Ability to generate unexpected solutions; think out of box	Originality of ideas generated

Table 2 Number of participants representing different majors

	Control group		Fixation group	
Major	Male	Female	Male	Female
Mechanical engineering	8	6	8	7
Civil engineering	8	6	6	8
Electrical and electronics engineering	7	8	6	8
Electronics and communications engineering	7	6	7	7
Computer science engineering	5	6	6	5
Information technology	3	5	4	3

skill set such as understanding the need and synthesizing alternative solutions which is common to creative problem solving in many domains. The task engages by providing autonomy, mastery, and purpose. Participants can design their solutions that reflect their values, views, and interests (autonomy); show creativity (mastery); and cherish results (purpose). Hence, the design has the intrinsic motivation for engagement. A diverse set of available solutions can satisfy the task.

The participants are told that the goal of the experiment is to generate as many solutions to the design problem as possible, where a prize will be given to participants with the greatest number of solutions. This prize is an extrinsic motivation for participants to devote serious effort to the design activity.

3.2 Control Group

Participants in the control group are provided with the design problem. They are not provided with any example solution or alternative representation of the design problem. A set of 150 engineering students is taken to conduct the task as shown in Table 2. Of them 75 are into the control group and the remaining into the fixation group.

3.3 Fixation Group

Participants in the fixation group are given the design task and a model solution as shown in Fig. 1. The solution is a colored piano, which highlights its features and parts and a short note stating: "This toy is enabled with braille for ease of use for a blind child. It is enabled with record and playback features which are fulfilled using microphone and speakers."

This solution is focused on the hearing and sensing aids, which are fulfilled using microphone, speakers, and braille on the keys of the musical instrument.

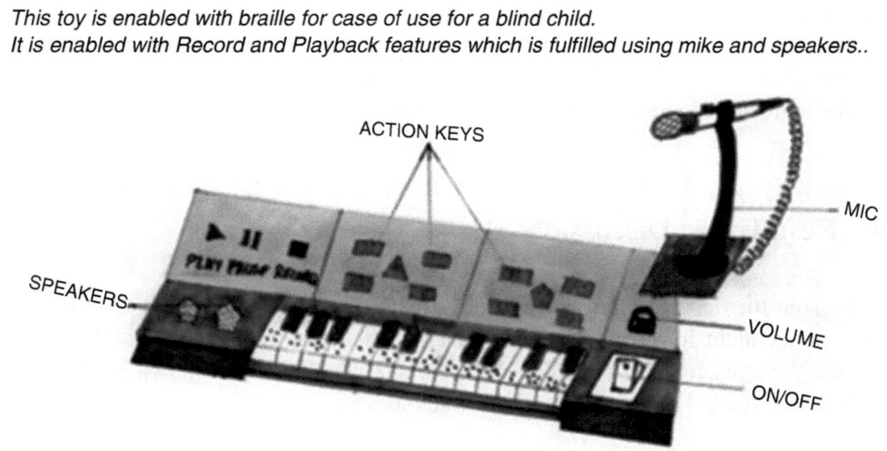

Fig. 1 Solution provided to the fixation group

3.4 *Hypothesis*

3.4.1 Hypothesis 1

Academic engineering freshmen are considered to show less evidences of design fixation and are likely to conceive more novel concepts. They exhibit less fixation compared to senior students.

It is expected that the creative problem-solving intelligence is highly active with less signs of fixation in freshmen students when compared to senior engineering students. There will not be any influence of curriculum and teaching pedagogy on engineering freshmen.

3.4.2 Hypothesis 2

Academic engineering students from mechanical trade can implement greater analogies and produce solutions with more design flexibility.

Design flexibility is an important skill in the context of product design. Designers with high design flexibility are open to acquiring the information and using it to adapt refine designs. They generate alternative solutions using the information and quickly evolve their designs.

To relate the given instructions and also prior design knowledge of putting them together to solve a problem is comparatively more intense and efficient in mechanical engineering students than in other trades. This is because mechanical engineering students are taught design-related approaches in their majors unlike other circuit trades such as electrical, electronics, and computer sciences. Thus, mechanical engineering students are exhibiting more design flexibility than students from other trades/branches of engineering.

4 Results and Discussion

Data from the design solutions and questionnaires is collected together, and a few important conclusions have been made in concurrence with the metrics of the experiment. Average scores of cognitive skills earned by students are shown in Fig. 2.

Following are the few results of the analysis:

- Academic freshmen designs are more original than those of senior engineering student whose designs are less fluent and flexible which supports our hypothesis 1.
- Designs of mechanical engineering students are far more fluent, flexible, and original than those of senior engineering students, supporting hypothesis 2.
- Senior students are more familiar with the design task given to them than freshmen students. Apart from the graphical representation, this result can also be supported by the analytical conclusion from the above statements, which show that as students are being aware of problems in the society, they start thinking analytically to solve them and thus were aware of the task given to them to some extent.
- The average scores earned by freshmen and seniors of both control and fixation groups based on the design parameters are shown in Table 3. It is evident from the results that freshmen show better overall performance than seniors in fulfilling the design parameters.

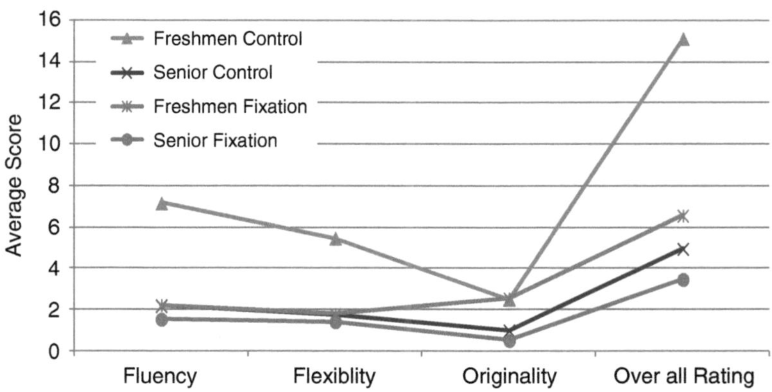

Fig. 2 Average scores of various cognitive skills

Table 3 Average design parameter scores

	Freshmen	Senior	Freshmen control	Senior control	Freshmen fixation	Senior fixation
Meeting the intent	0.764	0.690	0.885	0.750	0.655	0.628
Simplicity	0.964	0.678	0.962	0.773	0.581	0.966
User-friendly	0.855	0.678	0.760	0.750	0.931	0.605

Few more following observations were made by analyzing the questionnaire:

- Freshmen fixated group are highly influenced by the instructions given to them in the form of an example solution compared to the control group where there was no example solution provided. Comparing with the fixated group of freshmen students, seniors are less influenced by the instructions given to them.
- On average freshmen students feel that the assignment is moderately difficult (around 3.5 on 5) both in fixated and control groups. The same relative pattern is also followed in senior students. There is no difference in the mean rate of difficulty level in both senior and freshmen students.
- On average freshmen students felt their work to be highly creative giving 4.5 as the mean rating, and senior students both fixated and control together on an average rated 3.5 on 5.
- Freshmen students have rated their own performance and quality in the design task as 4 on 5 on an average. Senior engineering students have rated their performance in the task as 3.5 on 5 on the whole.

5 Conclusion

The experimental results show that freshmen engineering students exhibit fewer signs of fixation compared to seniors. These results highlight an immediate need for modifying the teaching paradigm for engineering education to increase the emphasis on innovation and creativity in the curriculum to keep the students competitive in the global economy.

The results also show differences in the level of fixation based on the discipline and identify mechanical engineering students are exhibiting more flexibility.

The paper also recommends that engineering faculty be conscious of fixation and its adverse effects and educate students. In the long run, as educators, we must reflect and reformulate the curriculum to balance fundamental understanding with the need to foster creative thinking skills so that the next generation designers not only analyze designs but also synthesize innovative solutions.

Acknowledgements The authors would like to acknowledge the support of the Department of Science and Technology, New Delhi, under Grant No: SR/CSI/130/2011(G) to carry out the present research.

References

1. Herzberger H. Lessons for student in architecture. Rotterdam: Uitgeverij; 1991.
2. Eckert C, Stacey M. Sources of inspiration: a language of design. Des Stud. 2000;21:523–38.
3. Eckert C, Stacey M, Earl C. Reasoning by references to past design. International workshop on studying designers. 2005.
4. Jansson DG, Smith SM. Design fixation. Des Stud. 1991;12:3–11.
5. Linsey JS, Tseng I, Fu K, Cagan J, Wood KL, Schunn C. A study of design fixation, its mitigation and perception in engineering design faculty. J Mech Des. 2010;132:41003.
6. Purcell AT, Gero JS. Design and other types of fixation. Des Stud. 1996;17:363–83.
7. Antonsson EK, Cagan J. Formal engineering design synthesis. Cambridge: Cambridge University Press; 2001.
8. Stone RB, Wood KL, Crawford RH. A heuristic method to identify modules from a functional description of a product. Des Stud. 2000;21:5–31.
9. Chakrabarti A. Engineering design synthesis: understanding approaches and tools. London: Springer; 2002.
10. Perttula M, Sipila P. The idea exposure paradigm in design idea generation. J Eng Des. 2007;18(1):93–102.
11. Moss J, Kotovsky K, Cagan J. The influence of open goals on the acquisition of problem-relevant information. J Exp Psychol Learn Mem Cognit. 2007;33(5):876–91.
12. Dahl DW, Moreau P. The influence and value of analogical thinking during new product ideation. J Mark Res. 2002;39(1):47–60.
13. Marsh RL, Landau JD, Hicks JL. How examples may and may not constrain creativity. Mem Cognit. 1996;24(5):669–80.
14. Simon HA. The sciences of the artificial. Cambridge, MA: MIT; 1968.

PPP: A Paradigm for Online Education in Engineering Colleges

Kalyana Chakravarthy Chilukuri and K.V.L. Raju

Abstract We propose a prepare, present, and publish (PPP) paradigm for online education in engineering colleges. We comprehensively survey various cost-effective avenues for digitizing and disseminating E-learning material based on the learners feedback. Most of these methods were implemented online and were evaluated based on survey polls and the learner's feedback. Since it is a difficult proposition for all engineering colleges to adopt a common method owing to their demographic, economic, and social differences, we analyze the strategies on factors such as the complexity and privacy. These online teaching strategies help a teacher save his valuable time by incrementally adding and revising his content, which allows him to recollect a previously taught course and gives him a larger scope for learning newer courses. In a student's perspective, learning becomes easier, particularly in case one is not very comfortable with the teacher's accent or has not completely understood or missed out on a particular topic, for instance.

Keywords PPP • E-learning • Online education

1 Introduction: Online Education, Advantages and Challenges

In the digital era, active teaching and learning using Internet and innovative techniques is slowly but surely taking over the conventional chalk and board system of education. For an experienced teacher, the plethora of tools allow quality content to be developed

K.C. Chilukuri (✉)
Department of CSE, MVGR College of Engineering, Vizianagaram, India
e-mail: kalyan@mvgrce.edu.in

K.V.L. Raju
Principal, Department of Mechanical Engineering, MVGR College of Engineering, Vizianagaram, India
e-mail: principal.mvgr@gmail.com

R. Natarajan (ed.), *Proceedings of the International Conference on Transformations in Engineering Education: ICTIEE 2014*, DOI 10.1007/978-81-322-1931-6_37,

Table 1 Tools for online education – a comprehensive survey

S.No.	Tool	Online?	Live/recorded	Advantages	Drawbacks
1.	PowerPoint with animations	No	Live (classroom)/ recorded narrations and animations	Small size Simple to install and set up	Needs a presenter
2.	Webinars	Yes	Live presentations, desktop sharing	Interactive	Presenter not visible, some issues with audio
3.	Skype	Yes	Live presentations, desktop sharing	Interactive	Issues with video, audio for increasing users
4.	VLC media player	Yes	Live video streaming, desktop sharing	Simple to install and set up	Issues with video, audio for increasing users
5.	Vimeo video hosting	Yes	Recorded videos	Easy to upload HD videos	User needs to sign up
6.	YouTube video hosting	Yes	Recorded videos	Easy to upload videos using existing Gmail IDs	Video access modes may effect popularity
7.	LENSOO video hosting	Yes	Recorded videos/ courseware	Easy to upload and market videos and courseware	No support for uploading PowerPoint with narration

in a short time and open sourced or distributed for a nominal amount to meet his efforts, whereas for a young, fresh teacher, they allow him to evaluate himself and correct himself for continuous improvement. Choosing effective tools to prepare or create, present, and publish or host (PPP) facilitates the entire digitization process for online teaching. Recent technological advances like videos, podcasts, wikis, and blogs and discussion forums can also help personalize the discipline and improve student's interaction with the teacher enabling them to see him as an individual who cares about their learning and success [1]. There are many ways to foster student engagement with faculty, peers, and content utilization in an online environment. Judicious and strategic use of humor, for instance, is an effective strategy instructors might employ to make their course content more interesting and enthralling to online students [2].

However, challenges still exist such as "which is the best online tool to teach a course" and assessment of student learning online. Moreover, required resources for the chosen online teaching strategy are a serious consideration. In spite of the challenges associated with online learning, online learning enrollments in advanced nations like the USA continue to grow at a much faster rate than overall enrollments in higher education, and nearly 30 % of US college and university students are now taking at least one online course [3]. It was predicted that this surge will not deplete but rather increase further owing to such factors as higher fuel costs, which make learning from home advantageous and practical (Allen & Seamen, 2008).

We present some of the tools that help a teacher better plan, prepare, organize, and disseminate his course content to students, promoting a congenial learning environment that caters to different learning styles.

2 Tools for Online Education

2.1 Creating Intuitive and Interactive Presentations

It is important for online and distance learning faculty to select strategies that are most appropriate for the intended outcomes of the course [4]. High-quality presentations are an important vehicle in making even the toughest of courses interesting and enjoyable for a learner. In particular, for courses that have a lot of logical explanation or theory, a simple diagram or a simulation would greatly help simplify things. The following are some of the tools for creating high-quality material.

2.1.1 PowerPoint with Audio and Animations

All versions of Microsoft PowerPoint after PowerPoint 2007 support numerous animations. In the animations, several effects such as entrance, emphasis, exit, and motion path can be chosen while making the slides. In particular, within motion paths, the custom path can be effectively used to move a desired object from one end along a path to another. We can also assign a trigger for actions. We may specify the object to move along the custom path (which we can draw using free hand) by clicking the mouse button or a key press. MS PowerPoint also has option to record an entire presentation along with audio, laser pointers, and animations. PowerPoint slides with audio and animations are tools that help the student recollect any missed out concepts and visualize sequences of actions in diagrams which are otherwise quite difficult explaining. We can take any scanned picture from a textbook or one that is available on Internet and, with some practice using PowerPoint, animate parts of the diagram. The most interesting fact with MS PowerPoint is that animations are triggered in the sequence they are created, which may be changed later. All we require for recording voice is a good quality microphone. The teacher may also choose to record the PowerPoint presentation with voice as he is presenting in the class, using an inexpensive collar microphone and a wireless receiver plugged into his laptop. Prezi is a nice alternative to PowerPoint, though comparatively limited in features. Earlier, Screen-Recording Software such as Camtasia and Adobe Captivate were popular tools for creating audio and video recordings from PowerPoint presentations. PowerPoint 2010 offers an option to directly save and send presentation as a video.

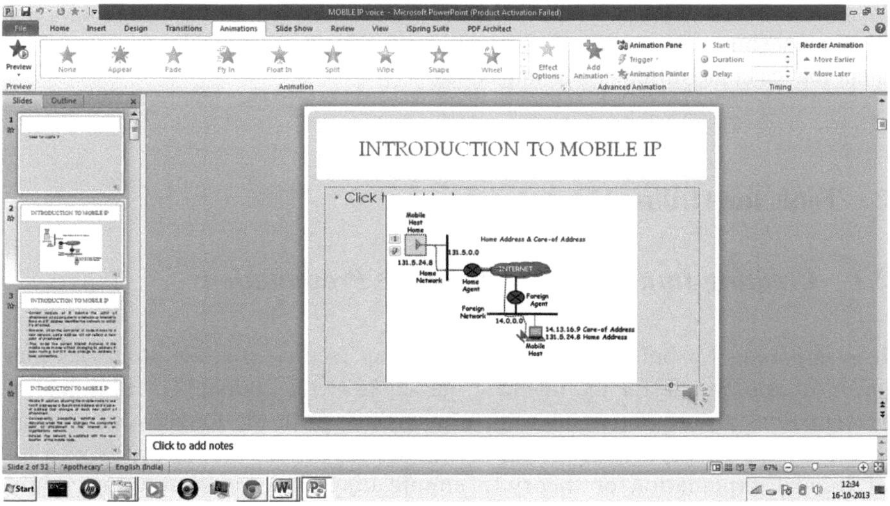

An example PowerPoint animation

2.2 *Presenting Online: Live Presentations*

While the influence of online teaching/learning in improving education standards is unquestionable, new issues confront instructors and students. Faculty, for example, have a tendency to cite an increase in workload as a challenge when teaching an online course [5]. In fact, preparing to teach in an online format for the first time generally requires at least 10 h of training outside of a regular teaching load according to Instructional Technology Council. Moreover, perception of social presence and interaction seems to be a concern common to both teachers and students. However, practices such as daily communication, periodic feedback, and response to inquiries will overcome these challenges, though it requires added commitment and responsibility on the part of the teacher. Video conferencing adds a new dimension of interaction and communication that was unavailable previously in the online learning environment [6, 7]. Issues that affect video conferencing include cost-effectiveness, access to recorded sessions, connectivity, scalability, and participant interaction [8–10].

2.2.1 Webinars for Live Presentations

Webinars are an extremely cost-effective solution, initially meant for conducting online meetings for large audience without them having to relocate from their place of work (or even the comfort of their home for that matter). They allow the presenter a wide range of options including PowerPoints, uploaded YouTube videos, or even a simple pdf or word document. The presenter can also mute or unmute his audience as he provides them opportunity to interact on microphone, indicate their mood, etc.

Two of the most interesting features in a webinar can however be to create survey polls while the webinar is conducted and to record the webinar to be published later either on YouTube or on some other platform such as the Learn Earn Network Socialize (LENSOO). Most popular webinar software includes GoToMeeting, Citrix, and AnyMeeting. All these software are useful for lecture-style classes where the instructor wishes to talk while showing students PowerPoint presentations, videos, or various websites. Though they are not designed specifically for the college classroom, they can easily be adapted for that purpose. Citrix's marketing material indicates that it has been used in such a setting [11], but it is difficult to find any scholarly reporting on such use. AnyMeeting provides the option for unlimited free webinars for a maximum of 200 audiences, with the only limitation stated being the webinar cannot be recorded live. Webinars allow foreign experts all over the world to present and share their courses for a nominal fee which includes the presentation and content hosting charges without them having to relocate to other countries. The Indo US Collaboration for Engineering Education (IUCEE) has been one organization using webinars most effectively as a medium for knowledge dissemination. It has been instrumental in connecting professors in the USA with other professors, faculty, and students in India.

Interestingly, most of the issues when conducting webinars were reported to be related to audio, not the video or even desktop sharing feature. Frequent voice dropouts and noise are quite common. It justifies the need for improving the so-called voice over IP technologies and integrating them with the webinar software. Another serious drawback with almost all webinar presentation software is that they do not allow a presenter to stream his videos unless he uploads them to YouTube. This is annoying because, while presenting, occasionally the presenter might like to stream his videos directly and show his audience rather than wait for long period uploading them. Moreover, the presenter might not always want to upload his proprietary content.

2.2.2 Skype for Presentations

Yet another tool for sharing presentations and videos online using the desktop sharing feature is Skype where in the audio quality is an issue when sharing with a number of users, though the quality of video is quite good. Shared online videos have been and remain quite popular. They represent 53 % of the Internet's traffic. However, the rapid advancements in wireless Internet resulting in even cheaper bandwidth costs for wired connections make video streaming absolutely smooth and seemless today. VLC includes a fairly easy-to-use streaming feature that can stream music and videos over a local network or the Internet and supports desktop capture, remote desktop sharing (screen casting), broadcasting, or multicasting video or audio in formats such as wma, mp4, and mp3 to remote locations. Although VLC is extremely good as a stand-alone player, its performance when streaming videos (particularly those with a high frame rate such as action videos), or screencasting, is debatable as the number of systems connecting to the streaming server increases.

2.3 *Video Hosting and Sharing*

One solution to sustain student's interest and to prevent unauthorized usage will be to host the videos periodically on a video sharing site. Mentioned below are some of the recent free video hosting solutions besides already popular Apple's iTunes U.

2.3.1 Vimeo for Video Hosting

Vimeo is a useful tool for uploading videos free so that one can only view videos without downloading. Free member gets 500 MB space/week after which his uploaded videos would expire. He would then be required to re-upload his videos. Nearly 3 h duration of video lectures in MP4 format with narrations and animations can be uploaded every week on Vimeo for students to make up for lost/miscellaneous classes. Though this method ensures videos are not downloadable and misused, it would require students to sign up and create accounts in order to upload or view videos. Using this method, in 4 week's time, a teacher may upload his entire course. He might also choose to upload more difficult units a number of times based on students' demand.

2.3.2 YouTube Videos for Video Hosting

Perhaps the most popular free video sharing option thus far is YouTube videos, which allow a teacher to upload and share his videos in the private, unlisted, and public modes. In the private mode, only users added by uploader would be granted access, in the unlisted mode, anyone with the link will be able to watch the videos, while in the public mode, any one could search for and watch the videos. The advantage of the method is that both the teacher and the student need not have a separate account and can watch or upload videos using their Google's Gmail account.

2.3.3 LENSOO for Course Development and Hosting

Learn Earn Network Socialize (LENSOO) is yet another platform that is gaining in popularity. The IUCEE uses LENSOO as a medium for hosting the recorded webinars by US experts. LENSOO offers several options that allow users to create and upload their courses including quizzes and pdf and video formats. It is quite similar to Moodle with several more exciting features such as allowing the provider to set a price for his course, to allow users a sneak preview of his course content, etc. It is in fact a first of its kind solution that combines content development and marketing at a go without having to do them separately. A developer can market his course right away after he develops it. However, support for streaming videos and uploading PowerPoint presentations with audio is still to be tested for success.

2.4 Outcomes

A part of course on mobile computing or students of CSE and IT, consisting of four units on Mobile IP Network Layer, Synchronization, Mobile Ad Hoc Networks, and Wireless LANs, was created using Microsoft PowerPoint 2010 with recorded narration and animations. Particularly for the first two units (Mobile IP and Synchronization) involving a lot of concepts to be memorized, a lot of animations, diagrams, and case studies beyond textbook have been added. The presentations were then converted to videos in wmv format, which were compressed into mp4 format and then uploaded to Vimeo and YouTube and were shared with 120 students. Videos were uploaded in private mode and links shared with the students. Students were asked to give thumbs up, thumbs down, or any comments and suggestions as feedback without any apprehensions. Before implementation of the method, students were critical in their feedback and were quite difficult to engage and deal with. In fact, the motive to this method is one of the students mentioning, "I am unable to understand when lecturers teach fast in English." This reflects the problems of students from rural engineering colleges in general. The response from students was overwhelming, against the one with PowerPoint without animations or voice. Some comments from students read "Now, I do not to have to go to my friend's house for clearing doubts." "This technology is very useful particularly for students who were absent for some classes." There were nearly 90 thumbs up for Unit Mobile IP and 30 for the Unit Synchronization, while other two units have recently been uploaded and comments are still awaited. There have been five comments, very positive for the first two units. There were not even single thumbs down or a negative remark by any of the students, indicating they all like the material.

A test webinar was conducted with thirty students of different percentage groups on the topic Mobile IP. AnyMeeting software was chosen to conduct a 1 h presentation. Except for three students for which uplink quality has been reported as A and downlink as A+, for the rest, these were reported as B or lesser. This is due to the fact that most of these students were attending the webinar from rural locations and the Internet speeds for these users are less than 2 Mbps. Twenty-eight students could see presentations without problems; however, many of them faced issues with audio reception and desktop sharing. Interestingly though the webinar was tested on a 16 Mbps leased line within the college, there were still some issues with audio; however, presentation and desktop sharing worked extremely well.

3 Conclusion

Though several tools are available today that make developing and sharing course material much easier, whether presenting online is as effective as teaching live in a classroom is still a topic of debate. The approach of traditional teaching followed by uploading content is the solution that works best in sustaining the interest and improving results of students. A teacher would be appreciated and widely followed if he adapts different styles of teaching using different tools. A teacher could conduct few webinars for add-on topics in one semester, while in another he could conduct a Skype conference with his students. He can use tools like Facebook, Twitter, or Google Hangouts for improving his interaction with students. Major challenge though is how to create an enabling environment for such teaching-learning techniques to be successful in a rural backdrop since students in rural engineering colleges typically need much motivation and preparation for setting up and persisting with the changing teaching methodologies, in spite of technological advances. Other challenges are the motivation levels of teachers themselves in preparing additionally to contribute to such enhancements.

References

1. Brinthaupt TM, Fisher LS, Gardner JG, Raffo DM, Woodard JB. What the best online teachers should do. MERLOT J Online Learn Teach. 2011;7(4):515–24.
2. LoSchiavo FM, Shatz MA. Enhancing online instruction with humor. Teach Psychol. 2005;32(4):246–8. doi:10.1207/s15328023top3204_8.
3. Allen IE, Seaman J. Staying the course: online education in the United States. Retrieved April 17, 2008, from Sloan-C Web site: http://www.sloanc.org/publications/survey/pdf/staying_the_course.pdf. 2008.
4. Lee MJW, McLoughlin C, Tynan B. Podcasting in distance learning: true pedagogical innovation or just more of the same? In: Lee MJW, McLoughlin C, editors. Web 2.0-based e-learning: applying social informatics for tertiary teaching. Hershey: Information Science Reference; 2011. p. 228–46. doi:10.4018/978-1-60566-294-7.ch012.
5. Sammons R, Ruth S. The invisible professor and the future of virtual faculty. Int J Instr Technol Distance Learn. 2007;4(1):3–17.
6. Palloff RM, Pratt K. Building online learning communities: effective strategies for the virtual classroom. San Francisco: Jossey-Bass; 2007.
7. Palloff RM, Pratt K. The excellent online instructor: strategies for professional development. San Francisco: Jossey-Bass; 2011.
8. Kear K. Online and social networking communities: a best practice guide for educators. New York: Routledge; 2011.
9. Smyth R, Zanetis J. Internet-based videoconferencing for teaching and learning: a Cinderella story. Distance Learning. 2007;4(2):61–70. Retrieved from http://www.usdla.org/assets/pdf_files/DL_4-2.pdf
10. White B. Using ICT to enhance curriculum opportunities for students in rural and remote schools. Aust Educ Comput. 2010;25(2):27–30. Retrieved from http://www.acce.edu.au/sites/acce.edu.au/files/pj/journal/AEC%20V25N2%20UsingICTtoEnhance.pdf
11. Mainstay Salire. OneMBA® program at UNC's Kenan-Flagler Business School. Retrieved from http://www.gotomeeting.com/fec/images/pdf/caseStudies/GoToMeeting_GoToWebinar_UNC_Case_Study.pdf. 2012.

Innovative Method of Teaching Digital Signal Processing Using Ubiquitous Learning Strategies

C.V. Suresh Babu, J. Vijay Daniel, and S. Joshna

Abstract Meaningful learning is often regarded as the ultimate learning status for a learner, regardless of the learning environment. Digital signal processing (DSP) is a core paper for students of Electronics and Communication. Unlike other subjects, DSP is considered to be a tough subject by students. In this paper, we provide suggestions for instructors and designers so that they can promote the quality of learning. This paper was inspired by various research outcomes in pedagogical teaching methods with respect to digital signal processing in which many researchers have discussed on ways to improve a classroom environment. Those ideas are discussed in this paper and then supplemented with general advice and specific suggestions from the experience of the faculties who handle digital signal processing for quite some time.

Keywords DSP • PBL • Pedagogical learning • Simulation software's and mini projects

1 Introduction

Digital signal processing has been a major specialty foundation course in the field of Electronics and Communication. This makes it vital that Communication Engineering students have a good understanding of the fundamentals of the subject and a practical appreciation of the possibilities it offers. This subject has a great

C.V.S. Babu (✉) • J.V. Daniel • S. Joshna
Department of Information Technology, VelTech MultiTech Dr. Rangarajan Dr. Sakunthala Engineering College, Chennai, India
e-mail: cvsureshbabu@yahoo.com; vijay.daniel@hotmail.com; joshna.cs@gmail.com

R. Natarajan (ed.), *Proceedings of the International Conference on Transformations in Engineering Education: ICTIEE 2014*, DOI 10.1007/978-81-322-1931-6_38,

scope in Electronics field, but it is still considered as one of the toughest subjects in engineering. DSP is popularly known among students as "Degree Stopping Paper" due to many reasons. The performance of students also varies between departments. For example, Electronics and Communication students will find it easier than other department students as they are fundamentally better than others. Students of Information Technology, Computer Science, and Electrical Engineering will find it quiet challenging compared with Electronics students. In this paper, we analyze the root cause of this problem, and we also suggest remedial steps based on various comparison studies between several colleges under Anna University, Chennai. When this subject is offered by ample number of universities across the globe, we take Anna University, Chennai, as a case study, and we are going to analyze the results, teaching methodologies, and problems with the existing system, and we also propose some innovative techniques to teach this subject in a much better manner. We also investigate whether problem-based learning (PBL) in combination with a lecture-based teaching style is able to enhance the learning outcomes of an introductory core course in digital signal processing. This paper reports on a pilot pedagogical project that combined for the first time problem-based learning with a lecture-based instructional style in digital signal processing. It is important to mention that our implementation of problem-based learning is different from the experiential learning, or the learning-by-doing approach that is usually implemented in engineering classes via laboratory assignments [1–3]. The problems to be solved in such assignments are specified in closed form by the course instructor, and they follow closely the theoretical content of the lectures [4–7] explains such concepts briefly. Our initiative aimed at providing a framework for solving multimedia-related DSP problems, where students were encouraged to formulate their own problems.

2 Digital Signal Processing

Digital signal processing can be explained simply as mathematical manipulation of real-world signals like voice, audio, video, temperature, pressure, or position that have been digitized. DSP applications include audio and speech signal processing, sonar and radar signal processing, sensor array processing, spectral estimation, statistical signal processing, digital image processing, signal processing for communications, control of systems, biomedical signal processing, seismic data processing, etc. DSP algorithms have long been run on standard computers, as well as on specialized processors called digital signal processor and on purpose-built hardware such as application-specific integrated circuit (ASICs). The hardware device where this DSP process takes place is called as DSP processor. A DSP contains some components like program memory, data memory, compute engine, and input/output.

Program memory stores the programs the DSP will use to process data. Data memory stores the information to be processed. Compute engine performs the math processing, accessing the program from the program memory and the data from the data memory. Input/output serves a range of functions to connect to the outside

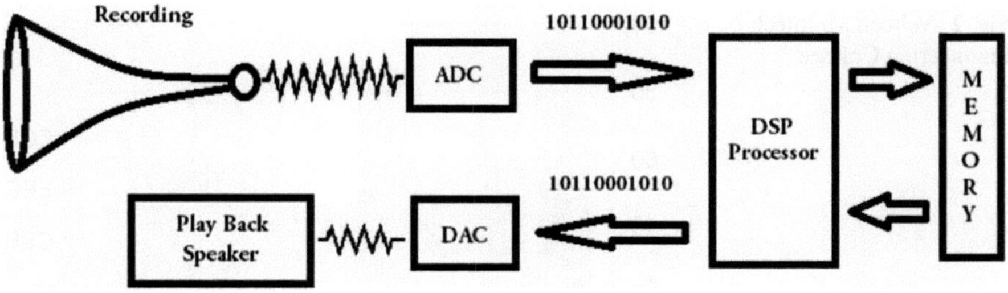

Fig. 1 A simple DSP process in an audio player

world. Figure 1 shows a simple DSP process happening in an audio player. The analog input data is fed inside the player using a microphone (recording), and the analog signals are sent to the ADC which converts the analog signals into digital format, and the digitized signals are fed into the DSP processor for further processing and then later they are stored in the memory. The reverse process of the above takes place to play that audio file from the memory. The DSP would perform other functions such as volume control, equalization and user interface, etc. Processing signals digitally provides the advantages of high speed and accuracy.

2.1 Learning Materials Followed Currently in DSP

As we are dealing with the syllabus of Anna University, Chennai, we take into account the structure of syllabus under Regulations 2004, 2008, 2009, and 2012. However, the Regulation 2013 is currently followed in all the engineering streams. We consider 4 departments for our study; they are Electronics and Communication, Electrical and Electronics, Computer Science, and Information Technology.

The subject has been widely divided into 5 units under these four streams. As we all know that DSP is 90 % mathematical oriented subject, finding a theory part would be rare. Most of the syllabus deals with derivations and other mathematical functions. We may have an assumption that students who are good at mathematics will excel in DSP, but it is a contrary. Most of the derivative and mathematical part included in the syllabus are complex compared with regular algebraic mathematics. These also include differentiation and integration of logarithmic and trigonometric functions. But deriving them using filter functions and window functions and understanding their terminologies differentiate DSP from other complex mathematics. Our initial findings with DSP are that students of ECE were performing much better than CSE, IT, and EEE. One of the major reasons is that Electronics students were strong in the basics of DSP, and they also had a paper called "Signals and Systems" prior to DSP which cover the fundamentals of DSP, and it also gives a clear picture about various concepts proposed in DSP. Unfortunately this subject was not a part of the syllabus for CSE and EEE. Henceforth these students found it challenging compared with students of Electronics and Communication.

Fig. 2 Veltech Multitech Engineering College

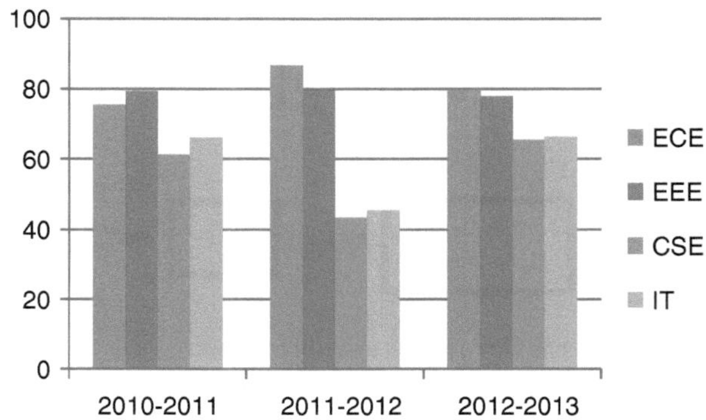

Fig. 3 Veltech Hightech Engineering College

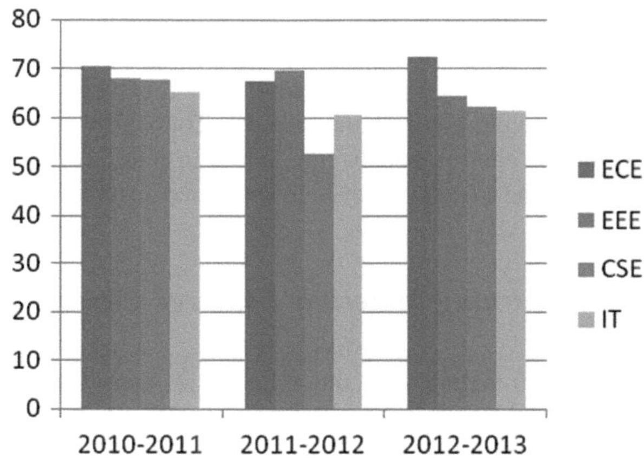

2.2 *Result Analysis*

The major reason for this paper is to increase the results of DSP. We have taken the results of various colleges into consideration, and we perform a detailed analysis of the same. Since the proposed paper is about syllabus followed by Anna University, Chennai, we consider the results of several colleges under Anna University, Chennai. For our pilot study, we randomly selected three engineering colleges from Chennai and analyzed their performance in their semester exams. We considered Veltech Multitech College, Veltech Hightech College, and S.A. Engineering College from Avadi. A total of 360 students handled by 6 different faculty members were taken into consideration. The overall analysis was found as below (Figs. 2 and 3).

The results of these colleges were based on the semester exams of the academic year 2010–2011, 2011–2012, and 2012–2013. The revaluation results were also considered. The general observation we could get from these analysis is that the results of ECE and EEE students were much better than CSE and IT. Another observation is that the results were comparatively higher when the same faculty member handles

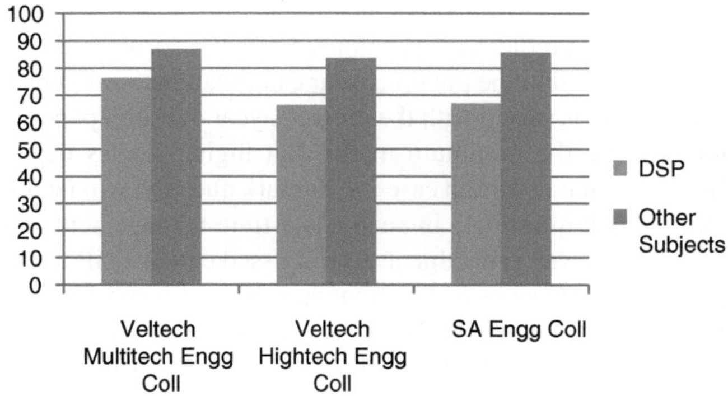

Fig. 4 DSP versus others

the subject in consecutive years. The subject was divided into 5 units. It was found that 3 cycle tests were conducted and students were asked to submit 3 assignments. Finally a model test consolidating all 5 units was also conducted. The mark in the semester exam was split up as 20+80. Twenty marks were awarded as internals based on their performances in the abovementioned tests, and 80 marks were extracted from the end semester exams. The end semester exam was for 100 marks and it will be converted to 80 marks and the 20 marks are the rest of the marks from cycle tests and model tests. It was found that the teaching level was slightly above average and most of the faculties were ME degree holders and only a few were doctorates. Compared with other subjects, the overall passing percentage of DSP was found to be low and there was a huge difference.

2.3 *Feedback from Students and Faculties*

We collected feedbacks from students regarding their performance in exams as well as about the subject and the faculties. Pilot study samples of students were 3 above average, 3 average, and 2 below average. Students gave genuine feedbacks about the subject and the faculties. Most of the students commented that they could not understand the concepts properly. When the faculty tendency was analyzed, most of the students felt that the faculties were teaching the subject based on questions asked in previous year exams and their teaching was not satisfactory (Fig. 4).

If teaching is marks oriented, then obviously the results should be better, but in our case the marks are low. This is a contradiction. The reason is that most of the current teaching techniques are marks oriented, in the sense the faculties would collect several previous year question papers and solve those problems and in certain cases they even omit several topics which were either asked comparatively less or never in the university exams. This technique is not a success always because the structure of the university question paper is mostly uncertain. Sometimes the question paper might be mostly problematic and sometimes it might be mostly theoretical. We cannot predict the questions, but we also noticed that in certain cases, the questions asked in the previous year examinations were repeated. In such scenarios if the student is strong with the previous year question papers, then he/she would definitely score the minimum marks, but higher grades depend on other skills. Another fact is that in certain cases, a 16-mark question will be divided as two 8-mark or four 4-mark questions; in such cases time management is a key issue. Moreover teaching this way is unethical. As discussed earlier DSP was a mathematical paper, and many students responded that they do not know why they solve these complex DSP problems and they were not educated well about the applications of the subject. So our overall observation from these feedback is that DSP was mostly syllabus centric and teachers were not in a position to take any chance in the form of implementing any new teaching or delivery technique and they were concerned too much only about the university results. Some of the contrasting comments were: We are being trained for exam rather than thought; most of us do not know the purpose of the course; few CSE students felt that the course "Signals and Systems" is a prerequisite of DSP, but it was just kept as a unit in their DSP syllabus; all students felt DSP was a difficult paper; this image has been created by many faculties and senior students; some students felt that it is other department's paper.

The faculty members expressed their comments over the subject, and they also sighted the reasons for the decline in the results. Most of the teaching strategies in the current trend are to be blamed. The college management on the other hand stresses the faculties to finish the syllabus on time. Pilot study samples of faculties were 2 from ECE, 2 from EEE, 1 from CSE, and 1 from IT. Many challenges were sighted by the faculties in handling this subject. A few comments from the faculties are summarized below; This paper is syllabus centric, time constraint, and has no cooperation from students, especially when a faculty from other department handles the subject, and students lack the basics and find it difficult to understand.

2.4 Remedial Steps from Faculty and Students

Based on all the analysis done and also considering the results, we asked both the students and the faculties to propose their own remedies for this issue. Most of the students felt that this subject is alienated and needs additional care which would increase their workload. They also felt that it might become an extra burden for them as the time stipulation to study that subject along with 5 other subjects (totally 6) was less than 6 months. Hence they felt that the syllabus could be made simple or reduced for CSE and IT students alone. It is not a convincing remedy as the syllabus was framed well, and our observation from senior faculty members is that the syllabus was well and good. Some students felt that more class tests need to be conducted.

The faculty members were also keen in proposing various strategies of improving the results in this subject. Many faculties felt that students of CSE and IT feel that it is a foreign (other department) subject and so they never give much importance to DSP compared with other subjects. Currently every college holds collective number of faculties in every department. To teach certain subjects like DSP, faculties from other departments are called in to teach those subjects. For example, DSP comes under Electronics Department and the Department of Computer Science need not require a faculty just to teach DSP alone; instead, they make faculties from other department (likely ECE) to handle the subject. But this could have a reverse impact with the students. Generally if other department faculties handle the subject, then the students would only give lesser importance to the subject and the faculty. This problem can be solved by making faculties from the same department to handle the subject. Some of the faculties also felt that the prescribed 45 h time duration by the university is not sufficient and they felt that 55–60 h time would be a fair deal.

3 Proposed Techniques to Enhance Performance

Based on the various studies and analysis done, we propose several innovative and unique schemes to enhance the performance of students, especially CSE and IT. We introduce problem-based learning (PBL) technique, and we also propose a pedagogical approach of learning DSP in an effective way. We also performed a test case by implementing this approach as a trial and error manner in an institution and found the results to be better than the earlier stages. The major aspect of our proposal is that instead of teaching theoretical concepts to the students, we make students learn problems based on real-time systems. Application-oriented learning is the major strategy of the proposed paper. It is important to mention that our implementation of PBL is different from the experimental learning or the learning-by-doing approach that is usually implemented in engineering classes via laboratory systems via engineering design projects. Projects represent the hands-on component of engineering courses, where students apply theoretical knowledge in the context of a specific application. However, formulating an appropriate project for an introductory DSP course presents challenges on how to avoid coursework overload. Since introductory DSP courses are typically very dense from a theoretical viewpoint, the PBL component needs to be carefully crafted in order to complement well the theoretical course content without overloading the student. Our hypothesis is that while observing how simple DSP components (e.g., first- or second-order filters, delays, and up/down samplers) work on familiar signals, students will grasp fundamental concepts such as superposition, convolution, frequency response, sampling, etc., more easily.

We must make the teaching method of this subject to be application oriented. The faculty before starting any topic under DSP must first explain the students about the applications of that topic in real time. Giving real-time examples would trigger interest from students and would also make them remember the topic forever. Giving activities would also help in boosting the student's desire toward the subject. Activities must be purely application oriented and should mainly focus on the practical rather than the theory part. For example, if a faculty is about to teach "Hidden Markov Process" he/she should not simply start what is Markov? Why Markov process? and get into the mathematical derivations. Such explanations would definitely bore the students and many dislike the course due to this approach. Instead if the faculty lets the students know about the applications of it, then it might trigger some sort of interest within the students. The hidden Markov process is one of the major concepts used in modern speech recognition systems. Definitely every student would have some knowledge about what a microphone is, so the faculty must explain it as a major necessity in speech processing, and examples like microphones, musical instruments or any device that captures sound can be cited as examples. Adding to this, an activity making the students to get to know the working principles of such speech processing devices can be employed. After the students finish this activity, then the faculties can get into the technical and mathematical stuffs about the Markov process in detail. Such type of teaching would help students to grasp things pretty quickly.

3.1 Pedagogical Approach

Pedagogical approach is the science and art of education, specifically instructional theory. An instructor develops conceptual knowledge and manages the content of learning activities in pedagogical settings. Students learn as they internalize the procedures, organization, and structures encountered in social contexts as their own schema. The learner requires assistance to integrate prior knowledge with new knowledge. Students must also develop metacognition or the ability to learn how to learn.

3.2 Problem-Based Learning

Problem-based learning (PBL) is a student-centered pedagogy in which students learn about a subject through the experience of problem solving. Students learn both thinking strategies and domain knowledge. Problem-based learning is a style of active learning. Learning is "student centered" because the students are given the freedom to study those topics that interest them the most and to determine how they want to study them. Students should identify their learning needs, help plan classes, lead class discussions, and assess their own work and their classmates' work [8, 9]. Problem-based learning addresses the need to promote lifelong learning through the process of inquiry and constructivist learning. Problem-based learning is becoming increasingly popular in engineering. Projects represent the hands-on component of engineering courses, where students apply theoretical knowledge in the context of a specific application. However, formulating an appropriate project for an introductory DSP course presents challenges on how to avoid coursework overload. Since introductory DSP courses are typically very dense from a theoretical viewpoint, the PBL component needs to be carefully crafted in order to complement well the theoretical course content without overloading the student. Our hypothesis is that while observing how simple DSP components (e.g., 1st- or 2nd-order filters, delays, and up/down samplers) work on familiar signals, students will grasp the fundamental concepts such as superposition, convolution, frequency response, sampling, etc., more easily. The programming environment enabled the direct perception of how filters change the input signals. The projects were formulated so that the students had to achieve a well-defined goal. This goal was reachable via multiple paths, i.e., by choosing among several multistep signal processing approaches. In our field study, two example problems were proposed by the instructor. Students were encouraged to specify their own problem within the area of audio/image processing.

Problem (A) Enhancement of Speech Signal: Speech signals are recorded using a microphone with unshielded cable. As a result, the signals are contaminated with 60 Hz interference. Also, due to the fact that the microphone's frequency response is not flat and that higher pitches are suppressed, the recorded voice sounds humming, with a background buzz. Students were to design a DSP system to enhance the signal-to-noise ratio of their speech signal and to modify its spectrum in such a way that it sounds as natural as possible. In the planning stage, students needed to select their signal processing blocks and to justify their selection. The implementation stage involves the interconnection of the blocks and the parametric adjustment of each block.

Problem (B) Enhancement of Print-Out Quality of Facial Images: Facial images are captured using a low-quality webcam. The images feature Gaussian noise and are overexposed, due to excessive lighting. The quality of the images needs to be optimized for printing purposes. The planning stage needs to identify the processes to be performed (such as de-noising, contour sharpening, histogram equalization, removal of specular regions, etc.). The implementation stage involves interconnecting the processes above, as well as tuning their parameters. All problems (either instructor or student formulated) were to be solved using simulators, a custom-designed software.

3.3 *The Mini Projects*

A series of mini projects use MATLAB to reinforce the concepts of the class.

Project 1: The projects start with a music synthesis project in which students are asked to find sheet music for a given tune (rhymes like jingle bells can be used) and then write a program to synthesize it, producing a .wav file that plays the song.

Project 2: In the second project, the students are given a .wav file containing a spoken message that has been masked by some loud sinusoids. Their task is to filter out the sinusoids.

Project 3: The weekly projects progress until the students are given a .wav file containing a recording of some instrument (a guitar and violin have been used in the past; in the future a tuba or trombone may be used). Their task is to produce a text file that tells what notes have been played, when, and for how long. This is just the reverse of the first mini project.

Project 4: Finally the students will be asked to create a simple program coding that would recognize these musical notes and would segregate them according to their theme or rhyming structure according to the preferences set by the students.

Other examples for mini projects are music synthesis, discrete convolution GUI, tone removal, tone removal via poles and zeros, note detection, simple song detection, etc.

4 The Simulator Environment

Advances in signal processing theory and its multi-domain applications make it increasingly difficult for educators to teach abstract mathematical concepts as well as illustrate their relevance in real-world system design in a single course. Learning a concept practically rather than theoretically makes students to easily understand the concepts. But in certain cases practical learning was also a huge challenge. In order to overcome these difficulties, the simulator environment came into picture. The advent of block-based design and simulation technologies like MATLAB, LabView, and Simulink enables educators to develop and employ visual and interactive models of real-world signal processing systems to meet the corresponding challenges. The paradigm that we propose is direct visual manipulation of discrete signals and systems. It is similar to the WYSIWYG (What You See Is What You Get) paradigm, a very popular choice for text editors. The graphical interface of simulators allowed students to manipulate the parameters of each incorporated block for digital signal processing and thus to gain understanding of how it affects the input signals. The development of a system in simulators consists of the following steps:

- Loading or creating the input signals
- Creating filters
- Interconnecting sources, filters, and arithmetic blocks
- Placing probes at desired points in the system for signal investigation
- System simulation, viewing and saving the outputs

Filter design has been implemented via dropdown menus and sliders for choosing filter type and parameters. Once the filter is designed, it can be visualized using a pop-up window. Students can investigate magnitude, phase, impulse, and unit step responses of their filter as well as zero-pole, group delay, and phase delay plots. These filter design options make the simulators more intuitive, and they enable novice users to locate, manipulate, and visualize their filters faster, without getting lost in advanced options. Simulators also provide the option to probe the signal at any desired location in the system by using pop-up windows or by saving it to a file. Probes are considered a signal sink and can be placed at the output of any block on the system. The simultaneous time (or space) domain and frequency domain visualization is useful for understanding the correlation between time (space) and frequency domain representations, as well as the trade-off between different constraints imposed on filters in either time, space, or frequency. The following are the popular simulators used in teaching DSP.

LabView: National Instruments' [10] LabView (short for Laboratory Virtual Instrumentation Engineering Workbench) is based on the concept of data flow programming and is particularly suited to test and measurement applications. The three important components of such applications are data acquisition, data analysis, and data visualization. LabView offers an environment which covers these vital components. Therefore, the full development version offers a number of signal processing functions, grouped under the headings of signal generation, time domain, frequency domain, measurement, windows, and filters.

MATLAB: MATLAB was introduced into digital signal processing as a teaching tool which can make students understand the abstract content intuitively, improving learning motivation and interest, strengthening the knowledge of integrated applications and creative ability, to achieve a multiplier effect, so they can achieve effectively understanding and mastery of digital signal processing class to improve teaching quality. These are the two most widely used simulators to test and work with DSP projects; however many other simulators are also available, but they are not commercially recognized as MATLAB and LabView.

5 Proposed Strategies

Our analysis and study about this problem led us to various strategies that could help students to improve their grades. We also took inputs from both students and faculties and developed a model or an approach that could help in improving the results. They are as follows; looking at the method of teaching followed currently we found that it's mostly syllabus oriented. This system should be changed. We must make the teaching method of this subject to be application oriented. The faculty before starting any topic under DSP must first explain the students about the applications of that topic in real time. Giving real time examples would trigger interest from students and would also make them remember the topic forever [11–13]. Giving activities would also help in boosting the student's desire towards the subject. Activities must be purely application oriented and should mainly focus on the practical's rather than the theory part.

For example, if a faculty is about to teach "Hidden Markov Process" he/she should not simply start what is Markov? Why Markov process? And get into the mathematical derivations. Such explanations would definitely boredom the students and many dislikes the course due to this approach. Instead if the faculty lets the students to know about the applications of it then it might trigger some sort of interest within the students. The hidden Markov process is one of the major concepts used in modern speech recognition systems. Definitely every student would have some knowledge about a microphone is, so the faculty must explain it as a major necessity in Speech processing and examples like microphones musical instruments or any device that captures sound can be cited as examples. Adding to this, an activity making the students to get to know the working principles of such speech processing devices can be employed. After the students finish of this activity then the faculties can get into the technical and mathematical stuffs about the Markov process in detail. Such type of teaching would help students to grasp things pretty quickly. Making students work with the simulators would be an added advantage for this course [5–7]. We consider giving more activities and mini projects to the students as this would help them visualize the DSP applications in real time scenarios. The teaching approach that we take is to start with a simple DSP system such as a small FIR filter, students will move from a (hopefully) familiar point and progress through the simulation and synthesis of the hardware. By leading them through their first complete system in a highly directed manner, students will be able to complete the entire design and implementation process within a few hours. The MATLAB–Xilinx interface toolbox is useful at this stage for allowing students to test the results of the hardware filter against those of a software implementation to verify correctness. Once students understand the entire process (i.e., the global perspective), the concentration can shift towards studying characteristics of hardware implementations such as quantization effects and resolution, area, power, and implementation optimizations. At this stage, verification of correctness becomes even more important but the toolbox also becomes useful for investigating engineering trade-offs such as the impact of coarser quantization.

5.1 The Mini Projects

A series of Mini Projects use MATLAB to reinforce the concepts of the class.

Project 1: The projects start with a music synthesis project in which students are asked to find sheet music for a given tune (Rhymes like Jingle Bells can be used) and then write a program to synthesize it, producing a .wav file that plays the song.

Project 2: In the second project, the students are given a .wav file containing a spoken message that has been masked by some loud sinusoids. Their task is to filter out the sinusoids.

Project 3: The weekly projects progress until the students are given a .wav file containing a recording of some instrument (a guitar and violin have been used in the past; in the future a tuba or trombone may be used). Their task is to produce a text file that tells what notes have been played, when, and for how long. This is just the reverse of the first mini project.

Project 4: Finally the students will be asked to create a simple program coding that would recognize these musical notes and would segregate them according to their theme or rhyming structure according to the preferences set by the students.

Other examples for Mini Projects

- Music Synthesis
- Discrete Convolution GUI

- Tone Removal
- Tone Removal via poles and Zeroes
- Note Detection
- Simple Song Detection

6 Conclusions

This paper depicts the importance of learning things with practical knowledge. We also conducted a pilot study on DSP and the issues related to the learning methods followed in DSP. We also conducted some test cases related to the results, and a detailed survey was conducted taking the results of 3 major engineering colleges in Chennai. Feedbacks were taken from both students and the faculty members, and we derived a summary of the same. Based on these statistics and feedbacks, we have proposed several strategies to teach DSP in an efficient way with the use of technology. The authors hope that more researches will be made under this domain to enhance the teaching methodologies.

References

1. Ausubel DP. Psychology of meaningful verbal learning: an introduction to school learning. New York: Grune & Stratton; 1963.
2. Chen C-M. Intelligent web-based learning system with personalized learning path guidance. Comput Educ. 2008;51(2):787–814.
3. Chen TS, Chang JS, Lin JS, Yu HL. Context-aware writing in ubiquitous learning environments. Res Pract Technol Enhanc Learn. 2009;4(1):61–82.
4. Kornecki AJ. Real–time systems course in undergraduate cs/ce programs, CD-ROM supplement. IEEE Trans Educ. 1997;40:9.
5. Wright CHG, Welch TB, Etter DM, Morrow MG. Teaching hardware–based DSP: theory to practice. In: Proceedings of the IEEE international conference on acoustics, speech, and signal processing, vol 4, Orlando, FL, May 2002, p. 4148–4151.
6. Chassaing R. Digital signal processing laboratory experiments using C and theTMS320C31 DSK. New York: Wiley; 1999.
7. Sorensen HV, Chen J. A digital signal processing laboratory experiments using the TMS320C30. Upper Saddle River: Prentice Hall; 1997.
8. Ho ML, Rad AB, Chan PT. Project-based learning. IEEE Control Systems Magazine, Oct 2004. p. 88–91.
9. Davenport D. Experience using a project-based approach in an introductory programming course. IEEE Trans Educ. 2000;43(4):443–8.
10. Chugani ML, Samant AR, Cerna M. LabVIEW signal processing. Upper Saddle River: Prentice Hall; 1998.
11. Mitra SK. Digital signal processing: a computer based approach. Boston: McGraw Hill; 2006.
12. Yin L, Yang R, Gabbouj M, Neuvo Y. Weighted median filters: a tutorial. IEEE Trans Circuit Syst II Analog Digit Signal Process. 1996;43(3):157–92.
13. Vieira J, Tome A, Rodrigues J. Providing an environment to teach DSP algorithms. In: Proceedings of the IEEE signal processing society's 2nd signal processing education workshop, Pine Mountain, GA, Sept 2002.

Transforming Engineering Students into Industry-Ready Professionals

Krishna Shastri

Abstract Employability of engineering graduates has been an area of concern for academia and industry alike. Amrita Vishwa Vidyapeetham has attempted to bridge the gap by getting feedback from industry and creating a course that will address these issues.

Keywords Employability • Soft skills • Skill development

1 Introduction

Every year close to 10,00,000 engineers graduate from the 3,000+ engineering educational institutions in India. Close to 30 % of them will not be placed in today's challenging market (The Economic Times, Jun 18, 2013) [1]. Adding to the market conditions is the fact that many of the students do not measure up to the requirements of the industry. Early in 2005, a study by NASSCOM revealed that only 25 % engineers were employable. The employability situation has only worsened since then. As per the 2011 National Employability Report conducted by Aspiring Minds [2], only around 20 % of the engineers are suitable for IT Services or IT Product companies. Companies run extensive training programmes for their fresh recruits. In addition to the costs incurred by the companies, the engineers are not productive for close to 1 year after they join.

The Government of India has set up the National Skills Development Corporation (NSDC) to fulfil the growing need in India for skilled manpower across sectors and narrow the existing gap between the demand and supply of skills. The mission statement of NSDC emphasizes the need to upgrade skills to international standards

K. Shastri (✉)
Corporate and Industry Relations, Amrita Vishwa Vidyapeetham, Coimbatore, India
e-mail: k_shastri@amrita.edu

R. Natarajan (ed.), *Proceedings of the International Conference on Transformations in Engineering Education: ICTIEE 2014*, DOI 10.1007/978-81-322-1931-6_39,

through significant industry involvement. Industry associations like NASSCOM for the IT-ITES Sector and FICCI for the Manufacturing Sector are participating in the Sector Skills Councils being set up by NSDC. In collaboration with industry, these associations have been outlining the need for making engineering curriculum industry relevant [3].

Amrita Vishwa Vidyapeetham (Amrita) has been interacting with industry for many years, and based on their feedback, a need was felt to incorporate Life Skills training in the engineering curriculum.

2 Design of Life Skills Course

Amrita has been working with industry to understand the competencies they seek in the students. Based on the inputs, the following areas need to be addressed:

- A degree is not an indicator of a person's employment skills. Companies are more interested in practical applications rather than just knowledge.
- Good problem-solving and creative skills are needed to solve practical problems.
- It is not individual capability but teamwork that is important to industry. Interpersonal effectiveness is very important to ensure success in career.
- Business is not local anymore. More and more work is being done across borders, languages and culture. As a result, good business communication skills are very important.
- Life in the industry is very different from a student's life. They need to be given inputs so that the transition to corporate environment is smooth.
- Technology is constantly changing. As a result, self-learning skills are very important to ensure that employees are constantly up to date with the latest advances and innovations.

The above inputs have come from companies across industries, though the IT industry has been in the forefront in defining the skills required.

The Life Skills course in Amrita [4] aims to complement the technical knowledge acquired by the students with skills that will become employable and enable them to have a successful career. The Life Skills Course in Amrita is handled by the Directorate of Corporate and Industry Relations (CIR). CIR adopts a multipronged approach towards development of Life Skills in the students. The Learning and Development wing of CIR imparts more than 240 h of training in the areas of verbal, aptitude and soft skills. These classes are spread over the entire duration of the course and are handled exclusively by well-qualified subject matter experts (SMEs).

The curriculum has been developed based on industry requirements and covers the following areas:

- Verbal and communication skills – spoken English, vocabulary, grammar, reading comprehension and business English

- Quantitative aptitude and logical reasoning skills – problem solving, quantitative ability, logical and analytical reasoning, data interpretation and data sufficiency
- Soft skills – attitude and behaviour, grooming for corporate, confidence building, presentation skills, resume writing skills, group discussion skills, team-building and leadership skills, interview skills
- Core competency – practical applications, guest lectures and workshops and participation in technical competitions

In addition, Amrita has many industry-designed electives increasing the relevance of the curriculum.

The board of studies for Life Skills courses includes members from industry. The syllabus is updated based on industry feedback, inputs from Human Resources Team of companies, alumni and students.

All students in their first year are briefed on the importance and the need for them to develop Life Skills. In addition, the following steps are taken to ensure that all students benefit from the courses conducted by CIR:

- In order to address learning needs of slow, moderate and advanced learners, the class size is kept to around 35 students. This ensures better coverage, monitoring, interaction and work assignment as per their needs.
- Remedial classes are conducted at the end of every year based on a student's performance in internal evaluations and in examinations.
- Records of each student are maintained by the appointed trainer. This record and the one-on-one interaction enable to address the development needs of each student on an individual basis.

The success of these courses is measured through the placement records and the feedback from the companies on the performance of students from Amrita. In addition, the AMCAT test conducted by external agency Aspiring Minds, a leading company in employability assessment, has validated student performance and placed Amrita in the top 10 % institutions in the country.

Some key data regarding the curriculum is given below:

- Life Skills course was started in 1998, with emphasis on soft skills.
- The course was enhanced in 2006 to include problem-solving and verbal skills.
- The course has been made a credit course from 2010 batch onwards.
- More than 2,000 students undergo the course from each engineering batch in Amrita.
- There are more than 30 full-time faculty in CIR to impart Life Skills training

3 Results

The placement records in Amrita are a testimony to the efficacy of the approach used to transform the engineering students into industry-ready professionals. The placement trend is given below.

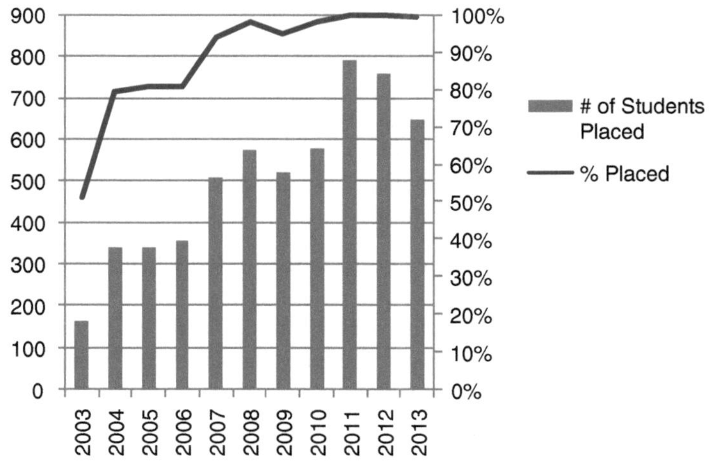

Fig. 1 Placement statistics over the years

Figure 1 clearly shows that the placement percentage has increased in line with the addition of Life Skills course components, even though the number of students to be placed in each batch has increased. In the years 2012 and 2013, the number of students planning for higher studies has substantially increased.

4 Conclusion

The approach taken by Amrita to address the issue of employability and industry readiness is exemplary. The investment made by the university has borne fruit based on the feedback from students and from the companies, which hire them. The university is committed to make necessary additions and changes to the Life Skills course based on the changing requirements of industry.

An area that would need further investigation would be the impact of internships, participation in world-class competitions and technical certifications on making the students industry ready.

References

1. Anumeha Chaturvedi, Rahul Sachitanand. A million engineers in India struggling to get placed in an extremely challenging market. *Economic Times*, 18 Jun 2013.
2. National employability report engineering graduates – annual report 2011. Aspiring Minds, Gurgaon.
3. Foundation skills in IT (FSIT) – guideline document. NASSCOM, India.
4. BTech curriculum and syllabus for SSK111, SSK112, SSK113. Amrita Vishwa Vidyapeetham, Coimbatore.

Impact Analysis of Students' Learning Styles in Effective Planning and Delivery of Courses: A Case Study of Data Mining Course

Basavaraj S. Anami and Deepa S. Garag

Abstract The learner style plays a pivotal role in pedagogy. In this paper, we have presented the adoption of Felder and Silverman model to analyze the learning styles of the students and use the same in the design and delivery of data mining course. One group is found to be verbal learners and the other is visual learners. The control groups have 52 students. The experiment shows that appropriately tailoring the teaching style has resulted in better results. The results reveal that 90 % of the students scored above 70 %. Thus, we have concluded that matching the teaching style and learning style gives good result.

Keywords Pedagogy • Learning style • Teaching style • Learning models • Delivery of courses

1 Introduction

Pedagogy is an art and also the science of teaching, which serves as a master plan that includes details of what is to be done by a teacher, the instructional strategies, instructional equipment, and the cardinal objectives of instructions. The most accepted criteria for measuring good teaching is the amount of student learning that occurs. An academician, Thomas Angelo, once said "teaching in the absence of learning is just talking." A teacher's effectiveness in teaching reflects the students' learning. The teacher (also referred as faculty) controls the instructional process. In other words, the teacher delivers the lecture content and the students listen to the instructions. Thus, the learning mode tends to be passive and the learners play a little role in their learning process. The learning abilities of the students are being ignored. Students learn in many ways, namely, by seeing and hearing; reflecting and

B.S. Anami (✉) • D.S. Garag
K.L.E. Institute of Technology, Hubli 580030, Karnataka, India
e-mail: anami_basu@hotmail.com; deepagarag@rediffmail.com

R. Natarajan (ed.), *Proceedings of the International Conference on Transformations in Engineering Education: ICTIEE 2014*, DOI 10.1007/978-81-322-1931-6_40,

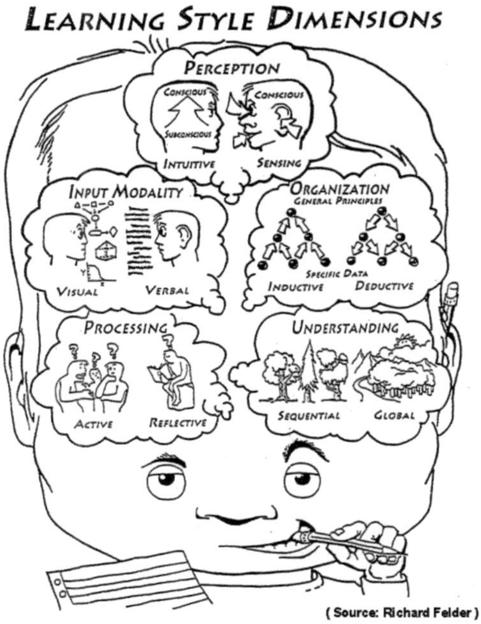

Fig. 1 Learning style dimensions

acting; reasoning logically and intuitively; and memorizing, visualizing, drawing analogies, and building mathematical models, steadily and in fits and starts [1]. Teaching methods also need to vary with learning styles. How much a given student learns in a class is governed by student's native ability, prior preparation about the topic and compatibility of his or her learning style and the instructor's teaching style. The gap between students' learning styles and teachers' teaching styles needs to be bridged [2].

A learning style is defined as the characteristics, strengths, and preferences in the way people receive and process information. There are several ways to identify students' learning styles. Some learning style models are (i) the Myers-Briggs Type Indicator (MBTI), (ii) Kolb's Learning Style Model, and (iii) Felder-Silverman Learning Style Model. We have selected the Felder and Silverman model to analyze the learning styles of the students because it has been modeled for engineering students [3].

The Felder and Silverman model classifies students as fitting into one of the following four learning styles, also called learning dimensions, which is depicted in Fig. 1. Types of learners include sensing learners or intuitive learners, visual learners or verbal learners, active learners or reflective learners, and sequential learners or global learners. Sensing learners are oriented towards facts and intuitive learners are oriented towards theories and meanings. Visual learners prefer visual representations and verbal learners prefer information and knowledge. Active learners like in trying things and reflective learners like working alone. Sequential learners are very systematic and global learners are able to learn in large leaps. A typical learning style profile obtained is shown in Fig. 2. These styles are not meant to be rigid

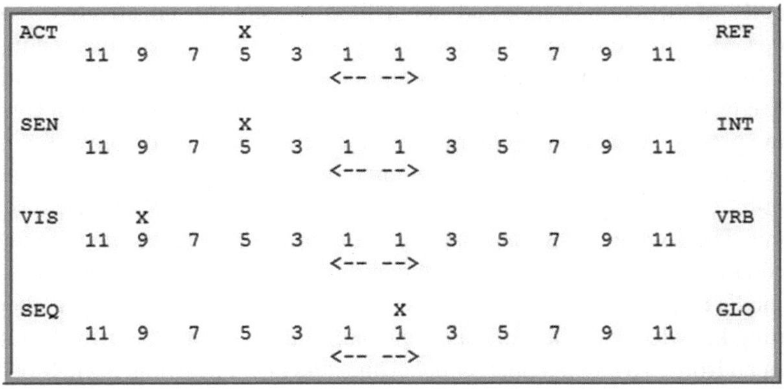

Fig. 2 A typical learning style profile

categories. A student may have a strong, moderate, or weak learning towards one end of a continuum or the other. Furthermore, one may lean in one direction for some tasks and in the other direction for other tasks. Teachers need to understand the learning styles of majority of the students and design their instructions. The latest developments in pedagogy treat teaching as an activity that conserves valued knowledge and skills. Pedagogy is adopted in technical education in India too [4]. Thus, in order to know the state-of-the-art practices, we have carried out a literature survey. The following is the gist of the literature survey carried out to know adoption of learning styles of students in instruction delivery.

Felder et al. [3] have defined two tiers of entering college students, the first consisting of those who go on to earn science degrees and the second those who have the initial intention and the ability to do so but instead switch to nonscientific fields. The authors have described learning and teaching styles of second category students. Franzoni et al. [1] have described the design of a personalized teaching method that is based on an adaptive taxonomy using Felder and Silverman's learning styles, which is combined with the selection of the appropriate teaching strategy and the appropriate electronic media. Students are able to learn and efficiently improve their learning process with such method. Bhowmik et al. [2] have tried to reflect the necessity of acquired knowledge of every teacher concerning pedagogy to perform his or her teaching profession effectively. Gilakjani et al. [5] have explained learning styles. The teaching styles match or mismatch between learning styles, namely, visual, auditory, and kinesthetic among Iranian learners are considered. The pedagogical implications are addressed for the EFL/ESL classroom. Entz et al. [4] have presented the seminal work of the Center for Research on Education, Diversity, and Excellence (CREDE). The five critical elements of effective pedagogy are summarized. The authors have demonstrated that consistent implementation resulted in greater student outcomes across the curriculum regardless of age and higher academic test.

Sajjad et al. [6] have determined the effectiveness of various teaching methods used at graduate level. Nearly 220 undergraduate students studying in 11 departments of faculty of Arts are interviewed about their perceptions on effective teaching methods. Most of the students rated lecture method as the best teaching method and group discussion the next. Students' perception and ratings about the interesting and effective teaching methods are considered helpful in the improvement in teaching-learning process. Prince and Felder [7] have reviewed most commonly used inductive teaching methods, including inquiry learning, problem-based learning, project-based learning, case-based teaching, just-in-time teaching, and discovery learning. The commonalities and specific differences for each method are defined. Damodharan et al. [8] have evaluated the traditional methods of teaching as well as multimedia teaching. The authors have suggested other useful teaching methods for imparting knowledge to the students. Hein and Budny [9] have described strategies developed from two independent models Dunn and Dunn Learning Style Model and Kolb Learning Style Model to teach physics for undergraduate engineering students. The basic elements of these two learning style models are compared and contrasted. Rutz and Westheider [10] have reported how engineering and technology students at the University of Cincinnati compared to other similar population showed significant differences in the learning styles. The educators who understand these various preferences and who have a good sense of the distribution of learning type have a better opportunity to enable all students to learn more fully. Kyun Lee et al. [11] have considered the students in two different mechanical engineering courses when surveyed using the VARK (visual, aural, read/write, kinesthetic) questionnaire to classify students' learning style and to determine their attitude towards PowerPoint lectures. The majority of students surveyed are found to be kinesthetic or read-write in their learning style. Livesay and Dee [12] have administered Index of Learning Styles (ILS) to all engineering freshmen on the first day of class and again 5 weeks later have examined the test-retest performance. On an average, individual students repeated greater than 75 % of their answers identically in the test and retest. Kaminski et al. [13] have aimed to explore engineering student's perceived learning characteristics in order to inform learning and teaching, carefully examined evidence of a link between student learning characteristics and student academic success. The results strongly suggest that recognizing this association between learning styles and academic success will necessarily lead to both more perceptive teaching and also more responsive learning.

From the literature survey, it is observed that the learning styles are considered in teaching physics, chemistry, and other subjects to technology education students. The results of adoption of learning styles are also reported. No case studies on the impact of learning styles in specific engineering subjects or courses are reported. Adapting different learning style models result in different outcomes in pedagogy. Suitability of learning style models to the given course delivery requires a study. Hence, a work on adoption of learning styles of students in teaching data mining course for undergraduate students of engineering is undertaken. We have adopted Felder's model in this work. The students of data mining course took test available on the link www.engr.ncsu.edu/learningstyles/ilsweb.html. The learning style profiles of the students are obtained and analyzed. The findings of analysis are used in tailoring the teaching style for the course. The analysis of the data obtained from the tests is given emphasis in this work and hence the work on impact analysis of the model proposed by Richard M. Felder.

The remaining part of the paper is organized into four sections. Section 2 deals with the proposed methodology where the different stages of the methodology are discussed in detail. Section 3 deals with the result and discussion. Conclusion of the work is referenced in Sect. 4.

2 Proposed Methodology

The work carried out comprises three stages, namely, Creation or Connection to Web Application, Storage, and Retrieval; Evaluation Analysis; and Information Generation. The stages are shown in Fig. 3.

2.1 Creation or Connection to Web Application

In this work, we have used a learning style model developed by Richard M. Felder and Linda K. Silverman to obtain the learning style profiles of each student in the class. They use the Index of Learning Styles (ILS) as an instrument to assess preferences in terms of four dimensions, namely, active/reflective, sensing/intuitive, visual/verbal, and sequential/global. A student's learning style profile provides an indication of probable strengths in academic settings. A student answers 44 questions and gets at the end the learning style profile. The profiles of the students of a class are analyzed for their learning styles of the entire class. Faculty adapts the majority of the learning styles of the class to match the teaching style. The Richard M. Felder and Linda K. Silverman learning style model is available for usage on the web. But, how it evaluates against the questionnaire is hidden. We have tried to explore the underlying process by analyzing the model and arrived at the pseudo code as given in the following algorithm.

2.2 Learning Style Profile Evaluation and Analysis

Every student is made to answer the questionnaire. The profiles of the students are obtained and are further evaluated using the algorithm.

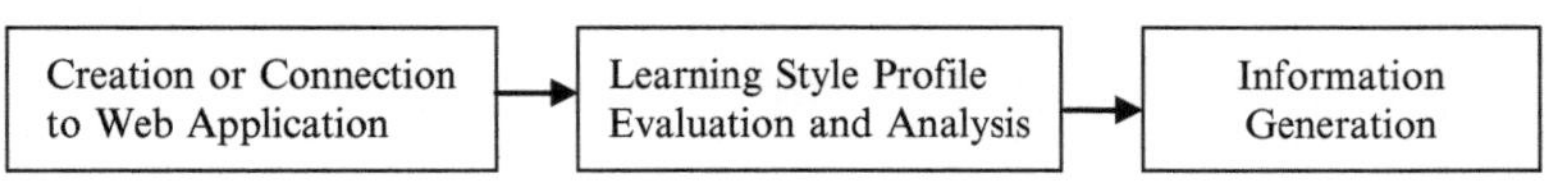

Fig. 3 Different stages of methodology

```
Algorithm. Students Learning Style Profile Evaluation
Input: Students Learning Style Profiles
Output: Learning Styles
Start
Initialize  the counters
/*Step1 to Step6 - finds the options for question no 1
to 44 and increment the appropriate dimension count*/
Step 1: For Question_no = 1 To 44 do
     Step1.1: If (Question_no mod 4 = 1)
                  If (option=a) then increment
                   Active_Learner_Count
                 else  increment
                  Reflective_Learner_Count
     Step1.2: If (Question_no mod 4 = 2)
                 If (option = a ) then increment
                  Sensitive_Learner_Count
                 else   increment
                  Intutive_Learner_Count
     Step1.3: If (Question_no mod 4=3)
                 If (option = a)  then increment
                  Visual_Learner_Count
                 else increment
                  Verbal_Learner_Count
     Step1.4: If (Question_no mod 4 = 0)
                 If (option = a )  then increment
                  Sequential_Learner_Count
                  else increment
                  Global_Learner_Count
Step2: If  (Active_Learner_Count>Refletive_Learner_Count)
            x1_score_active=abs(Active_Learner_Count -
            Reflective_Learner_Count)
            else
            x1_score_reflective=abs(Active_Learner_Count
            - Reflective_Learner_Count)
Step3: If  (Sensitive_Learner_Count> Intu-
            tive_Learner_Count)
            x2_score_sensitive=abs(Sensitive_Learner_Coun
            t -   Intutive_Learner_Count)
            else
            x2_score_intutive=abs(Sensitive_Learner_Count
            -   Intutive_Learner_Count)
  Step 4:If  (Visual_Learner_Count > Ver-
            bal_Learner_Count)
            x3_score_visual=abs(Visual_Learner_Count -
            Verbal_Learner_Count)
            else
            x3_score_visual=abs(Visual_Learner_Count -
            Verbal_Learner_Count)
```

```
  Step5:If   (Sequential_Learner_Count > Glob
al_Learner_Count)
            x4_score_sequential=abs(Sequential_Learner_Co
            unt - Global_Learner_Count)
            else
            x4_score_sequential=abs(Sequential_Learner_Co
            unt - Global_Learner_Count)
  Step6: If  (x_score is in the scale 1-3)
            Declare " student is fairly well balanced  on
            the two dimensions".
  Step7: If  (x_score is in the scale 5-7)
            Declare " student have a moderate preference
            for one dimension of the scale and will learn
            more easily in a teaching environment which
            favors that dimension".
  Step8: If  (x_score is in the scale 9-11)
            Declare " Student have a very strong prefer-
            ence for one dimension. The student may have
            real difficulty learning in an environment
            which does not support that preference".
  Stop
```

2.3 *Information Generated*

It shows that majority of the students are visual learners in the class. The students with balanced preference generally adopt to the teachers teaching style. The data mining course is planned and delivered by using PowerPoint presentations. We have designed quiz, showed video lectures on some topics, and discussed case studies. The same method is tried second time to another group of students. This time students are found to be verbal learners. Hence, new teaching style is followed to teach the same course. The course is taught by using chalk and talk method with moderate use of PowerPoint presentation. Audio lectures are played on few topics with explanation on board. The case studies on different data mining techniques, group discussions and seminars, are dealt by students and teachers using black board. In this group also, 90 % of the students have scored 70 % and above. Thus, we opine that the learning styles drive the teaching styles and resonance in results occurs upon match of the two.

3 Results and Discussion

We have carried out an experiment on a control group of 52 students (25 male and 26 female), the entire class of VII semester in the age group of 19–20 years, for data mining course to study their learning styles. The result obtained for an individual is shown in Fig. 4. The students have a very strong preference towards reflexive learning. The

Fig. 4 Learning styles of students

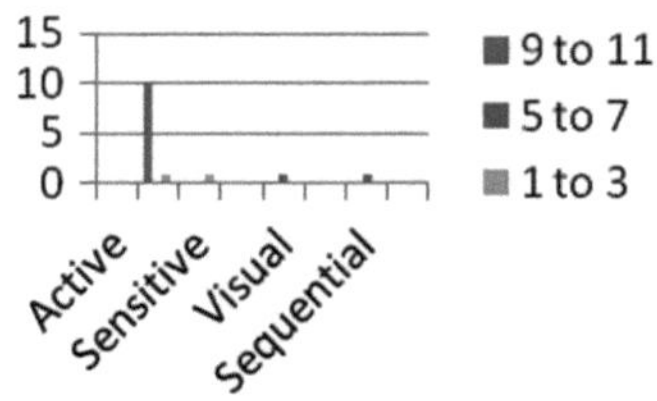

Fig. 5 Active/reflective learners

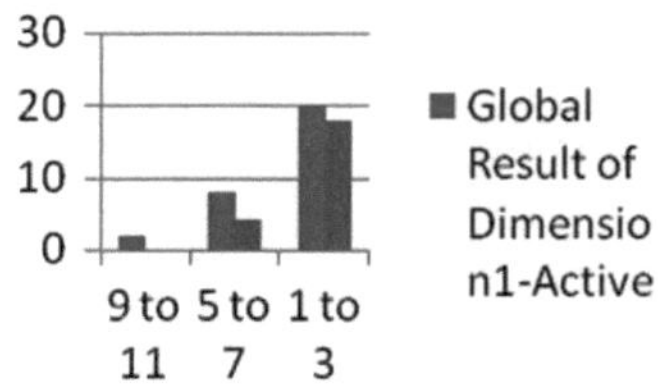

Fig. 6 Sensitive/intuitive learners

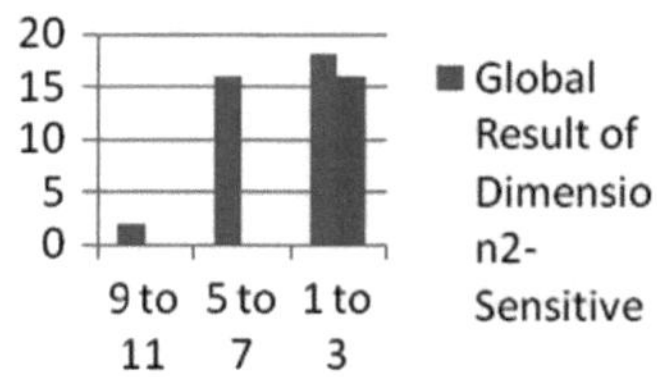

Fig. 7 Visual/verbal learners

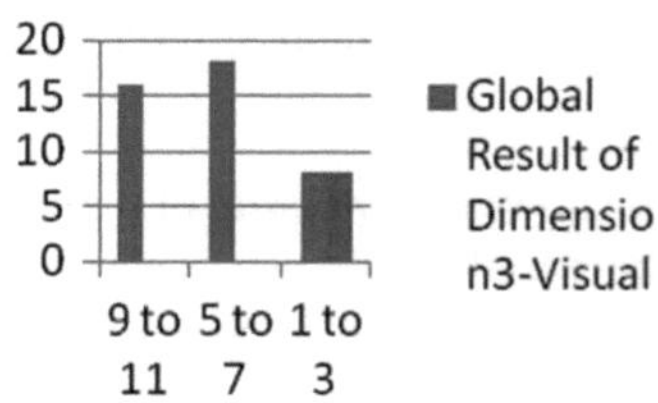

Fig. 8 Sequential/global learners

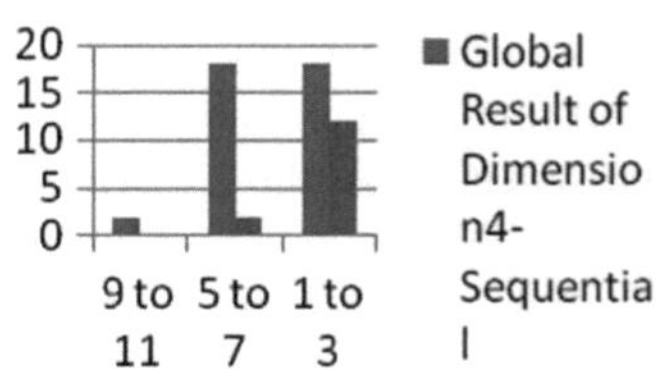

number of active/reflective learners is shown in Fig. 5. Nearly 70 % of the students have balanced preference to both styles. The rest 30 % of students have moderate preference towards active learning. The number of sensitive/intuitive learners is shown in Fig. 6. Nearly 70 % of the students are balanced. Nearly 30 % of the students have moderate preference towards sensitive learning. The number of visual/verbal learners is shown in Fig. 7. Majority of the students are visual learners. The number of sequential/global learners is shown in Fig. 8. Fifty percent of the students have moderate preference towards sequential learning and the rests are balanced. The balanced learn equally well either way. The result of entire class strength of 52 is shown in Fig. 9.

Fig. 9 Entire class behavior

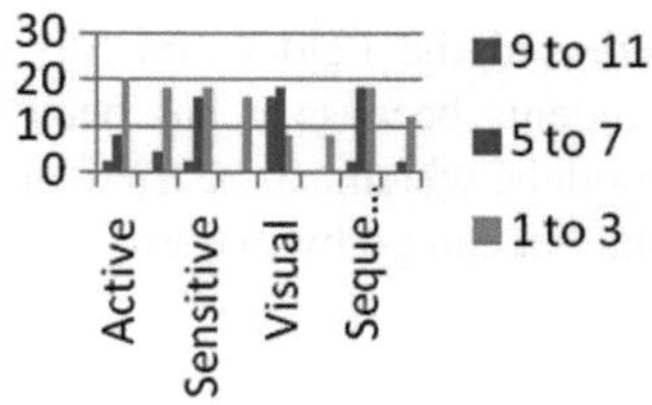

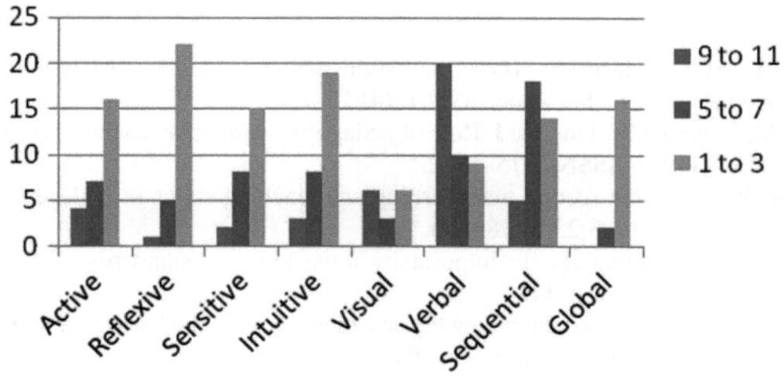

Fig. 10 Entire class behavior using traditional chalk and talk approach

The same data mining course is taught for the second time to different group of students. The students have taken online test and their students learning style profile is evaluated and analyzed. We have found that the majority of the students are verbal learners as shown in Fig. 10. The teaching style is used for visual group of students which could not be used to the group of verbal students, as their learning style is different. Hence, new teaching style and course delivery is adapted to this group of students in teaching data mining course. The course is taught using blackboard approach with moderate use of PowerPoint presentation whenever necessary. The audio lectures are played on few topics.

The case studies on different data mining techniques are presented for group discussions. The seminar's topics are assigned to the students. We tailored the course delivery to suit both groups of students. We have changed to the course delivery and not the design of the course. In a typical affiliated system, there is no scope for changing the design of the course. However, one has freedom to adopt the necessary teaching style. There is always scope for research on the teaching styles to achieve good results.

4 Conclusion

The learning styles must be taken into consideration while teaching any course at any level. Analyzing the learning styles of students becomes helpful to the teacher in adapting his or her teaching style, which in turn benefits the students. We have selected the Felder and Silverman model to analyze the learning styles of the students because it has been modeled for engineering students. There are other models, which can be tried and compared. The teacher cannot satisfy each individual but can go by majority and address specific types on case-to-case basis.

References

1. Franzoni AL. Student learning styles adaptation method based on teaching strategies and electronic media. Educ Technol Soc. 2009;12(4):15–29.
2. Bhowmik M, Banerjee B, Banerjee J. Role of pedagogy in effective teaching. Basic Res J Educ Res Rev. 2013;2(1):1–5. ISSN 2315–6872.
3. Felder RM. Reaching the second tier: learning and teaching styles in college science education. J Coll Sci Teach. 1993;23(5):286–90.
4. Entz S. Why pedagogy matters: the importance of teaching in a standards-based environment. In: Forum on public policy. 2011.
5. Gilakjani AP. A match or mismatch between learning styles of the learners and teaching styles of the teachers. J Mod Educ Comput Sci. 2012;11:1–60.
6. Sajjad S. Effective teaching methods at higher education level. 2010.
7. Prince MJ, Felder RM. Inductive teaching and learning methods: definitions, comparisons, and research bases. J Technol Educ. 2006;95(2):123–38.
8. Damodharan VS. ACCA, AICWA, Rengarajan V. AICWA. Innovative methods of teaching. 2004.
9. Hein TL, Budny DD. Teaching to students' learning styles: approaches that work. In: Proceedings of the frontiers in education conference – 1, San Juan, Puerto Rico. 2000.
10. Rutz E, Westheider V. Learning styles of engineering & engineering technology students – similarities, differences and implications for effective pedagogy. In: Proceedings of the 2006 American Society for Engineering Education annual conference & exposition copyright © 2006, American Society for Engineering Education. 2006.
11. Kyun Lee B, Leiffer P, William Graff R. Understanding cognitive style and learning style and their application to teaching and learning in engineering. In: Proceedings of the 2009 American Society for Engineering Education annual conference & exposition copyright © 2009, American Society for Engineering Education. 2009.
12. Livesay GA, Dee KC. Test-retest reliability of the index of learning styles for first-year engineering students. In: Proceedings of the 2005 American Society for Engineering Education annual conference & exposition copyright © 2005, American Society for Engineering Education. 2005.
13. Kaminski DA. Exploring the link between student learning styles & grades in an introductory thermal-fluids course: a three-year study of associations. In: Proceedings of the 2005 American Society for Engineering Education annual conference & exposition copyright © 2005, American Society for Engineering Education. 2005.

WiMAX in Education: A Wireless Networking Lab Design

Akashdeep and K.S. Kahlon

Abstract WiMAX is one of the latest broadband wireless technologies that supports applications like voice, video and surfing over very large geographical areas. WiMAX is seen as a technology that can solve the reachability problem in countries like India because of its wider reach. Connectivity and education is vital to any country to usher in economic growth, health care and improved entertainment services. Broadband wireless technologies like WiMAX have started providing cheap connectivity to schools and colleges. Telecommunication industry is also smart enough to quickly adapt to this new technology. In order to bridge the gap between industry and academia, there is a need for more flexible delivery and extensive study of such courses to satisfy the needs of telecommunication industry. The changing trends in telecommunication companies towards WiMAX deem it necessary to pod it to every engineering student. This paper presents a WiMAX hands-on lab for graduate and postgraduate students aimed to awaken students to this technology. The contents of lab have been designed based on the use of simulator like Qualnet, ns-2 or OPNET. We present a brief overview of experiments that can be performed on WiMAX using Qualnet and can be included in the subjects of wireless technologies. Learning outcomes of this lab are also presented.

Keywords WiMAX • Wi-Fi • Wireless communication • Broadband

Akashdeep (✉)
UIET, Panjab University Chandigarh, Chandigarh, India
e-mail: akashdeep@pu.ac.in

K.S. Kahlon
Department of CSE, Guru Nanak Dev University, Amritsar, India
e-mail: karanvkahlon@yahoo.com

R. Natarajan (ed.), *Proceedings of the International Conference on Transformations in Engineering Education: ICTIEE 2014*, DOI 10.1007/978-81-322-1931-6_41,

1 Introduction

India is the second most populous country in the world, with an annual GDP per capita of around $1900 billion [1] and is placed at the eighth position worldwide. Indian government in its recent budget has given top priority to education sector (Rs 65,867 crores) as a means of building a reservoir of competent leaders and skilled personnel who will guide and sustain the region's current pace of development [2]. As a result, both the government and private sectors are directing huge technological inputs towards improving educational content and delivery systems and in upgrading infrastructures. Recent BWA allocation done in 2012 [3] indicates that WiMAX will emerge as the quintessential answer to these issues given its superior performance and lower costs as compared to other similar technologies like 3G, LTE, etc. Therefore, WiMAX is also going to play a major role in the field of education as more and more universities are going to adopt it as has been done by Michigan University in the USA [4]. The emergence of WiMAX and its backing by telecoms will help set new norms in providing education to students of all ages [5].

There have been a number of universities offering course on wireless networking in India and abroad [6, 7]. Most of the networking lab experiments focus on either Internet protocols or MAC and PHY mechanisms in Wi-Fi or fixed LANs. Work on designing labs that teaches concepts of cellular or wireless communication is rather sparse. Dawy et al. [6] have designed a list of experiments which also includes experiments on WiMAX, but these have been covered and aimed at analysis and resource planning. There have been recent attempts by [7, 8, 9, 10] to design wireless labs based on WiMAX networks. Alqudah et al. [7] have proposed a high-level design of wireless lab on WiMAX but have not provided any actual implementation. Marasevic et al. [8] and Pati [9] have also proposed some experiment for WiMAX lab, but they are very limited. As per the authors' knowledge, there are no universities in India where experiments on concepts of WiMAX technology are provided to students. IITs have just now started the inclusion of WiMAX networks in their courses [10], but experiments listed are very basic, and concepts such as throughput, QoS provisioning and MAC layer operations are not included. Experiments proposed in this paper are very novel and informative. The upcoming sections will highlight the lab structure and experiments to be performed.

2 Lab Outcomes and Experiments

This lab is a research-oriented graduate-level course and can be included with subjects like computer networking/data communication or wireless networking laboratories. The main focus of this lab will be on the functionalities of PHY and MAC of IEEE 802.16 standard. The objective is to provide students with understanding of network design using WIMAX technology, protocols and proposed algorithms.

There are at least 60–120 students each year in an engineering course who study course on wireless networks having background of computer, electronics and electrical engineering to whom this course can be offered. The course may also be offered online on the lines of universities like Stanford, MIT and IITs. The experiments can be performed on world-class simulators like Qualnet, OPNET, ns-2, ns-3, Winprop, etc. [11], or funded institutes can opt for the purchase of WiMAX test bed. We had designed this course by using simulations on Qualnet 5.2, and students are expected to learn and develop a simulation of IEEE 802.16d/e standard while evaluating its performance (throughput, collision probability) in different scenarios. This provides them with an intuition on random-access MAC protocols and allows them to confirm the results. The experiments in this course are aimed at the following learning outcomes:

1. Learn how to use professional simulators like Qualnet, OPNET, etc.
2. Understand the basic design and operation of the wireless technologies like WiMAX.
3. Understand how QoS is affected by variation in PHY parameters.
4. Understand the basic principles of radio network planning and optimization.
5. Understand the importance and basics of Wi-Fi, fixed and WiMAX networks and their co-sitting and coexistence.

2.1 Experiments

This section describes a total of 6 experiments which can be added partially or fully to the course of wireless communication/computer networks already studied by engineering students. The authors had proposed only six experiments as WiMAX is not studied as an independent subject; however, the number can be increased or decreased as per the requirements of the students. Table 1 provides the summary of learning outcomes along with the list of experiments. These experiments have been performed by authors using Qualnet 5.2 from Scalable Technologies [12]. Qualnet is a world-class simulator that can model IEEE 802.16d/e WiMAX networks. Same experiments can be performed with ns-2 simulator which is based on an open-source platform.

Experiment 1: Designing and Configuring WiMAX Network: WiMAX operates in two modes: PMP and Mesh modes. The working of two modes is entirely different. As starters, students will be required to design a WiMAX network consisting of one BS or more BS and multiple SS (Fig. 1). Students shall learn how to configure a node as BS and SS, how to connect them with wired or wireless subnet and what the effect of changing different parameters on BS and SS is.

Experiment 2: Study Support for Various Kinds of Traffic: Students in this experiment are required to generate different kinds of traffic that WIMAX supports like real-time, non-real-time or simple web traffic without any basic requirements. Traffic can be generated by changing the features of simulators. Value of precedence

Table 1 List of experiments

No.	Experiment title	Learning outcomes
1.	Designing and configuring WiMAX networks	(a) Learn how to design a WiMAX network with one/two BS and multiple SS (b) Learn how SS interconnect with BS (c) Learn how different SS can communicate among themselves in WiMAX operation
2.	Study support for various kinds of traffic	(a) Learn WiMAX support for various application like the Web, video, FTP, etc. (b) Learn the basics of WiMAX traffic analysis
3.	Maximum throughput measurement	(a) To observe difference between maximum throughput obtained by SS at different locations (b) Study the influence of physical parameters on modulation scheme selected by BS
4.	Quality of service measurement	(a) To measure QoS through the use of parameters like throughput and delay (b) To identify which SS was configured to use services like UGS, rtPS, nrtPS, etc.
5.	WiMAX as backbone technology	(a) How WIMAX can be a solution to interconnectivity problem (b) Observe the performance of each network in WiMAX environment
6.	WiMAX radio network planning	(a) Learn to plan a fixed network using WiMAX (b) Learn the basics of WiMAX coverage and planning

field of IP protocol header can be mapped to direct traffic to specific service class (Fig. 2b). Students are required to learn how WiMAX system supports various kinds of traffic.

Experiment 3: Maximum Throughput Measurement: The objective of this experiment is to learn how WiMAX responds to SS based on their distances. As per IEEE 802.16 standard, WiMAX will choose different modulation schemes for SS which are at different distances. The experiment will be performed by placing SS at different distances from BS and observing QoS parameters (Fig. 3a).

Experiment 4: Quality of Service Measurement: The goal of this experiment is to identify which SS was configured to use UGS service class. Students are required to observe the behaviour of WiMAX system under load conditions. Traffic is varied by increasing the size of packets that are sent. Since allocations to QoS class are fixed, SS with UGS class will incur less delay and better throughput. Students will be asked to reason about the values of parameters obtained and determine which SS is using UGS class.

Experiment 5: WiMAX as Backbone Technology: Assuming that at this stage the students have enough background about the working of WiMAX, they are made to learn how WiMAX can fill the role of backbone technology giving connectivity to Wi-Fi nodes. Scenarios of Fig. 3b illustrate how WiMAX can provide connectivity over a geographical area.

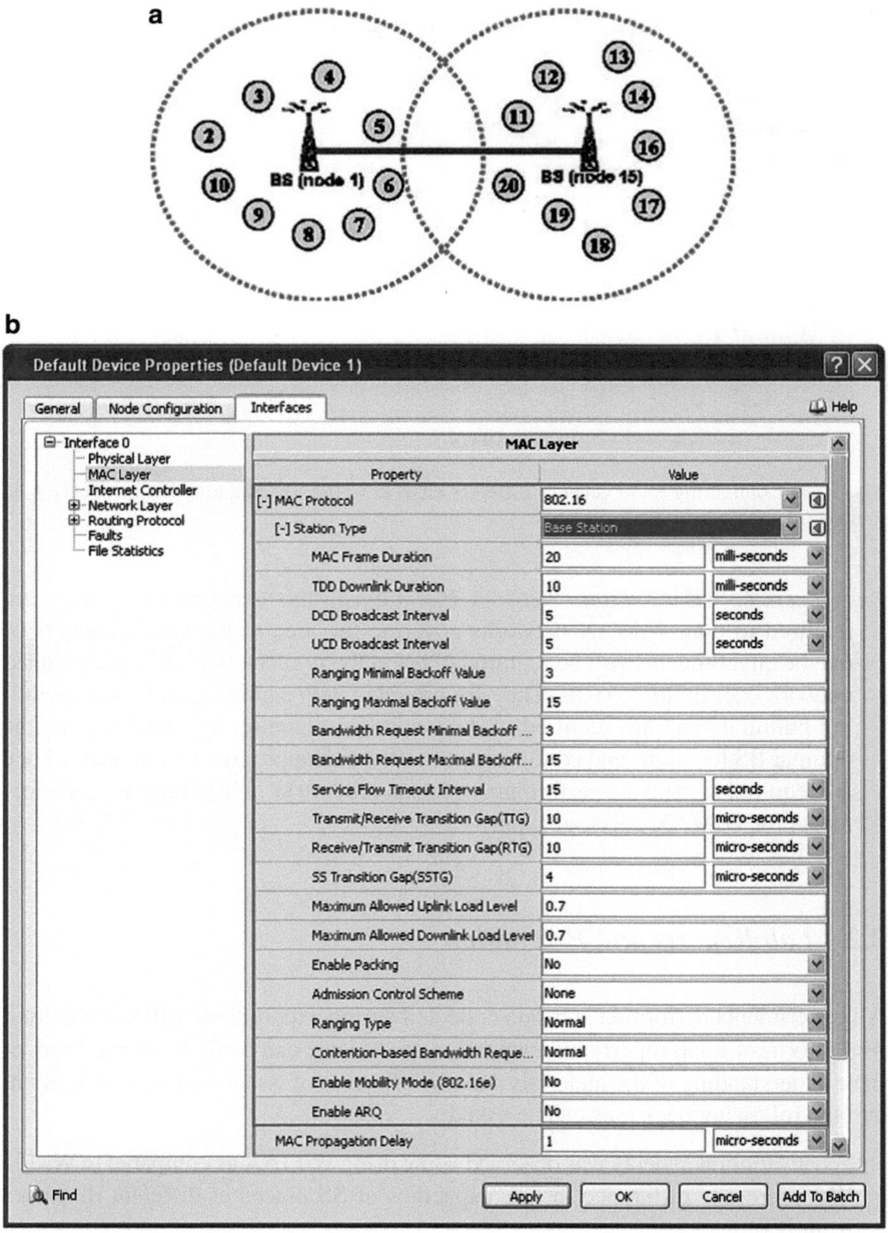

Fig. 1 Design of sample scenario in Qualnet. (**a**) Picture of scenario. (**b**) Sample properties for base station

a

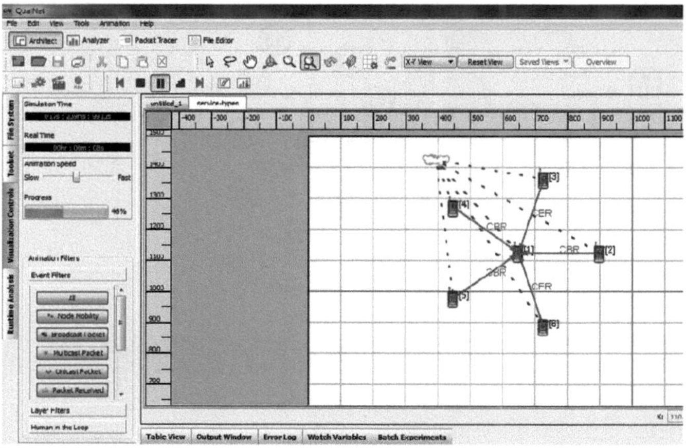

b

WiMAX Service class	Precedence field
UGS	7,5
ertPS	4
nrtPS	6,2,1
rtPS	3
BE	0

Fig. 2 (**a**) Configuring SS to generate different kinds of traffic. (**b**) Mapping IP precedence field to traffic classes

Experiment 6: WiMAX Radio Network Planning: In this experiment, students are required to plan WiMAX networks covering the area of the whole campus or maybe city. Students will be required to begin the implementation of the planned network over multiple steps. They are provided with a geographical map including buildings and are required to do a nominal planning by searching for the optimal BS locations and configuring the BSs with appropriate parameters. They are required to generate performance plots in order to be able to propose enhancements for WiMAX coverage plan.

2.2 Lab Reports and Evaluation

All experiments performed and data collected during experiments will be submitted in the form of a lab report, and evaluation of students can be done on the basis of their understanding of the materials and identification of issues in hands-on session, and the following questions can be asked:

- What variation in delay was observed while using WiMAX as compared to Wi-Fi?
- Was there any difference in link capacities of SS placed at different distances from BS?
- Determine which SS was working with real-time and which on non-real-time data. How?
- Which traffic class observed the highest and lowest throughput and why?
- Was WiMAX able to provide the minimum required throughput to all traffic classes?
- Was the plan generated by the students good enough to provide coverage over the entire region?

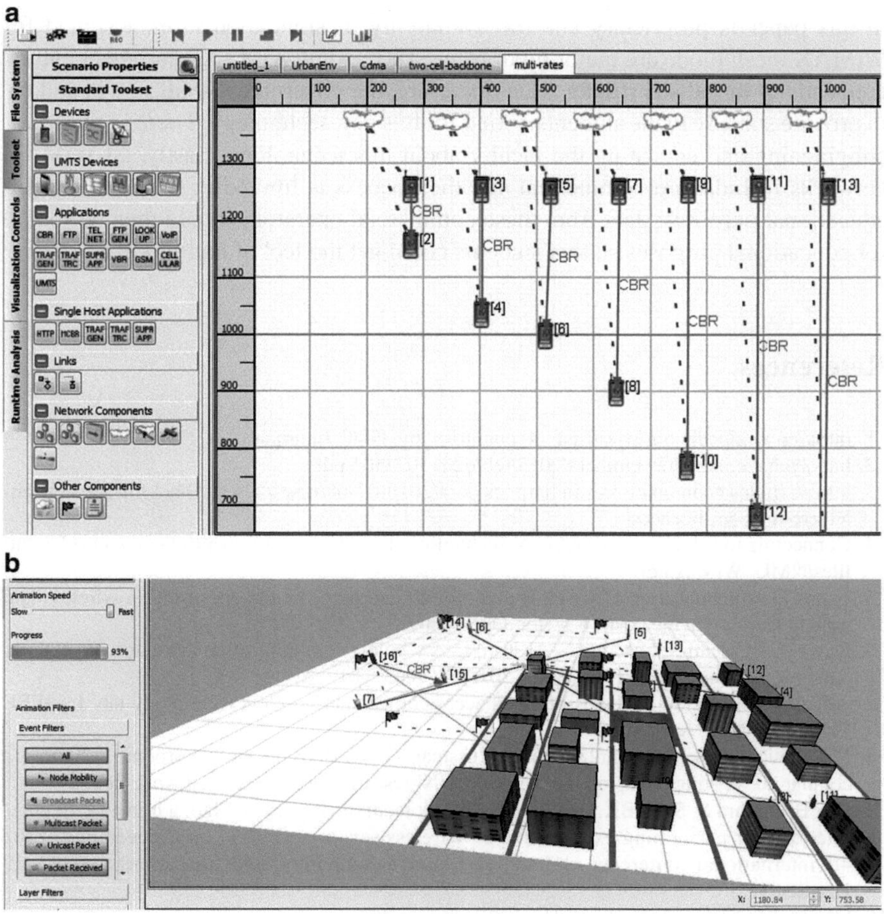

Fig. 3 (**a**) Different SS generating different traffic and placed at different distances from BS (Experiments 3). (**b**) WiMAX offering connectivity over defined geographical area (Experiment 5)

3 Conclusion

This paper has proposed a one of a kind hands-on lab on WiMAX designed for a course in wireless and mobile networking. This paper is aimed to bridge the gap between academic curriculum and industry standards. Since WiMAX is an ideal choice for telecommunication companies for providing high-speed wireless network at affordable prices, a student is expected to be aware of this futuristic technology and this is just a tiny attempt in this direction. Although the experiments given in this paper are very comprehensive, still there is considerable space for improvements. First is the availability of simulators for students as the simulator proposed in this paper is proprietary and the availability of simulation tools designed for WiMAX mesh mode are major hindrances. The cost of WiMAX equipment and its connectivity in India is still an issue; therefore, the government shall step in and try to provide subsidy to the students opting for this new technology. There is a need for enlightening and educating the faculty about this technology. Lastly, all WiMAX networks in India are commercial and their access is limited or not available for educational purposes; data from these commercial operators should then be shared for educational purposes so that students could get the feel of real-world situation.

References

1. http://en.wikipedia.org/wiki/List_of_countries_by_GDP_(nominal)
2. http://pib.nic.in/archieve/others/2013/feb/d2013022806.pdf
3. http://articles.economictimes.indiatimes.com/2010-07-08/news/27631294_1_bwa-spectrum-augere-telecom-licences
4. Connecting to Education by WiMAX Technology. http://www.k12blueprint.com/sites/default/files/NMU_WiMax.pdf
5. Evans D. Potential uses of wireless and mobile Learning. Landscape study in wireless and mobile learning in the post-16 sector, Open University, April 2005.
6. Dawy Z, Husseini A, Yaacoub E, Al-Kanj L. A wireless communication laboratory on cellular network planning. IEEE Trans Educ. 2010;53(4):653–61.
7. Alqudah Y, Cooklev T. Hands-on open access broadband wireless technology lab. In: IEEE EDUCON 2012, April 2012.
8. Marasevic J, Janak J, Schulzrinne H, Zussman G. WiMAX in the classroom: designing a cellular networking hands-on lab. Second GENI research and educational workshop. 2013.
9. Pati B, Mishra S, Saha BK. Advanced network technologies virtual lab: a human-computer interface for performing experiments on wireless sensor networks. In: Proceedings of the 4th international conference on intelligent human computer interaction, Kharagpur, 27–29 Dec 2012.
10. http://iitkgp.vlab.co.in/?sub=38&brch=121&sim=580&cnt=1
11. WiMAX Simulation tools. http://4g360.com/profiles/blogs/wimax-simulation-tools
12. http://web.scalable-networks.com/content/qualnet

Creating an Integrated Learning Experience Within Curriculum Threads Through Mini-projects

Uma Mudenagudi, Ashok Shettar, M. Meena, C. Nalini, K. Sujata, B.L. Desai, and R.M. Banakar

Abstract This paper discusses details of the curricular reform effort to design an integrated learning experience in curriculum threads through mini-project in Electronics and Communication undergraduate. Traditionally, capstone projects because of their integrative nature are expected to provide an opportunity for the student to synthesize and demonstrate the learning that has happened across the program. Often it becomes difficult to choose problems for capstone projects that call for the application of the knowledge and skills acquired in all the curriculum threads. Mini-projects vertically integrated with each of the curriculum threads of the program are carried out by the students in the prefinal year and provide an opportunity to integrate knowledge and skills acquired in a set of courses belonging to the curriculum thread to solve complex engineering problems. Introduction of mini-projects also provides sufficient resolution to directly assess student learning outcomes in each of the curriculum thread. As these projects are carried out in teams, students are able to develop and demonstrate several professional

U. Mudenagudi (✉) • K. Sujata • B.L. Desai • R.M. Banakar
Department of Electronics and Communication Engineering, BVB College of Engineering and Technology, Vidya Nagar, Hubli, Karnataka 580031, India
e-mail: uma@bvb.edu; sujatask@bvb.edu; bldesai@bvb.edu; banakar@bvb.edu

A. Shettar
Department of Civil Engineering, BVB College of Engineering and Technology, Vidya Nagar, Hubli, Karnataka 580031, India
e-mail: ashok@bvb.edu

M. Meena
Department of Information Science and Engineering, BVB College of Engineering and Technology, Vidya Nagar, Hubli, Karnataka 580031, India
e-mail: msm@bvb.edu

C. Nalini
Department of Instrumentation Engineering, BVB College of Engineering and Technology, Vidya Nagar, Hubli, Karnataka 580031, India
e-mail: nalinic@bvb.edu

R. Natarajan (ed.), *Proceedings of the International Conference on Transformations in Engineering Education: ICTIEE 2014*, DOI 10.1007/978-81-322-1931-6_42,

competencies that are critical for engineering practice. The paper demonstrates that the theme-based mini-projects provide an integrated learning experience in a curriculum thread and help to develop curriculum design and modifications.

Keywords Student outcomes • ABET • OBE • Capstone • Curriculum design • Mini-projects • Tollgate courses • NBA

1 Introduction

This paper provides details of creating an integrated learning experience within curriculum threads using theme-based miniprojects. Typically capstone projects are used to provide an integrated learning experience. They also play significant role in assessing technical and professional competencies attained by the students. By design, undergraduate engineering programs are broad in scope with two or three vertically integrated curriculum threads, which are largely independent. It is important that the students get an opportunity to connect learning that has happened in various courses under a curriculum thread to solve complex engineering problems. The Department of Electronics and Communication Engineering introduced miniprojects in its undergraduate program to strengthen integration of learning in program verticals. The themes of the mini-projects are carefully chosen by the department undergraduate committee to ensure that students while carrying out these projects are required to integrate and apply the knowledge and skills learnt in sequence of courses belonging to a particular curriculum thread. Prior to capstone projects students are exposed to solving of unstructured, open-ended problems. Mini-project can act as tollgate course in the assessment of student outcomes and provide better resolution due to the use of themes, which helps in the curriculum design and modifications.

Often, the capstone projects done near the end of the program are used to evaluate attainment of student outcomes a to k, in an outcome-based education framework proposed in ABET and NBA [1–3, 16, 17]. Though integrative in nature it is very difficult to design capstone projects to cover all the curriculum threads of the program and they can be used to initiate entrepreneurial and innovative thinking [4]. There are attempts to use open-ended experiments and design experiments to evaluate student learning in a group of courses [5–8]. It is difficult to design open-ended projects and design experiments to give an integrated experience covering curriculum thread. Few experiments are carried out to assess the student outcomes using representative course of a group of courses called tollgate course [9–13]. There are few attempts in introducing mini-projects at earlier levels of the program to strengthen the outcome evaluation process [4, 5].

Each of these methods has weaknesses if the primary objective is curriculum improvement. Author in [14] discusses the mini-projects for senior students as tool for curriculum modifications and improvement. Mini-projects can be a means for students to apply the learning from cluster of courses belonging to a curriculum thread and also act as evidence of student learning in that vertical like tollgate course. Attainment of student outcome in miniprojects provides better resolution than capstone projects and can help to design and modify the curriculum. This paper presents details of mini-projects as a tool in providing integrated experience in the curriculum threads of ECE undergraduate program.

Paper demonstrates that theme-based mini-projects:

1. Provide integrated learning experience in a curriculum thread to the student prior to capstone project.
2. Act as tollgate course to evaluate outcomes in a curriculum thread, which acts as an input to the continuous improvement of teaching learning process and the curriculum design and modifications.
3. Help to develop and reinforce professional competencies that have not been sufficiently emphasized in the earlier courses.

Section 2 discusses the ECE curriculum structure and framework for theme-based mini-projects. Assessment of mini-projects is provided in Sect. 3. Impact of student learning and continuous improvement with curriculum design/modifications are described in Sect. 4. Conclusions of the paper are given in Sect. 5.

2 Curriculum Threads and Framework of Miniprojects

Broad curricular components of ECE program are shown in Fig. 1a. The curricular components of the program are: (1) Basic science, Mathematics, Engineering, and Humanities, (2) Program Foundation Courses, (3) Program Verticals, (4) Integrated Experiences. As stated earlier in an undergraduate education program verticals or curriculum threads are largely independent. Hence, it is necessary that opportunities are created in curriculum to demonstrate and also to assess student learning in each of these verticals.

Assessment of mini-projects provide, useful student, course and curriculum assessment data, which are used as inputs for continuous curriculum improvement. There are several ways of carrying out course assessments like teacher/course evaluations, homework and test results, and minor, major exams. These methods do not provide any platform for assessing student's capability to integrate his learning in different courses

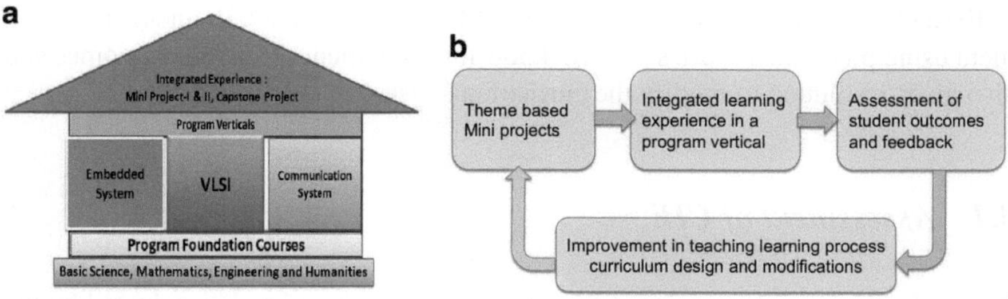

Fig. 1 Theme-based mini-projects: (**a**) Curriculum structure with threads, (**b**) frame work of mini-projects

to solve complex problems. Mini-projects can be of great help in overcoming this problem. The assessment data that is obtained in mini-projects at prefinal year of the UG program can be used effectively for corrective actions. Hence, to facilitate continuous improvement in the curriculum, a need for early evaluation of integrated learning in program verticals prompted us to propose mini-projects at the prefinal stage of UG program. Mini-projects are also used as tollgate courses to evaluate the learning in a group of subjects. The main objectives of the mini-projects are (1) to develop problem-solving skills by applying and integrating the knowledge base acquired, (2) to learn modeling/design/develop prototype/algorithm for a solution, (3) to develop implementation skills and team interaction, and (4) to learn technical documentation and presentation.

In the academic year 2007–2008, mini-projects were introduced for the first time in ECE curriculum; no themes were given to students. The spectrum of topics selected for projects was much wider and implementations were similar to miniature version of capstone projects. The process followed did not ensure that all the students were able to demonstrate competency pertaining to a particular vertical which made assessment much more complicated. To overcome this problem in 2009–2010 department came with an idea of fixing themes for the mini-projects. The themes for the projects were chosen in such a way that the student will be able to integrate and apply the knowledge and skills learnt in a set of subjects belonging to a particular vertical to solve complex engineering problem. While choosing the themes inputs from employers and immediate graduates were taken to ensure that the experience student gains is relevant to industries. For example, following few themes that are defined for each of the verticals.

1. Embedded System: (i) Automotive, (ii) Assistive applications.
2. Communication system: (i) Signal acquisition and processing, (ii) Information processing.
3. VLSI: (i) FPGA implementation of applications, (ii) Digital VLSI, Analog VLSI, or mixed signal VLSI design.

3 Assessment of Miniprojects

Student assessment consists of two components: (1) Continuous Internal Evaluation (CIE) and (2) Semester End Examination (SEE). Evaluation of CIE and SEE assessment using predefined rubrics helps to know the attainment of student outcomes and also gives resolution to modify the curriculum.

3.1 Assessment of CIE

The evaluation of CIE consists of four reviews and each review is conducted by the guide and a evaluation/review team (group of 3–4 faculty). The rubrics followed for the CIE evaluation are:

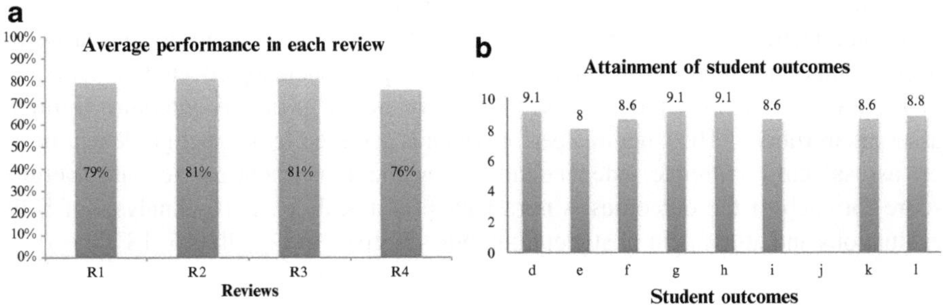

Fig. 2 Assessment of student outcomes: (**a**) average performance in each review in CIE (**b**) attainment of student outcomes

- Review-1: Identification of the problem, understanding of professional ethics (copy right, plagiarism), problem definition and literature survey, identifying multiple solutions, selection of the optimal solution with justifications.
- Review-2: Distribution of work among team members by leader and team work, specification and identification of input and output, functional block diagram relating input and output, design on paper with listing of the required components, simulation of the design using any EDA tool.
- Review-3: Detailed block diagram with all specifications/algorithm, integrating the functional blocks, debugging details, partial demonstration of results.
- Review-4: Hardware implementation, demonstration of results, analysis of results, report submission, paper presentations, awards, budget planning, and management.

3.2 Assessment of SEE

Semester end examination (SEE) includes submission of the project report, demonstration of the mini-projects and viva-voce conducted by the external and internal examiner. Performance criteria or rubrics used for evaluation of SEE are (numbers inside the bracket represent the maximum percentage of marks for the SEE):

1. *Problem statement (10)*: understanding of the theme, coverage of subjects, problem definition, design approaches.
2. *Design methodology (15)*: applying theoretical knowledge in solution paradigm, capability of integrating knowledge in the given solution, design specifications, and implementation methodologies.
3. *Demonstration of results and analysis (15)*: Representation of results, analysis of results, comparison, and conclusion.
4. *Report presentation, viva/voce (10)*: Outline flow of presentation, clarity, usage of language.

Average performance of students in CIE for each of the reviews is shown in Fig. 2a. The performance is comparatively low in Review-1 and Review-4. The Review-1 data shows that student needs to improve skills in problem identification, formulation, and conceptualization of alternative solutions. Further, student's capability in hardware implementation and project management needs improvement as reflected in the scores of Review-4. The assessment process is also related to program outcomes through assessment rubrics. The student's performance is assessed by project guide and project assessment committee independently using the assessment rubrics and average score for each of the outcomes is normalized on a scale of 1–10. Analysis of SEE evaluations and attainment of student outcomes is given in Fig. 2b [15, 18].

4 Impact on Student Learning and Continuous Improvement

The perception of students about the impact of mini-projects on their learning is carried out through a survey. The survey questions are designed to understand how students perceive their performance in meeting each of the performance criterion specified for the implementation of the project and they are:

Q1. To develop problem-solving skills
Q2. To understand modeling of a problem into realizable blocks
Q3. Usage of learnt theory in solving a problem
Q4. Time management and implementation of the project
Q5. Learning programming skills, coding standards, and analysis of results
Q6. Ability to work in a team
Q7. To develop technical writing skills
Q8. To develop presentation skills

Distribution of the survey questionnaire according to the performance criteria (Sect. 3.2) is:

1. Problem statement – 3 questions (1, 2, 3)
2. Design Methodology – 2 questions (4, 6)
3. Demonstration of results and analysis – 1 question (5)
4. Report presentation and viva voce – 2 questions (7, 8)

Figure 3 shows the result of the survey of mini-projects using eight survey questions covering the four performance criteria. The survey shows that students' perception is not very good in Q3 and Q4. The student survey reinforces the student weakness in problem identification, formulation, and conceptualization of alternative solutions and also student's capability in hardware implementation and project management (Q4). Further, the survey brings out some of the weaknesses associated with design methodologies.

Based upon the assessment data the department carried out following major improvements in the curriculum to improve the student attainment of the program outcomes. The modifications address both the weakness identified from qualitative analysis of attainment of student outcomes as well as through perception of student and they are:

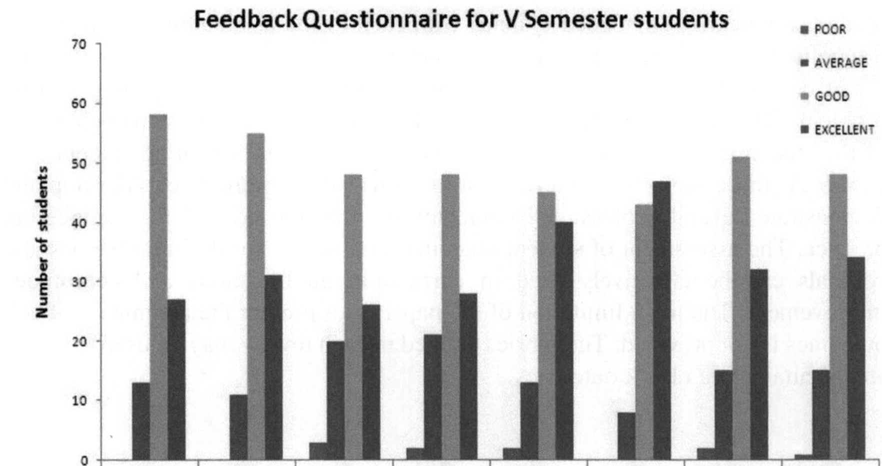

	1	2	3	4	5	6	7	8
POOR	0	0	3	2	2	0	2	1
AVERAGE	13	11	20	21	13	8	15	15
GOOD	58	55	48	48	45	43	51	48
EXCELLENT	27	31	26	28	40	47	32	34

Questionnarie

Fig. 3 Results of the survey of V semester mini-projects (2008–2012)

1. Introduction of 'Engineering Design' course: Both the assessments have brought out the weakness of students in problem identification, formulation, and conceptualization of alternate solutions. This is due to the fact that ECE students lack the exposure to engineering design process. The department designed and introduced a core course on 'Engineering Design' at 4th semester level from the present year. The course which is more of learning by doing takes the student through whole process of design.
2. Department designed few modifications to the curriculum to address the problem associated with hardware implementation and project management and they are:
   - Basic microcontroller course is introduced in IV semester in place of microprocessor.
   - State of art microcontroller like ARM is introduced in V semester in place of advanced microprocessors.
   - Embedded systems course is converted from elective to core course.

5 Conclusions

This paper discussed details of the curricular reform effort to design an integrated learning experience in curriculum threads through mini-projects in undergraduate program. The paper demonstrated that mini-project can be vertically integrated (demonstrated with embedded system) with each of the curriculum threads of the program and provide an opportunity to integrate knowledge and skills acquired in a set of courses belonging to the curriculum thread to solve complex engineering problems. The analysis of student attainment with evaluation rubrics and perception of the students using survey provided better resolution in evaluating student outcomes. As these projects are carried out in teams, students are able to develop and demonstrate several professional competencies that are critical for engineering practice. The assessment of student outcomes and survey for different themes and verticals can be effectively used in curriculum modifications and continuous improvement. The main limitation of the paper is, at present the attainment of a–k outcomes is not provided. The rubrics defined in each review may be used to evaluate the attainment of a–k outcomes.

References

1. Accreditation Board for Engineering and Technology (ABET). Engineering criteria 2000: criteria for accrediting engineering programs, effective for evaluations during the 2002–2003 accreditation cycle, Baltimore, MD, 2001.
2. http://www.nba-aicte.ernet.in/manual.html
3. Bradley University Electrical Computer Engineering Department. Capstone project: evaluation relative to professional level. http://cegt201.bradley.edu/~gld/prof_cap.html. 2002, August.
4. Soares FO, Sepúlveda MJ, Monteiro S, Lima RM, Dinis-Carvalho J. An integrated project of entrepreneurship and innovation in engineering education. Proc Sci Direct Mechatr. 2013;23(8):987–96.
5. Dempsey GL, Huggins BD, Anakwa WKN. Use of senior mini-project for electrical and computer engineering curriculum assessment. In: Proceedings of the 2001 American Society for Engineering Education annual conference and exposition, Albuquerque, NM, June 2001.
6. Dempsey GL. EE451: mini-project. http://cegt201.bradley.edu//~gld/miniproj.html. 2002, August.
7. King RH, Gosink JP. Meeting ABET EC 2000 criterion 3 out- comes with a laboratory course. In: Proceedings of the 2001 American Society for Engineering Education annual conference and exposition, Albuquerque, NM, June 2001, Session 2526.
8. Newcomer JL. Reassessing design goals: using design projects to meet assessment goals. In: Proceedings of the 2001 American Society for Engineering Education annual conference and exposition, St. Louis, MO, June 2000, Session 1625, [Online].
9. Roslan Hashim, Mokhtar Azizi Mohd Din. Implementing outcome based education using project based learning. University of Malaya, Eur J Sci Res. ISSN 1450-216X. 2009;26(1):80–6.
10. Mechanical Engineering Department. Measurable course outcomes to support the program outcomes and objectives. Ohio University Mechanical Engineering Department, Assessment and CI process. 2001–2002.
11. Johnson ME, Dubikovsky S. Introduction of Lean Six Sigma Principles into aeronautical engineering technology courses. In: Proceedings of the 2008 IAJC-IJME international conference. ISBN 978-1-60643-379-9, Paper 047, ENT 203, Year 2008.
12. University Report, Ohio University Mechanical Engineering Department. Assessment and CI process review, 2002. Review of the department's constituencies and the plan for soliciting and incorporating their input. 2002.
13. Svensson T, Gunnarsson S. Using a project model for assessment of CD skills. Queens University Kingston, Ontario, Canada, June 2005.

14. Powell PC. Assessment of team-based projects in project-led education. Eur J Eng Educ. 2004;29(2):221–30.
15. Perez GL, Shuman L, Wolfe H, Besterfield-Sacre M. Measuring continuous improvement in engineering education programs: a graphical approach. In: Proceedings of the 2001 American Society for Engineering Education annual conference and exposition, Albuquerque, NM, June 2001.
16. Basterfield-Sacre M, Shuman L, Wolfe H, Atman C, McGourty J, Miller R, Olds B, Rogers G. Defining the outcomes: a framework for EC-2000. IEEE Trans Educ. 2000;43:100–9.
17. Felder RM, Brent R. Designing and teaching courses to satisfy the ABET engineering criteria. J Eng Educ. 2003;92(1):7–25.
18. Nicoletti D, Orr J. An implementable/sustainable outcomes assessment process for an electrical engineering program. In: Proceedings of the 2001 American Society for Engineering Education annual conference and exposition, Albuquerque, NM, June 2001, Session 2532.

Experience of Using Felder–Soloman Index of Learning Styles

Vasim Shaikh and Pradeep Waychal

Abstract Education brings about holistic development of an 'individual' and depends on learning, which is engendering changes in knowledge, skill, and attitude. Both education and learning are essentially subjective processes. The developmental goals and needs and ways to realize them differ with almost every individual. The modalities of acquiring knowledge, skill, and attitude also vary widely. Therefore, it is important to understand individuals in the realm of education and learning. While there are many possibilities, Learning Style that describes relatively stable preferences of students to receive and process information appears to be more appealing to understand students. There are umpteen number of learning style models and instruments that are available. We surveyed them and zeroed on Felder–Soloman Index of Learning Styles (ILS). It is based on Felder and Silverman's model and assesses preferences on four bipolar dimensions: Active–Reflective, Sensing–Intuitive, Visual–Verbal, and Sequential–Global. The paper discusses our experience of using it in a variety of courses.

Keywords Felder–Soloman index of learning styles • Liberal learning • Fundamentals of computer programming • Software testing and quality assurance

1 Introduction

Education helps develop an individual in every aspect of life. It allows him to realize his full potential and contribute toward larger goals. It depends on learning, i.e., developing knowledge, skill, and attitude. It is important to realize that every

V. Shaikh • P. Waychal (✉)
Department of Computer Engineering and Information Technology,
College of Engineering, Pune, Shivajinagar, Pune 411005, Maharashtra, India
e-mail: Pradeep.waychal@gmail.com

R. Natarajan (ed.), *Proceedings of the International Conference on Transformations in Engineering Education: ICTIEE 2014*, DOI 10.1007/978-81-322-1931-6_43,

individual maintains his distinct identity in terms of development goals and potential and ways of learning.

The engineering education is unique in many ways. It is becoming increasingly complex across all branches from traditional civil engineering to new computer, space, and genetic engineering. The rising number of patents is making a wide range of new technologies available to engineers which are further adding to the complexities [1]. Richard Riley has beautifully articulated the current situation in the statement: "we are currently preparing (engineering) students for jobs that don't exist using technologies that haven't been invented in order to solve problems we don't even know are problems yet." While we cannot list technologies and problems, we certainly can prepare our students to learn to learn or lifelong learning. We can leverage knowledge of learning preferences of individuals in the process.

While there are many possibilities to bring in such understanding, learning style appears to be more apt. There are close to 100 different learning style models that have been formulated, and instruments to assess preferences on some of those models are available either from open sources or from commercial vendors with millions of users accessing them each year. We surveyed various models and decided to use Felder–Soloman Index of Learning Styles (ILS) that has been developed by engineering educators. The model helps students identify their strengths, nature, and thought process of learning and also helps the faculty understand their students better and adopt appropriate pedagogical techniques.

Major contribution of the paper is to share our experience and utility of using the Felder–Soloman Index of Learning Style (ILS) in a variety of courses. Further, the paper shows that balance in the learning styles can be brought in during a semester using some innovative elements in a course.

The next section of the paper explains the Learning Styles and elaborates on the ILS. That is followed by our experience of using it on various courses. We then enlist our observations and provide concluding remarks and point out future scope for the study.

2 Learning Styles

Learning styles are relatively stable preferences students have for ways of receiving and processing information. Hawk and Shah [2] claim that the use of learning style instruments should allow students and faculty to seek out the activities that are conducive to more effective and deeper learning. The concept of learning styles also has detractors in the psychology community. Coffield et al. [3] studied 13 different models and asserted that in the current state of research-based knowledge about learning styles, there are real dangers in commending detailed strategies to practitioners, because the theories and instruments are not useful. There is no consensus about the recommendations for practice too, they added. Pashler et al. [4] argue that very few studies have even used an experimental methodology capable of testing the validity of learning styles applied to education and the ones that have used to contradict the popular meshing hypothesis according to which instruction is best provided in a format that matches the preferences of the learner.

Felder [5] has made an interesting observation. He says that every 2 years or so, some academic psychologists conduct a literature review and conclude that no research supports the use of learning styles in teaching. Journal reviewers and editors treat this conclusion as a new revelation that once and for all debunks learning styles. These pronouncements, he has added, have never had the slightest effect on the world academic community's extensive and continually growing use of learning styles models and assessment instruments, but that has never deterred others from repeating the exercise 2 years later. Felder [5] further points out that learning style instruments are neither infallible guides to student behavior nor made-up constructs that have no basis in reality. He opines that they are useful descriptions of common behavior patterns and have been used frequently and successfully to help teachers design effective instruction, help students understand their own learning processes better, and help both teachers and students realize that not everyone is like them and the differences are often worth celebrating. We have assessed learning styles of many students and found that they are used and appreciated by both students and teachers.

2.1 *Felder–Soloman Index of Learning Style (ILS) Instrument*

We surveyed the available instruments and chose the (ILS) [6] for our class. It assesses preferences on four learning style dimensions on a model developed by Felder and Silverman [7]. The model defines learning style as "the characteristic strengths and preferences in the ways individuals take in and process information" and originated in the engineering discipline. The instrument is freely available for academic institutes and is very well documented. The lead author of the model and instrument, Prof Richard Felder, is an active researcher, is approachable, and is immensely helpful.

The Felder–Silverman model asserts that individuals have preferences along four bipolar dimensions: Active–Reflective, Sensing–Intuitive, Visual–Verbal, and Sequential–Global. Active learners prefer doing things, particularly in groups. Reflective learners work better alone, with time to think about the task before doing it. Sensing learners like facts, data, and experimentation and work well with detail. Intuitive learners prefer ideas and theories, particularly when they can grasp the new ideas. Verbal learners like to hear their information and engage in discussion, especially when they can speak and hear their own words. Visual learners like words, pictures, symbols, flow charts, diagrams, and reading books. Sequential learners prefer linear reasoning, step-by-step procedures, and material that come to them in a steady stream. Global learners are strong integrators and synthesizers, making intuitive discoveries and connections to see the overall system or pattern. Both innate personality traits and prior experiences can affect students' preferences on each of those scales.

The Index of Learning Styles provides scores showing the strengths of an individual's preference for one category or the other on each of the four dimensions. The instrument is a free, 44-item questionnaire [6] that requires choosing one of two options that focuses on some aspect of learning. The choices result in a score of 1, 3, 5, 7, 9, or 11 for each of the preferred categories, with a 1 signifying a very slight preference for the category and an 11 a very strong one.

Felder and Spurlin [8] have cited various studies that support the reliability and validity of the ILS. Reliability is widely analyzed using Cronbach's coefficient alpha, which is an average of all possible split pair correlations of responses to items assessing preferences on a particular scale. Tuckman [9] and Sekaran [10] propose a minimum alpha value of 0.5 for assessments of preferences such as those done by the ILS. Felder and Spurlin [8] have quoted four studies in which alpha for each of the four dimensions exceeded 0.5 except for one global–sequential value that fell slightly below it, indicating that the instrument has satisfactory internal consistency reliability, and other measurements have shown satisfactory test–retest reliability. Assessed preferences at 10 engineering institutes were found to be quite similar, with the subjects consistently being predominantly active, sensing, visual, and sequential. Only at one institute, a batch of students was found to be a little more global than sequential [8].

While it is important that students establish balance in all the dimensions of learning styles, two dimensions are more important in engineering education – sensing–intuition and global–sequential. It has been observed that precollege education tends to push people toward a sensing and sequential orientation. In order to be successful professionally, engineering graduates must establish balance and develop intuitive and global preferences.

3 Our Study

Our main goals are to enhance learning experience of every student and make them better lifelong learners. Toward that, in every course we assess learning styles and use a variety of pedagogical techniques. We are presenting our finding of three different courses – a first year course on Fundamental of Computer Programming, a second year course on Liberal Learning, and a final year course on Software Testing and Quality Assurance.

3.1 Fundamental of Computer Programming (FCP)

This was a first year course. The faculty found the students not responding well to his questions and assignments and not progressing satisfactorily. We then introduced the faculty and students to the learning style concept. We administered the Index of Learning Styles (ILS) assessment to the students and are presenting the

Table 1 Summary of the ILS assessment of FCP class

ACT	REF	SNS	INT	VIS	VRB	SEQ	GLO
5.44	5.53	6.09	4.91	7.78	3.22	5.72	5.28

Table 2 Feedback of the ILS assessment of FCP class

	Usage of the assessment results	Benefits thereof	Recommendation for carrying out assessment of other students
Absolutely	1	20	44
To some extent	64	43	21
Not at all	1	2	1

results in Table 1. The faculty made changes to the teaching style like introducing group assignments, providing holistic picture, leveraging more visual aids based on the assessment results. The steps brought in significant improvement in the class performance.

We carried out a survey – a month after the assessment – to ascertain efficacy of the ILS by asking usage of the assessment results, benefits thereof, and their recommendation to use the assessment in other classes. Their response is given in Table 2 and was overwhelmingly positive. The faculty's performance was appreciated by the students in a senate meeting.

3.2 *Software Testing and Quality Assurance (STQA)*

This final year course provides an excellent platform for experimentation. Both testing and quality assurance activities can be executed using both the polarities of all the dimensions especially sensing–intuitive and global–sequential. The course is a part of Software Engineering which is a young engineering discipline. In fact, some critics even do not recognize Software as an engineering discipline. These disagreements provide a very good base for debate and opportunities to encourage thinking. We included many open-ended problems during the course and facilitated their solutions. Some open-ended problems were a part of open-book regular tests and the final examination. The class was also assigned to write wiki articles, underwent a daylong innovation workshop, and used many active learning strategies.

We started by explaining fundamentals of learning and learning styles to the students and administered the ILS assessment and helped them interpret its results. At the end of the course, we carried out another assessment. We are presenting results of pre- and postassessments in Table 3.

At the middle of the semester, we carried out a survey in the class to ascertain efficacy of the ILS by asking usage of the assessment results, benefits thereof, and their recommendation to use the assessment in other classes. Their response is given in Table 4 and was quite positive.

Table 3 Summary of the ILS assessments of STQA class

	ACT	REF	SNS	INT	VIS	VRB	SEQ	GLO
PRE	5.80	5.20	6.49	4.48	8.27	2.71	5.92	5.07
POST	6.10	5.50	5.86	5.49	8.60	5.50	5.16	5.50

Table 4 Feedback of the ILS assessment of STQA class

	Usage of the assessment results	Benefits thereof	Recommendation for carrying out assessment of other students
Absolutely	14	18	29
To some extent	31	26	16
Not at all	3	4	3

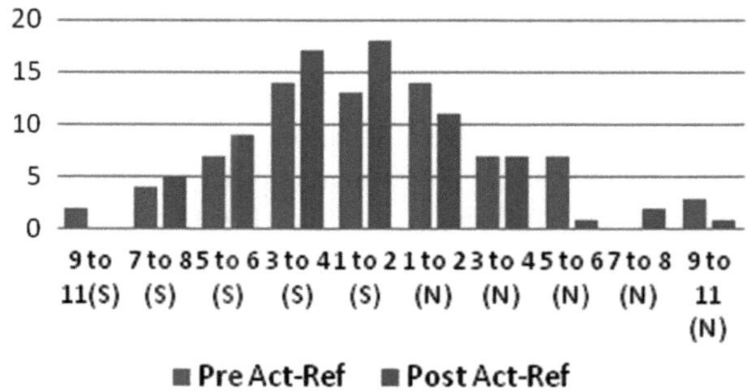

Fig. 1 Changes in active–reflective dimension

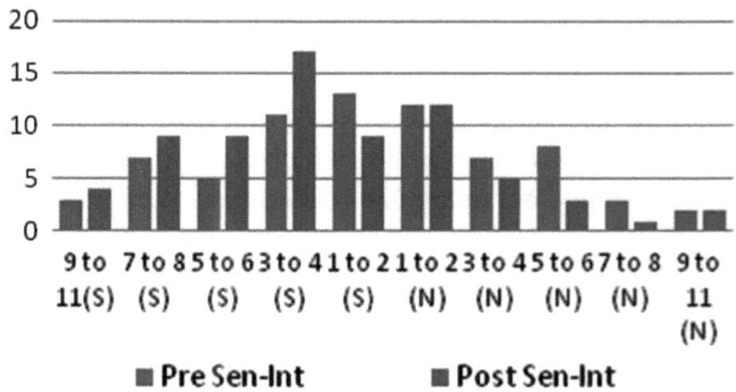

Fig. 2 Changes in sensing–intuitive dimension

At the end of the course, the students were asked if understanding of their learning styles has helped them to become a better learner using the 5-point Likert scale with 5 being strongly agree and 1 being strongly disagree. The question received an average of 4.3. The changes in various dimensions of the learning style are shown (Figs. 1, 2, 3, and 4):

We saw a noticeable pre-post shift from sensing toward intuition and from sequential to global styles. Paired t-tests indicated the pre-post change in the SNS-INT and SEQ-GLO average preference scores as statistically significant ($p = 0.004$

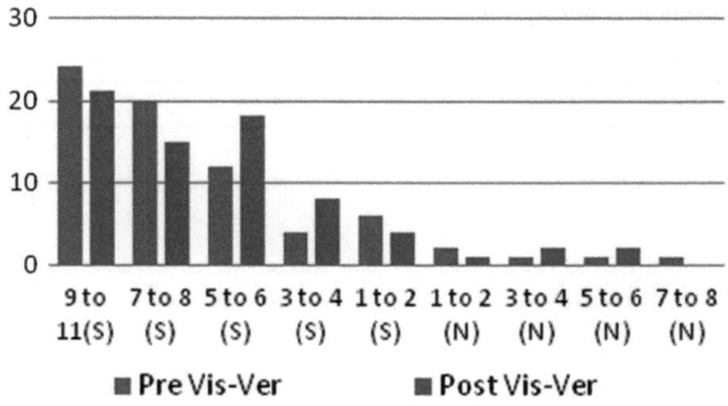

Fig. 3 Changes in visual–verbal dimension

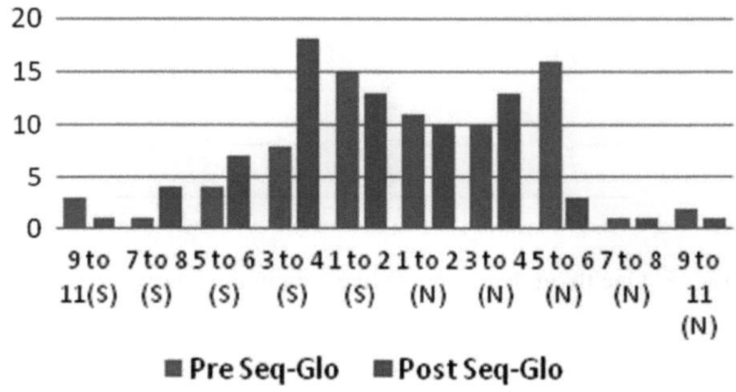

Fig. 4 Changes in sequential–global dimension

and $p = 0.002$). We did not see statistically significant change in the other two dimensions. The course and faculty secured the best rating of 4.7/5 and received excellent qualitative comments.

3.3 *Liberal Learning Course (LLC)*

We ran a liberal learning course for 328 second year students of noncircuit branches [11]. The primary goals of the course were to inculcate a lifelong learning process that helps students to extend knowledge base beyond engineering, appreciate the interplay of engineering with other disciplines, and make students better learners. The course did not have any contact hours in the timetable and expected students to learn the new topics on their own. Their awareness of learning style concepts and own assessment results was very useful for the students. We conducted an orientation class for all the students at the start of the semester and carried out the ILS assessments therein. We are presenting the results in Table 5.

At the end of the course we asked students to fill in the online feedback consisting of two questions: (1) one thing that they liked about the course and (2) one thing they disliked. Total of 123 students responded. Figures 5 and 6 present the consolidated feedback:

Table 5 Summary of the ILS assessment of LLC class

ACT	REF	SNS	INT	VIS	VRB	SEQ	GLO
5.94	5.06	5.45	5.55	7.91	3.08	5.73	5.26

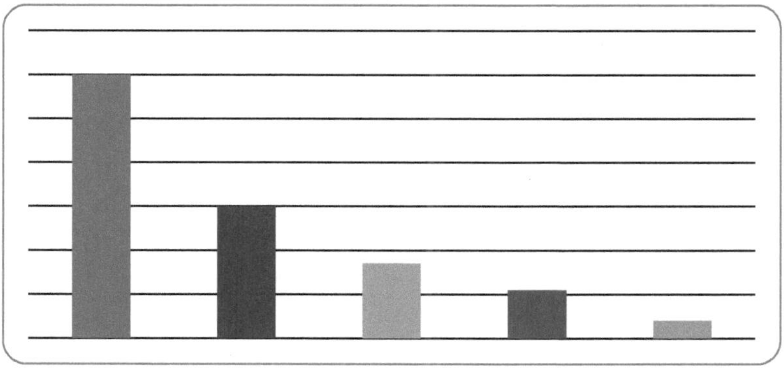

Fig. 5 One thing that students liked about the course

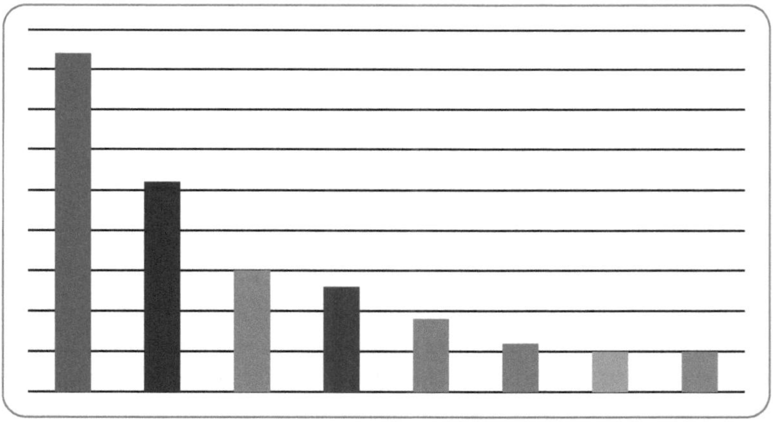

Fig. 6 One thing that students disliked about the course

The students liked the course and its approach. Besides overall theme, they liked freedom offered to choose topics and knowledge/learning they gained out of the course. Interestingly, one third of the students who provided feedback did not dislike anything.

Table 6 Summary of assessments of all the three courses

	ACT	REF	SNS	INT	VIS	VRB	SEQ	GLO
FCP	5.44	5.53	6.09	4.91	7.78	3.22	5.72	5.28
LLC	5.94	5.06	5.45	5.55	7.91	3.08	5.73	5.26
STQA PRE	5.80	5.20	6.49	4.48	8.27	2.71	5.92	5.07
STQA POST	6.10	4.90	5.80	5.20	8.62	2.36	5.14	5.84

4 Concluding Remarks

We found that the ILS assessments can be very helpful, across three significantly different courses. In these courses, we have essentially treated education and learning as an individualized self-discovery process and focused on lifelong learning skills. While we have added modern, enhanced, personalized, and current methods of teaching, we have maintained the subtlety that every student will learn and progress using a different trajectory. The students became aware of their preferences and became better learners. While most of the students expressed this thought, we are citing responses of a couple of students from their course-end feedback:

Now I am pretty aware of environment which helps me improve my learning. My outlook toward studying has changed a lot and I also find myself more confident.

Before knowing my learning styles, I used to study things without any interest, Knowing the styles has increased my interest and changed my thinking.

The summary of all the assessments is given in Table 6.

It is clear that most of the students across all the classes are highly visual. Barring LLC, we notice that students tend to be more sensing than intuitive. On sequential–global dimension, students seem to be tilted toward sequential dimension. Except for FCP, students seem to prefer active learning.

In the case of STQA, between the PRE and POST assessments, we see clear shift toward Intuitive and Global polarities, the two critical ones for engineering profession. While use of Learning Styles is a reasonably well accepted learning aide, changes in the styles in a semester are not yet reported. We understand that we require fortification of our claims and carry out more experiments on larger population covering different cultural and geographic settings but the initial results suggest that the approach has some merits and is worth exploring.

Acknowledgement We would like to thank Prof. Richard Felder who helped us to understand learning style concepts and its proper usage and provided many useful comments on various iterations of the paper. We also acknowledge all the students of the class who participated in various exercises which allowed us to test our hypothesis and Mr. Abhay Joshi for reviewing the early manuscripts.

References

1. US Patent and Trade Marks Office. U.S. patent activity calendar years 1790 to the present. http://www.uspto.gov/web/offices/ac/ido/oeip/taf/h_counts.pdf. 2012.
2. Hawk TF, Shah AJ. Using learning style instruments to enhance student learning. Decis Sci J Innov Educ. 2007;5(1):1–19.
3. Frank C, Moseley D, Ecclestone K. Learning styles and pedagogy in post-16 learning: a systematic and critical review. London: Learning and Skills Research Centre; 2004.
4. Harold P, McDaniel M, Rohrer D, Bjork R. Learning styles concepts and evidence association for psychological science. Psychol Sci Public Interest. 2009;9(3):105–19.
5. Felder RM. Are learning styles invalid? (Hint: No!). North Carolina State University. On-Course Newsletter, September 27, 2010. http://www.oncourseworkshop.com/Learning046.htm. 2010.
6. http://www4.ncsu.edu/unity/lockers/users/f/felder/public/ILSpage.html
7. Felder RM, Silverman LK. Learning and teaching styles in engineering education. Eng Educ. 1988;78(7):674–81.
8. Felder RM, Spurlin JE. Applications, reliability, and validity of the index of learning styles. Int J Eng Educ. 2005;21(1):103–12.
9. Tuckman BW. Conducting education research. Wadsworht Group. 1999.
10. Sekaran U. Research methods for business: a skill building approach. 2nd ed. New York: Wiley; 2003.
11. Pradeep W, Anil S. Liberal learning framework. In: Proceedings of the ASEE annual conference, Atlanta 2013.

Teaching Reform Through Model Building in Theory of Machine Course

G.R. Chalageri and G.U. Raju

Abstract Theory of machine is an important course of mechanical engineering, which develops students' ability of system and concept design. The concept design of mechanical motion is the most crucial part in mechanical engineering design that plays a vital role in developing students' creative ability. In the teaching process of this course, kinematic kit and modern design tools are used to guide the students to realize the concepts and it is believed that this lays a solid foundation for understanding of design of machine elements course. Model building has been done using elements available in kinematic kit. At the end of the course, analysis was done to measure the effectiveness of students learning and it is found that physical construction of a model helped students to generate ideas and resulted in better visualization and understanding of mechanisms. This pedagogical practice also helped students to understand the concept of motion change by introducing different kinematic elements.

Keywords Mechanisms • Kinematic kit • Modern tool

1 Introduction

In theory of machine course, the concept design of motion is the most creative content, in developing students' ability to design. This course not only imparts professional knowledge to students, but also develops problem solving and analyzing abilities among students [1, 2]. According to Jianguo [3], developing and improving students' creative thinking ability is an important task in higher education reform.

G.R. Chalageri (✉) • G.U. Raju
Mechanical Engineering Department, BVB College of Engineering and Technology, Vidya Nagar, Hubli, Karnataka 580031, India
e-mail: gireesha@bvb.edu; raju_gu@bvb.edu

R. Natarajan (ed.), *Proceedings of the International Conference on Transformations in Engineering Education: ICTIEE 2014*, DOI 10.1007/978-81-322-1931-6_44,

For many years, traditional teaching method was adopted in teaching theory of machine course. Later, animated videos were shown to realize the working of mechanisms, which improved the level of understanding marginally. Further, we thought of providing hands-on experience to the students that could significantly augment learning of the course. This paper brings up about developing students' creative thinking ability through pedagogical practices such as building up of mechanisms, in learning the course.

Zhang Meilin [4] reported that knowledge development is a continuous iterative process which contains old knowledge, innovation, and new knowledge. Accordingly, teachers must constantly provide new knowledge to enable students to search for alternate methods to solve the problems. Kinematics of machines deals with the study of the relative motion of machine parts. It involves study of position, displacement, velocity, and acceleration of machine parts. Knowledge of motion and kinematic elements is important to build a mechanism. In teaching the theory of machine course, many of the concepts/mechanisms need to be explained in two-dimensional images. Students should have good imagination skill to understand these concepts. If students lack this imagination skill, then it is hard to imagine the movement of elements and force transmission through them. Use of visual aids enriches the student's imagination and thus enhances understanding the concepts to some extent. For example, in the explanation of the planar linkage mechanism, it can be demonstrated by animation of dump truck, engine slider–crank mechanism, and so on. Further, in our institute students build the physical model of the same using a kinematic kit which provides hands-on experience. Its effect is often beyond the teacher's expectation. The hands-on experience motivated students' creative ability and desire to learn effectively.

2 Course Project Through Model Building

Theory of Machines course is one of the essential courses of Mechanical Engineering undergraduate curriculum. Previously, this course was taught by traditional instruction-based pedagogy. In order to enhance the learning, this course is restructured on project-based learning approach. The purpose of course project in the theory of machines course is to enhance the learning of basic principles. The lack of imagination ability in students to understand the kinematic mechanisms was the impetus for our investigation of model building as an instructional strategy. In addition, although previous researchers studying the kinematics have utilized sketches, drawings, verbal accounts, and written communications, no literature is available investigating whether hands-on model building can contribute to a better understanding of the kinematics of machines. In view of this, the present paper gives insight about whether a hands-on, model-building activity can contribute to students understanding of the course. The topic of course project was designed by the teacher. Students determine the kinematics motion parameters, draw kinematics diagram of the mechanisms, and then build physical models using kinematic kits shown in

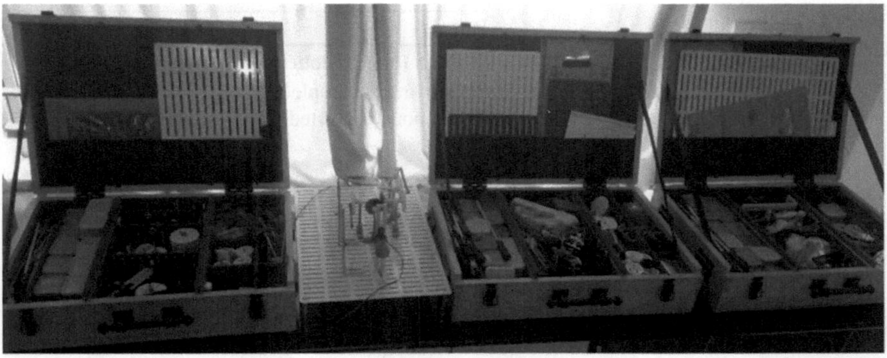

Fig. 1 Kinematic kits

Fig. 1. To analyze the effectiveness of the course project, analysis was carried out. In the analysis, data from 150 mechanical engineering students was collected.

3 Methodology

It is not possible to view cognitive processes during a model-building task. Other way to know what students are thinking during a design task is to ask them. Course project analysis (CPA) was carried out to know about the process of development of solutions for a given problem. During CPA data collection, students are asked to think aloud while performing a task. Think aloud involved students thinking aloud when they are working on a course project. Students are asked to say whatever they are thinking and doing as they go about model building task. This enabled course instructor to observe the process of model building rather than looking only the final product. From the course project analysis, we get insight into how they utilize subject knowledge to develop a solution to the given problem.

A total of 30 groups were made out of 150 students and the objective of each group is to build a physical model of a mechanism to convert rotary motion into linear motion for an application. Use of elements in part A is compulsory and they use one or more elements from part B as given in Table 1. Finally, they are required to simulate the localized mechanism using multibody dynamic software.

The following procedure was followed in building a mechanism.

1. Initially students have to select the appropriate links and gear elements from part A and need to assemble them with element(s) selected from part B with the help of different supporting elements available in kinematic kit.
2. Then, students need to check the feasibility of mechanism by giving input through motor. If they are unable to get the required results then add/remove few elements from the mechanism or change the combination of elements in order to get the required output as linear motion. This task was accomplished by using

Table 1 List of kinematic elements

Part-A	Part-B
1. Links (binary/slotted)	1. Cams
2. Gear trains (simple/ compound/reverted)	2. Geneva mechanism
	3. Coupling
	4. Belt drive
	5. Spring

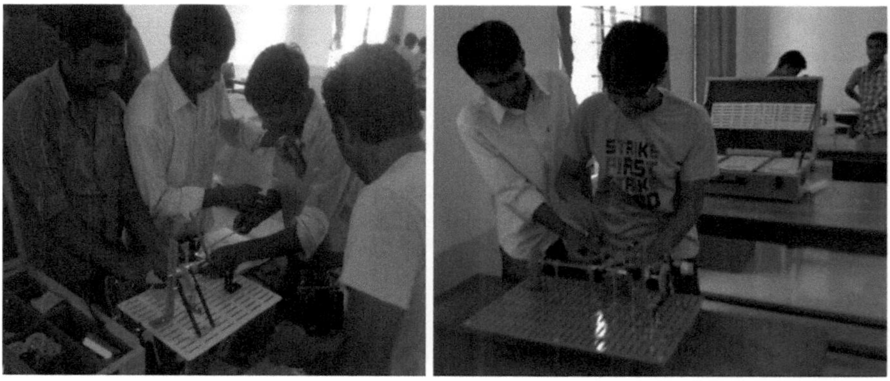

Fig. 2 Students working in a group to develop a mechanism

tools and methods dealt in theory classes. Once they get a required output, demonstrate the physical model.

3. After demonstrating the physical model, students make a complete velocity analysis of kinematic links and calculate velocity ratio of gear trains used. Also calculate the mobility of mechanism in order to justify the number of inputs required to control the mechanism [5].
4. Students simulate the localized mechanisms using multibody dynamic software ADAMS. Compute the velocity of the output link using relative velocity method and compare with ADAMS results [6].

Students working in a group to develop a mechanism for an application are shown in Fig. 2. Physical models of the course projects are shown in Fig. 3.

4 Assessment

Theory of machine course has a total of 100 marks, in which 50 marks is for Continuous Internal Evaluation (CIE). In the CIE, 40 % weightage, i.e., 20 marks was given to model building through course projects. For evaluation, the 20 marks are split as given in Table 2.

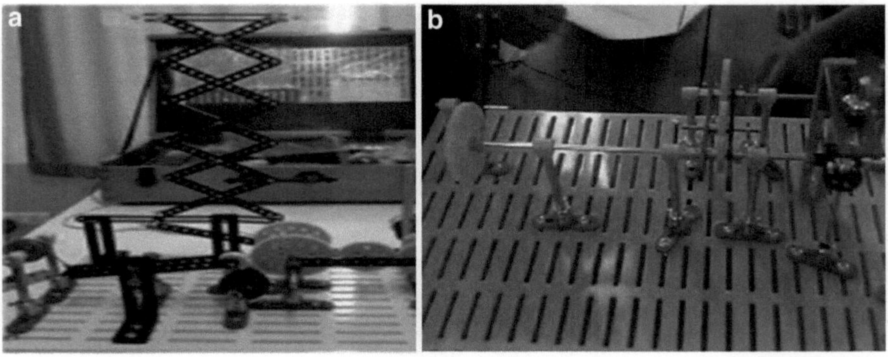

Fig. 3 Physical models of the course projects. (**a**) Lifting mechanism model (**b**) Pulverizing and printing mechanism model

Table 2 Evaluation of course projects

Details	Marks
Development of mechanism for an application	10
(a) Only one output = 06	
(b) Two outputs = 02	
(c) More than two outputs = 02	
Simulation using ADAMS	05
Analysis using relative velocity method	03
Report containing details as mentioned in methodology	02
Total	20

Following questions have been asked and analyzed as feedback questionnaires

1. Use of kits has improved students participation in learning theory of machine course
2. Use of kits helped the students in learning working of different mechanisms and motion concepts
3. The course project has developed critical thinking ability among the students in building the mechanisms
4. The course project has developed ability to work in teams and improved presentation skill
5. Better understanding of the course by the use of kits has helped to improve performance in minor exams
6. Co-relation of graphical and multibody simulation results of working models helped in better understanding of mechanisms.

The course project feedback analysis is shown in Fig. 4. SA&A: strongly agree and agree, N: neutral, and D&SD: disagree and strongly disagree.

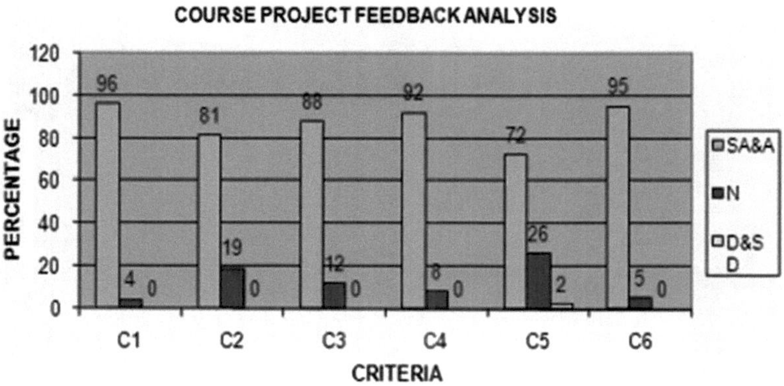

Fig. 4 Course project feedback analysis

5 Conclusions

The methods and tools which were discussed in the theory are well demonstrated through model building in the course projects. This in turn helped students for the better learning of rudiments of the course. Students have justified that motion-related problems are better structured through course projects. It also helped students to expose and familiarize with team-based activities. The present course project addresses criterion (d), i.e., an ability to function on multidisciplinary teams and criterion (k), i.e., an ability to use the techniques, skills, and modern engineering tools necessary for engineering practice of ABET 3a–3k program outcomes. It also addresses the performance indicator of mechanical engineering department, i.e., MEPI (a)-2- An ability to apply laws of kinematics and dynamics. Finally with this course project the Blooms taxonomy level has increased from L3 to L4, i.e., apply to analyze.

References

1. Wang Ping, Shen Xiaoyang. Mechanical engineering machinery foundation series practice teaching reform of the course. J Coll Educ Technol. 2005;24:57–8.
2. He Guoqi, Hu Chengwu, Wu Jiping. Mechanical principle in the course design of training students innovative ability exploration. J Zhuzhou Eng. 2006;20:145–6.
3. Lou Jianguo. Mechanisms and machine theory course design teaching exploration and practice. J Shaoxing Univ (Nat Sci). 2003;23:114–6.
4. Zhang Meilin. Mechanical create design. Beijing: Chemical Industry Press; 2005.
5. Shigley JV. Theory of machines and mechanisms. Delhi: McGraw Hill International; 2003.
6. Multi body dynamics software ADAMS user manual.

Application of Design of Experiments (DOE) in Course Project

V.N. Gaitonde, B.B. Kottershettar, and S.B. Burli

Abstract Accreditation Board for Engineering and Technology (ABET) in its format has put a new spin on experimentation skills in engineering education. The National Board of Accreditation (NBA) in India has adopted the graduate attributes (GAs) in line with the program outcomes (POs) of ABET. Specifically, outcome (b) states that engineering graduates must have "an ability to design and conduct experiments, as well as to analyze and interpret data." While the ability to conduct experiments, as well as the ability to analyze and interpret data has been addressed by traditional laboratory courses, the ability to design an experiment presents a new challenge for teachers and students alike. The paper first discusses steps involved in design and conduct of experiments and analysis and interpretation of data/results, then a general process for experimental design and finally, presents an example of how this process is used to teach design of engineering experiments in a postgraduate program. The students were made to work in laboratory on an open-ended experimental design as a course project in addition to regular classroom activity. In the tutorial component of this course, students work through 7 multiweek modules that have been developed with a robust framework. Integrating design of experiments (DOE) into a course project builds teamwork, communication, and use of statistics in machining process in addition to enhancing the learning experience.

Keywords DOE • Statistics • Teaching design of experiments

V.N. Gaitonde (✉) • B.B. Kotturshettar • S.B. Burli
Department of Industrial and Production Engineering, BVB College of Engineering and Technology, Vidya Nagar, Hubli, Karnataka 580031, India
e-mail: vngaitonde@bvb.edu; bbkshettar@gmail.com; sbburli@bvb.edu

R. Natarajan (ed.), *Proceedings of the International Conference on Transformations in Engineering Education: ICTIEE 2014*, DOI 10.1007/978-81-322-1931-6_45,

1 Introduction

Accreditation Board for Engineering and Technology (ABET) in its format has put a new spin on experimentation skills in engineering education. The National Board of Accreditation (NBA) in India has adopted the graduate attributes (GAs) in line with the program outcomes (POs) of ABET. Specifically, outcome (b) states that engineering graduates must have "an ability to design and conduct experiments, as well as to analyze and interpret data." While the ability to conduct experiments, as well as the ability to analyze and interpret data has been addressed by traditional laboratory courses, the ability to design an experiment presents a new challenge for teachers and students alike. The department has made several attempts to understand the requirements of the outcome (b) so that our processes could be designed or redesigned to fulfill the requirements. This paper emphasizes our learning by way of first discussing the general process for experimental design followed by an example of how this process is used to teach design of engineering experiments in a postgraduate course.

2 Experimental Design Process

Inquiry-based learning as outlined in reference [1] provides a framework to understand the process and the skills needed to design an engineering experiment. As we move from the 'conduct' type of experiments to student 'design'ed experiments under the supervision of a teacher, the responsibility for the various tasks involved in doing so, gradually shifts from the teacher to the student. This is a very important observation because research has shown that taking responsibility for one's own learning is one of eight conditions that must be satisfied in order to master a task or subject matter [2]. It is also an essential condition for the development of students as lifelong learners. Hence, it must be understood that without an opportunity to take responsibility for the decisions about the various tasks of an experiment, students cannot master the process of experimental design. The design of experiment concepts can be applied to one or more of the experimental and/or testing scenarios as described below that the students come across during their study of engineering.

- An experiment has to be conducted to verify the known relationship between two or more variables
- An experiment has to be conducted to establish the relationship between two or more variables which is unknown
- A test has to be conducted to confirm its meeting the design specifications for the newly developed product, before production.

The first two types of experiments are shared between engineering and all science. The third type is unique to engineering. Just like with the design of an engineering product, it is desirable to have a general process that one can follow to design an experiment under any circumstances. This process can also serve as a tool for teaching students experimental design. An attempt to create such a process involves seven steps. They are:

1. Define the goals and objectives of the experiment.
2. Research any relevant theory and previously published data from similar experiments. The purpose of this step is to have an idea about what to expect from the experiment.
3. Select the dependent and independent variable(s) to be measured.
4. Select appropriate methods for measuring these variables.
5. Choose appropriate equipment and instrumentation.
6. Select the proper range of the independent variable(s).
7. Determine an appropriate number of data points needed for each type of measurement.

Naturally, additional skills are needed to meet the other three components of outcome (b). For example, to conduct an experiment, an engineer should be able to:

- Familiarize himself/herself with the equipment.
- Calibrate the instruments to be used.
- Follow the proper procedure to collect the data and/or measure the performance of the product.

To analyze a set of experimental data an engineer should be able to:

- Carry out the necessary calculations.
- Perform an error analysis.
- Tabulate and plot the results using appropriate choice of variables and software.

Finally, to interpret the data an engineer should be able to:

- Make observations and draw conclusions regarding the variation of the parameters involved.
- Compare with predictions from theory or design calculations and explain any discrepancies.

The following sections provide an example of how production management (PG) students of industrial and production engineering department at BVB College of Engineering and Technology (BVBCET), Hubli are taught experimental design. Our approach to educating students with regard to experimentation has evolved ever since our thought process started to meet the requirements of outcome (b). Prior to 2010, all "experiments" used to be predefined measurement exercises. Since 2010, we have begun to introduce open-ended experiments including a course project, where the definition, execution, and documentation of engineering experiment is left entirely up to the student. An example of a course project by one batch of students from the academic year 2012–2013 is summarized below in light of the seven-point framework outlined above. The course provides students with the fundamental knowledge and principles in modern material removal processes. A physical understanding of the machining processes including forces, power consumption, temperatures, and other machinability aspects such as tool wear, tool life, and surface finish has been illustrated in this course. This core course consists of lecture (4 h/week) and tutorial (2 h/week), which includes course project. The modules developed in the course project of this course build teamwork, communication skills, and use of statistics and DOE in machining process along with the hands-on experience of the course.

3 Statistics and Design of Experiments

Statistics and DOE were cited as skills our industry personnel felt were weak in engineering graduates they were currently hiring. The employers also felt these skills were critical to a process engineer's success. They emphasize statistics and DOE as two of the top skills needed for new college graduates to be hired as process engineers. There are frequently stand-alone classes on statistics and design of experiments. However, integrating the statistics and DOE directly into a laboratory class stemmed from a desire for the students to gain a more hands-on understanding of statistics. Six lab modules were designed to provide learning experiences in a range of basic topics in DOE and statistical analysis of data. Table 1 lists the modules and the schedule for a 14-week semester. The laboratory experiments are open ended, where teams of students design their own experiments to answer given questions. The modules increase in complexity of the DOE and statistics covered as well as the experimental planning required of the students [3].

The first week of each lab module is reserved as a dry lab session. During this week, the fundamentals of the DOE or statistical analysis of data needed for that module are taught. Students gain mastery of these skills by working through given dry lab exercises (numerical problems related to machining process). Then as a

Table 1 Modules used in course project

Course project module	No. of weeks	Schedule
Module 1: Dry lab session: overview of DOE and its importance in engineering applications.	01	7th week
Module 2: Dry lab exercises: analyzing the DOE principles through various numerical problems.	01	8th week
Module 3: Planning phase (*Design of Three Factor Experiment*): detail experimental plan for the course project, problem statement, execution, and documentation of the course project for each team.	01	9th week
Module 4: Conducting phase: performing the experiments as per experimental plan.	01	10th week
Module 5: Analyzing phase: analyze and interpret the experimental results.	02	11th and 12th week
Module 6: Assessment phase: evaluation of students' performance.	01	13th week

Each week of 2 h tutorial session

Table 2 Example of learning objectives used in module: design of a three factor experiment and analysis of variance

Sl. no.	Learning objectives
1.	Write clear objectives and statement of problem for an experiment.
2.	Identify controllable and noncontrollable factors in an experimental setup.
3.	Choose factors for ANOVA based on expected outcome.
4.	Determine an appropriate level to be researched for factors.
5.	Design an experiment using proper replication, randomization, and control of variables.
6.	Develop a regression model to explore the effect of machining parameters.
7.	Calculate F statistic to ascertain adequacy of the model.
8.	Plot data of all the levels and factors to show variation between levels.
9.	Organize technical information into a clear and concise formal laboratory report.

team they design an experimental plan to be carried out in the following weeks. The plan should be based on a specific assigned question and generate the needed data to be analyzed according to that module's DOE and statistics principles. The students and faculty are provided with a list of learning objectives for the course. The list utilizes Bloom's taxonomy and represents skills that can be assessed in the lab's outcomes. They include technical skills on the theory and laboratory equipment as well as mastery of the DOE and statistical analysis of data principles. Table 2 lists the learning objectives for Module 4: Design of a Three Factor Experiment and Analysis of Variance as an example.

4 Module Contents

This course project is open ended, where teams of students design their own experiments for work-tool and process parameter combinations to perform the analysis on various machinability aspects in turning process.

- The first module session is a dry lab session. During this week, the overview of DOE and statistical analysis of data needed for the course project is taught. For the current program of the same semester, simultaneously the students are already exposed to statistical experimental designs in the Robust Design course (Theory core course) for a period of 6 weeks from the beginning of the semester. Hence, the first module commences from the 7th week from the beginning of the semester and only 1 week is sufficient for this module as the students are already exposed to practical exercises of DOE in Robust Design course.
- In the second module session, students gain mastery of skills by working through dry lab exercises by working numerical problems related to various engineering applications. The students begin to use the statistics and design of experiments they have been learning in the other modules to evaluate results from complex sample sets, that is multiple lots (runs) that each contain multiple samples to determine the variation within a process. In the dry-lab portion of the exercise,

students review the F-test and how it can be applied to multiple runs and samples. Students then plan and carry out an experiment to statistically evaluate the repeatability of a process. Each team is assigned a machining process. They design a process flow through the equipment.

- The third module is the planning phase, which is the heart of the course project that gives opportunities for the students to develop the experimental design skills. The students of each team work out a detailed plan of experiments. Prior to beginning of the lab experiment, each team reviews the experimental plan with the course instructor, discusses the various aspects such as proper control of factors, measurement of expected outcome (machinability aspect), and time management to complete the experiments in allotted session with the lab instructor. During this week, students of each team write clear objectives and statement of the problem, identifying the factors (controllable variables) and responses (expected outcomes) in an experimental setup. Each team determines the appropriate levels (ranges) of the identified factors based on response, control, and precision of machine tool and measuring instrument and time constraints. Finally each student team is assigned the lab equipments as per the experimental plan and the specific goal to be investigated.
- Module 4 is the conducting phase, wherein each team carry out the experiments as per the experimental plan. In this session, the students learn overview of machining process and the measurement techniques of various machinability aspects. Teamwork and communication skills are broadly utilized in this class.
- In Module 5, DOE and statistics are directly integrated into the experimental results. Each student team constructs the mathematical model of the proposed outcome. The ANOVA is used to test the statistical adequacy of the developed model [4, 5]. The coefficient of multiple regression is determined to test the goodness of fit of the model. The developed model is then used to predict the proposed machinability aspect by substituting the values of process parameters within the identified ranges. Two-factor interaction effects of process parameters are plotted for analyzing and interpretation of results using modern engineering tools/software. Students show final mastery of the concept by utilizing ANOVA in their final report to prove whether their variable was statistically significant.
- Module 6 is the assessment phase. The evaluation of each student team performance is done through dry labs, development of experimental plans and management of time in the experiments in the lab, communication of the experimental results through statistical analysis and also in the form of reports.

5 Case Study: Course Project

Example of students' course projects carried out by postgraduate Production Management program is outlined below.

Table 3 Machining parameters and their levels

Parameter (factor)	Level		
Cutting speed, *v* (m/min)	34	51	68
Feed, *f* (mm/rev)	0.05	0.125	0.16
Depth of cut, *d* (mm)	0.5	1	1.5

(a) Goal and objectives of the course project: Machinability analysis in turning for a specified work–tool combination to explore the effects of process parameters.

(b) Research on literatures: Each team investigated various sources of information such as internet sources, published research articles on machinability aspects, visit to local suppliers for the availability of work and tool materials, reference of machine tool design data handbook, etc.

(c) Selection of materials, machine tool, and measuring instrument: After investigating various sources of information along the way of designing the experiment, the following were selected.

- Machine tool: Conventional centre lathe (Machine shop at BVBCET, Hubli)
- Work material: Stainless steel (SS 316)
- Tool material: Cemented carbide (K10)
- Measuring equipment: Tool dynamometer for measurement of cutting forces in turning (Machine shop at BVBCET, Hubli)

(d) Identifying the turning process variables: The key variables selected were

- Independent variables (Factors): Cutting speed, feed, and depth of cut
- Dependent variables (Responses): Machining force (for Team 1); Cutting power (for Team 2); Specific cutting pressure (for Team 3).

(e) Selection of levels for the factors: The ranges of the identified process parameters were chosen based on the range of speeds and feeds available in conventional lathe and data available on machine tool design handbook. The proposed parameters are kept as wide as possible to obtain the variation in the response parameters. Table 3 presents the machining parameters and their levels selected in the current course project.

(f) Developing the experimental plan: Based on 3 levels, 3 factors, design of a 3 level full factorial experiment has been selected which involves 27 experiments. The experimental layout plan for the current course project is presented in Table 4.

(g) Experimental details: The experiments were carried out in 30 mm diameter and 100 mm length SS316 workpieces using cemented carbide tool (K10) without coolant. The strain gauge type tool dynamometer was used to measure the various forces, namely, cutting force (F_c), thrust force (F_t), and radial force (F_r). The machinability is assessed through:

$$\text{Machining force, } F_m = \sqrt{F_c^2 + F_f^2 + F_d^2} \tag{1}$$

Table 4 Experimental design and machinability aspects

	Experimental design					
Trial no.	v (m/min)	f (mm/rev)	d (mm)	F_m (N)	P (W)	K_s (MPa)
1.	34	0.05	0.5	410.26	177.89	12,556.80
2.	34	0.125	0.5	387.84	205.68	5,807.52
3.	34	0.16	0.5	421.94	227.92	5,027.63
4.	51	0.05	0.5	321.94	208.46	9,810.00
5.	51	0.125	0.5	367.84	291.85	5,493.60
6.	51	0.16	0.5	434.97	358.56	5,272.88
7.	68	0.05	0.5	308.04	300.19	10,594.80
8.	68	0.125	0.5	399.45	433.60	6,121.44
9.	68	0.16	0.5	434.09	478.07	5,272.88
10.	34	0.05	1.0	436.19	227.92	8044.20
11.	34	0.125	1.0	674.90	378.01	5,336.64
12.	34	0.16	1.0	719.62	405.81	4,475.81
13.	51	0.05	1.0	387.09	316.86	7,455.60
14.	51	0.125	1.0	713.98	600.37	5,650.56
15.	51	0.16	1.0	775.73	650.40	4,782.38
16.	68	0.05	1.0	407.91	433.60	7,651.80
17.	68	0.125	1.0	699.54	778.26	5,493.60
18.	68	0.16	1.0	763.92	844.97	4,659.75
19.	34	0.05	1.5	548.66	277.95	6,540.00
20.	34	0.125	1.5	1,010.48	539.22	5,075.04
21.	34	0.16	1.5	1102.92	600.37	4,414.50
22.	51	0.05	1.5	657.64	483.63	7,586.40
23.	51	0.125	1.5	1224.56	942.25	5,912.16
24.	51	0.16	1.5	1,325.77	1,033.97	5,068.50
25.	68	0.05	1.5	583.18	589.25	6,932.40
26.	68	0.125	1.5	1,112.39	1,189.63	5,598.24
27.	68	0.16	1.5	1,203.56	1,267.45	4,659.75

$$\text{Cutting power, } P = F_c v \tag{2}$$

$$\text{Specific cutting pressure, } K_s = \frac{F_c}{f * d} \tag{3}$$

The computed values of machining force (F_m), cutting power (P), and specific cutting pressure (K_s) are summarized in Table 4.

(h) Analysis of experimental results through DOE: It involves the following steps:

- Development of mathematical models: The second order mathematical models have been developed to explore the effects of three process parameters, namely, cutting speed (v), feed rate (f), and depth of cut (d) on three responses, machining force (F_m), cutting power (P), and specific cutting force (K_s). The mathematical model for three factors considering the two factor interactions is given by

Table 5 Summary of ANOVA for proposed machinability models and R^2

#	Sum of squares		Degrees of freedom		Mean square			
	Regression	Residual	Regression	Residual	Regression	Residual	F-ratio	R^2
F_m	2,456,224	49,937	9	17	272,914	2,937	92.91*	0.98
P	2,384,949	31,487	9	17	264,994	1,852	143.07*	0.99
K_s	95,585,775	7,472,151	9	17	10,620,642	439,538	24.16*	0.93

*Significant at 99 % confidence interval

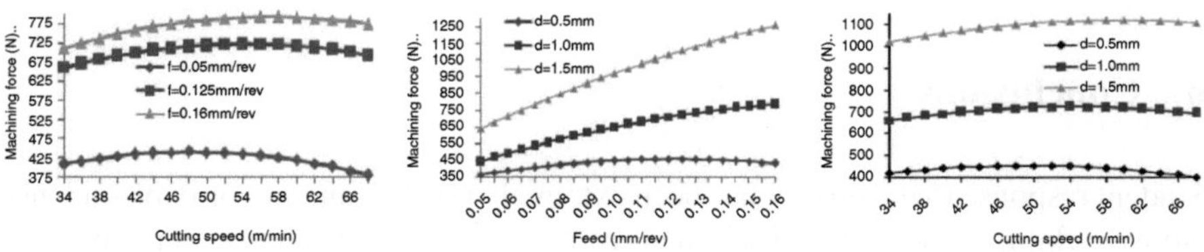

Fig. 1 Interaction effects of process parameters on machining force

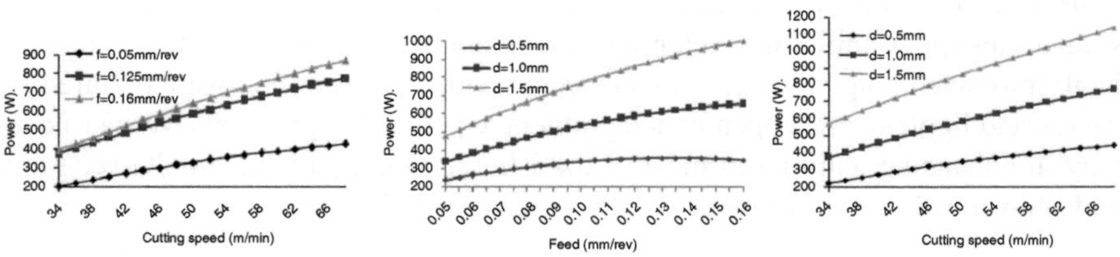

Fig. 2 Interaction effects of process parameters on power

$$Y = b_0 + b_1 v + b_2 f + b_3 d + b_{12} vf + b_{13} vd + b_{23} fd + b_{11} v^2 + b_{22} f^2 + b_{33} d^2 \quad (4)$$

where, Y: desired response and $b_0, \ldots\ b_{33}$: regression coefficients to be determined for each response. Each team determined the regression coefficients using modern engineering tool/software.

- Adequacy of the model was tested by each team through ANOVA and R^2; the results are summarized in Table 5.

(i) Interpretation of the experimental results through DOE: The machinability characteristic was analyzed by each team through developed prediction model by generating the plots considering two parameters at a time while keeping third parameter at center level. Figures 1, 2, and 3 exhibit the plots of interaction effects of process parameters on machinability for various teams. The variation tendency for each of the characteristics has also been analyzed and results were interpreted.

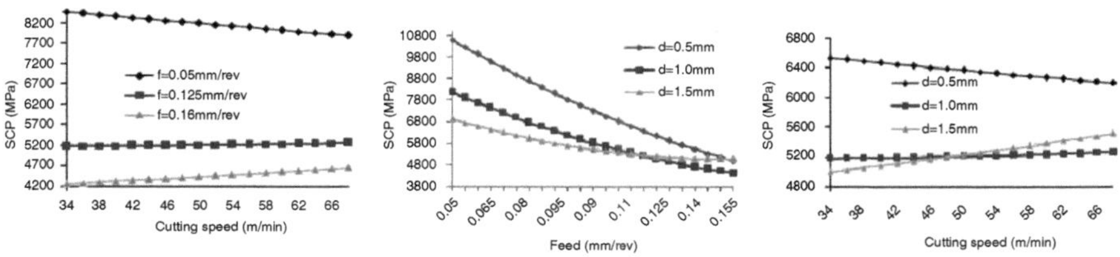

Fig. 3 Interaction effects of process parameters on specific cutting pressure

6 Conclusions

Student response to the project has been extremely positive and majority of students enjoy working on the project. By its nature, the project lends itself to a more involved discussion of the concepts as they are presented. The concepts no longer exist as isolated cases, but rather are parts of a whole. Six modules were designed for a course project in Analysis of Machining Processes Theory core course of PG program in Production Management. The main aim of this course project is to give students an opportunity for developing experimental design skills and then integrate DOE into machining process. Students were divided in teams and were required to design and implement an open-ended problem. Open-ended problems have a role to play in enhancing learning in most of the laboratory experiments in UG program and show a path toward research for the students of PG program.

References

1. Du WY, Burford JF, Nikos JM. On the ability to design engineering experiments. In: Proceedings of the 8th UICEE annual conference on engineering education, Kingston, Jamaica, 7–11 February 2005.
2. Mourtos NJ. From learning to talk to learning engineering: drawing connections across the disciplines. World Trans Eng Technol Educ. 2003;2(2):1–6.
3. Gleixner S, Young G, Vanasupa L, Dessouky Y, Allen E, Parent D. Teaching design of experiments and statistical analysis of data through laboratory experiments. In: Proceedings of the 32nd ASEE/IEEE frontiers in education conference, Boston, MA, 6–9 November, 2002.
4. Montgomery DC. Design and analysis of experiments. New York: Wiley; 1997.
5. Gaitonde VN, Karnik SR, Davim JP. Design of experiments, chapter 5. In: Ozel T, Davim JP, editors. Intelligent machining: modelling and optimization of the machining processes and systems. London: ISTE-Wiley; 2009. p. 215–42. ISBN 978-1-84821-129-2.

SUCC Tool Kit: An Activity in Data Structure Laboratory

Vidya S. Handur, Jayalakshmi G. Naragund, and Priyadarshini D. Kalwad

Abstract The laboratories play a significant role in the professional career of engineers. They strengthen the concepts learnt by the students in the classroom. Data Structure (DS) laboratory course in 2nd year of undergraduate Computer Science and Engineering aims to think logically to develop good programming skills by considering real-time applications. DS laboratory course has impact on learning of courses in higher semesters.

The paper presents *S*ample, *U*se DS, *C*hoose DS, and *C*hoose Problem (*SUCC*) tool kit to describe the activities for different categories' like demonstration, exercises, structured enquiry, and open-ended experiments. The platform required to conduct the laboratory and usage of data structure for a given application is achieved through Sample and Use DS activities, respectively. Choose DS experiments focus on recognizing the suitable data structure for a given application. Depending on the data structure given, the students identify a suitable problem and implement as Choose application. These activities enhance the ability to identify and define data structure and problem formulation and design a program. The SUCC tool kit is compared with traditional method and study shows the higher learning curve in students. The SUCC tool activities are also analyzed and results show that 90 % of students are successfully able to use and choose DS in implementation of different applications.

Keywords Stack • Queue • Linked list • Trees and dynamic memory allocation

V.S. Handur (✉) • J.G. Naragund • P.D. Kalwad
Department of Computer Science and Engineering, BVB College of Engineering and Technology, Vidya Nagar, Hubli, Karnataka 580031, India
e-mail: vidya_handur@bvb.edu; jaya_gn@bvb.edu; priyadarshini@bvb.edu

R. Natarajan (ed.), *Proceedings of the International Conference on Transformations in Engineering Education: ICTIEE 2014*, DOI 10.1007/978-81-322-1931-6_46,

1 Introduction

Engineering is a profession of practicing. It is a hands-on profession where the buzzword is 'doing.' The overall goal of engineering education is to prepare the students to practice engineering that can be obtained in laboratories, an integral part of engineering curriculum [1]. According to a Chinese proverb "Tell me and I'll forget; show me and I may remember; involve me and I'll understand." Therefore, laboratories play a vital role in acquiring the practical knowledge.

The laboratory courses in Computer Science and Engineering discipline are designed to focus on Software Development. The choice of data structure impacts the efficiency of software development process. Therefore, Data Structure laboratory course is a foundation for the students to apply the appropriate data structure in software design. It is a base for follow-up courses like Operating System, Computer Networks, Web applications, Compiling Principles, etc. and also a necessary subject for interviewing employee. The challenges faced by students in dealing with this course are the selection of suitable data structure for a given application, developing logic to solve a problem, following coding standards, and write syntactically correct program. To address these challenges, authors propose novel activities to conduct DS laboratory (lab).

The activity-based conduction of laboratory increases the learning ability more than 90 % according to [2]. The authors propose SUCC tool to conduct the DS lab course according to the structure given by [3]. The Sample activity focuses on instruction of basics for the laboratory. In this activity, the course instructors demonstrate an open source software tool like Code::Blocks [4], which is used to implement DS experiments based on advanced concepts of C language. In Use DS activity students implement an application based on the given DS. The Choose DS activity requires the students to identify the suitable DS for given applications. The Choose Problem encourages the students to formulate the problems for a given specific DS. Thus SUCC tool motivates the students to implement the application by using appropriate DS.

The article is systemized in the following sections: Section 2 narrates the survey of conduction of DS laboratory course. The proposed tool SUCC is explained in Sect. 3 along with evaluation process. The student feedback analysis is presented in Sect. 4. The conclusion and future work are revealed in Sect. 5.

2 Literature Survey

Data Structures is a foundation course for Computer Science and Engineering students. A laboratory course associated with theory makes the students conceptually strong and also enhances their programming skills. Following are some of the contributions toward conducting the course.

Ramon Lawrence [5] proposes a project-based learning for data structure course on the idea of competitive programming. Competitive programming motivates student learning by allowing students to evaluate and improve their programs throughout an assignment by competing their code against instructor-defined code and the code of other students in a tournament environment.

Wanggen Li et al. [6] describe the study of linear data structure in DNA computer, which can help to improve the application capability of DNA computer. The stack and queue are considered as an example to design and implement double-stranded DNA molecules. Data structure can help to organize the information processed by DNA computer correctly and efficiently, and make DNA computer available for practical application.

Shuling Di et al. [7] opined that some modes like project description, application-based teaching contents such as open teaching, tutor system, scientific research, programming contest motivate the students' will, ability of self-directed learning, and innovation ability.

Liu Chengxia et al. [8] coined a new reformed experiment teaching methods of data structure. After completion of each chapter a case study is given for a group of students possessing different abilities. Once the task is completed, the students present the results to the tutor. With this the student's study positivity is stimulated or in other words the quality and the efficiency of study are improved greatly.

Ping Zhang et al. [9] introduced a group cooperation model. The model is to conduct teaching activities by dividing students into several groups according to factors like teaching contents, students' capabilities, and preprepared knowledge. In group teaching, students can conduct group discussions about the thought provoking questions, make complementary remarks, and perfect their opinions. In group experiments, tasks assigned by teachers need students' clear division of work, mutual cooperation, related material consultation, and programming. Group cooperation model cannot only solve the difficult problem of class teaching in accordance with students' various levels, but also create very good learning environment for students' cooperative study.

Sujatha et al. [10] describe the new approach for teaching the data structure lab course, which focuses on active learning, ability of learning data structure concepts, applying them to solve the real-world problems, formulate the problem, and implement the solution using programming skills. The lab exercises are designed based on data structures topics such as stack, queues, lists, binary trees, and few algorithm design techniques such as divide and conquer and dynamic programming. All the exercises are open ended of level 3, where the problem, ways and means, and answers are open. In other words, the problem statement, algorithm, or solutions are not known, but only the concept is given to the students. This motivated them to think of applying the suitable data structures in different courses of their curriculum and also in other domains or fields.

The proposed SUCC tool focuses on activity-based learning of Data Structure lab course and it is discussed in the next section.

3 Proposed Work

Data Structure laboratory is a second year course where the experiments are conducted using C programming language. The Course Learning Objectives are set as following:

- Implement standard data structures like stack, queues, lists, and trees.
- Choose appropriate data structure to effectively model the information in a given problem statement.
- Demonstrate testing and debugging skills using standard data structures.

The stated objectives of the laboratory are to expose the students for demonstrating the basic operations of different data structures, to choose suitable data structure for a given application, to formulate a suitable problem for a given data structure thereby enhancing the programming, debugging, and testing abilities.

Structure of the lab plan includes different categories such as demonstration, exercise, structured enquiry, and open-ended experiments [3]. SUCC tool is used to conduct experiments under these categories as explained in Sect. 3.1.

3.1 SUCC Tool Kit

The SUCC tool kit is used to conduct the activities in the laboratory. The tool operates as follows:

Sample

To start the course, the learners need to learn the basics which are achieved through this activity. The first 2 weeks of the laboratory are used to conduct demonstration experiments [3] by the course instructor. The students are introduced to the open source software tool Code::Blocks [4], coding standards to be followed throughout the semester, and few sample C programs using this tool. There is no weightage of marks given for this phase. The goal of this phase is to make the students aware of the importance of the coding standards and Code::Blocks tool.

Example

Define a structure called CRICKET that includes the following information: player name, team name, batting average. Using CRICKET declare an array *player* with 50 elements and write a program to read the information about all the 50 players and print a team wise list containing names of players with their batting average.

Use DS

A set of exercises are given to students wherein they have to demonstrate the basic operations of the various data structures like stack, queues, linked list, trees. They do these exercises by writing program for the given application and executing the same by applying all the standards demonstrated in Sample activity. This phase emphasizes on sound knowledge of usage of all the data structures and their applications. These exercises enhance the programming, testing, and debugging skills. This activity illustrates exercise category [3] of experiments.

Example

Part I: Perform the following primitive operations using singly linked list for Employee information:

Insert Item at end of list, Display the list content and free all the nodes

Part II: For the above problem add the following operations

1. Delete the employee record that is at the end.
2. Find sum of annual expenditure on all employees.
3. Display all the Employee information whose salary is less than 10,000.

Choose DS

This activity is conducted by giving a set of applications. Each student has to identify the suitable data structure for the given application, write C program, and execute the same by applying the knowledge gained in activity Sample & Use DS. The students instead of simply presenting a solution, they must justify the suitability of DS for a given application. The objective is to choose appropriate data structures to effectively model the information in a problem. This activity enhances the thinking ability in the students. Choose DS is used to conduct structured enquiry [3] experiments.

Example

Write a C program to simulate a simple multiuser computer system as follows. Each user has a unique ID and wishes to perform a number of transactions on the computer. However, only one transaction may be processed by the computer at any given time. Each input line represents a single user and contains the users ID followed by a starting time as a series of integers representing the duration of each of the transaction. The input is sorted by increasing starting time, and all times and durations are in seconds. Assume that the user does not request time for a transaction until the previous transaction is complete and that the computer accepts transaction on a first come first serve basis. The program should simulate the system and print a message containing the users ID and the time whenever a transaction begins and ends. At the end of the simulation, it should print the average waiting time for a transaction. Use appropriate data structure to mold the information given in the above problem.

Choose Problem

This activity includes the student to formulate a problem which suits the given data structure. This task makes use of all the concepts educated in the above activities. The learner presents a solution with justification. The solution is to identify a problem using given data structure, recognize the functional requirements, write a C code, and execute the same. This activity makes the students to think creatively and apply the depth knowledge of usage of data structures. Choose Problem demonstrates open-ended experiment [3].

Example

Following are the some of the sample applications for given DS.

1. Circular doubly linked list
   - Simulation of slide show of photos.
   - Simulation of web pages.
   - Flipping of Adobe reader pages.
2. Singly linear linked list
   - Simulation of ticket reservation
   - Simulation of Common Entrance Test for seat allotment
   - Simulation of group discussion

The exercises conducted as a part of this course meet all the objectives that are set before the commencement of the semester as given in the beginning of this section. Figure 1 shows the overview of SUCC tool to attain the desired outcomes. The lab assignments are carried out in four phases. In each phase a different activity is conducted and the skills gained in the previous phases turn out to be the inputs to provide the solution in the next phase.

3.2 *Implementation*

The laboratory is conducted as a set of 10 slots. Each slot is 3 h in length. Table 1 depicts the systematic approach to implementation of SUCC tool.

In these lab slots, the students work individually. The goal of Use DS activity is to introduce the students to different data structures such that they are comfortable with them and gain confidence in their own abilities. Hence more weightage is given for Use DS activity. The Choose DS and Choose Problem activities are based on applying and analyzing the concepts of DS, so the weightage is not as much of Use DS activity.

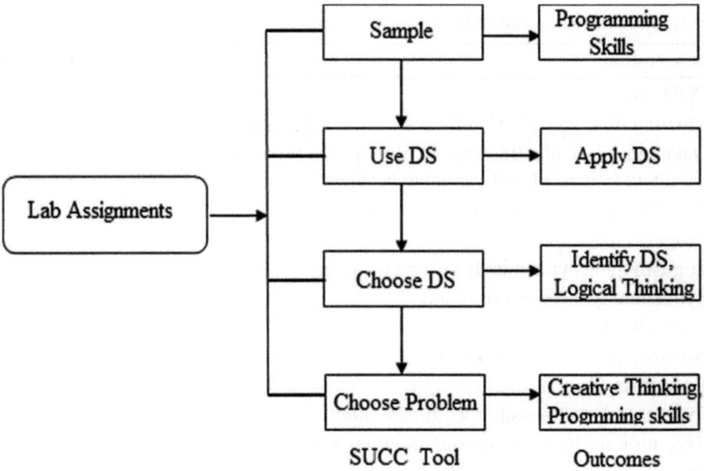

Fig. 1 Outcomes of SUCC tool kit

Table 1 Organization of activities

Slot no.	Demonstration of activity	Weightage of marks (in %)
1–2	Sample activity	0
3–8	Use DS activity	75
9	Choose DS activity	13
10	Choose problem activity	12

The assessment rubrics for the SUCC tool activities as a part of Continuous Internal Evaluation (CIE) are given in Table 2. The Program Outcomes (POs) such as applying engineering knowledge (a), identifying the required solution (b), design and implement program (c), which are set by the Computer Science and Engineering department of authors' institute based on Graduate Attributes (GAs) are given in [11]. Table 2 also shows the mapping of rubrics to POs.

4 Result Analysis

The traditional approach of conducting DS lab is based on implementing a given problem for a specific DS. So this approach does not ensure identifying and applying suitable DS for real-world applications in students. This method also lacks in enhancing the ability of learners to identify the suitable problem for a given DS. The proposed SUCC tool kit activities are designed to incorporate the deficiency of traditional approach. Figure 2 depicts the comparison study of these two methods of strength around 160 students. The traditional approach is followed in the previous

Table 2 Assessment rubrics and POs

Sl. no.	Description	Marks	PO mapped
I	**Write up (04)**		
	Abstract data type (ADT)/identify DS/formulate problem (2)		a
1.	Data definition, identification of all required prototypes/correct DS identified/identify suitable problem	2	b
2.	No ADT/incorrect DS/not suitable problem	0	
	Program write up (2)		a
1.	Logically and syntactically correct program	2	
2.	Syntactically incorrect program	1	
3.	Logically incorrect	1	
4.	No write up/incorrect program	0	
II	**Execution (06)**		a, c
1.	Execution for all possible legitimate inputs	6	
2.	Execution for basic operations and two operations	4–5	
3.	Execution for basic operations	2–3	
4.	Logical error	1	
5.	No execution/execution of irrelevant program	0	
	Total	**10**	

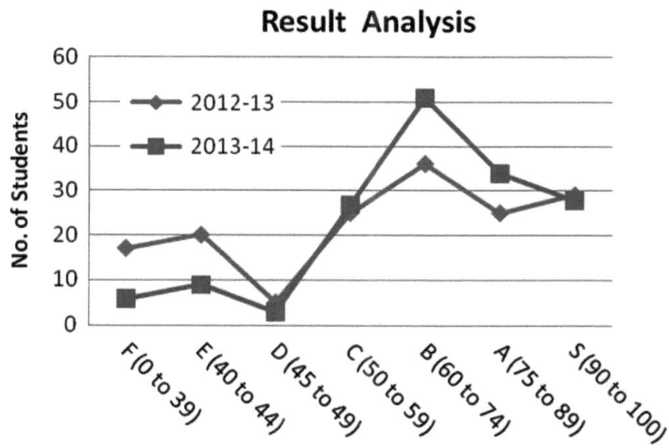

Fig. 2 Comparison study of traditional approach and SUCC tool kit

academic year 2012–2013, where 73 % of students have scored higher grades (S, A, B, C) and 10 % of student are unable to clear the course. SUCC tool kit is used in the current year 2013–2014, where 87 % of students have scored higher grades and 4 % of students are unable to clear the course. The practice of SUCC tool shows the higher learning curve.

The SUCC tool activities are also analyzed based on the feedback taken from the students of strength around 70. Analysis is categorized according to the SUCC tool activities as explained in the further part of this section.

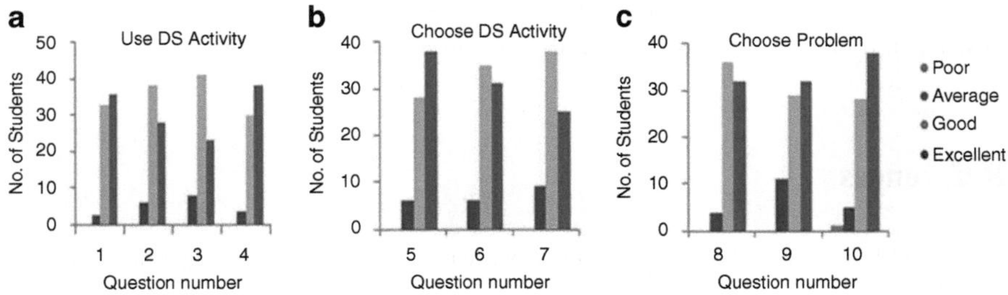

Fig. 3 Analysis of SUCC tool. (**a**) Use DS activity (**b**) Choose DS activity (**c**) Choose problem activity

Since Sample activity is a demonstration activity and has no weightage, analysis is not considered. In the following graphs, X-axis represents the feedback question numbers and Y-axis shows number of students.

Use DS activity is analyzed based on 1–4 questions depicting the usage of DS, improvement of programming and debugging skills, objective of each assignment, correlation of theory and lab assignments, respectively. According to the analysis in Fig. 3a, 92 % of the students are favored the proper usage of DS in writing the program by this activity.

Choose DS activity is analyzed using feedback questions 5–7 representing identifying suitable DS, improvement in thinking ability and effort needed for preparation, respectively. According to the Fig. 3b analysis, 90 % of the students are able to choose appropriate DS by applying the skills developed in the Use DS activity.

Choose Problem activity is analyzed with the feedback question 8–10 which illustrate the formulation of problem spontaneously, innovative thinking, and enhancement in self-learning, respectively. Figure 3c shows that 90 % of the students are able to choose problems spontaneously for a given DS.

5 Conclusion

Conduction of DS lab at the 2nd year of undergraduation is really a challenge for course instructor. Identifying data structure for a problem and vice versa is an issue to be addressed. The proposed SUCC tool in this article is one of the approaches to overcome this. The skills developed in each activity will turn out to be the input to the next activity to achieve the outcomes like applying DS knowledge, identifying the required solution, design, and implement program. The analysis results also depict the improvement in these skills. The comparison study of SUCC tool kit with traditional method shows that higher grades obtained by students are increased by 14 %. Since any programming lab in Computer Science and Engineering curriculum follows the steps specified in this article, SUCC tool can be applied to conduct these labs. Any activity-based learning requires active coordination between the timing of the lectures and the timing of the laboratories. Few extra hours of teaching are needed to implement SUCC tool.

References

1. Feisel LD, Rosa AJ. The role of the laboratory in undergraduate engineering education. J Eng Educ. 2005;94:121–30.
2. Surgenor B, Firth K. The role of the laboratory in design engineering education. Can Eng Educ Assoc. 2011;23(2):287–93.
3. Laboratory Plan Format. AMEC cell BVB College of Engineering and Technology. 2013.
4. Code::Blocks. The open source cross platform IDE. 2010.
5. Lawrence R. Teaching data structures using competitive games. IEEE Trans Educ. 2004;47(4):459–66.
6. Li W, Ding Y, Huang Y, Ye X. Linear data structure for DNA computer. In: Fuzzy systems and knowledge discovery. Los Alamitos: IEEE; 2009;7:141–5.
7. Di S, Liu L, Sun S. Study on after class teaching of data structure for undergraduates. In: Proceedings of the 2nd international conference on education technology and computer (ICETC). Piscataway: IEEE; 2010;1:216–18.
8. Chengxia L, Ying C, Ning L, Tiefeng W. The study of experiment teaching in data structure. In: Proceedings of the the 5th international conference on computer science and education. Piscataway: IEEE; 2010;1:631–4.
9. Jiang L, Zhang P, Liang R, Hu Y. The application of group cooperation model in teaching data structures. In: Proceedings of the circuits, communications and system (PACCS), 2011 third Pacific Asia conference. Piscataway: IEEE; 2011. p. 1–3.
10. Jayalaxmi GN, Suvarna GK, Sujatha C. An innovative approach carried out in data structures and algorithms lab. In: Proceedings of the engineering education: innovative practices and future trends (AICERA), 2012 IEEE international conference. Piscataway: IEEE; 2012. p. 1–4.
11. National Board of Accreditation New Delhi-110003. Manual for accreditation of undergraduate engineering programs, I edn. March 2012. Available in www.nbaind.org

Open End Activity (OEA): An Experience in Computer Graphics Laboratory

Meenaxi M. Raikar, Shantala G. Giraddi, and Jayalakshmi G. Naragund

Abstract The paper discusses the Open End Activity (OEA) initiated in the computer graphics laboratory course. The course includes assignments on basic concepts such as 2D/3D object creation, clipping, transformation, and open-ended experiments. The open end means no protocols to execute the task. The students are given the openness to choose the problem statement in computer graphics area. The different areas in which the open-ended experiments explored are gaming, simulation, and animation, such as car race game, story animation, and wireless network simulation. OEA is used to conduct open-ended experiments. It is a group activity, which is carried out by different teams. The activity is conducted in four different phases such as problem identification, design, implementation, and uploading video to the Internet. The rubric-based assessment is followed to evaluate the open-ended activity. OpenGL, an open source library, is used for the implementation of the assignments/open-ended experiments.

The OEA induces the students to develop problem identification, problem-solving skills, independent and creative thinking, willingness to explore new ideas, self-learning, and teamwork experiences. The authors' observations indicate that the open-ended experiments achieve a, c, g, and k Program Outcomes (POs) of ABET. The results show that the average attainment of these POs is 7.6 on the scale of 10.

Keywords Open-ended experiment • Texture mapping • OpenGL • Animation

M.M. Raikar (✉) • S.G. Giraddi • J.G. Naragund
Department of Computer Science and Engineering, BVB College of Engineering and Technology, Vidya Nagar, Hubli, Karnataka 580031, India
e-mail: mmraikar@bvb.edu; shantala@bvb.edu; jaya_gn@bvb.edu

R. Natarajan (ed.), *Proceedings of the International Conference on Transformations in Engineering Education: ICTIEE 2014*, DOI 10.1007/978-81-322-1931-6_47,

1 Introduction

Nowadays, the educationalists have begun to question the traditional, or "cookbook," method of learning in the engineering field. The traditional laboratory (lab) activities are characterized by high degree of memorization. The students in these labs retain little of what they learn and have difficulty in applying what they know. To overcome the drawback of traditional rote learning, the authors explore activity-based learning in the computer graphics lab [1].

Computer graphics (CG) is one of the most exciting and rapidly growing fields in computer science. It has become common element in user interfaces, virtual reality, data visualization, television commercials, motion pictures, medical field, fine arts, and many other applications. The CG applications have changed the way of using technology by the society. The computer gaming industry is one of the major industries that depends on CG and currently its annual turnover is an order of magnitude greater than the whole of the film industry [2]. The user interface based on CG has contributed tremendously for the growth of computer industry. Hence, CG is one of the important courses in the computer science and engineering curriculum.

The challenges in learning CG are representing real-world objects or system in the coordinate system, imagination of building the object, and obtaining a realistic view of the objects. These issues encourage the authors to implement an innovative OEA.

All these activities were implemented using openGL [3], which is an open-source library that is supported on every gaming platform including Mac, Windows, Linux, and iPhone. Popular applications which are using openGL as renderer are Adobe Photoshop CS3, 3D Studio Max, Autodesk Maya, Google Earth, SAP2000, and Scilab.

The OEA is experimented for three successive batches. The first batch uses texture mapping and second batch applies imaging operation along with texture mapping. Along with these concepts, third batch uses the feature of importing the 3D objects, which are created using other tools like 3D MAX. Thus, the percentage of openness is increased in the successive batches to motivate the students.

The paper is organized in the following sections. Section 2 deals with literature survey of conducting computer graphics lab. The proposed OEA description and implementation is discussed in Sect. 3. The analysis of this activity is presented in Sect. 4. The article is concluded in Sect. 5.

2 Literature Survey

Several innovations have been tried during the recent years in teaching computer graphics [4]. Some of the innovations are described in the further part of the section. Two themes of computer graphics, namely, computer-generated color and computer-generated visualization, are used while teaching an interdisciplinary course,"Creativity and Technology" [5]. The authors have used an educational software tool SIMBA for teaching CG using breadth-first approach. This approach enhanced students' understanding and the relevance of the topic.

In [6] the authors focus on the use of new tools to improve the learning of CG. The authors have explored the union of two areas: the educational strategy named Problem-Based Learning (PBL) and the use of Interactive Embodied Pedagogical Agents (IEPAs) for education and training tasks. They have explored usage of Maxine, a script-directed engine for the management and visualization of 3D virtual worlds. The learning process becomes highly interactive and flexible using their method.

In [7], the authors have used Microsoft XNA as a development platform to teach CG for game development. XNA is a collection of tools from Microsoft designed to ease the burden on game programmers by offering built-in functionality commonly used in games. According to the authors, any game programming course needs to include programming, software engineering, algorithms, and computer graphics. They found that successful completion of projects increased the students' confidence.

In [8], the authors explain the approach followed at the University of Lisbon for teaching CG. The authors opine that CG is far from being restricted to 3D graphics, no matter their increasing importance today.

3 Proposed Work

The open-ended experiments pose several challenges to the learners in identifying the problem and implementing the real-world scenario. However, the advantages of these experiments are multifold, such as:

- Provide a forum for students to express creative ideas.
- Encourage students to be open-minded when analyzing and solving most real-life engineering problems which have multiple solutions.
- Give students an opportunity to work in a team environment and apply the concepts learnt in the classroom.

Thus, these issues inspire the authors to conduct OEA. The further subsection discusses the objectives, implementation, and assessment strategy of OEA.

3.1 Objectives of OEA

The OEA focuses on enhancing the self-learning ability of the students. The students' perception about CG lab course is different from practicality. Considering these issues, the authors propose OEA with the following objectives:

- Perform animation of the objects.
- Explore texture mapping techniques to provide realistic view of the scene.
- Position and orient the camera using openGL [9].

Table 1 Assignment on 2D and 3D objects creation

2D objects assigned	Implemented model	3D objects assigned	Implemented model
	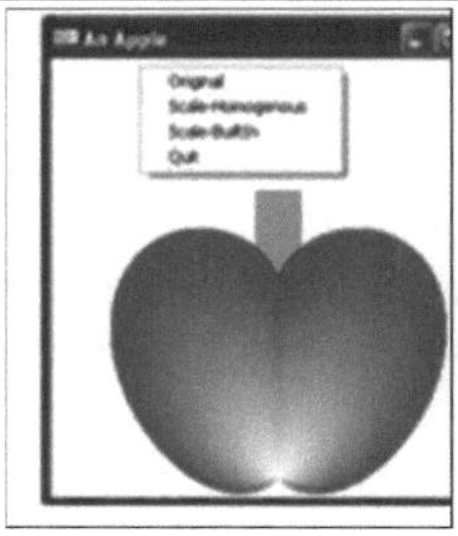		

Table 2 Assignments on texture mapping

3D objects assigned for texture mapping	Implemented model	3D objects assigned for texture mapping	Implemented model
	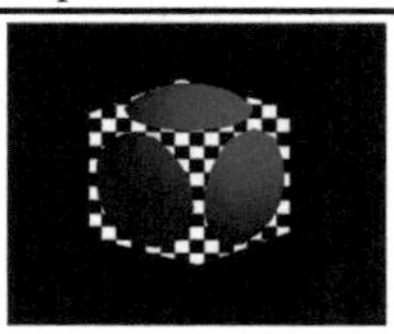		

- Explore the imaging operations.
- Importing 3D objects from other tools like MAYA.
- Improve thinking ability, communication skills and work together in a team.

3.2 *Implementation*

The authors recommend conducting some basic experiments on the creation of 2D/3D graphics primitives, transformations of 2D/3D objects, and application of lighting/shading effects for the objects and parallel/perspective projections, to prepare the students to implement open-ended experiments. Table 1 illustrates some of the sample 2D/3D assignments and Table 2 depicts the sample assignments on texture mapping.

Along with these experiments, course instructor can motivate the students to do various assignments on these concepts. These assignments create awareness about the coordinate system in students and improve their programming skills. All these assignments provided the platform for the students to take up the OEA.

Thirty-six batches, each batch comprising four students, are involved in OEA. The activity is conducted in four different phases such as problem identification,

Table 3 Rubric-based assessment of the OEA

Problem identification phase (7 M)	Design phase (6 M)	Implementation phase I (7 M)	Implementation phase II (10 M)
1. Problem identification (04 M)	2D/3D object creation (2 M)	Appropriate transformations applied (3 M)	Texture mapping (2 M)
If the problem suits the emerging trends and appropriate objects chosen for animation (3–4 M)	ADT/data structure used (2 M)	Camera position (2 M)	Lighting/shading (2 M)
If the problem suits the emerging trends and without proper objects chosen for animation (0–2 M)	Animation (2 M)	User interface (2 M)	Video of the animation along with the explanation of the openGL concepts used (2 M)
2. Presentation (03 M)			Uploading the video (4 M)

design, implementation, and uploading video to the Internet. The explanation of these phases is given below:

- In the first phase, students are intimated to identify a minimum of three different problem statements on various applications such as social issues, story animation, games and computer science concepts, etc. All the batches have presented the identified problems. The course instructors have directed to continue with the better one based on feasibility, innovativeness, and ability of the batch.
- In the design phase, the students are expected to show the identified 2D/3D objects, data structures, and definition of ADT (Abstract Data Type).
- In the third phase, students are expected to demonstrate the transformations applied, various camera positioning, user interface, importing objects from other tools, and basic operations of animation.
- In the last phase, students are supposed to show the complete implementation of the open-ended experiments along with uploading the video to a website like youtube.com [10–15].

3.3 Assessment Strategy

The authors have used rubric-based assessment for the evaluation of OEA. The weight age given for the lab assignment is 50 % and for the OEA is 50 %. Table 3 summarizes the rubrics used for this activity. The observation is that after being aware of the rubrics defined, the students start improving their animation models.

The effectives of the OEA activity and lab assignments are analyzed in the next section

4 Result Analysis

The Program Outcomes (POs) enlighten the skills developed by the students in curriculum. ABET a–k [16] specifies a to k outcomes for the program. In this section, the authors narrate the outcomes achieved by OEA in teaching CG.

To perform open-ended experiments, students start exploring tools like Blender, an open source tool; Photoshop for editing the images; and languages like C# and WebGL. Video of the animation is created using the open source tool, Camtasia Studio. Thus, OEA induces a lot of self-learning among the students.

The student performance in implementation of OEA is illustrated in Fig. 1. The score in Figs. 1 and 2 is in a scale of 10. As the implementation process proceeds, students are more comfortable in implementing the models. On an average more than 75 % of the students successfully implemented the chosen models.

Table 4 shows the achievement of POs through different activities conducted by the authors in CG lab. Through the activities specified in Table 4, the authors have

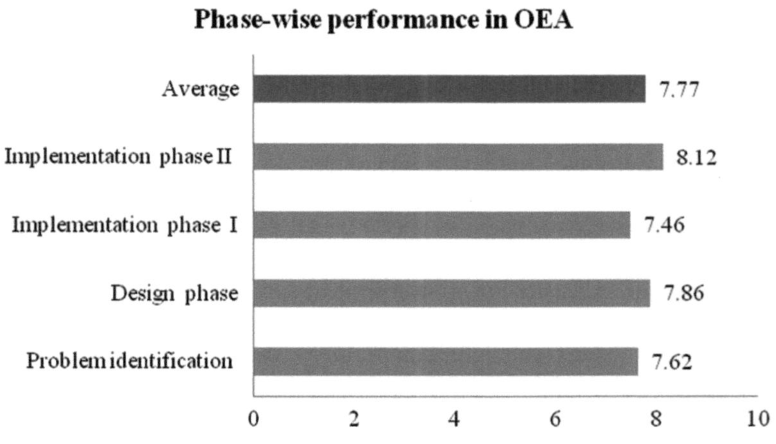

Fig. 1 Analysis of OEA implementation

Fig. 2 Attainment of POs

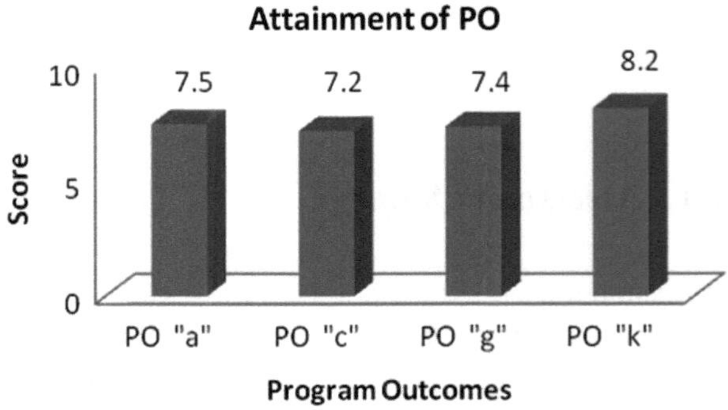

Table 4 Mapping of the POs

Transformation	a, c
Imaging	k
OEA	a, c, g, k
Texture mapping	k

Table 5 Sample screenshots of open-ended experiments

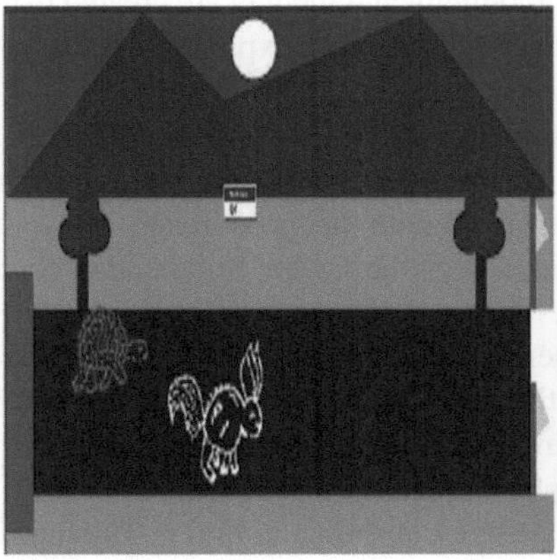 **Fig. 3** Story animation using point cloud technique	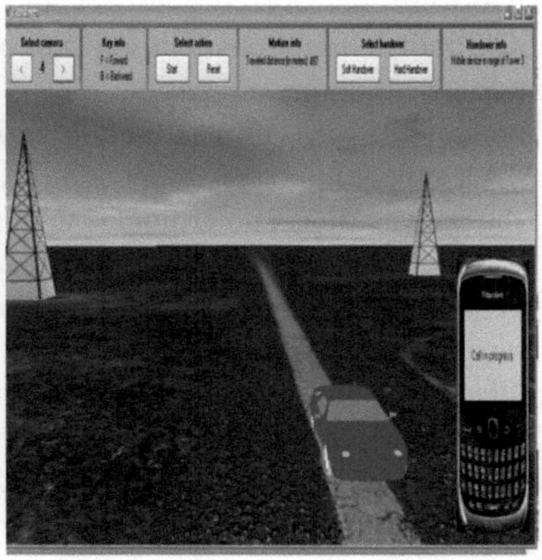 **Fig. 4** Handoff management, importing 3DMAX object.
 **Fig. 5** Maze game using basic primitive spheres	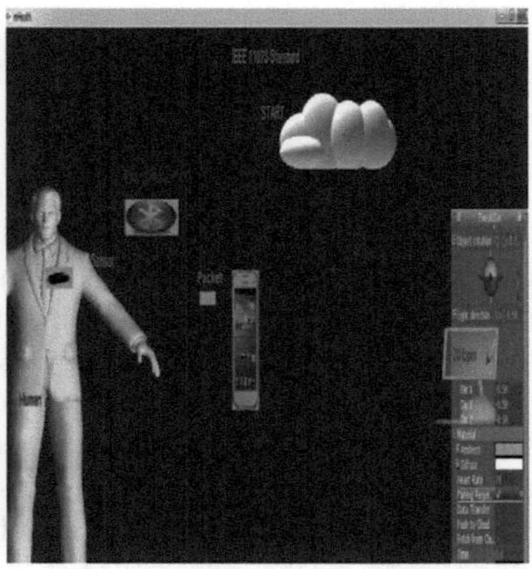 **Fig. 6** WBAN animation importing,'.obj' objects.

attempted to address ABET EC 2000 outcomes a, c, g, and k. These are now considered important outcomes of an engineering education. OEA addresses more outcomes compared to other activities. Hence, OEA helps to improve the basic concepts of CG, designing aspects, communication skills, and usage of open source tools.

The attainment of POs a, c, g, and k is shown in Fig. 2. The different activities conducted by the authors strongly improved the basics of CG and component design aspects. The learning of open source tools by the students is almost 80 %.

The students have developed games, operating system concepts, networking scenarios, moral stories, etc. Table 5 shows screenshots of the sample experiments conducted through OEA.

The animation of the story "Hare and Tortoise" is implemented as a part of the OEA. A point cloud is used as shown in Fig. 3, to represent the object hare as well as the tortoise. The objects lane, tree, moon, mountain and the flag are created using openGL built-in functions. The concept of double buffering is explored in the experiment to get smoother animation.

The handoff management animation is shown in the Fig. 4. The car shown is a 3D MAX object imported in openGL [10]. The pyramid primitive is used to represent the base station. The ground is texture mapped with the grass image. The camera position could be changed to get the aerial view of the screen. Soft hand over and hard hand over are illustrated by animation.

The maze game is implemented using openGL [17]. The user has to find the path from the start point to the end point using the arrow as user interface as shown in Fig. 5. The background is collection of spheres arranged to represent a maze. As the user traces the path with arrow keys a red sphere appears on the path.

The animation of the wireless body area network is shown in Fig. 6. The human in the figure is an imported ".obj" file into the openGL program. Texture mapping is applied to represent the mobile. Tweak Bar is used as the interface to change the attributes like color, camera position and lighting effects.

5 Conclusion

The conduction of OEA in the CG lab encourages the students to utilize higher-order thinking skills, creativity, innovation, and usage of open source tools. Also cooperative learning and teamwork are achieved through OEA. The four phases in the implementation of OEA have helped the students to understand the goals of this activity. An open source tool, used by the students, enlightens the suitability of the tool depending on the context. For example, they find that graphical user interface forms can be created with much ease using C# and WebGL. Thus, the active learning that happened through OEA is intrinsically motivating and engaging.

The different activities such as OEA help to achieve the POs a, c, g, and k. The analysis given by the authors depict the effective learning of open source tools through OEA. Applying the basic concepts, formulating the problems, programming, and communication skills are also improved among the students with OEA.

References

1. Foley JD, Van Dam A, Feiner SK, Hughes JF. Computer graphics. Reading: Addison-Wesley; 1997.
2. calipso.ugr.es/dmartin/Research/Vindication
3. Angel E. Interactive computer graphics – a top-down approach with OpenGL. 5th ed. Reading: Addison-Wesley; 2008.

4. Hearn D, Baker P. Computer graphics – OpenGL version. 2nd ed. Boston: Pearson Education; 2003.
5. Domik G, Goetz F. A breadth-first approach for teaching computer graphics, Education paper, Eurographics. 2006. Available in: Pdf.aminer_org/000/590/225/approaches_to_teaching_computer_graphics_abstract.pdf
6. Seron F, Cerezo E, Baldassari S. Computer graphics: problem based learning and interactive embodied pedagogical agents, Education paper, Eurographics. 2001. Available in: giga.cps.unizar.es/affectivelab/publications/E9-08.pdf
7. Mihail RP, Goldsmith J, Jacobs N, Jaromczyk J. Teaching graphics for games using Microsoft XNA. In: 18th international conference on computer Games, IEEE; 2013.
8. Santos Manuel Prospero dos. Computer graphics in the scope of informatics engineering education. Comput Graph. 2001; 25(5):909–16, Elsevier.
9. Hill FS. Computer graphics using OpenGL. 2nd ed. Upper Saddle River: Pearson Education; 2001.
10. Prashant MN, Shamanth MN, Divya K, Samarath B. Eco friendly network management. http://www.youtube.com/watch?v=DTlECsiQnbI&feature=youtu.be
11. Rane T, Athani P, Madival A, Arakeri N, Gouranna U. Handover mechanism in cellular networks. http://www.youtube.com/watch?v=IBfl-GZHA2I
12. Hiremath N, Kattimani P, Sunagar V, Padasali V. Wireless sensor network for obstacle detection. https://www.youtube.com/watch?v=1miVHh99NOc
13. Pagad S, Kalasannavar V, Hubli S, Channagoudar A. FIFO scheduling queue. http://www.youtube.com/watch?v=ss7xFhFpR6E
14. Shetty S, Prabhu R, Harihar S, Kolli N. Wireless sensor network Testbed animation. http://www.youtube.com/watch?v=068GKuimaRo
15. Raj R, Prabhu S, Singh P, Shetty N, Purad G. Snake and ladder game. http://www.youtube.com/watch?v=jR3jOH83eZs&feature=em-share_video_user
16. http://www.abet.org/uploadedFiles/Accreditation/Accreditation_Process/Accreditation_Documents/Current/eac-criteria-2012-2013.pdf
17. Wright RS, Lipchak B, Haemel N. OpenGL superbible comprehensive tutorial and reference. 4th ed. Upper Saddle River: Addison Wesley; 2007.

Leveraging Technology in Outcome-Based Education

B. Kanmani and K. Mallikharjuna Babu

Abstract In this work, we attempt to leverage technology in the conventional teaching methods, to result in improved teaching–learning process. The improved teaching–learning process happens due to the shift from an input-based education system to outcome-based education (OBE) system. In an OBE system, the focus is on the learning by the student, and is measured by the competencies/skills developed by the student at the end of every course, and contributes to the overall attainment of skills at the time of graduation. The National Board of Accreditation (NBA) has specified through the Graduate Attributes (GAs), the abilities every engineering graduate needs to have acquired. The usual delivery methods in a course are lecture, tutorial, and laboratory. We discuss possible modifications in these traditional teaching methods by leveraging technology and include another self-study component. Every delivery method is discussed with the GAs being addressed. The methods are general and may be incorporated in any course.

Keywords Outcome-based education • Delivery methods • Graduate attributes

1 Introduction

Outcome-Based Education (OBE) is penetrating the Engineering Education system at an increasing pace. Today, every activity, every course, is measured by the Outcomes being addressed. The outcomes of a program are the set of competencies

B. Kanmani (✉)
Department of Telecommunication Engineering, BMS College of Engineering, Bangalore 560019, Karnataka, India
e-mail: bkanmani.tce@bmsce.ac.in; bkanmani@rediffmail.com

K.M. Babu
BMS College of Engineering, Bull Temple Road, Bangalore 560019, Karnataka, India
e-mail: principal@bmsce.ac.in; drkmbabu@gmail.com

R. Natarajan (ed.), *Proceedings of the International Conference on Transformations in Engineering Education: ICTIEE 2014*, DOI 10.1007/978-81-322-1931-6_48,

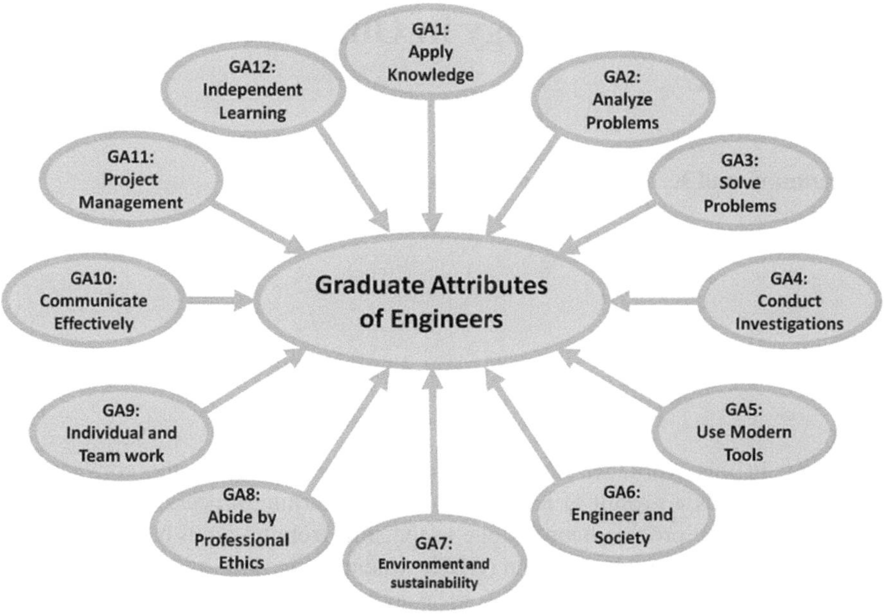

Fig. 1 The twelve graduate attributes (GAs) represented by their key words

acquired by the student at the time of completion of the program. The broad guidelines for desired competencies are defined by the accreditation bodies [1–4]. The National Board of Accreditation (NBA), the accreditation body for India, has defined the desirable competencies through the 12 Graduate Attributes (GAs) [1]. Figure 1 gives the 12 graduate attributes of the NBA represented by their key words. It can be observed that GA1–GA4 develops the application of basic science/ engineering concepts to solve, design, analyze systems/subsystems in the core engineering domain. These attributes are usually addressed by a significant number of courses in a curriculum. GA5 deals with the competency to use a modern engineering tool, and this component is addressed by the laboratory component of the curriculum. GA6 and GA7 develop the awareness, concern, and responsibility toward health, society, and environment. GA8 emphasizes the importance of abiding by professional ethics, and the student is introduced to the norms of professional practice. GA9 measures the ability to perform as an individual/as a member of a diverse team. This attribute helps develop the competency to respect diverse cultures and perform in diverse teams, and thus prepares the skills necessary to pursue a professional career. GA10 emphasizes the need for effective communication, both with the engineering community and nonengineering community. GA11 takes the student through the norms of planning and implementing a project. GA12 attempts to measure the ability to engage in independent learning and the desire to continuously upgrade the knowledge.

An OBE system of education designs a curriculum comprising a set of courses [1–5]. Each course has a specific Course Outcomes (COs), each of which addresses one or more GAs through the Program Outcomes (POs). The POs are aligned to the GAs and are specific to a program. The COs are sometimes known as the Student Outcomes (SOs). Through the courses of the curriculum, the student at the end of the program needs to have acquired all the GAs. Given the content for a course, it is possible to design the COs, the delivery methods, and assessment tools for each of the COs, so that a larger number of POs/GAs are addressed. In this work we present some possible delivery methods that can contribute to improved learning abilities of the student and hence leading to an effective implementation of OBE system of education. In the next section, we discuss some delivery methods, and how technology can be leveraged into every delivery method.

2 Delivery Methods

Various delivery methods through formal allocation of credits are given in Fig. 2. We now discuss each of these formal delivery methods, and how including some additional changes leads to an increase in the outcomes addressed by a course. The methods suggested are general and be incorporated in any course.

2.1 Lecture

Lecture has been an accepted and essential form of delivery method since many years. Let us consider a teaching–learning environment that existed few centuries ago. The teacher used a 'chalk,' delivered the 'talk'; students listened, and got evaluated and were finally declared as having gained knowledge. In this age-old classroom environment, if students 'spoke,' they were punished, it was treated as indiscipline, if he could not reproduce he was considered incompetent to study/learn. Now this was the teaching–learning process centuries ago. It is surprising to note that the then education system could not identify the genius, the innovative talent present in scientists like Albert Einstein or Thomas Alva Edison, during their early schooling. There are many more great personalities who have achieved significant success in their life time and yet were individuals

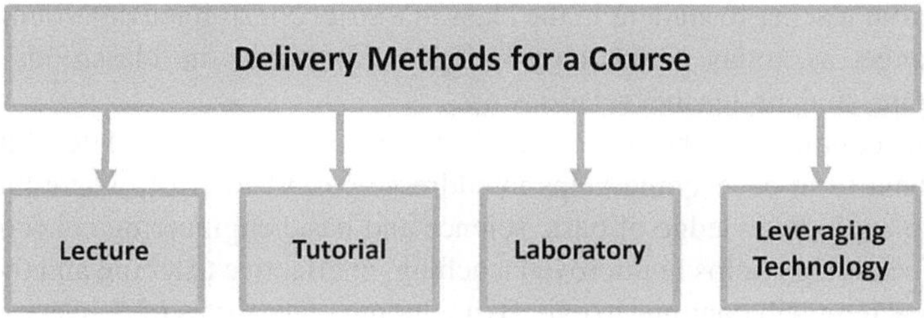

Fig. 2 The possible delivery methods in a course through credit allocation

Fig. 3 Classrooms during the fifteenth, seventeenth, and the eighteenth century [6–8]

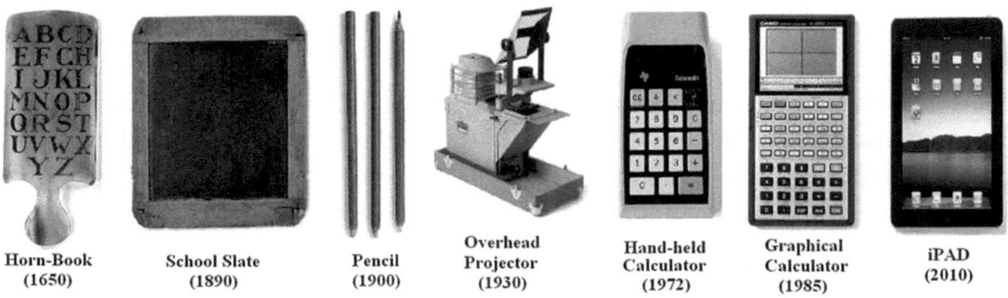

Fig. 4 The evolution of learning aids [9]

who miserably failed in the then educational platform. Some pictures of classrooms, some years ago is given in Fig. 3.

How, has time changed this classroom environment? Accessories that have found place in the classroom today are: the white board, the computer, the LCD projector, the digital board, the tablets, digital television, the video-conference facility. But, surprisingly, this invasion of new learning aids has not brought about change in the classroom environment. The evolution of teaching aids is given in Fig. 4. Maybe we have the blackboard being been replaced by the white board, the chalk being replaced by the marker pen, maybe in addition we have an audio system, we have the LCD projector, maybe we have facility to put a video, etc. But still the flow of knowledge is from teacher to student. We still maintain the 'no talking,' 'no discussion,' 'no eating,' 'no laughing,' disciplined environment. This model of wisdom flowing from teacher to student in the class in a strict constrained environment calls for a change, as, today, technology has penetrated into the classroom, without changing the student–teacher relationship.

Lecture continues to be an essential and accepted method of teaching. This conventional form of teaching helps in addressing GA1 to GA4, where there is an attempt to apply knowledge of basic science and basic engineering concepts to the course. Technology helps in improved teaching, in effective teaching and the lecture form of teaching will continue to exit. By including other forms of delivery methods, we can improve the teaching–learning process.

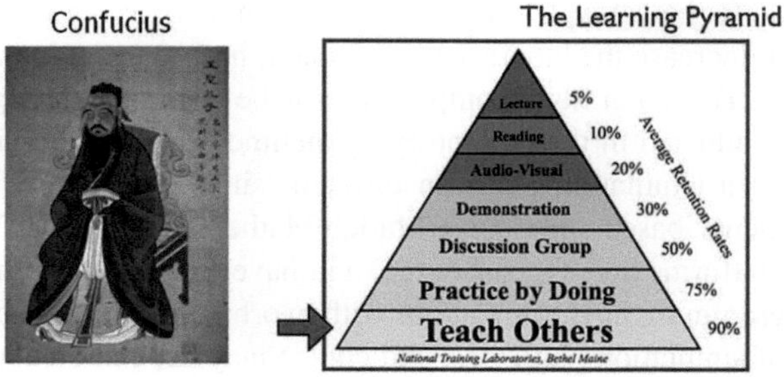

Fig. 5 Confucius, the person behind the learning pyramid [11, 12]

2.2 *Tutorial*

Tutorial is another component of the teaching–learning process, and it usually involves solution to problems with partial guidance by the course instructor in the classroom. This delivery method once again is likely to address GA1 to GA4. Usually, the tutorial session is included after delivery of lectures related to the concept. The number of students per teacher in a tutorial session is less compared to that of a regular classroom, and hence the student gets necessary guidance in attempting to solve the problem.

This conventional form of regular teaching in a classroom, followed by tutorial sessions in the classroom, can be attempted using the '*Flipped Classroom*' [10]. Flipped classroom model is one that delivers instruction at home through interactive, teacher-created videos and moves "homework" to the classroom. This model is presently gaining wide acceptance, as the student has the freedom of listening to a lecture in the home. In this model, the typical 'classwork' happens at home and the typical 'homework' happens in the class. The student listens to the class lecture available as a video at home, and in the classroom, the course instructor can have a quick recap of the concepts and proceed with the tutorial session. The tutorial sessions now become effective and lead to efficient utilization of time and enables addressing higher order GAs like GA3 and GA4. This delivery method can be introduced in significant courses. It can be observed that we are leveraging technology into the conventional tutorial method of course delivery.

2.3 *Laboratory*

This laboratory component in the teaching–learning process involves 'hands-on' experience. The significance of the laboratory component has been expressed, as long ago as 451 B.C., by Confucius, "What I hear, I Forget; What I see, I Remember; but What I do, I Understand." His words are better represented by the learning pyramid Fig. 5, which shows that 'listening to a lecture' results in just 5 % of learning, while 'reading' leads to 10 % learning. Demonstration enhances the learning to

30 %; however, 'Doing' results in learning of as high as 75 %. Hence, there is a need to increase the laboratory component in the curriculum.

The laboratory component can be included through 'Hands-on' experiments conducted in the laboratory. Sometimes it is also possible to conduct experiments on a simulation platform through 'simulation tools,' like programming environments based on C/C++/Matlab/Mathematica/LabVIEW. While some simulation platforms are 'Open-Source' and have free access, others provide wonderful programming environment but with exorbitant costs. Hence, there exists a wide range of simulation platforms, and choice may depend on application and other deciding factors. This delivery method addresses GA5 and GA9, as the student performs experiments using a 'Modern Tool' as an individual or as a member of a team. Usually, the student performs a fixed set of experiments, as designed by the teacher, and under the guidance of the teacher. A small modification to the usually practiced existing laboratory course model is to introduce ONE experiment as a self-designed experiment by the student in addition to the usual fixed set of experiments. The self-designed experiment may be conducted under the guidance and with prior approval of the course instructor. This miniscule modification introduces in the student the ability to formulate, design, implement, demonstrate, and document the process, without any additional investment from the institution. Hence introduction of 'Open-Ended' experiment addresses GA1, GA2, GA3, GA4, GA5, GA9, GA10, and GA12. Today, technology has brought laboratory into the home, through the initiative of Ministry of Human Resource Development (MHRD) project Under the National Mission on Education through ICT (www.vlab.co.in), wherein the student can perform an experiment through remote login in the virtual labs. Most of the IITs are part of this initiative and have made available state-of-the-art labs to students across the county through use of technology.

2.4 Leveraging Technology

The next delivery method is the component that attempts to leverage technology in the delivery methods of a course [13–15]. Today technology has made available videos on various concepts in mathematics, physics, and chemistry. Taking cue from here and combining the fact that most students today possess a mobile with an inbuilt camera, the student may be asked to prepare a 'Video' on the topic assigned by the teacher or as decided by the student, with prior consent of the course instructor. Officially permitting the usage of the mobile phones for developing a video/ audio presentation, helps develop soft skills in the student, in addition to the technical skill development. Allowing the student to bring out his hidden priced possession and do an assignment, completely transforms the class, and tremendously increasing the learning percentage. It can be observed that including this delivery method addresses GA8, GA10, GA11, and GA12. It can be seen that in this model, we are moving away from the conventional classroom environment. Technology leverage can also occur through the student taking up Massive Open

Online Course (MOOC), through independent selection by the student or based on recommendations by the course instructor. Effective MOOC can become a part of the course through introducing a suitable assessment tool. This delivery method develops the ability of the student to engage in independent learning, and hence addresses GA12.

3 Conclusion

The attempt in this paper is to present few different delivery methods for a course. All the methods presented are general and can be included in any course. Most methods do not call for additional investment in space or time or finance. It includes small changes within the classroom and increases the number of GAs addressed by the course. We have proposed different delivery methods in an OBE system. It cannot be mentioned which model is the best model, as it depends on the course, the teacher, the infrastructure available in the college. A good model may be one which introduces all the above methods in small proportions in different courses.

Acknowledgments The authors acknowledge IUCEE (Indo-US Collaboration for Engineering Education) for the series of webinars conducted on Outcome Based Education (www.iucee.org)

References

1. http://www.nbaind.org/Files/General%20Manual%20of%20Accreditation.pdf
2. http://www.abet.org/changed-definitions-for-program-educational-objectives/
3. http://www.washingtonaccord.org/APEC/Documents/Singapore_A3.pdf
4. https://www.ieee.org/education_careers/education/accreditation/ctaa/ieee_role.html
5. www.iucee.org
6. http://commons.wikimedia.org/wiki/File:15th.century.classroom.jpg
7. http://lifetakeslemons.wordpress.com/2011/06/15/portrait-of-an-english-schoolboy/
8. http://en.wikipedia.org/wiki/School_of_Salamanca
9. http://www.nytimes.com/interactive/2010/09/19/magazine/classroom-technology.html. The learning machines from New York Times.
10. http://en.wikipedia.org/wiki/Flip_teaching
11. http://davidtjones.wordpress.com/2009/10/11/the-learning-pyramid-true-false-hoax-or-myth/
12. http://plannetplc.wordpress.com/2010/02/02/37/
13. Felder RM, Silverman LK. Learning and teaching styles in engineering education. Eng Educ. 1988;78(7):674–81.
14. Faculty development course material on 'effective teaching' by Rich Felder, North Carolina State University and Rebecca Brent, Education Designs, during the 2009 Faculty program organized by IUCEE at Mysore. http://iucee.org/iucee/login/index.php
15. Kanmani B, Mallikharjuna Babu K. Towards improvement in quality of higher education through activity based teaching-learning process. In: Proceedings of the national conference on "modern trend & importance of quality assurance in higher education & role of stake holders", conducted by Guru Shree Shantivijai Jain College for Women at Chennai, February 22–23, 2013.

Multidisciplinary Approach to Product Design and Realization

Ashok Shettar, Uma Mudenagudi, Sanjay Eligar, Arunkumar C. Giriyapur, and Nitin Kulkarni

Abstract This paper discusses development of project-based 'product design and realization' course for undergraduate students. The course offered at an early stage of the curriculum is aimed at providing engineering design and product realization skills to the students. Creating an appropriate learning experience in product design is challenging owing to its multidisciplinary nature. An innovative multidisciplinary design-to-realization approach is adopted in this course and student teams are required to design and build working prototypes for predefined products. This course brings a new perspective to the multidisciplinary approach to teaching product design. Introduction of project-based design experience at an early level provides students with an opportunity to develop capabilities to design complex systems in the future. Further, this approach facilitates meeting challenging requirements of several ABET-based educational outcomes: technical as well as professional.

Keywords Multidisciplinary approach • Experiential learning • Engineering education • Product design • ABET outcome

---

A. Shettar (✉)
Department of Civil Engineering, BVB College of Engineering and Technology, Vidya Nagar, Hubli, Karnataka 580031, India
e-mail: ashok@bvb.edu

U. Mudenagudi
Department of Electronics and Communication Engineering, BVB College of Engineering and Technology, Vidya Nagar, Hubli, Karnataka 580031, India
e-mail: uma@bvb.edu

S. Eligar • N. Kulkarni
BVB College of Engineering and Technology, Vidya Nagar, Hubli, Karnataka 580031, India
e-mail: eligar@bvb.edu; nitin@bvb.edu

A.C. Giriyapur
Department of Automation and Robotics, BVB College of Engineering and Technology, Vidya Nagar, Hubli, Karnataka 580031, India
e-mail: aaron@bvb.edu

R. Natarajan (ed.), *Proceedings of the International Conference on Transformations in Engineering Education: ICTIEE 2014*, DOI 10.1007/978-81-322-1931-6_49,

1 Introduction

This paper addresses the development of product design and realization (PDR) course offered as an open elective to pre-final year undergraduate students from engineering disciplines of Electrical and Mechanical Sciences. Product design is a complex process requiring a cross-functional team consisting of design engineers from different engineering disciplines. An experience of multidisciplinary nature usually presents itself during the professional career after graduation. To create such a learning experience at undergraduate level is a challenging task. Approaches to address the interdisciplinary challenge are presented in [1–3]. Attempts have been made at the post graduate level [4] with students from different disciplines. The Department of Electronics and Communication and Department of Automation and Robotics made an effort to design and deliver a course that allows the undergraduate student to experience the complete product design and realization process, working in a multidisciplinary environment.

The primary objectives of PDR course are: the student should be able to (1) design multidisciplinary projects culminating in a finished product, (2) engage in a systematic approach towards design, (3) develop a new product or improve an existing product. The students undergoing this course will also be well placed to explore the possibilities of entrepreneurial ventures, since they would be well equipped to handle all the aspects of product design and realization.

The scope of PDR in providing integrated learning experience in the undergraduate program is presented here and makes the following main contributions.

- The PDR course is designed to provide experience of 'product' using multidisciplinary teams at the undergraduate level.
- The course delivery uses innovative pedagogical approaches and they are
  - Processes of reverse engineering, user survey, need analysis, conceptual design and advanced tool learning are integrated into student training to provide a hands-on experience before the student initiates the process of PDR.
  - Sample case studies are facilitated by the course instructor to provide experiential learning.
  - Interactions with local entrepreneurs at their design and production facilities are organized to provide real exposure to the PDR process.
- Attainment of student outcomes, in particular outcome (e), assessed using continuous review process by an expert review committee comprised of faculty from the disciplines of both mechanical and electrical engineering sciences, is used for continuous improvement of the teaching–learning process.

In Sect. 2 the curriculum design process is illustrated that meets the course objectives. The pedagogical practices used in the course delivery and rubrics used for assessing student learning are discussed in detail. Section 3 illustrates the evaluation of the course and Sect. 4 lists the steps taken for continuous improvement. The conclusion is presented in Sect. 5.

2 Curriculum Design and Course Delivery

The process of curriculum design began by acknowledging the need for experience of product design at an undergraduate level, followed by interaction with prominent academicians and industry personnel. An evaluation of gaps in the skill sets of the graduate engineers was done, and an effort was made to address the issues. The conventional design flow using isolated disciplines/teams would result in increased product design life cycle thereby delaying the final product/prototype. It is important that in this course students become aware of the problems in optimizing the concurrent design processes involving coordination of the multidisciplinary technical functions of design to improve productivity. An innovative multidisciplinary design-to-realization approach is adopted in this course and student teams are required to design and build working prototypes for predefined products.

The entire course spanned 6 weeks during the summer vacation for the student. The first 2weeks comprised of:

- Interactive lecture sessions on product design process and basics of engineering design
- Active learning sessions on relevant topics of product design: Reverse engineering, User survey, Need analysis, Product planning, CAD tool usage, open-source tools
- Case studies incorporating idea generation, conceptual design, detailed design and prototype verification
- All assignments carried out as group activity involving teams made up of students from diverse engineering disciplines
- Industry visit to enable interaction with entrepreneurs and get acquainted with state-of-the-art industrial production equipments and product design and development
- Interactive sessions on 3-D printing for prototyping

The teams consisting of two students from electrical sciences and two students from mechanical sciences background were formed to ensure that each team has skill sets from both the disciplines. The products to be designed were carefully chosen to enhance the experiential learning process, rather than implementation of hi-tech products. The two products given for the student teams to choose from were:

1. Remote control for elderly people
2. Position sensitive Musical Toy

The institutions adopting the Outcome Based Education framework in engineering education are accredited worldwide as per ABET engineering criteria 2000 (EC 2000) [5] and as per new NBA accreditation in India [6]. Often, the capstone projects are used to evaluate student attainment of technical and professional outcomes [7]. Due to the integrative and multidisciplinary nature, the course helps the student to attain few of the challenging program outcomes specified by ABET. The programs can also use this course as a tollgate course to assess attainment of student outcomes by using proper assessment rubrics. The design process of a remote control for elderly people is presented here as a case study.

Table 1 Need analysis chart

Sl. no.	Requirement	Importance
1.	Provision to change channels, volume, power on and off	10
2.	Light weight	8
3.	LED indication for buttons	1
4.	Easily operated by both the hands	9

Table 2 Morphological chart

Functions	Option 1	Option 2	Option 3	Option 4
Accepting force	Press button	PC mouse button	Switch	
Processing signal	By standard IC	By micro controllers	By discrete components	
Holding the base and upper pieces	Adhesives	Screws	Snap fitting	Mechanical lock (notch, sliders)

2.1 Case Study of Remote Control for Elderly

2.1.1 Need Analysis

Need Analysis is the first phase of the course. The need is to design a product which can be used by elderly to handle different television operations with ease. The analysis of the constraints, like difficulty in handling small buttons, frequent fall of remote, trembling hands, confusion while handling more buttons, is documented and listed in the form of a need analysis table as shown in Table 1, which enables the designer to arrive at the primitive specifications of the product.

2.1.2 Product Planning and Market Analysis

The next phase is to estimate the time required for the product design and realization in the form of a Gantt chart highlighting the multiple tasks which could be either sequential or concurrent. Apart from user survey, the students conduct a market survey in the resident city and arrive at an estimate for the market consumption and approximate pricing.

2.1.3 Conceptual Design

After finalizing the specifications of the product to be designed, the student demonstrates the solution-neutral design. This involves listing of all the functions and subfunctions of the product in the form of a morphological chart. Table 2 shows a sample set of the concepts.

The morphological chart provides the student with an opportunity to select multiple options for each function implementation. The final concept selection is done

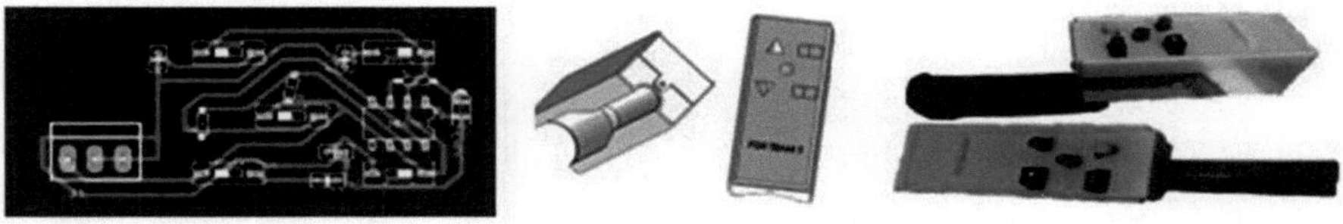

Fig. 1 PCB layout, CAD design and working prototype

Table 3 Evaluation rubrics sample set

Rubrics	
1.	Product planning
2.	Need analysis
3.	Identifying constraints, attributes, goals
4.	Team dynamics
5.	Morphological chart
6.	Detailed design
7.	Need mapping
8.	CAD modelling
9.	PCB design
10.	PCB implementation
11.	3-D printing of CAD model
12.	Demonstration

based on the needs and constraints analysis. This involves various trade-offs against size, cost, ease of use, availability of materials, simulation models and design complexity to name a few.

2.1.4 Detailed Design

In this phase the student performs the functionality test using Proteus® simulation software after finalizing the electrical circuit schematic. The culmination of this phase is the design of PCB using Eagle® CAD tool and implementing it using the PCB prototyping machine.

The mechanical design process is carried out using Solid Works® software. The CAD model for the enclosure and all other functionalities provided by the mechanical subfunctions is shown in Fig. 1 along with the working prototype that has been realized. The unique feature of the product is that it adds the functionality of a remote control to a torch available in the market and uses the same battery source.

3 Evaluation

The evaluation is carried out based on a set of carefully designed rubrics specific to each area of design and realization process quantifying the degree of attainment of ABET outcomes (a–k). A sample set of rubrics is shown in Table 3.

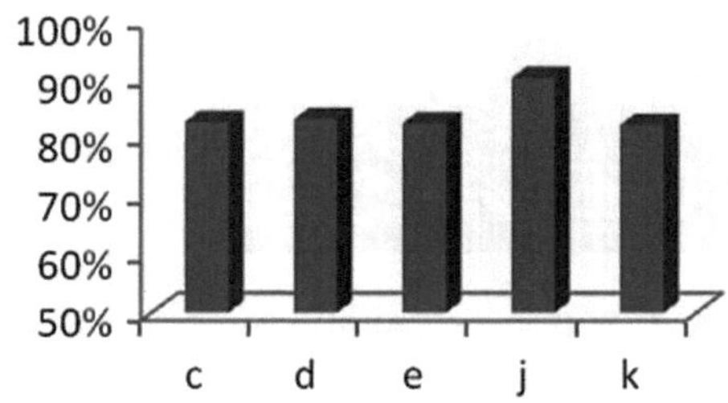

Fig. 2 Attainment of ABET outcomes

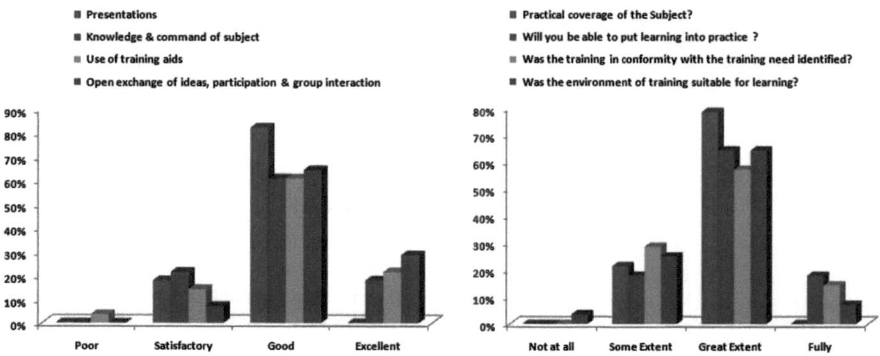

Fig. 3 Feedback analysis

The evaluation consists of two categories, namely Continuous Internal Assessment and Semester End Examination. The assessment of the PDR is carried out regularly twice a week, and teams' progress is evaluated and guided by an expert review committee comprising of faculty from both the disciplines of mechanical and electrical engineering sciences. The interactions are carried out with the intent of highlighting the issues in multidisciplinary design and team work. The experts provide the bigger picture of product design and realization without getting lost out in the technicalities of the design. The attainment of ABET outcomes is shown in Fig. 2, which indicates the attainment of both technical and professional outcomes adequately.

4 Continuous Improvement

Curriculum design is a continuous process and must evolve with the needs of the industry and skill levels of the students for a multidisciplinary course like PDR. The need to redesign the curriculum arises from many factors such as: Evolving global ecosystem of product design and development: Emerging technologies that enable multiple solutions to the problems encountered during product design and development: Redeployment of skills training in evolving regular courses in the undergraduate engineering curriculum: Overcoming the shortcomings of the previous courses based on the feedback from all the stakeholders.

The feedback analysis of the overall course experience is summarized in Fig. 3, which addresses the overall PDR course delivery. The analysis highlights the need to strengthen some aspects of course design and delivery.

5 Conclusions

An innovative multidisciplinary design-to-realization approach is adopted in this course and student teams build working prototypes for predefined products. This course brings a new perspective to the multidisciplinary approach to teaching product design. Introduction of project-based design experience at an early level provides students with an opportunity to develop capabilities to design complex systems in the future. An attempt is made to bridge the gap between the skill level of the graduates and the industry expectations, so that the students are industry ready. The outcome assessment meets the requirements of most of the ABET outcomes adequately and provides inputs for continuous improvement of the course. The future work is to share the experiences with the industry stakeholders and incorporate the suggestions into subsequent courses. At present the rubrics do not address the applicability as a tollgate course, however it can be extended as a tollgate course to assess attainment of student outcomes. Using this course as an early intervention for the students desiring to take up entrepreneurship opportunity is also explored.

References

1. Bronet F, Eglash R, Gabriele G, Hess D, Kagan L. Product design and innovation: evolution of an interdisciplinary design curriculum. Int J Eng Educ. 2003;19(1):183–91.
2. Limaa RM, Carvalhoa D, Floresb MA, Van Hattum-Janssena N. A case study on project led education in engineering: students' and teachers' perceptions. Eur J Eng Educ. 2007;32(3):337–47.
3. Lattuca LR, Trautvetter LC, Codd SL, Knight DB, Cortes CM. Promoting interdisciplinary competence in the engineers of 2020. In: Proceedings of the conference of the American Society for Engineering Education (ASEE), 2011.
4. Soares FO, Sepulvedaa MJ, Monteiroa S, Lima RM, Dinis-Carvalho J. An integrated project of entrepreneurship and innovation in engineering education. Mechatronics. 2013;23:987–96.
5. Accreditation Board for Engineering and Technology (ABET). Engineering criteria 2000: criteria for accrediting engineering programs, effective for evaluations during the 2002–2003 accreditation cycle, Baltimore, MD. http://www.abet.org/images/Criteria/2002–3EACCriteria.pdf. November 2001.
6. http://www.nba-aicte.ernet.in/manual.html
7. Bradley University Electrical Computer Engineering Department. Capstone project: evaluation relative to professional level. http://cegt201.bradley.edu/~gld/prof_cap.html. August 2002.

Contextual Integration of Industrial and Production Engineering Curriculum

B.B. Kotturshettar and Gururaj Fattepur

Abstract In an effort to increase the richness and scope of students' learning experience, the Industrial and Production Engineering program has recently upgraded its curriculum with manufacturing technology as the major thread. It is proposed to make the curriculum context based on ASME #150 Gate Valve. The purpose of this project is to provide a real-world, product-based framework to integrate and expand the learning experiences from different courses and lab exercises. As students are exposed to the design, manufacturing, and quality challenges presented by the project, they are better able to see the limitations and constraints that would otherwise be simply theoretical and easy to ignore. This paper gives an overview of the project and outlines some of the possible benefits as well as challenges that may have to be encountered while integrating the curriculum in this way.

Keywords Curriculum integration • Manufacturing • Product realization

1 Introduction

The use of mini and major projects to integrate and apply student knowledge is common practice and with a known reason. The benefit obtained when students experience their knowledge in action is hard to understate. It is this kind of experience that has driven the development of integrating curriculum on context

B.B. Kotturshettar (✉) • G. Fattepur
Department of Industrial and Production Engineering, BVB College of Engineering and Technology, Vidya Nagar, Hubli, Karnataka 580031, India
e-mail: bbkshettar@gmail.com; gururaj_f@bvb.edu

R. Natarajan (ed.), *Proceedings of the International Conference on Transformations in Engineering Education: ICTIEE 2014*, DOI 10.1007/978-81-322-1931-6_50,

that is discussed in this paper [1, 2]. The goal of this project will be to distribute some of the capstone design experience throughout the Industrial and Production Engineering program curriculum at BVBCET. The integrating context is, in this case, a product – specifically, an ASME #150 Gate Valve. Through a sequence of several existing courses, students design, document, and build a number of complete valves. The project is being implemented by strengthening existing course delivery and lab exercises. This paper discusses the development of the project and how it is being used to integrate various coursework.

2 Background

The Industrial and Production Engineering Department at BVBCET consists of 12 full-time faculty members. The student population comes from a diverse background and with a reasonable ranking in the common entrance test (CET) conducted by the state and Comed-K test conducted by the Consortium of Private College Managements. About 20 % in a class of 75 comprises of students with diploma background making lateral entry to the program. The diversity of the group in terms of prior knowledge can be a hindrance or an advantage depending on the approach used in the classroom. The integrated context makes the most of this situation because it lends itself to a team approach, leveraging the knowledge possessed by the more experienced members, while providing a real-world platform for developing the skills of the less experienced. Another benefit of the integrating project has been the incorporation of previously under-utilized equipment into the coursework. For example, the project relies on rapid prototyping, sand casting, and CNC machining. Rapid prototyping and lean product development have been the subjects of much research recently and have been incorporated into the curriculum first time. The Industrial and Production Engineering Department, along with Mechanical Engineering recently acquired a rapid prototyping machine. The models for the valve project, the parts and assembly will be produced squarely in the production environment. The models are not the end product, but become casting patterns that must function, and produce cast metal parts that must function as well.

As they progress through their coursework, students will experience the valve project from various perspectives. One of the major inputs to the development of valve is adherence to ASME standards. Students refer the internet to gather information on ASME standard codes on gate valves. This will help them acquire the knowledge on sizes, materials to be used, operating pressure range, etc. and also create a curiosity among them to know various other products that need to be developed as per standard codes. They prepare working drawings that will be used to create CNC machining programs, and they assign tolerances that will later drive quality characteristics. Although the work may occur in different semesters, each class will be treated as if it is occurring simultaneously with the other activities, resulting in a tangible experience of concurrent engineering.

If the progress on student learning is satisfactory, it would be feasible to incorporate some of these activities into the curriculum of other courses.

3 Method

The design is reverse engineered from a commercially available #150 gate valve. The valve consists of 12 components that will be made by students. In addition to these, fasteners, and a few other components (considered purchase items) will have to be purchased separately. The student's exposure to the project is determined by the course sequence. This means that normally, their first involvement is through the second level manufacturing processes course. In this course (Manufacturing Technology I – Foundry and Welding), there is a laboratory exercise in sand casting in addition to earlier experiments on sand quality testing. The integrated lab will use a split pattern of the valve block made on the rapid prototyping machine. The pattern will be scaled up to allow for shrinkage and includes an interlocking sprue, riser, and runner system. The goal of this type of integration will be to go beyond understanding the technology to adaptation to leveraging of emerging technology. Figure 1 shows the casting of a gate valve body.

During the same semester, students will typically take a materials course (Engineering Materials) in which they perform mechanical testing of various materials. Previously, the labs had been done in isolation of unrelated materials, now the course draws on the integrating product to develop material specifications and perform tests related to the specifications for the valve materials.

In the same semester, students use solid modelling tool Pro-E to reverse engineer the valve. They will be provided un-parameterized models of all of the valve parts. The students create working drawings from dimensions they measure on these models and use their working drawings to create parametric models of the components in the Engineering Design-CAD lab. For some of the components, only functional parameters will be given, and students are required to develop designs

Fig. 1 Gate valve body

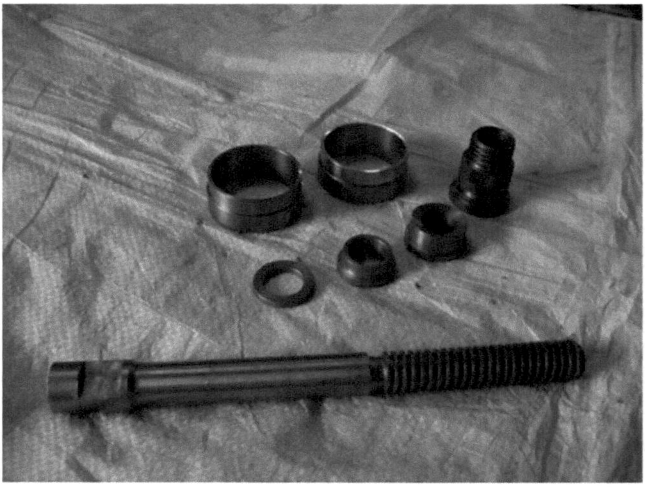

Fig. 2 Machined components

that function in their fourth semester course (Design of Machine Elements) and will be manufactured by the equipment available. The context provides its own limitations and gives students a more realistic idea of concepts such as design for manufacture. Additionally, students will determine acceptable tolerances for certain components that are manufactured in the Manufacturing Technology II – Manufacturing Processes class and will be evaluated by the Metrology and Quality Engineering class.

In what is usually their fifth and sixth semester, students will take several courses that incorporate the valve. Most of the production will take place in the Manufacturing Technology III course. This course requires students to take the parametric models created by the CAD class and create code to machine the various components. Some of the labs require writing G-Code manually, such as the lathe operations on the body and bonnet, while others will incorporate the manufacturing application of Pro-E, while some will use Master-CAM. In the Quality Engineering lab, the process of machining the stem and other machined parts will be examined. Using the tolerances defined in the CAD course, students will study the process capability and try to reduce variability. Mechatronics and PLC course helps students identify the automation components – sensors, actuators, etc. for valve operation. Several of the machining fixtures would be designed as part of the Tool Design course, and further plan is to use Manufacturing Technology IV – Forming and Form Tool Design course to develop punch-press tooling to create gaskets and brackets that otherwise would be treated as purchase parts. The course on Fluid Power Control would help students design circuitry for automation of the opening and closing of the valve and implement the same. Figure 2 shows some of the machined components used in the valve.

4 Results

Student response to the project is extremely positive and majority of students enjoy working on the project. The project, by its nature lends itself to a more involved discussion of the concepts as they are presented. The concepts are rather parts of a whole as they no longer exist as isolated cases. A change in the working drawing as students can now see has an effect that permeates throughout the entire production environment. At the same time, the challenges presented by the project have been real challenges that require real solutions, or else the entire project fails. Figure 3 shows the gate valve assembly.

4.1 Success

The integration of the curriculum on context has advantages that include an increased emphasis on team-based, experiential learning; increased opportunities to improve problem-solving skills. The nature of the project, with its self-defining limitations, makes it a good fit for team-based learning. Teams will be given an objective and will be limited by the equipment and material available. The project, as it continues to mature, becomes more formalized, and economic aspects of the project will become more realistic. For example, in the Manufacturing Technology II course, the current emphasis will be on successfully manufacturing the parts. Each team will be assigned a specific group of components and fixtures, and they are required to successfully machine each one of them. The project is flexible enough that it can

Fig. 3 Gate valve assembly

be integrated at varying levels in any of the courses where it is used. Student interest in the project is the key and it is expected to be extremely encouraging. Students may even choose to develop some elements of the valve as part of their mini/major project. Since the students are involved in the design and production of the valve, they will take ownership of the ideas behind the work. The department hopes that students live up to the challenges they are presented by the project and exhibit a good degree of pride in their workmanship.

5 Conclusion

Although the project is not without its problems, the overall benefit to student understanding would be gratifying. The students' level of confidence to solve complex engineering problems would be enhanced as they have learnt the concepts in context and hence become more rational in applying the knowledge that they have acquired. In its current form, the integrating project will be incorporated into coursework for second and partially third year, and the results will justify further work to complete the project and further integrate it into the course and lab work. Although it requires more effort both from the students and faculty members, the ability of this kind of project to bring together so many of the techniques used in the field of manufacturing will make it an ideal solution to many of the problems facing faculty in our program and create profound learning experience for students.

References

1. Hewitt RW. Using a two-cycle engine to integrate manufacturing engineering technology curriculum. In: Proceedings of the ASEE southeast section conference. 2007.
2. Ziemian CW. A systems approach to manufacturing as implemented within a mechanical engineering curriculum. Int J Eng Educ. 2001;17(6):558–68.

Automotive Electronics: Learning Through Real-World Problem-Based Case Studies

P.C. Nissimagoudar, Nalini C. Iyer, B.L. Desai, M. Uma, C.D. Kerure, V.R. Mane, M.R. Kiran, and Ramakrishna Joshi

Abstract Automotive electronics is a course that requires skills from multiple disciplines including, but not limited to, mechanical, control, computer science, and electronics. The course is introduced to address the needs of embedded and automotive industries, hence providing the necessary knowledge and skills required for those industries. The objective of the curriculum is to enhance learning and improve student's implementation skills. In this paper, we propose to introduce the exercises including real-world case studies and experiential learning. The major challenge of teaching this course was to teach mechanical concepts for electrical science students and to develop electronics for mechanical systems. The practical demo sessions by automobile labs gave the desired foundation for the course. The engine management concepts were taught using a very popular simulation software, AT Electronics tool, which is a combination of electronics and diagnostics. This activity gave a real feel of engine management systems to learn how complex systems work and to diagnose faults with them. The paper also discusses another major activity in the form of course projects. The course projects resulted in the application of domain knowledge and improvement of skills by using appropriate tools. In addition to these activities, all regular classes included animations and video presentations to make the concepts clearer. Special lectures by industry experts were also arranged to give the students a

P.C. Nissimagoudar (✉) • M. Uma • C.D. Kerure • V.R. Mane • M.R. Kiran • R. Joshi
BVB College of Engineering and Technology, Vidya Nagar,
Hubli, Karnataka 580031, India
e-mail: sn_asundi@bvb.edu; uma@bvb.edu; cd_kerure@bvb.edu; mane@bvb.edu

N.C. Iyer
Department of Instrumentation Technology, BVB College of Engineering
and Technology, Vidya Nagar, Hubli, Karnataka 580031, India
e-mail: nalinic@bvb.edu

B.L. Desai
Department of Electronics and Communication Engineering, BVB College of Engineering
and Technology, Vidya Nagar, Hubli, Karnataka 580031, India
e-mail: bldesai@bvb.edu

R. Natarajan (ed.), *Proceedings of the International Conference on Transformations in Engineering Education: ICTIEE 2014*, DOI 10.1007/978-81-322-1931-6_51,

wide perspective of the subject. The paper discusses the impact of these activities in the form of student feedback, placement results, and participation in technical events. This experiential learning helped the students to improve comprehensive application ability and innovative consciousness.

Keywords Program outcomes • Bloom's taxonomy • Course project • Course design • Delivery methods • Student feedback

1 Introduction

Automotive electronics is highly multidisciplinary requiring the knowledge of various engineering domains like mechanical, electronics, and computer engineering. The said field is highly application oriented and mainly deals with embedded systems applied to various domains of automotives such as power train, chassis, body, infotainment, multimedia, and human-machine interface (HMI) with an emphasis on engine management systems, automotive safety systems, automotive communication protocols, automotive sensors, and automotive software development methods. The topics by themselves are highly multidisciplinary requiring relevant real-world case studies to understand the concepts and correlate with suitable application domains [1].

There is a potential growth and scope for innovations in various domains of automotive research, resulting in rising demand for qualified employees. This is a major driving force for introducing the course on automotive electronics in the electrical engineering discipline.

BVB College of Engineering and Technology, an autonomous institute, affiliated to Visvesvaraya Technological University, offers the automotive electronics course at the sixth semester level, for all the programs of the electrical science stream involving instrumentation technology, electronics and communications engineering, and electrical and electronics engineering wherein the student has a sufficient background of embedded systems development, design of electronics systems, and basics of mechanical systems.

The curriculum has been designed by seeking inputs from industry and academia experts in the Board of Studies (BOS) of each engineering discipline every year. The BOS is responsible for approving the program curriculum and educational experiences and assessment pattern for each program as per the norms of autonomous college. The BOS committee guides the board about current trends in the instrumentation and its allied industries and the skills that are required for the graduates to succeed. The BOS meets yearly on campus and reviews the proposed changes in the academic program. The BOS also discusses new activities and initiatives of the department. The curriculum structure of each program has been broadly classified as basic science, mathematics and engineering, program foundation course, and program verticals (specializations) along with integrating experience.

The common program vertical for all the disciplines of electrical science stream is embedded systems. Automotive electronics is one of the core courses belonging to this vertical and courses such as Infotainment and AUTOSAR being the electives.

This vertical helps in enhancing the skills needed to develop and design embedded products and meet the requirements of the society. The concept of communication protocols, concept of electronic control unit (ECU), and engine/fuel management covered in the automotive electronics course along with the knowledge of processors, controllers, and programming skills cater to the design of automotive embedded systems.

The contents of the course are framed considering the inputs of the various automotive embedded systems organizations like KPIT Cummins and Robert Bosch Engineering and Business Solutions India (RBEI) and ARM Limited to cater to the needs of the industry. The teaching-learning process has been made more effective by adopting various activity-based learning apart from classroom teaching. These methods include practical exposure to automotive mechanical systems, simulation of engine management systems, expert lectures to introduce students to the current technology-related issues of industries, animations, and video presentations to make the concepts clearer.

These delivery methods gave a required foundation for a student to frame a problem statement for their course project, wherein the student deals with the need and functional requirement analysis and proposes a suitable solution for real-world problems of automotive domain [2].

The automotive electronics is of three credits with 42 h of teaching. Assessment techniques involve the continuous internal evaluation (CIE) and semester-end evaluation (SEE). CIE involves minor exams and course project with weightages of 80 % and 20 %, respectively. The questions of minor/semester-end examinations have been framed to cater to the learning levels of Bloom's taxonomy. The course along with the delivery and evaluation methods addresses both the technical and professional outcomes of the program.

The organization of the rest of the paper is as follows: Section 2 deals with the design of the course; Sect. 3 deals with delivery methods; Sect. 4 deals with course projects, evaluation methods, and their effectiveness; and Sects. 5 and 6 discuss the conclusion.

2 Course Design

The course has been designed with the following objectives:

- To provide a comprehensive overview about the existing and future automotive electronics systems
- To illustrate the distinctive features of the automotive world in terms of requirements and technologies
- To demonstrate the role of electronics for the areas like in-vehicle architectures, networking, engine management systems, vehicle safety systems, and infotainment systems
- To analyze the current status of software in the automotive industry and present the specifications elaborated within the AUTOSAR consortium in terms of standardization

The course mainly focuses on developing the student's ability to analyze the electronics systems in present-day automobiles and their specific requirements to become acquainted with the new applications that are being developed for future automobiles. To cater to these objectives, the course contents were framed by considering the inputs of various automotive industries like KPIT Cummins, RBEI, and ARM Limited. The contents included both the basic topics like automotive mechanical systems and advanced topics like automotive communication protocols and automotive open system architecture (AUTOSAR) with the equal weightage to hardware and software concepts.

3 Course Delivery Methods

This section deals with the details of teaching-learning methods adopted to teach automotive electronics. As the course deals with two very diverse fields of engineering, mechanical and electronics, the delivery of the course demands different delivery methods to be practiced along with the conventional classroom teaching [3]. The methods adopted include the following:

- Practical exposure to automotive mechanical systems through demonstrative laboratory sessions
- Augmenting concepts of engine management systems through simulation tool
- Exposure to current technological requirements and advances through expert lectures by industry personnel
- Providing design experience through course projects

The details of each method are summarized below.

3.1 Practical Exposure to Automotive Mechanical Systems with the Demo Lab Sessions

The course is taught for electrical science stream students who have restricted knowledge about mechanical concepts. As the course deals with the development of embedded systems for mechanical systems, it becomes sheer important to make the students understand the mechanical concepts in a short time. The conventional methods of teaching only provide an abstract learning. The demo and hands-on sessions provided by automobile labs present them knowledge of real working models of an automobile. Students very easily understand the working principles in a very short time. The lab sessions include hands-on sessions on basic automotive mechanical systems consisting of different types of engines (gasoline, diesel), transmission systems, braking systems, steering systems, and suspension systems. The

Fig. 1 Practical exposure to automotive mechanical systems with the demo lab sessions

labs provide the demos and hands-on experiences using cut sections of all these models. The snapshot of one such model is given in Fig. 1.

3.2 *Augmenting Concepts of Engine Management Systems Through Simulation Tool*

Simulation tools provide an opportunity to test and understand the behavior of the complex systems through experimentation. Engine management system is one such similar system whose behavior and performance depend on various input parameters. There are various control actions to be provided to control many outputs. The performance of the system depends on tuning various control parameters. The behavior of such systems can be very easily explained with the use of simulation tool. AT Electronics simulation tool helps in analyzing the behavior of engine management system. Here, the user can vary different control parameters like air-fuel ratio, ignition timing, and exhaust gas recirculation (EGR) valve position to monitor the system behavior in terms of torque produced, fuel efficiency, and emission control. The snapshot of such simulation is shown in Fig. 2.

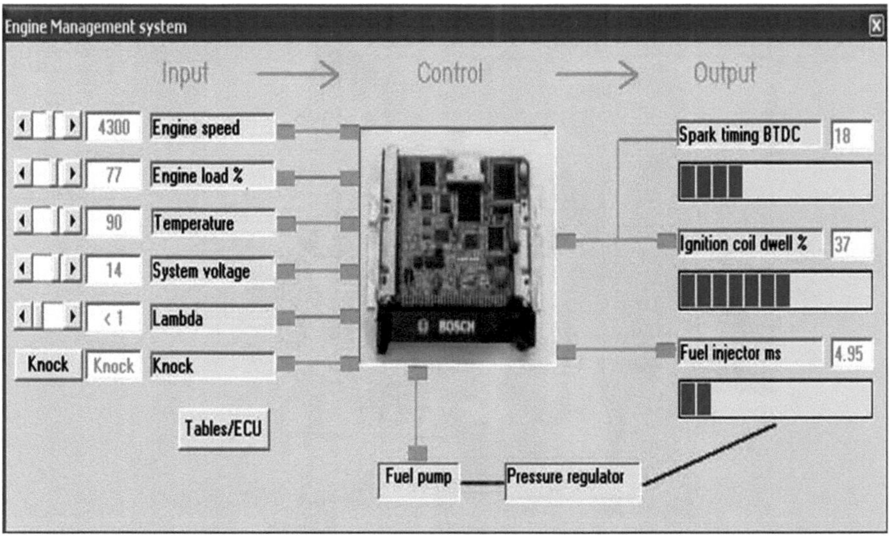

Fig. 2 Simulation of engine management systems

3.3 Exposure to Current Technological Requirements and Advances Through Expert Lectures by Industry Personnel

To make sure that the students get to know about the latest applications, technologies, and requirements in automotive electronics, expert lectures by industry personnel from automotive companies like KPIT Cummins and RBEI were included in the curriculum. With this activity, the student can correlate the theoretical concepts learned to the applications of the industry. The gaps and scope for innovation can also be identified.

3.3.1 Providing Design Experience Through Course Projects

The course projects are introduced where the concepts can be well understood and higher-level learning can be achieved by developing an application or a product or a process. The activities previously mentioned facilitate the students to identify and formulate the problem. The delivery methods discussed above help the students to learn the concepts of more advanced subject like automotive electronics in a better way.

Along with these activities, a lot of the underpinning theory is covered using the conventional methods and using the instructor powerpoints, animations, and videos.

4 Course Projects

The main objective of the course projects is to give more hands-on experience and an integrated experience to the students in addition to that knowledge gained in the regular course [2]. Integrated experience is a multidisciplinary, design project-based approach emphasizing on engineering design and incorporates fundamentals from several engineering disciplines such as instrumentation, electronics, electrical, computer science, automobile and mechanical involving science, mathematics, humanities, and engineering components. It will improve both practical and academic skills of individuals to prepare them for a wide range of opportunities in either employment in industries, R&D, or higher education.

The students will be given topics/themes related to the subject; they have to explore an interesting problem of their choice in the context of the course. Projects can be done individually or in teams of two/three students. The students will be given various themes related to automotive electronics like the realization of control algorithms required for specific automotive applications using MATLAB/Simulink and the implementation of communication protocols; automotive sensors; emission control systems; safety, security, and driver assistance systems; etc. The activity involves different phases like the following:

- Literature survey and problem definition: The students need to formulate the problem based on the requirement analysis, explore alternate approaches through survey, and select the appropriate one.
- Concept-level design: Identifying input/output variables, establishing the relationship between them, and proposing the functional block diagram are carried out. The development of circuit diagram and algorithm development are also part of this phase.
- Implementation: This phase includes the selection of appropriate simulation and modeling tools, implementing the design on breadboard, debugging, testing the design, and finally implementing on PCB.
- Demonstration and report writing: Here, the student needs to present their work in both oral and written form along with the demonstration of their work.
- The assessment is carried out according to the rubrics mentioned in Table 1 and is in line with the review required for different phases of the activity.

Based on the teaching-learning methods followed for automotive electronics, the course outcomes for the course projects are as follows.

The student will be able to:

- Perform need analysis and identify the problem from the given theme and propose the solution ("a," "c," "f," "h," "i," "j," "l")
- Analyze and record the requirements for the identified problem ("a," "c")
- Transform the fundamental knowledge gained in the curriculum to model ("c")

Table 1 Course project rubrics

Rubrics	Weightage (%)
1. Literature survey and problem definition	20
Problem statement	
Identify the resources to meet the desired needs	
Exploring alternate approaches through literature survey	
Selection of an appropriate approach	
Team formation and defining the roles	
2. Concept-level design	20
Identification of inputs and outputs	
Establishing the relationship between input and output	
Circuit design/algorithm	
Verification of obtained results with standard results	
3. Implementation	40
Functional block diagram	
Choose an appropriate simulation and modeling tool	
Implementation, debugging, and demonstration of results	
4. Demonstration and report writing	20
Preparation for report, poster, use of visual aids	
Clear and well-organized report	
Presentation skills	

- Design/develop prototype/system to meet the identified requirements ("c," "k")
- Develop technical writing skills and presentation skills and build the ability to work in team ("d," "g")

Course outcomes are also mapped with program outcomes of ABET. Some of the selected course projects are listed here.

4.1 Fast Tire Puncture Detection System

This project intends to design a system that automatically detects any foreign body that pierces the tire and may cause its puncture. The system outputs the message to the user regarding the interruption caused in the tire.

4.2 A Simulation of Speed Variation Based on Accelerator Pedal

This project deals with regulation of throttle position depending on the accelerator position which is proportional to vehicle speed.

4.3 Modeling of Cruise Control System Using MATLAB

The objective of the project is to model cruise control system of vehicle for different conditions.

4.4 Brake Signal Modulator

While traveling, when the driver applies the brake, a continuous brake light is turned on, which does not catch the attention of the vehicle behind. Hence, the solution to the above problem can be given by developing an electronic module that converts ordinary brake light into an attention-getting visual alerting device.

4.5 Lambda Probe Readout for Carburetor Tuning: Improving the Combustion Efficiency by Lambda Sensor

In addition to the regular reviews and evaluation by the course instructor, each course project has been also evaluated by the industry personnel. The projects carried out by students gave them a real-time exposure to the subject. Many of the problem statements addressed the contemporary issues and provided an innovative solution for the problem.

The method of assessing the effectiveness of the activity includes student performance assessment, student self-assessment, feedback from industry experts, and student feedback. Contributions to the activity can be assessed in terms of individual deliverables and group deliverables.

5 Evaluation Methods, Their Effectiveness, and Discussion

This section discusses the various evaluation methods adopted for the course and the discussion of the effectiveness of the activities realized and their mapping toward program outcomes. Equal weightage has been given to continuous internal evaluation (CIE) and semester-end evaluation (SEE). The CIE includes two minor exams and the activity in the form of course project with the weightage of 40 % and 10 % and the SEE with 50 % weightage. The question paper pattern involves the questions addressing different learners' levels according to Bloom's taxonomy [4, 5]. Below is a graph showing the attainment of different learners' levels for a model question paper and course activity. Questions of CIE and SEE mainly addressing Bloom's taxonomy levels L1 to L4 and the course activity in the form of course project addressing higher-learning levels are illustrated in Fig. 3.

The effectiveness of the course project as part of a core subject is mapped to the learning outcomes "a" to "k" defined according the Accreditation Board for Engineering and Technology (ABET) [6]. Table 2 presents that the outcomes "a" to "e" belong to technical outcomes and "f" to "k" belong to professional outcomes. It can be seen that the course project addresses both technical and professional outcomes defined by the program.

5.1 *Reflections of Course Project with Continuous Monitoring and Feedback*

With this activity, different teams were able carry out course projects addressing different themes. Around 25 different projects with different themes were carried out. This activity resulted in building a strong knowledge of fundamentals, realizing

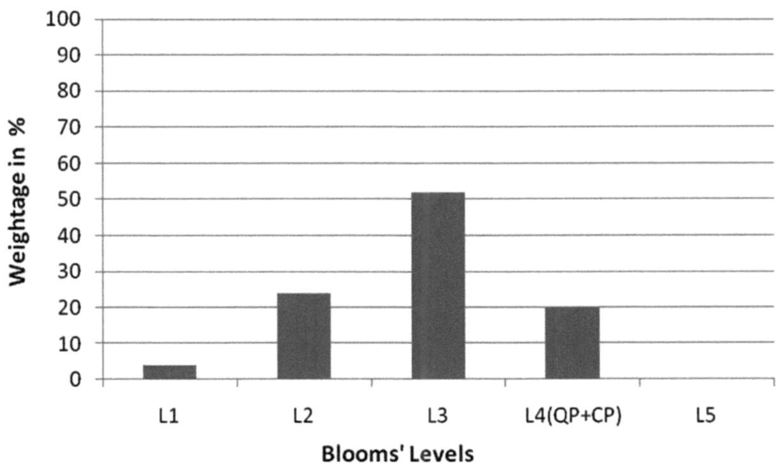

Fig. 3 Question paper pattern

Table 2 Activity outcomes mapping to program outcomes

Activity phases	Outcome elements	ABET program outcomes addressed
Literature survey and problem definition	An ability to identify needs An ability to formulate problem An ability to find alternate solutions and choose the best one Ability to work effectively in project teams, as both a member and leader, with different skills	c, d, i, j
Concept-level design	Understand environmental, political, economical, aesthetic, and social impacts of engineering work An ability to use established method to design a system/process Knowledge of contemporary issues in the field of instrumentation technology	c, j
Implementation	An ability to use modern engineering tools for modeling and simulation Ability to engage in independent and lifelong learning in the broadest context of technological changes	i, k
Demonstration and report writing	Oral and visual communication skills appropriate to the profession of engineering	g

the various aspects of engineering analysis and hence the problem-solving ability of the student enhanced [7]. Finally, the assessment based on student feedback has been collected by each team as detailed in the Appendix, and the statistics shows that the objectives of the course project were satisfactorily met as mentioned in Fig. 4. Questions 1 and 3 relating to the understanding of concepts and applying the same to the application have made an impact on learning. Question 2 relating to the scope of understanding beyond curriculum is satisfactory. Suggestions by the review committee and participation in technical events reflected in questions 4 and 5 need a scope for improvement.

5.2 *Reflections of Activity in Participation in Technical Events and Placement Activities*

A total of ten project batches have participated in different project competitions conducted by automotive industries and three batches have won the top prizes. The projects have been appreciated by industry personnel and few ideas were having research attributes in them. The number of automotive embedded industries visiting the college campus has been increased and the number of students placed in those industries is also increased considerably.

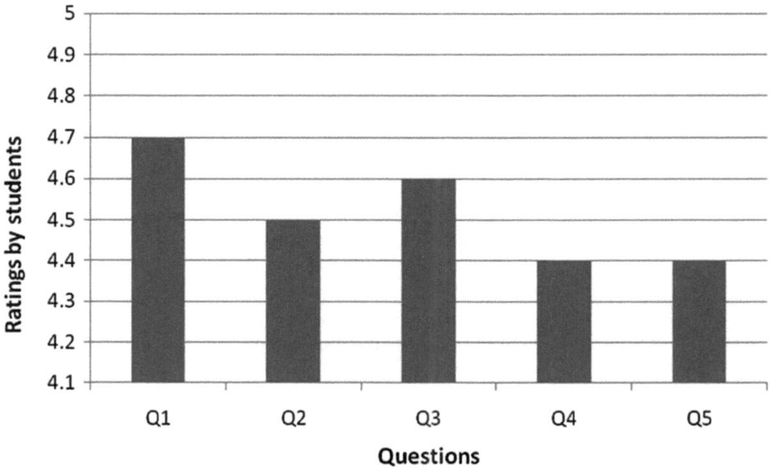

Fig. 4 Feedback questionnaire response

6 Conclusion

The details of the course design and delivery methods and the activity planned and executed for the course automotive electronics as a part of an innovative approach to enhance the learning outcome of the students have been presented. The metrics and the techniques adopted for the assessment of the learning outcome have been listed and the results are presented.

The overall outcome as seen from the result analysis clearly indicates that the approach adopted has significantly been encouraging in terms of the students' overall growth. The students' participation in technical events and their placement and the students pursuing career in automotive embedded domain have been significantly increased.

A majority of the students have expressed their satisfaction over the course contents, delivery methods, and course project. The activity provided the necessary platform to showcase the innovative ideas of the students.

Acknowledgment The authors would like to acknowledge Dr. Ashok Shettar, Principal, BVBCET, for providing constant support and encouragement and Dr. Anil Badiger, Head, Dept. of Automobile Engineering, and his faculty for providing practical exposure to automotive mechanical systems with the demo lab sessions. We also would like to acknowledge the industry personnel from automotive industries KPIT Cummins and RBEI for their contributions in framing the curriculum and its implementation.

7 Appendix

The questionnaire consists of five questions; the students are asked to rate on a scale of 1–5.

1. Did the course project help in understanding the automotive systems better?
2. Did the course project boost your knowledge in the area of automotive domain?
3. Are you able to apply your knowledge of instrumentation and control in real-time embedded automotive applications?
4. Did the review committee give you the right feedback to guide you for the implementation of the course project?
5. Did the course project help you to participate in technical events and placement activities?

References

1. Lehtovuori A, Honkala M, Kettunen H, Leppavirta J. Interactive engagement methods in teaching electrical engineering basic courses. In: Proceedings of the global engineering education conference (Educon), IEEE, 2013. doi:10.1109/Educon.2013.6530089. Berlin, Germany. p. 75–84.
2. Adamo OB, Guturu P, Varanasi MR. An innovative method of teaching digital system design in an undergraduate electrical and computer engineering curriculum. IEEE Department of Electrical Engineering University of North Texas, Denton, TX 76207, USA.
3. Case JM, Light G. Emerging methodologies in engineering education research. J Eng Educ. 2011;100(1):186–210.
4. Azuma M, Coallier F, Garbajosa J. How to apply the Bloom taxonomy to software engineering. In: Proceedings of the eleventh annual international workshop on software technology and engineering practice, Washington: DC, USA 2004.
5. The University of Wisconsin-Madison. http://teachingacademy.wisc.edu/archive/Assistance/course/blooms.htm
6. http://www.abet.org/special-reports/
7. Redish EF, Smithg KA. Looking beyond content: skill development for engineers' unpublished.

Pedagogical Transformation in Heat and Mass Transfer Laboratory Course of Undergraduate Mechanical Engineering Program

N.R. Banapurmath, P.P. Revankar, and M.B. Gorawar

Abstract The heat and mass transfer is an important application-oriented course taught for working in the mechanical engineering arena and the other allied courses like automobile engineering and chemical engineering. The pedagogy for thermal engineering courses like heat transfer has undergone changes over the years through novel ways of industry participation, discussion of application-driven issues, facilitation of problem-solving skills, performance of analysis/design of thermal systems, and the use of computational tools. Albeit to these initiatives, the process has complicated due to the urgent need to update the course material with new technology issues and reduction in credit hours available to the thermal courses [1].

A segment of the students have an apprehension to this course owing the prerequisite of calculus knowledge and analytical skills as evidenced in thermal courses. In view of this, the heat and mass transfer course needs implementation of certain pedagogical tools to facilitate active learning by the student community. The implementation of course delivery based on Bloom's taxonomy has improved the modus operandi of delivery of this lab course. The categorization of the experiments as demonstration, exercise, structured enquiry, and open-ended enquiry has given ample opportunity for the faculty to address the ABET 3a, 3b, 3c, 3d, 3e, 3g, and 3k criterion to a greater extent.

Keywords ABET criteria • Heat transfer • Structured enquiry and open-ended problems

N.R. Banapurmath (✉) • P.P. Revankar • M.B. Gorawar
Department of Mechanical Engineering, BVB College of Engineering and Technology, Vidya Nagar, Hubli, Karnataka 580031, India
e-mail: nr_banapurmath@bvb.edu

R. Natarajan (ed.), *Proceedings of the International Conference on Transformations in Engineering Education: ICTIEE 2014*, DOI 10.1007/978-81-322-1931-6_52,

1 Introduction

The teaching-learning activities in the technology-enabled world today are a real challenge to conventional mold teaching. The advent of web technologies has led to information flooding in every field of engineering that has extended the teaching-learning process beyond the confines of a closed classroom interaction. The innovations in pedagogical approaches have been reported in several streams of study as early as the 1990s when modules were designed for interactive learning that proved to be delivering greater learning gains than the traditional lecture format. The measures were implemented based on the observations of poor student attentiveness and performance even to lectures delivered by eminent teachers despite of the good math skills possessed by the students. The traditional lectures, although excellent for many purposes, did not communicate the concepts well because of their passive nature. The meticulously designed assessment techniques have enhanced gains amongst student groups constituted by a heterogeneous mix thereby establishing effectiveness of interactive student engagement [2].

The engineering profession harnesses resources like energy, materials, and information to contribute usefully to mankind. The engineering education paradigm in the early days gave more emphasis to theory courses and teaching methods with very little attention (<5.2 %) to the application segment. The engineering education in the present context has the prime objective to mold students and enable them to practice engineering to utilize meaningfully the resources of nature. This demands a good understanding of nature that goes beyond theoretical knowledge based on traditional educational laboratories [3].

Engineers are involved in the compilation of experimental data that guide them to develop products and ensure products to perform as intended through methodological measurements. Their performances are compared with supplier specifications to demonstrate compliance or indicate scope for changes to overcome deviations. There is a need for "design-based effort" to practice activities towards the attainment of excellence in the assigned task though the practicing engineers by default apply conventional pedagogy concepts.

The engineering education system has proved to be a vital social fabric and has made it mandatory for it to be responsive to the changing societal pattern. The present-generation engineers operate in a challenging environment that poses seven deterrent factors: information proliferation, multidisciplinary technology, environmental concerns, growing social responsibilities, globalized markets, participatory corporate modules, and rapid rate of change. Therefore, the paradigm shift is mandatory in educating future-generation engineers whose components include domain knowledge, work skill sets, and goal-driven attitudes. The major impediments to bring this paradigm shift of engineering education policy are the fear of loss of control that can lead to diversions into areas that the teacher is unaware of. However, governmental regulation provides succor in reviewing the transformation process towards the better. The conventional instructional methods are inadequate to equip engineering graduates with the knowledge, skills, and attitudes they will need to meet the demands likely to be placed on them in the coming decades, while alternative methods like problem-based learning have been time tested to offer good prospects to future engineering and technology [4].

2 A Paradigm Shift: Teacher-Centric to Student-Centric Practice

The mechanical engineering students as part of the UG curriculum learn the heat transfer course in the third year of the program. This course addresses questions that emphasize the underlying principles of heat transfer and focuses hands-on experiments on principles that relate to conduction, convection, radiation, boiling, and condensation. This section tries to explain the experiences of stakeholders in two different methodologies of learning and teaching the heat transfer course at UG level in mechanical engineering. The brief account of two methods is explained in the context of the learning process, evaluation criteria, and benefits realized.

Approach 1: The traditional approach of delivery in the heat transfer course was based on the execution of the prescribed set of course modules based on "teacher-centric: follow-the-instruction method" that specified steps to be executed to get the final expected result. The students opting the course could be grouped into exquisite band and repulsive band based on their extent of involvement in course sessions. The standard approach did make a limited part of the class involved seriously, while the remainder part was left into disillusion by monotony of the following steps to get the result. The students as part of the course were to complete a specified set of chapters and take up CIE exams that determined their performance grade in the course.

On the evaluation front, the heat transfer course posed challenges to course instructors in terms of devising methodology that would investigate the depth of student understanding. The standard practice was to assess student performance through minor exams and assignments based on stereotyped questionnaires. The overall exercise for the student was more of a memory teaser rather than knowledge exploration as the course provided minimum space for the student to think beyond the prescribed set of exercises.

Therefore, the course in heat transfer, which involved more numerical calculations, turned out to be a nightmare for a substantial part of the class. The student performance in the semester end examination reflected the lack of ingenuity in their performance due to improper understanding of the course. The course thus turned out to be a "tough-to-digest course," though it had a wide scope for practical applications. Thus, the heat transfer course was able to reach only a small part of the exquisite student layer, leaving the intermediate and repulsive student layers puzzled on why it was important for mechanical engineering graduates. The traditional approach of course delivery lead to disdain to heat transfer course which otherwise would be mandatory course for any graduate in Mechanical engineering. This is a brief overview of how the course was taught based on the teacher-centric traditional approach that kept the student comfort level too low with a drooping learning curve that led to a phobia towards thermal engineering courses.

The experiments designed in the conventional approach of teaching the course are indicated in Table 2. The contents clearly reveal that the experiments listed do not involve more concepts of active learning and are therefore referred to as the teacher-centric approach.

Approach 2: The approach of delivery is based on flexible learning that has no rigid prescribed set of recipe, making it more student participative, "student-centric: do-it-yourself method." The approach stressed on knowledge exploration through teamwork and practical experimentation without barriers on student thought process.

The sessions were conducted to promote better learning through group learning activities that also fostered group dynamics apart from learning the course content. The activities were designed to promote better student participation through group tasks that promoted better understanding of concepts without being only memory teasers but involved higher levels of learning skills. The tasks performed by students included the following:

- Using computational tools for heat transfer through simple illustrations
- Investigating applications of different modes of heat transfer
- Performing heat transfer analysis of thermal systems such as hot box, refrigerator systems, and 2D heat transfer analysis
- Delivering course seminar on contemporary heat transfer issues

The course delivery led to the realization of higher levels of learning as evidenced through a better student performance in the CIE and SEE. The experiments were designed in four types, namely, demonstration, exercises, structured enquiry, and open-ended experiment as indicated in Table 3. The proposed system of delivery of the course clearly reveals the proper blend of teacher-designed experiments and the student-developed experiments. The proposed delivery approach promotes a more active participation of the student, thereby realizing a higher level of learning as per Bloom's taxonomy. The Course Articulation Matrix: Mapping of Course Learning Objectives (CLO) with ABET 3a to 3k criterion (program outcomes) indicated that the course delivery addressed 3a, 3b, 3d, 3e, 3g, and 3k criterion to a medium level of realization.

This proposed revision in the lab course gave the students a better learning experience compared to the earlier approach of standard "stepwise procedure-based experiments" as evident in the feedback indicated in Fig. 1a, b.

The students found the new method to be a more active means of assimilating knowledge in heat transfer and learned additional computational skills as against the teacher-centric approach that made very little room for independent thinking or learning.

After the end of each experiment, each group is advised to duplicate the recorded data in a soft copy; write a theoretical background of each experiment using primary, secondary, and electronic sources; and do the specimen calculations for their

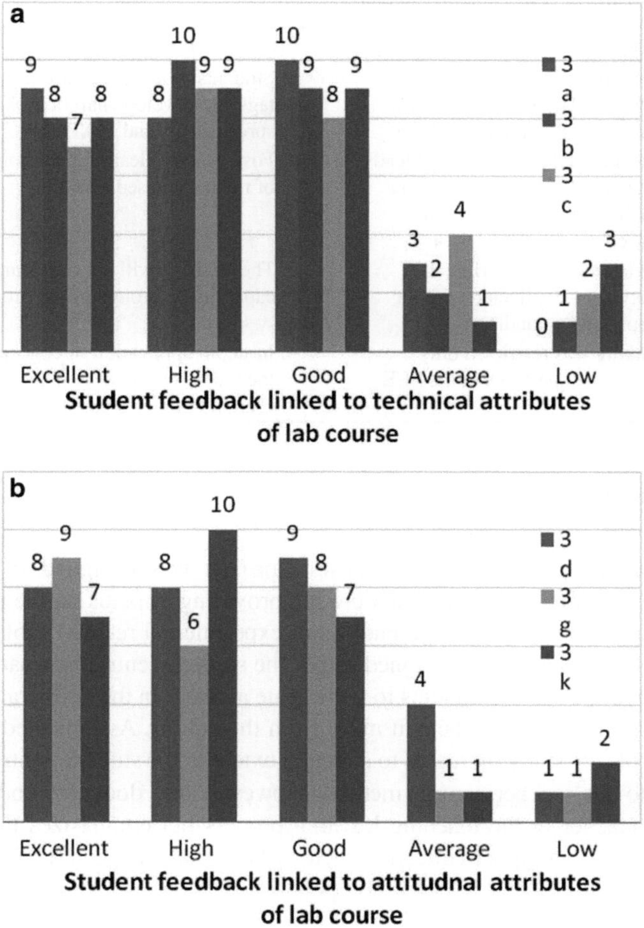

Fig. 1 (**a**) Student feedback linked to heat and mass transfer lab course (criteria a, b, c, and e). (**b**) Student feedback linked to heat and mass transfer lab course (criteria d, g, and k)

group data with the help of an equation editor. Each student of the batch is suggested to analyze the recorded data of the whole batch and to plot graphs using a modern computer tool and write the conclusion (ABET 3a, 3b, 3g, and 3k).

3 Comparison Between Different Pedagogical Approaches

This section presents the comparison between the teacher-centric approach and student-centric approach with reference to the heat transfer course. The study is presented in terms of the content, delivery, and evaluation strategies adopted for the heat transfer course taught to mechanical engineering graduates at sixth semester level.

Table 1 Comparison of pedagogical approaches

Teacher-centric	Student-centric
1. Every student performs prescribed experiments, standard set of specimen calculations with little lateral thinking	1. Students motivated to use *what-if* and *why* tags and develop analytical skills through spreadsheets and MATLAB
2. Simple experimental procedure that leads to "memorization by repetition" or rote learning as the evaluation was mainly memory-based	2. Fosters active learning environment instead of memory-based learning
3. Less active learning as experiments were stereotyped—conducted without much in-depth understanding	3. The students will develop teamworking capabilities through group projects
4. Lateral thinking was restricted only to a few viva voce questions in the SEE	4. In-depth application of concepts learned in the theory course as demanded by sections C and D of new syllabus

3.1 Content

The syllabus as per the teacher-centric approach was designed to perform a stereotyped set of experiments that were not providing impetus on the use of computational tools to make detailed analysis of experimental results. Table 2 given in Annexure provides syllabus designed as per the student-centric approach that gave an adequate scope for the students to participate actively in the deliberations during the lab session and thereby benefit more from the course. As indicated earlier, the component of self-study increases to a greater extent in the student-centric approach compared to the teacher-centric method. However, this does not undermine the role of the teacher in the teaching-learning process but emphasizes the changing trends in the present higher education sector where students should be prepared for self-learning.

3.2 Course Delivery

The comparison between course deliveries through the two approaches is presented in Table 1, highlighting the benefits of the proposed scheme. It is clear that the student-centric approach creates a better ambience for assimilation and retention of the course content. It also alleviates the teaching-learning process to higher levels of Bloom's taxonomy as compared to the teacher-centric traditional method. In the traditional approach, students were made to strictly follow the instructions and obtain the results, while the proposed method provides a 50 % (15 h) duration for the course instructor to provide the necessary technical support as the remainder of the allotted duration is utilized to develop competency through the self-learning component.

3.3 *Course Evaluation*

The teacher-centric approaches involved a lesser degree of student involvement due to the limited number of assessment levels that were again memory dominated rather than honing the skill sets of students. On the contrary, the student-centric evaluation scheme provided myriad milestones for the evaluation as it included computational skills and also addressed contemporary issues for which the student may be motivated to read research publications from reputed journals in the discipline of thermal engineering. The evaluation of the teaching-learning process followed in two approaches clearly distinguishes the student-centric approach from its counterpart as being more effective, leading to better performance and satisfaction for the student.

4 Additional Benefit of Pedagogical Reforms

The process of learning should engage the imagination—both of students and of the faculty. The passive lecture mode was transformed into an intense, active, personalized, and highly collaborative active learning. The more flexible mode of learning stimulates discovery and improves understanding of conceptual material. The student-centric teaching practice resulted in better clarity in concepts among students as well as a higher level of realization in Bloom's taxonomical levels. The active learning ensured greater student participation in the teaching-learning process. The evolved experience through implementation of Bloom's taxonomy over the years improved modus operandi of delivering this course as a theory and practical course and motivated us to present our experience in heat and mass transfer through a well-researched textbook on "heat and mass transfer." The proposed textbook gives a methodological approach through the concepts of structured enquiry and open-ended problems in heat and mass transfer.

5 Conclusions

The pedagogical ordeal clearly indicates that outcome-based education is the need of the hour for the higher education sector to address the challenges posed by the industrial sector to the academia. The technology-enabled teaching environment with more avenues for knowledge exploration demands a more interactive teaching-learning environment compared to the earlier generation. The techno-savvy student community does not restrict learning only to the portals of the classroom but expects a more activity-based interactive teaching-learning process. The courses like heat transfer provide adequate scope to address most of the 3a–3 k criteria specified by ABET.

Acknowledgment We are grateful for the support by our beloved principal, Dr. Ashok Shettar, in taking initiatives to promote outcome-based education on campus.

6 Annexure

Table 2 Course syllabus: teacher-centric approach

Course code: MEL610	Course title: heat and mass transfer lab	
L-T-P: 0-0-1	Credits: 1	Contact hours: 02 h/week
CIE marks: 50	SEE marks: 50	Total marks: 100
Teaching hours: 30 h	Exam duration: 3 h	

1. Determination of thermal conductivity of an insulating powder
2. Determination of thermal conductivity of metal rod
3. Determination of thermal conductivity of liquid
4. Determination of thermal conductivity of composite wall apparatus
5. Determination of heat transfer effectiveness of a fin
6. Determination of heat transfer coefficient for the free convection
7. Determination of heat transfer coefficient for the forced convection
8. Determination of emissivity of a surface
9. Parallel/counter flow heat exchangers: determination of LMTD and effectiveness
10. Demonstration experiments on
    (a) Condensation heat transfer
    (b) Refrigeration and air conditioning 15 h

Scheme of continuous internal evaluation (CIE)
Major experiments: 25 marks
Minor experiments: 15 marks
Journal and viva voce: 10 marks
Total marks: 50 marks
Scheme of semester end examination
One question from major experiments: 20 marks
One question from minor experiments: 20 marks
Viva voce: 10 marks
Total: 50 marks

Table 3 Course syllabus: student-centric approach

Course code: MEL610	Course title: heat and mass transfer lab	
L-T-P: 0-0-1	Credits: 1	Contact hours: 02 h/week
CIE marks: 80	SEE marks: 20	Total marks: 100
Teaching hours: 30 h	Exam duration: 3 h	
A. Demonstration (2 sessions)		
1. Construction and calibration of thermocouple junctions		
2. Air and water flow measurements		
3. Thermal conductivity of metals, insulating materials, and liquids		
4. Determination of free and forced heat transfer coefficient		
5. Determination of emissivity of a surface		
B. Exercise (4 sessions)		
1. Temperature dependence of thermal conductivity		
2. Investigation of free/forced convection on effectiveness of fin		
3. Flow dependence of heat transfer coefficient		
4. Investigation of fluid dependence on heat transfer in HEs		
C. Structured enquiry (6 sessions)		
1. Design "cylindrical container insulation" to minimize heat loss		
2. Investigation of combined modes of heat transfer		
3. Experimental investigation of 2D temperature distribution in a heated plate and comparison with analytical solution		
D. Open-ended enquiry (6 sessions)		
1. Analyze different designs available for exhaust gas recirculation and radiators in passenger cars		
Evaluation scheme for CIE and SEE		
Assessments	Weight-age in marks (CIE)	Weight-age in marks (CIE)
Section A	10	05
Section B	20	05
Section C	30	10
Section D	20	10
Total	**80**	**20**

References

1. DeWitt DP, Figliola RS. Heat transfer education: integration of the thermal sciences stem. Purdue heat transfer celebration, West Lafayette, 3–5 Apr 2002, pp. 61–6
2. http://web.mit.edu/edtech/casestudies/teal.html
3. Feisel LD, Rosa AJ. The role of the laboratory in under graduate engineering education. J Eng Educ. 2005;94:121–30.
4. Rugarcia A, Felder RM, Woods DR, Stice JE. The future of engineering education – a vision for a new century. Chem Eng Educ. 2000;34(1):16–25.

Enhanced Learning Through Self-Study Component in Engineering Education

Saroja V. Siddamal, Suneeta V. Budihal, Ujwala Patil, Uma Mudenagudi, Rajeshwari Banakar, Vijaya Eligar, Shivaraj Hublikar, and Shivashankar Huddar

Abstract The paper presents Self-Study (SS) component as a pedagogical tool for enhanced learning in a set of identified course. The objective of the SS component is to enhance the understanding of the course beyond the classroom teaching. The traditional modes of course design, course delivery and course assessment provide less scope for in-depth learning of courses. Outcome-Based Education (OBE) is an effort to overcome limitations of traditional education by using many progressive pedagogical models, ideas, to meet *a-k* student outcomes of ABET and NBA. To meet the expectations of present engineering education, few courses of Electronics and Communication engineering programme are modified using SS component. Typical methods of SS component are course project, term paper/seminar, field work, modelling and simulation, case studies, proof of concept/prototype development, etc. The attainment of depth of learning is demonstrated using two activities, namely, course project and term of SS component. The student outcomes are assessed by using designed set of rubrics for each course-SS combination. The tradition mode of delivery facilitates the attainment of technical outcomes of ABET, whereas along with technical outcomes professional outcomes are also achieved through SS component. The assessment of attainment of student outcomes and student feedback of SS component act as inputs for the continuous improvement of teaching-learning process.

Keywords Self-study (SS) component • a-k student outcomes of ABET • Term paper • Course project • Evaluation rubrics

S.V. Siddamal (✉) • S.V. Budihal • U. Patil • R. Banakar • V. Eligar
S. Hublikar • S. Huddar
BVB College of Engineering and Technology, Vidya Nagar, Hubli, Karnataka 580031, India
e-mail: sarojavs@bvb.edu; suneeta_vb@bvb.edu; ujwalapatil@bvb.edu; banakar@bvb.edu; vijaya@bvb.edu; shivaraj@bvb.edu; shivashankarhuddar@bvb.edu

U. Mudenagudi
Department of Electronics and Communication Engineering, BVB College of Engineering and Technology, Vidya Nagar, Hubli, Karnataka 580031, India
e-mail: uma@bvb.edu

R. Natarajan (ed.), *Proceedings of the International Conference on Transformations in Engineering Education: ICTIEE 2014*, DOI 10.1007/978-81-322-1931-6_53,

1 Introduction

The paper address in-depth learning in a set of courses using SS activity as a pedagogical tool in Electronics and Communication engineering programme. Traditional course design and delivery includes theoretical concepts, tutorials and supported by a set of experiments. However, these activities fail to provide depth of learning and assessment is limited to technical outcomes. In most cases, capstone projects carried out in the final of are used to evaluate attainment of student outcomes a to k in an outcome-based education framework proposed in ABET and NBA [1–4]. Miniprojects [5–8] carried out in the prefinal year of graduation are used to enhance the depth of learning and to achieve professional outcomes. Due to broad base of the curriculum, these projects may not provide in-depth learning in many courses and also spectrum of identified problems is much wider which poses difficulty in the assessment of student outcome [9–11]. These methods may not provide sufficient resolution at the course level for the curriculum improvement. There are several ways of engineering course assessments like teacher/ course evaluations, homework, tests, minor and semester end exams. These methods provide less scope for assessing the extent of in-depth learning in the course. SS components provide useful student assessment data at course level resolution which are used as inputs for continuous curriculum improvement. SS component is introduced at III-VI semester of ECE programme to strengthen the course learning objectives as a part of course design, which can provide better assessment of student outcomes.

The paper demonstrates the following:

- In-depth learning in four courses using course project and term paper as SS activity.
- Attainment of professional outcomes along with technical outcomes.
- Analysis of attainment of SS and survey of SS activity can be used for continuous improvement and curriculum modifications.

The framework of SS component is discussed in Sect. 2. Continuous improvement of teaching-learning process is discussed in Sect. 3. Conclusions are provided in Sect. 4.

2 Framework of the SS Component

The primary objective of the SS component is to enhance the understanding of the course beyond the classroom teaching. The self-study is categorized as course project, term paper/seminar, field work, modelling and simulation, case studies, proof of concept/prototype development. In 2011, the department designed curriculum structure and self-study activity is offered for three courses per semester. Depending on the course content, the SS activity and associated rubrics to evaluate student outcomes are identified. The activity identified for Pulse Digital Circuits and Data Structures with C of III semester is course project. For Digital Signal Processing (DSP), Information Theory Coding (ITC) of V and VI semester is term paper. Continuous evaluation of the SS component is carried out by the respective course instructor in 2–4 reviews. The evaluation of each phase is done based on the rubrics. The review process in the continuous evaluation measures the student learning using defined rubrics

2.1 *Guidelines for the SS Component*

The activity begins 2 weeks after the commencement of the semester and continues till the end of the semester. The implementation of the self-study component was spread over 2–4 phases during the entire semester. The SS component needs to encompass the concepts learnt in the course. Each batch is given a problem statement based on the application areas of the course. At the beginning of SS component, the students are made familiar with the evaluation rubrics. SS component is designed equivalent to 1 credit with a time effort of 10–12 h.

2.2 *Evaluation of SS Component*

The evaluation of CIE consists of 2–4 reviews that are conducted by the course instructor. The basic criteria for evaluation include an ability to understand the problem and carry out the literature survey, and ability to work as a team member. It also includes quality of work, documentation, report writing and presentation skills. The example rubrics followed for Pulse and Digital Circuits are as follows:

- Review 1: Understanding the problem definition, block diagram representation.
- Review 2: Identifying multiple solutions, selecting the best suited solution with justifications, simulation of any five digital circuits using available open source tools.
- Review 3: Design on paper with listing of the required components, simulate the same using available open source tool.
- Review 4: Hardware implementation, presentation and demonstration of results, report submission.

The basic purpose is to assess the student competencies in understanding the problem definition, identifying the alternate solution and the selection of optimal solution. More specifically to assess the student's individual contribution to the SS component, to establish the level of understanding of basic theoretical knowledge relevant to the course.

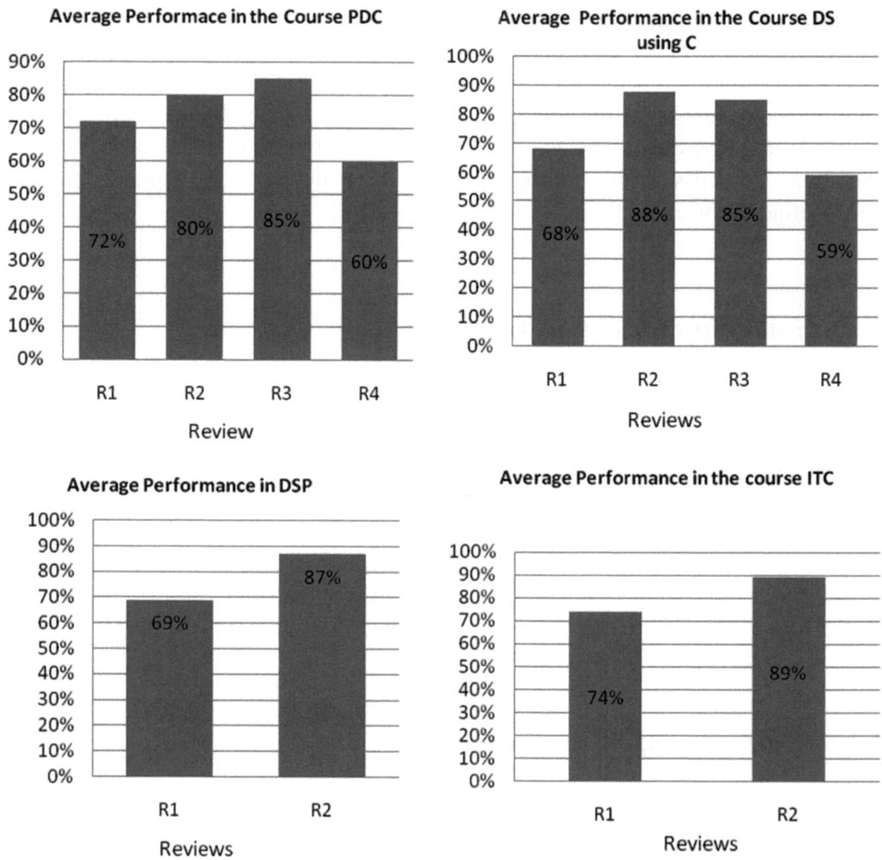

Fig. 1 Performance of students in SS activity offered in four courses

The course instructor has the responsibility of educating the students about the criteria/standards being used for course project development and implementation. The course instructor adopts a clear and consistent pattern of asking questions from general to specific aspects of the course project.

Performance of students in SS activity offered in four courses is shown in Fig. 1. For example in the PDC course, the student performance in Review-1(R1) and Review-4(R4) is comparatively low. The Review-1 data shows that student needs to improve skills in problem identification and block diagram representation. As reflected in Review-4, student's capability in hardware implementation, presentation, demonstration of results and documentation needs improvement. Review-2(R2) and Review-3(R3) data show student's performance is good in identifying multiple solutions, simulation and design on paper.

3 Continuous Improvement

The continuous improvement of the course and curriculum modification can be achieved through analysis of attainment of student outcomes and student survey.

3.1 Evaluation of Student Outcomes

Evaluation rubrics are used to assess student outcomes. The review process in the continuous evaluation measures the student learning against the student outcomes as per the a-k program outcomes of ABET. The rubrics are framed to assess the professional outcomes along with technical outcomes which are otherwise difficult with regular course design. The student's performance is assessed by course instructor using the assessment rubrics and average score for each of the outcomes is normalized on a scale of 1–10. Figure 2 demonstrates the attainment of students outcomes g, i, j, k, g which is otherwise difficult to achieve in regular course without SS component. SS component enhances the professional outcomes like lifelong learning, communication skills and leadership qualities of students.

3.2 Student Survey

A set of survey questions are given to students to know the effectiveness of SS activity. The survey questions are framed to measure the perceptual or subjective achievement in Pulse and Digital Circuits course and they are as follows:

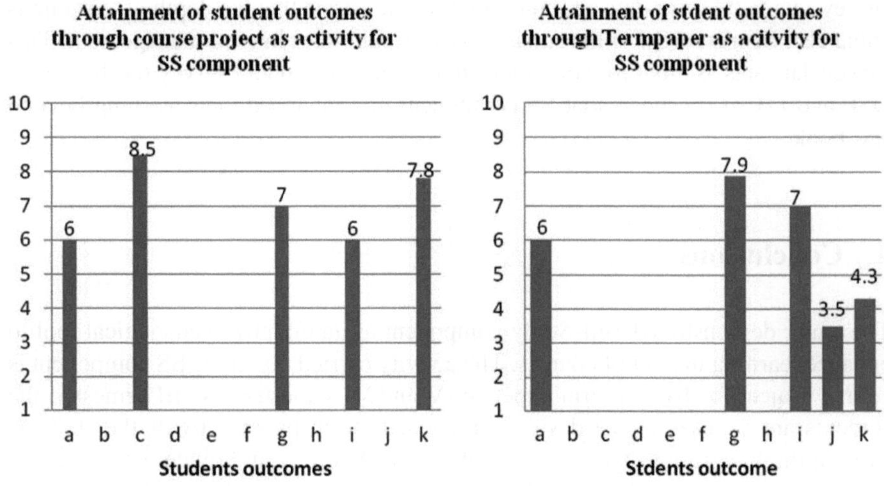

Fig. 2 Attainment of student outcome through SS component

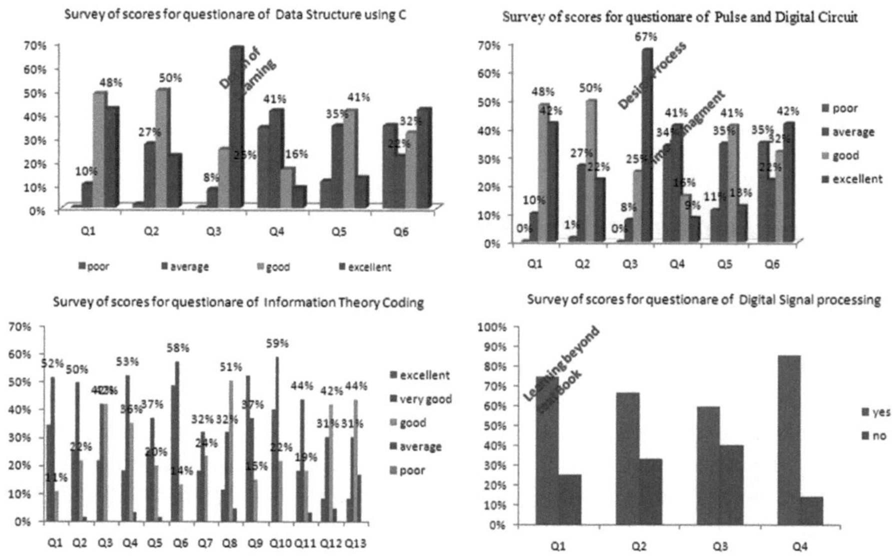

Fig. 3 Survey of SS activity offered in four courses

1. To use learnt theory in solving a problem
2. To gain extra knowledge about the course
3. To explore modern engineering tools
4. To utilize time management to carry out the project
5. To develop the ability to work in a team
6. To develop technical writing skills and presentation skills

The survey of SS activity offered in four courses is presented in Fig. 3. The survey scores (Q3) for data structure, PDC demonstrate that the depth of learning is enhanced. Scores of Q4 of the same survey reveal issues in time management. This survey data acts as input for continuous improvement. Student survey for the course DSP and ITC (Q1) shows that SS component has enhanced their learning beyond text book.

4 Conclusions

The paper demonstrated Self-Study component as an effective pedagogical tool to enhance learning in a set of courses. The activity carried out in the SS component is course project for III and term paper for V and VI semesters. At III semester, the students are exposed to the design approach. In term paper activity, the students explore the literature to know state-of-the-art technology and relate the classroom learning to the practical issues. The evaluation rubrics are designed to meet the objectives. The assessment was carried out as per the rubrics designed for attainment of a-k. The criteria a, c, i, g, k were achieved through SS component catering to professional outcomes along with technical outcomes. The student feedback indicates good evidence of effectiveness of the approach. The student outcomes and feedback act as inputs to the continuous improvement of teaching-learning process.

References

1. Accreditation Board for Engineering and Technology (ABET). Engineering criteria 2000: criteria for accrediting engineering programs, effective for evaluations during the 2002–2003 accreditation cycle, Baltimore. http://www.abet.org/images/Criteria/2002–03EACCriteria.pdf. November 2001.
2. http://www.nba-aicte.ernet.in/manual.html
3. Bradley University Electrical Computer Engineering Department. Capstone project: evaluation relative to professional level. http://cegt201.bradley.edu/~gld/prof_cap.html. August 2002.
4. Felder RM, Brent R. Designing and teaching courses to satisfy the ABET engineering criteria. J Eng Educ. 2003;92(1):7–25.
5. Dempsey GL, Huggins BD, Anakwa WKN. Use of senior mini-project for electrical and computer engineering curriculum assessment. In: Proceedings of the 2001 ASEE annual conference exposition, Albuquerque, June 2001, Session 3232. http://www.asee.org/conferences/search/01078_2001.pdf
6. Dempsey GL. EE451: mini-project. http://cegt201.bradley.edu//~gld/miniproj.html. August 2002.
7. Basterfield-Sacre M, Shuman L, Wolfe H, Atman C, McGourty J, Miller R, et al. Defining the outcomes: a framework for EC-2000. IEEE Trans Educ. 2000;43:100–9.
8. Bryden KM, Flugrad DR. Implementing a program of continuous assessment and improvement for a new sophomore design course. In: Proceedings of the 2000 ASEE annual conference exposition, St. Louis, MO, June 2000, Session 1625. http://www.asee.org/conferences/search/20294.pdf
9. Nicoletti D, Orr J. An implementable/sustainable outcomes assessment process for an electrical engineering program. In: Proceedings of the 2001 ASEE annual conference exposition, Albuquerque, June 2001, Session 2532. http://www.asee.org/conferences/search/00475_2001.pdf
10. Ashok Shettar UM, Desai BL. Mini-projects: an impact on student learning and curriculum design. In: Proceedings of the 1st world summit on accreditation, National Board of Accreditation, 25–28 Mar 2012, New Delhi.
11. Junge B, Quiñones C, Kakietek J, Teodorescu D, Marsteller P. Promoting undergraduate interest, preparedness, and professional pursuit in the sciences: an outcomes evaluation of the SURE program at Emory University. PMC2879378.

Mapping Graduate Attributes of NBA with the Program Outcomes of the ABET/OBE to Establish Consistency Between the Two

B.B. Kotturshettar and A.S. Shettar

Abstract The National Board of Accreditation (NBA) in India has come up with a new requirement in its document published in January 2013 – which is a complete overhaul of what it was a few years ago. The expectations of NBA for an engineering educational institution are now driven by graduate attributes – the qualities that a graduating student has to demonstrate for having possessed through the engineering education. The graduate attributes mentioned in the NBA document are comparable with the more familiar program outcomes (POs) – the term used in Outcome-Based Education (OBE) framework and also in the ABET document to measure the attainment status of competence of professionals. The competence is also measured subsequently through Program Educational Objectives (PEOs) after 3–5 years of graduation which is beyond the present scope. Many engineering institutes in India have started work on implementing OBE framework with program outcomes as the measure of attainment of attributes expected of a graduating engineer. Such institutes are in an ambiguous state as to how they can switch their processes to meeting the new NBA requirement of addressing graduate attributes from the program outcomes that they have acquired familiarity with. The paper shows how the graduate attributes can be mapped to POs in institutes offering undergraduate engineering education so that one can continue with the established processes of measuring the attainment status of program outcomes and also meet the requirements of NBA.

Keywords NBA • Graduate attributes • Program outcomes • OBE framework

B.B. Kotturshettar (✉)
Department of Industrial and Production Engineering, BVB College of Engineering and Technology, Vidya Nagar, Hubli, Karnataka 580031, India
e-mail: bbkshettar@gmail.com

A.S. Shettar
Department of Civil Engineering, BVB College of Engineering and Technology, Vidya Nagar, Hubli, Karnataka 580031, India
e-mail: ashoksshettar@gmail.com

R. Natarajan (ed.), *Proceedings of the International Conference on Transformations in Engineering Education: ICTIEE 2014*, DOI 10.1007/978-81-322-1931-6_54,

1 Introduction

With India after having become signatory to the Washington accord, the National Board of Accreditation in India had to shift the accreditation paradigm from erstwhile input focus to emphasizing on outcome based. It was not easy in the contemporary scenario in the country owing to the complexities of institutions offering engineering education (offering Diploma, UG, and PG programs). There are affiliated colleges – government, aided, and unaided with varying degree of delivery potential, autonomous institutions – a few of them with good length of experience while many of them novice requiring time to switch over to the new found freedom in the academic space, deemed universities and the private universities. The NBA had daunting task to arrive at an accreditation framework that could encompass the spectrum of institutions. The final document that was published in January 2013 has addressed this issue to an extent by identifying tier-I and tier-II categories for autonomous institutions, university departments, and the affiliated institutions respectively. While engineering educators in India are embracing, although with some apprehensions, the substantial change set forth by NBA in the light of Washington accord, there is much concern as to how to best operationalize each graduate attribute for use within one's own institution. Graduate attributes form a set of individually assessable outcomes that are the components indicative of a graduate's potential to acquire competence to practice at the appropriate level as specified in criteria 2. The discussions in the following paragraphs will be on the implementation details associated with the tier-I category institutions.

2 Graduate Attributes

All graduates of any higher education programs are expected to have identified with technical/functional, generic, and managerial competencies. The competencies that a graduate of a program should have are called graduate attributes. The attributes that a graduating engineer should have are generally identified by the accreditation agency for engineering and technical education, namely, National Board of Accreditation in India. The graduate attributes of engineering programs are as follows [1].

1. Engineering knowledge: Apply the knowledge of mathematics, science, engineering fundamentals, and an engineering specialization to the solution of complex engineering problems.
2. Problem analysis: Identify, formulate, research literature, and analyze complex engineering problems reaching substantiated conclusions using first principles of mathematics, natural sciences, and engineering sciences.
3. Design/development of solutions: Design solutions for complex engineering problems and design system components or processes that meet the specified needs with appropriate consideration for the public health and safety, and the cultural, societal, and environmental considerations.

4. Conduct investigations of complex problems: Use research-based knowledge and methods including design of experiments, analysis and interpretation of data, and synthesis of the information to provide valid conclusions.
5. Modern tool usage: Create, select, and apply appropriate techniques, resources, and modern engineering and IT tools and prediction and modeling to complex engineering activities with an understanding of the limitations.
6. The engineer and society: Apply reasoning informed by the contextual knowledge to assess societal, health, safety, legal and cultural issues and the responsibilities relevant to the professional engineering practice.
7. Environment and sustainability: Understand the impact of the professional engineering solutions in societal and environmental contexts, and demonstrate the knowledge of, and need for sustainable development.
8. Ethics: Apply ethical principles and commit to professional ethics and responsibilities and norms of the engineering practice.
9. Individual and team work: Function effectively as an individual, and as a member or leader in diverse teams, and in multidisciplinary settings.
10. Communication: Communicate effectively on complex engineering activities with the engineering community and with society at large, such as, being able to comprehend and write effective reports and design documentation, make effective presentations, and give and receive clear instructions.
11. Project management and finance: Demonstrate knowledge and understanding of the engineering and management principles and apply these to one's own work, as a member and leader in a team, to manage projects and in multidisciplinary environments.
12. Life-long learning: Recognize the need for, and have the preparation and ability to engage in independent and life-long learning in the broadest context of technological change.

The graduate attributes as stated by NBA are in the spirit of program outcomes as stated in Washington Accord (*graduate attributes are referred to as program outcomes by Washington Accord*).

3 Program Outcomes

The POs are observable and measurable manifestations of applied knowledge. They describe what students are expected to know and be able to do by the time of graduation. These relate to the knowledge, skills, and behaviors that students acquire as they progress through the program. Hence, the 11 outcomes serve as a foundation for all engineering programs, but each program must then define itself by adding its own specificity to the outcomes [2]. The POs for the Industrial and Production Engineering Program are defined to be in alignment with the department objectives. They are as follows.

Graduates of the Industrial and Production Engineering Program are expected to have the following:

(a) Ability to apply knowledge of mathematics, science, and engineering, to model and analyze manufacturing and industrial engineering problems.
(b) Ability to design and perform laboratory experiments for manufacturing and allied systems as well as to analyze and interpret data.
(c) Ability to design systems, components, or processes to meet customer needs.
(d) Ability to participate effectively in multidisciplinary team.
(e) Ability to identify, formulate, and solve manufacturing/industrial engineering problems.
(f) Ability to demonstrate his understanding of professional and ethical responsibility while assessing a situation.
(g) Ability to communicate effectively in both oral and written forms and to become proficient in working with diverse teams.
(h) Broad understanding of the impact of engineering in a global, economic, environmental, and societal context.
(i) Ability to engage in lifelong learning that will help him/her for growth in professional career.
(j) Knowledge about contemporary issues in engineering.
(k) Ability to use modern modeling and simulation techniques, and computing tools.
(l) Ability to engineer projects from P to P (*i.e., from plan to prototype/product*) with adept management of 5 Ms (*i.e., man, material, machine, money, and minute*).

4 Methodology

As we see, the POs are much the same in spirit of the graduate attributes of NBA. The NBA document further expects that the POs must foster the attainment of PEOs. It is here that there is a need to establish clarity and an unambiguous relationship between the GAs and POs so that we meet the expectations of NBA unequivocally.

Much before the release of new NBA document, the department in line with the institutional strategy followed through a process to define the POs for the program and also implemented the assessment mechanism to measure their attainment.

While defining POs it was ensured that PEOs were taken note of so that the attainment of POs leads to the attainment of PEOs.

After the NBA came up with desirable GAs, the already defined POs were revisited, an additional PO to address the attainment of knowledge and skill on project management and finances was introduced as PO (l) and mapping of GAs and POs was carefully done to ensure that all the POs are in alignment with the GAs, the details of which are discussed in the following sections.

Table 1 Graduate attributes and expected competence from graduate attributes

GA-#	Graduate attribute	Expected competence
GA-1	Engineering knowledge	Solve complex engineering problems
GA-2	Problem analysis	Analyze complex engineering problems and reach substantiated conclusions
GA-3	Design/development of solution	Develop solution to complex engineering problems
GA-4	Conduct investigation of complex problems	Provide valid conclusions to the complex problems after investigation
GA-5	Modern tool usage	Model complex engineering activities and predict the solution
GA-6	The engineer and society	Assess societal, health, safety, legal, and cultural issues and the consequent responsibilities relevant to the professional engineering practice
GA-7	Environment and sustainability	Demonstrate the knowledge of and need for sustainable development
GA-8	Ethics	Commit to professional ethics, responsibilities, and norms of the engineering practice
GA-9	Individual and team work	Accomplish as a team member/leader the intended task of solving complex problems
GA-10	Communication	Communicate effectively on complex engineering activities with the engineering community and with society at large
GA-11	Project management and finance	Manage projects in multidisciplinary environments
GA-12	Life-long learning	Understand the requirement and offer solution to complex problems in the broadest context of technological change

The expected competence in graduate on attainment at the time of graduation from each of the graduate attributes is extracted and listed in Table 1.

As a first step, each GA is subdivided into plausible elements, i.e., “Attribute Elements” representing an ability expected from a graduate to attain the competence as shown in Table 2.

The next step includes subdividing POs into plausible elements, i.e., “Outcome Elements (OEs)” for each of the POs defined by the IPE department. Table 3 lists elements representing abilities expected from graduates in POs. The abbreviation of OE is used in our explanation to indicate the outcome element of the POs defined for the department.

The process of dividing GAs into “Attribute Elements” and POs into “Outcome Elements” allows for sufficient resolution to map POs with GAs.

The last step of mapping of “Attribute Elements” with “Outcome Elements” was carried out and documented as in Table 4 to ensure that the defined POs address all the GAs.

From the abovementioned example in Table 3, it is evident that all the elements of the GAs are addressed by the POs of the IPE program.

Table 2 Expected competence from graduate attributes and attribute elements

GA-#	Attribute elements
GA-1	*The graduates are expected to attain this competence, when they have the ability to* Apply the knowledge of mathematics; apply the knowledge of science; apply the knowledge of engineering fundamentals and domain knowledge
GA-2	Identify the engineering problems; formulate the problem; research relevant literature, use principles of mathematics, science, and engineering fundamentals
GA-3	Appropriately consider specified needs, give due considerations to public health, safety, and cultural and environmental issues; design systems, components, or processes that meet the specified needs
GA-4	Use research-based knowledge and research methods; carry out design of experiments, analyze and interpret data; synthesize the information
GA-5	Create, select, and apply appropriate techniques to predict and model complex engineering activities; allocate appropriate resources; use modern engineering and IT tools; understand the limitations of tools and techniques
GA-6	Apply knowledge to assess societal, health, safety, legal, and cultural issues related to engineering practice; understand professional responsibilities of engineering practice
GA-7	Understand the impact of professional engineering solutions in societal and environmental context; demonstrate the knowledge of sustainable development
GA-8	Apply ethical principles in his endeavors; commit to professional ethics, take up responsibilities and practice norms of the engineering
GA-9	Work effectively as an individual and as a team member in diverse teams in multidisciplinary settings; function as effective team leader
GA-10	Communicate effectively with engineering community and society; comprehend and write effective reports and design documents; make effective presentations; give and receive clear instructions
GA-11	Demonstrate knowledge and understanding of the engineering and management principles; apply knowledge and understanding to one's own work, as a member and leader in a team, to manage projects and in multidisciplinary environments
GA-12	Recognize the need for lifelong learning; have the preparation and ability to engage in independent and life-long learning

Figure 1 helps visualize the process of measurement of the attainment of GAs. Table 5 shows the mapping of GAs with the POs at a glance through the corresponding elements. Now it is clear that as we assess the courses for accomplishing the course outcomes by measuring the extent of attainment through the corresponding outcome elements of program outcome, the assessment of program outcomes in turn happens by combining the attainment in one or more contributing outcome elements. Thus, one can realize that the expected attainment of desired graduate attribute has happened with the attainment of correspondingly mapping program outcome.

A methodology for assessing the program outcomes has been established by the IPE program accreditation committee. For each of the courses a set of course outcomes (COs) are defined and mapped with the POs. A strategy for assessing individual COs is developed by choosing appropriate assessment tool. Overall success in attainment of each outcome is identified by analyzing combination of individual course reports.

The college NBA executive committee carries out detailed analysis of each GA specified by the NBA for the implementation process and attainment status.

Table 3 Program outcomes and outcome elements

PO(#)	Outcome element(s)
a	Formulate and solve mathematical models that describe the behavior and performance of physical systems and processes of manufacturing and industrial engineering
	Use basic scientific and engineering principles to identify applications, explain and analyze the performance of processes and systems of manufacturing and industrial engineering
b	Design an experiment to verify the conceptual understanding
	Conduct (or simulate) an experiment and report the results
	Analyze a set of experimental data
	Interpret the results
c	Analyze the need and understand the customer expectations
	Develop system/process for the problem as part of solution
	Build prototype of a design and demonstrate that it meets performance specifications
d	Participate in team activities as member(s) in discussions and consolidate the ideas
e	Identify gaps in a process or product in the domain
	Formulate the identified problem
	Solve the problem
f	Assess the situation that requires a decision on ethical implications and professional acumen
g	Develop written and graphical communication skills appropriate to the profession of engineering
	Demonstrate oral and visual communication skills appropriate to the profession of engineering
h	Degree of awareness of the global, economic, environmental, and societal impact of engineering solutions
i	Find relevant sources of information about a specified topic and meet the challenges for growth in career
j	Knowledge of current events, developments, and issues in the technical and nontechnical space at the regional, national, and global level
k	Competence to use techniques, skills, and modern engineering tools
l	Competence to adopt technical knowledge and managerial skill in planning projects and deployment of resources
	Lead projects of diverse nature demanding multidisciplinary team efforts to completion on time, within estimated cost and to the satisfaction of stakeholders

Table 4 Mapping of graduate attributes through attribute elements with program outcomes through outcome elements: an example

GA-1	Engineering knowledge: solve complex engineering problems		
AE	*Attribute elements*	*OE*	*Outcome elements*
AE-1.1	Apply the knowledge of mathematics	OE (a)-i	Formulate and solve mathematical models that describe the behavior and performance of physical systems and processes of manufacturing and industrial engineering
AE-1.2	Apply the knowledge of science	OE (a)-ii	Use basic scientific and engineering principles to identify applications, explain and analyze the performance of processes and systems of manufacturing and industrial engineering
AE-1.3	Apply the knowledge of engineering fundamentals and the domain knowledge	OE (a)-iii	

Table 5 Mapping at a glance

	OE(a)-i	OE(a)-ii	OE(a)-iii	OE(b)-i	OE(b)-ii	OE(b)-iii	OE(b)-iv	OE(c)-i	OE(c)-ii	OE(c)-iii	OE(d)-i
AE-1.1	✓										
AE-1.2		✓									
AE-1.3			✓								
AE-2.1											
AE-2.2											
AE-2.3											
AE-2.4											
AE-3.1								✓			
AE-3.2									✓		
AE-3.3										✓	
AE-4.1				✓							
AE-4.2					✓						
AE-4.3						✓					
AE-4.4							✓				
AE-5.1											
AE-5.2											
AE-5.3											
AE-5.4											
AE-6.1											
AE-6.2											
AE-7.1											
AE-7.2											
AE-8.1											
AE-8.2											
AE-8.3											
AE-9.1											✓
AE-9.2											✓
AE-10.1											
AE-10.2											
AE-10.3											
AE-10.4											
AE-11.1											
AE-11.2											
AE-12.1											
AE-12.2											

OE(e)-i	OE(e)-ii	OE(e)-iii	OE(f)-i	OE(f)-ii	OE(g)-i	OE(g)-ii	OE(h)-i	OE(i)-i	OE(j)-i	OE(k)-i	OE(l)-i	OE(l)-ii
✓												
	✓											
	✓											
		✓										
										✓		
										✓		
										✓		
										✓		
							✓					
				✓								
							✓					
									✓			
			✓									
			✓									
			✓									
					✓							
					✓							
						✓						
						✓						
											✓	
												✓
								✓				
								✓				

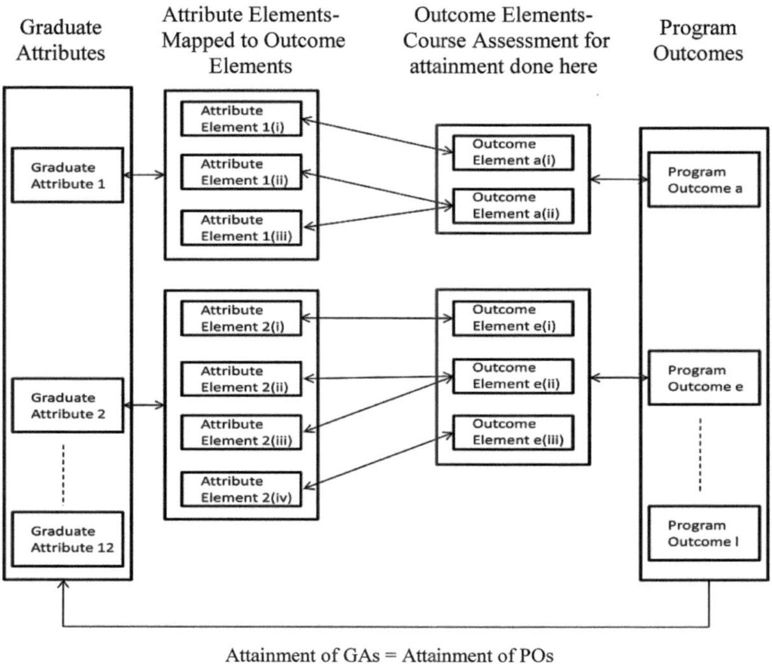

Fig. 1 Pictorial representation of attainment of graduate attributes through program outcomes

5 Conclusion

The present approach is expected to allay the apprehensions about establishing or adapting to the established processes towards attainment of GAs. This is enabled through mapping GAs with POs through corresponding elements giving more clarity for the faculty members to design an appropriate delivery mechanism and select proper tools for assessment. The entire exercise of outcomes' assessment would be more enjoyable and satisfying for faculty members as they understand the intent of the whole exercise and witness their students demonstrate the acquired knowledge and skills more convincingly by way of solving complex problems that they encounter even during their graduating period and also after graduation.

References

1. National Board of Accreditation, India. Accreditation manual for UG engineering programmes (Tier-1). 2013.
2. Besterfield-Sacre ME, Shuman LJ, Wolfe H, Atman CJ, McGourty J, Miller RL, Olds BM, Rogers GM. Defining outcomes: a framework for EC-2000. IEEE Trans Educ. 2000;43(2):100–10.

Integrated Experience: Through Project-Based Learning

R.M. Shet, Nalini C. Iyer, P.C. Nissimgoudar, and S. Ajit

Abstract Project-based learning (PBL) is an instructional method in which students learn a range of skills and subject matter in the process of creating their own projects. Sometimes, these projects are solutions to a real-world problem. But what is most important in project is that students gain the design experience. They work in groups and bring their own experiences, abilities, learning styles, and perspectives to the project. Students work in teams under the direction of a faculty advisor to tackle an engineering design project. Engineering communication, such as reports and oral presentations are covered. We emphasize practical, hands-on experience, and integrate analytical and design skills acquired in companion senior-level core courses. In this paper, we propose to introduce the project-based learning approach throughout the course at various levels and its impact at learning ability and employability. This experiential learning helped the students to improve comprehensive application ability and innovative consciousness.

Keywords Program outcomes • SEE • CIE • Course project • Mini projects • Capstone project

1 Introduction

The dominant pedagogy for engineering education was "chalk and talk," but due to the constant change in the innovative product-driven society the approach toward the self-study or implementation is picking up. Project-based learning is the highly comprehensive domain of engineering which requires the knowledge and skill of

R.M. Shet (✉) • N.C. Iyer • P.C. Nissimgoudar • S. Ajit
Department of Instrumentation Technology, BVB College of Engineering and Technology, Vidya Nagar, Hubli, Karnataka 580031, India
e-mail: raghu@bvb.edu; nalinic@bvb.edu; sn_asundi@bvb.edu; ajit_sam@bvb.edu

R. Natarajan (ed.), *Proceedings of the International Conference on Transformations in Engineering Education: ICTIEE 2014*, DOI 10.1007/978-81-322-1931-6_55,

project design and development at various levels of engineering program for domains like mechanical, electronics, and computers [1].

In project-based learning, students work in groups to solve challenging problems that are open-ended, curriculum-based, and often interdisciplinary. Firstly, students decide how to approach a problem and look for a solution based on their prior knowledge. They gather information from a variety of sources and synthesize, analyze, and derive knowledge from it. The main characteristic of a project is to make the learning environment student-centered. Teamwork is regarded as a way to construct knowledge. Therefore, projects are designed to demonstrate their acquired knowledge and understanding of concepts to correlate with the application domains or relevant real-world case studies.

This approach is introduced at the various levels of engineering in Instrumentation Technology program wherein the student has a sufficient background of electronics systems and basics of design and development. The design and assessment are framed to test the learning levels of students according to Bloom's taxonomy.

Firstly, at individual course level students are exposed to the course project where the main objective is to build a module or implement related principle of the course. This course project also forms a part of continuous internal assessment. Secondly, the knowledge gained through the cluster of courses at lower semester is used to design and develop a subsystem level problem in mini projects. This course and mini project in turn leads to the capstone project at the final year where a system level implementation is carried out. PBL integrates the knowledge gained through various levels of design and development. This also addresses a–e program outcomes in achieving the problem solving skills and f–k program outcomes which addresses the professional skills.

Organization of this paper is as follows. Section 2 deals with PBL and design of project at various levels, Sect. 3 deals with assessment and evaluation methods, and Sect. 4 deals with effectiveness, discussion, and conclusion.

2 Project-Based Learning

Problem-based learning was first developed in the medical field in mid-1950s, which was used to replace the traditional lecture-based approach. It was also adopted by variety of educational disciplines including Business, Law, Education, and Engineering. This involves students first learning the fundamentals and then utilizing "total recall" to apply in solving problems, learning objectives are set by the instructor and principles are presented to the students through lectures, but often students learn with repeat back approach to satisfy the instructor [2].

In contrast, PBL approach employs a problem as a driving force for learning the fundamentals and principles required to approach the solution. It permits to develop the knowledge based on the fundamental principles of science and mathematics.

A project is a sequence or series of smaller tasks to be performed at various levels of design and development to reach a desired objective. The objective is to break out the traditional learning of "chalk and talk" and make the students to have a product or solution-oriented approach.

PBL experience begins with the students identifying performance requirements, formulation of design objective, list of functional requirements, and conceptual design solution. Potential design solutions are analyzed from a system level perspective which explores the inter-relationships of components, including how they interact with each other and their operating environment. Next a detailed design solution is developed and specifications are established that will enable the design to be fabricated and tested. Throughout this process the students are challenged to learn how to work in teams, practice system level thinking when integrating technologies. The key elements of PBL are as follows:

1. Develop and encourage team dynamics in a group. Team building exercise should be utilized to facilitate the development of trust and communication within the team.
2. Team develops necessary knowledge and skills in identifying the real problem statement by analyzing its functional requirement and constraints.
3. Details all the parameters necessary to solve the problem with limited tolerance.
4. Identify various alternate solutions for the problem within the group and settle on to final solution through constructive brainstorm among the teammates.
5. Action plan with component level, system level development. Implementation and prototyping of the product by analysis of results and conclusions.
6. Summarize the results in both oral and written reports. Team should also be encouraged to present their work to the classmates, conferences, and competitions.

The program aims at achievement of high level learning through PBL used at various levels of courses. Students will be carrying out following types of projects during their study in the IT program.

2.1 Course Projects

The course projects are introduced at the individual course level where the concepts can be well understood and higher level learning can be achieved by implementing any principle or concept. The courses are identified based on its complexity and ability of implementing various concepts and principles for better understanding and clarity. Normally course instructor suggests the theme/s of the project and the students carry out the projects in the group of 3 or 4. The main objective of the course projects is to give more hands on experience to the students in addition to that knowledge gained in the regular course.

2.1.1 Course Project: Digital Electronics Course

A problem definition was formulated to group of students to perform and understand the following theoretical concepts:

- Display numbers from 00 to 99 using 7-segment display
- 1-bit serial adder
- 2-bit ALU

The implementation of the problem statement and its review was made during assessment period based on solution proposed, method of implementation, components used, and its correlation with theoretical concepts. This activity helped the instructor and students in teaching learning process of the course for better understanding of combinational and sequential digital circuits.

2.2 *Mini Projects*

Mini projects I and II are carried out in fifth and sixth semesters. The mini projects give the students an opportunity to demonstrate the knowledge and skills learnt in cluster of core courses. They give students design, prototyping experiences and also enable them to deal with subsystem level problem solving capability. The mini projects are carried out in a group of three or four students. The themes for mini projects are decided by the departmental faculty committee.

The course outcomes of mini projects: The student will be able to:

- Perform need analysis and identify the problem from the given theme and propose the solution (a, c, f, h, i, j, l).
- Analyze and record the requirements for the identified problem (a, c).
- Transform the fundamental knowledge gained in the curriculum to model/design/develop prototype/system for component/system to meet the identified requirements (c, k).
- Develop technical writing skills, presentation skills, and build ability to work in team (d, g).

2.3 *Capstone Projects*

The projects executed by the students in their final year have a primary objective of giving an integrated experience to solve complex engineering problems. Typical expectations from these projects are identification of the problem, researching the literature, problem definition, concept level design, and functional level design, implementation, demonstration, and report writing. The students have the flexibility in selecting an academic project, industry-sponsored project, or community-based project.

All the projects are evaluated continuously and at the semester end through the assessment rubric developed for the purpose. The projects are not only assessed for technical competency but also for professional competencies required by the outcomes: f–k.

The course outcomes of Capstone projects: The student will be able to:

- Identify the needs and formulate the problem statement through surveys (a, e, f, h, i, j, l).
- Develop problem solving skills by applying and integrating the knowledge base acquired (a, e).
- Design/develop the prototype/algorithm for a solution (e, k).
- Develop team building and leadership qualities (d).
- Prepare technical documents and presentations (g)

3 Assessment and Evaluation Methods

3.1 Course Project

The main objective of the course projects is to give more hands on experience to the students in addition to that knowledge gained in the regular course. Assessment is carried out as a part of CIE with a 20 % weightage. Table 1 shows the detailed review for the course project assessment with evaluation.

3.2 Mini Project Rubrics

The knowledge gained in cluster of subjects is incorporated to formulate and solve a real-life problem which is assessed by the instructor allotted and also by an expert review committee at various levels of project. Table 2 shows the detailed reviews for the mini project assessment.

3.3 Capstone Project Rubrics

The final project to be carried out during the final year of program is in two semesters. Table 3 shows the detailed review for the capstone project assessment.

Table 1 Rubrics for course projects

Rubrics	PO attained	Weightage (%)
Literature survey and problem definition	c, d, i	20
Problem statement		
Identify the resources to meet the desired needs		
Exploring alternate approaches through literature survey		
Team formation and defining the roles		
Concept level design	c, k	20
Identification of inputs and outputs		
Establishing relation between input and output		
Circuit design/algorithm		
Implementation	c	40
Functional block diagram		
Choose an appropriate simulation and modeling tool		
Demonstration and report writing	g	20
Preparation for report, poster, use of visual aids		
Clear and well organized report		
Presentation skills		

Table 2 Reviews for mini project assessment

Reviews	Stages of the project	PO attained	Weightage (%)
Review 1	Preparation of requirement form	c, f, h, i, j, g, l	20
	Identification of functional and nonfunctional requirement		
	Problem statement		
	Exploring alternate approaches through literature survey		
	Selection of an appropriate approach		
	Road map of the project		
Review 2	Team interactions and discussions	c, d, g, k, l	50
	Identification of inputs and outputs		
	Design on paper with listing of the required components		
Review 3	Circuit design/algorithm	c, g	20
	Demonstration of obtained results with standard results		
	PCB implementation, debugging, and validation of results		
Review 4	Preparation of report, use of visual aids	g	10
	Clear and well-organized report		
	Presentation and communication skills		

Table 3 Reviews for capstone project assessment

Reviews	Stages of the project	PO attained	Weightage (%)
Review 1	Identify the problem	d, e, f, h	20
	Ability to arrive at problem statement/scope for future enhancement		
	Alternate approaches		
	Presentation and communication skills		
Review 2	Identification and establishing relation between input and output through block diagram representation	d, e, l	20
	Component list with specifications		
	Road map for the project		
	Budget estimation		
	Roles and responsibilities of team members		
Review 3	Functional block diagram	d, e, j	20
	Use of appropriate formula/method		
	Suggestions from experts and discussion of the latest issues		
	Integration of subfunctional blocks for final solution		
Review 4	Proper selection of simulation tool	e, i, k, l	30
	Identification and optimal use of available resources		
	Validation of results		
Review 5	Preparation for presentation, use of visual aids	e, g, i	10
	Conclusion and future scope		
	Participation in technical events		
	Presentation and communication skills		

4 Effectiveness, Discussion, and Conclusion

The effectiveness of PBL can be determined in terms of five components: ownership, creativity, collaboration, critical thinking, and fun. The measure of these components is reflected in assessment rubrics. This PBL experience has encouraged the students in participation of project competitions to showcase the work carried out at various levels. These project competitions include various state and national levels like Impress-IT, JED-I, etc.

Learning through projects has a positive effect on student content knowledge and the development of skills such as collaboration, critical thinking, and problem solving which increases their motivation and engagement. It also gives them an opportunity to see design from a system perspective and develop an appreciation for technical challenge in the context of global, societal, economical, and environmental requirements. The PBL process requires students to be very self directed in their learning and to take ownership of their own education. It provides a framework for embedding experiential and rich learning activities, integrated with discipline-based curriculum that improves employment and career outcomes.

References

1. Chandrasekaran S, Littlefair G, Joordens M, Stojcevski A. Cloud-linked and campus-linked students' perceptions of collaborative learning and design based learning in engineering. Int J Digit Inf Wirel Commun (IJDIWC). 2014;4(3):1–9. The Society of Digital Information and Wireless Communications, ISSN: 2225-658X.
2. Richard NS, Chen KC, Linda V. Integrating project-based learning throughout the undergraduate engineering curriculum. J STEM Educ Innov Res. 2007;8(3/4):15.

Problem Identification Through Literature Survey: A Course Project Activity to Satisfy Accreditation Requirements

Jyoti Bali and Arunkumar C. Giriyapur

Abstract This paper is intended to discuss the active learning practices followed for students of Automation and Robotics as a course project activity, evolved through a structured literature survey under subjects like Mechatronics System Design [MSD] and Real-Time Embedded Systems [RTES]. Activity-based learning has always proved that it can foster faster learning by students with enhanced skills essential for their engineering career. Thus aiming at increased student participation and effective learning in line with ABET accreditation requirements, a literature survey-based course project activity was introduced to enable students to acquire some of the prominent competencies like designing, identifying, formulating and solving engineering problems, communicating effectively, professional ethics, using modern engineering tools and lifelong learning capability. Thus the activity involved two important steps: (i) Literature survey of papers for the allotted areas in a group of two to three members for the entire class of students. (ii) Extending the literature survey in order to solve problem, build solution, and demonstrate using software and hardware models in the laboratory. This is the activity proposed to achieve higher levels of learning and thus address majority of a–k criteria, which otherwise cannot be achieved through traditional lecture-based teaching. Hence the methodologies adopted for guiding students, evaluation of developed models, and the analysis of feedback got from students before and after undergoing this activity are discussed.

Keywords Automation and Robotics • Mechatronics system design [MSD] • Literature survey • Real-time system design [RTES] • ABET accreditation requirements

J. Bali (✉) • A.C. Giriyapur
Department of Automation and Robotics, BVB College of Engineering and Technology, Vidya Nagar, Hubli, Karnataka 580031, India
e-mail: jyoti_bali@bvb.edu; aaron@bvb.edu

R. Natarajan (ed.), *Proceedings of the International Conference on Transformations in Engineering Education: ICTIEE 2014*, DOI 10.1007/978-81-322-1931-6_56,

1 Introduction

Under engineering discipline, practical work allotted to students boosts their interest in learning the subject and enable them to draw parallels between theoretical concepts and practice [1]. Being under autonomous curriculum, teachers have the advantage of running activity-based courses under theory and laboratory with improved participation and learning by student which is usually lacking in lecture-based teaching practices. Activity-based learning helps students overcome the monotony of routine tests and assignments [1, 2]. The key objective is to promote effective learning and achieve higher levels of skills required in engineering education by addressing c, e, f, g, i, and k factors of ABET criteria [3]. Hence the course project activity involving literature survey-based identification of problem in the allotted area, finding solution to the selected problem, and implementing a working model in the laboratory is proposed. In Automation and Robotics program, which is a multidisciplinary field of engineering, it is required that students undergo much of hands on activities in order to observe intricacies of cross-functional mechatronics elements which can be very well realized through project work activity [4, 5]. Introducing interesting group activities based on literature survey and project-based activity is one of ways by which students can be motivated to go few steps beyond the prescribed syllabus and learn faster [3, 4, 6]. Hence the activity was introduced for two batches of students of V and VII semester. The activity complements the deficiencies in the traditional theory and laboratory sessions to a very good extent which is evident from the student feedback response and performance of students [1, 2].

2 Proposed Objectives

The objectives decided for the proposed activity are to enable students to achieve competencies as stated under ABET criteria a–k. The standard a–k criteria are listed as follows:

(a) An ability to apply knowledge of mathematics, science, and engineering
(b) An ability to design and conduct experiments, as well as to analyze and interpret data
(c) An ability to design a system, component, or process to meet desired needs
(d) An ability to function on multi-disciplinary teams
(e) An ability to identify, formulate, and solve engineering problems
(f) An understanding of professional and ethical responsibility
(g) An ability to communicate effectively
(h) The broad education necessary to understand the impact of engineering solutions in a global and societal context
(i) A recognition of the need for and an ability to engage in life-long learning
(j) A knowledge of contemporary issues

(k) An ability to use the techniques, skills, and modern engineering tools necessary for engineering practice

Under the proposed activity, the target is to address ABET criteria c, e, g, and k to a higher level of competency whereas f, i, and j only to the basic level of competency. In line with ABET criteria, it was decided to concentrate mainly only six objectives listed as follows:

1. To demonstrate problem-solving skills by identifying the problem in the assigned area, formulating and solving using structured literature survey methodologies (criteria – c, e, i, and j)
2. To develop the ability to build system by selecting appropriate system components and modeling using simulation or real time hardware/software tools (criteria – c and k)
3. To exhibit professional ethics by getting approval of their project implementation plan and ensuring that problem identified and solutions built by each team is unique (criteria – i and f)
4. To undergo a unique research experience by referring the latest research papers and preparing a survey report along with results of implementation of solution, analysis, and verification (criteria – g, i, and j)
5. To build competence in the usage of tools in laboratories, by exploring various capabilities of software and hardware to build models (criteria – k)
6. To enhance peer learning through demos and presentation by other teams (criteria – g)

In order to achieve the abovementioned objectives, the plan, methodologies followed, implementation details, and the achieved results are discussed in further sections.

3 Methodologies and Implementation

The proposed activity is being conducted for V and VII semester students for theory subject MSD and RTES. As both of these subjects are associated with regular practical sessions, the activity of literature survey for identification of the problem in the allotted area could be combined with that of building a solution for the same through project activity in the laboratory. The activity was announced at the start of semester along with details regarding the team formation, schedule of completion of each of stages, submission dates, evaluation attributes, etc. The activity of literature survey for identification of the problem and solution in the assigned area along with survey report generation is proposed in theory subject with a weightage of 40 % of internal marks. The activity of building solution for the identified problem under project activity is planned in the associated laboratory and evaluated again for 40 % of total marks. Application and verification of theory using practical work is an integral part of engineering curriculum and can be very effectively done through the proposed activity [5, 7].

Students are encouraged to select their partners in a group of two to three members and thus experience the team spirit aspect in the activity. It was even ensured that students contribute individually by limiting the size of team to a maximum of three members only. Then teams were asked to select topics from the assigned chapters of the subject through literature survey. Thus all students together are made to address content of each chapter for the subjects under consideration, i.e., MSD & RTES. They were guided to survey current state of developments in the allotted areas using latest IEEE research papers and later identify and formulate a problem [8].

A set of guidelines and demo was given to students for gathering the recent and related papers, short-listing most relevant three to four papers for further study, which in turn can help them to identify the problem to work on. Students were given a time duration of 30 days for doing literature survey in order to identify the problem. Students surveyed and selected on an average of six to eight IEEE papers in the assigned area and further shortlisted around two to three in order to identify the problem. Thus all 32 teams together were able to collect around 250 IEEE papers. Later they were asked to do literature survey for the possible solutions for the selected problem. Students were guided to analyze problems from multiple perspectives [7, 8].

Students were further guided to select the best possible solution for the problem as they had options of using different hardware and software tools. Thus students could get an experience of identifying a problem, following the design steps and experimenting to find solution to the problem in team. This was a step taken to improve their interpersonal skills, attitudes, and professional ethics required in teamwork [1, 4, 6].

Students were evaluated for written and oral communication skill through survey report writing and presentations, respectively. Report writing and communication skills are considered to be very prominent competencies required by a graduating engineer in his profession. Students have submitted the project reports with the discussion of survey of literature, problem identified, and results of implementation of solution. Further, around 30–40 % of the project reports were identified for converting as papers for publication in some reputed journals and conferences [1, 2]. Thus the students involved in those projects are now being guided for writing the paper following the standard guidelines. Evaluation attributes are designed carefully to address each of the phases in the activity, so that the learning levels are measured [3].

3.1 Implementation Plan

The simplified process flow chart shown in Fig. 1 depicts the stepwise approach followed in practicing the proposed literature survey-based project activity. Even though the activity seems to be a sequential process, students can always go back and rework to correct the methods adopted previously before moving to the next

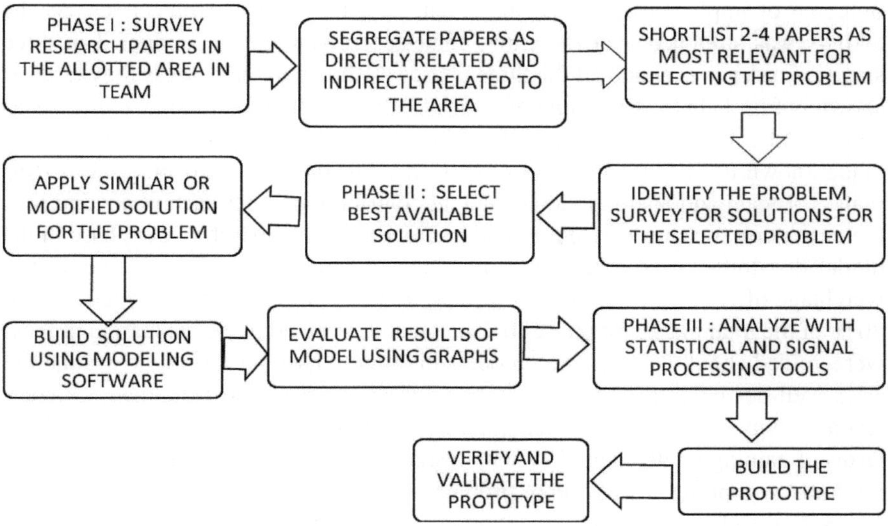

Fig. 1 Process flow chart of the proposed activity

step in process [5]. Students were guided about the importance of undergoing each of the step in the process so that they undergo the activity consciously [9].

Students adopted the Kolb's Experiential Learning model while implementing the project solution undergoing the four steps of learning, i.e., Doing, Reviewing, Concluding, and Planning [1, 5]. Thus they were able to refine the solution by iterating through the Kolb's learning cycle. The activity was purely student-centric enquiry-based learning where students were made to explore various library resources including IEEE literature for identifying real-world problems and solutions for them [5, 7, 8]. Hence the activity provided them the opportunity to develop competencies like teamwork, self-learning, experimental design, analysis, and innovation which are very essential attributes of a good engineer [4, 6, 7].

4 Assessment

Assessment methods have to be designed in order to collect the facts about the student learning based on the data collected through rubrics for the activity. Even the assessment process can be iterative so that the methods used for assessment can be refined in order to measure student learning accurately [10–12]. Thus the measurement methods introduced should be able to measure the level of attainment of objectives set for the activity [3, 6]. This in turn will help teachers to monitor and improve student learning [4].

Based on the objectives set for the proposed activity as listed in Sect. 2, the evaluation methods and feedback questionnaire were designed to measure the outcome of implementation and the extent of achievement of objectives. As the activity was stretched over the entire semester, reviews were conducted in three stages in order to evaluate all the three phases as shown in the process flow chart as shown in Fig. 1. In the beginning of semester, the schedule for all three reviews was announced and the requirements of attainment of objectives for each phase were made known to the students. The rubrics was designed for measuring the performance of team and also individual student contribution under each phase and thus the evaluation formats were developed in order to evaluate each of the phase in the model development. The level of attainment of objectives was evaluated with a weightage of 25 % of total marks for phase I, 35 % of total marks for phase II and 40 % of the total marks for phase III. The faculty involved in the evaluation process were guided about the evaluation process in order to have uniformity in evaluation.

A well-designed questionnaire is being used to assess the outcomes of undertaken activity in meeting the competencies as required by ABET a–k criteria in terms of gaining knowledge of system design approach, using of modern engineering tools, equipping with skills of problem identification and solving, professional ethics, oral and written communication [10, 11]. Later the data collected from student feedback as well their performance was analyzed. Pre- and post-activity surveys were planned to collect the data regarding expectations of outcomes and accordingly compare the views of students before and after conducting the activity.

5 Results

The outcome of implementation of the proposed activity for the present batch was very impressive; as around 4–5 of stated objectives could be fulfilled satisfying c, e, f, g, i, and k factors of ABET criteria to a good extent. This is evident from the feedback collected from students regarding the competencies achieved by them in accordance with ABET a–k criteria before and after completion of the activity as shown in Table 1 as pre- and post-survey data columns. There are eight questions in the questionnaire in order to measure the level of attainment of a–k criteria for six proposed objectives listed in Sect. 2. Students were asked to rate for the attainment level under each question using attributes such as Strongly Agree (SA), Agree (A), Disagree (D), Strongly Disagree (SD), and Neural having weightage of marks from 5 points to 1 point, respectively.

Here the numbers given indicate the number of students answering for a specific question using any of the attributes among SA, A, D, SD, or Neutral. It was noted from the data gathered as per Table 1 that average rating has improved in post-survey as compared with pre-survey data for each of the competency under objectives and also an improved positive response of students toward the activity. The same analysis is shown using bar charts plotted for pre- and post-survey data in Figs. 2 and 3, respectively, where y-axis indicates the number of students in

Table 1 Pre- and post-survey done for 62 students for the levels of attainment of six objectives

Abilities to develop/	Q1.0 system design approach		Q2.0 identify, formulate/ solve problem		Q3.0 professional and ethics		Q4.0 communication oral and written		Q5.0 lifelong learning		Q6.0 use of modern engineering tools	
attributes	Pre	Post	Pre	Post	Pre	Post	Pre	Post	Pre	Post	Pre	Post
SA	11	15	8	12	18	19	19	21	16	18	11	15
A	31	36	27	35	23	31	25	30	33	35	27	33
D	11	7	12	8	8	5	8	6	4	3	12	8
SD	7	3	13	6	7	4	9	4	6	4	8	4
Neutral	2	1	2	1	6	3	1	1	3	2	4	2
Average (5)	3.7	**4.0**	3.4	**3.8**	3.7	**4.0**	3.8	**4.0**	3.9	**4.0**	3.5	**3.9**
Average in (%)	74	**80**	68	**76**	73	**80**	77	**81**	77	**80**	71	**78**

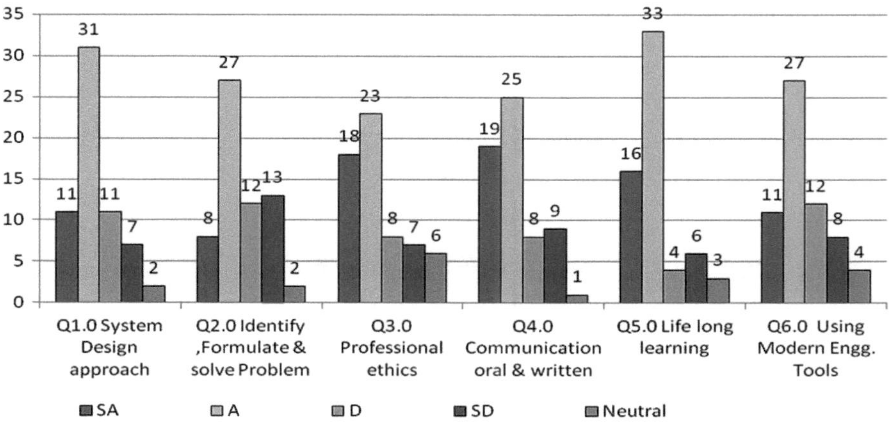

Fig. 2 Feedback analysis as per the data gathered in Table 1 (From pre-activity survey)

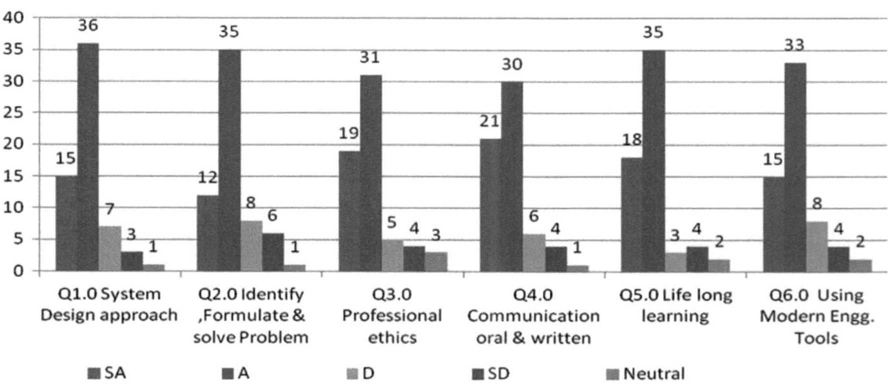

Fig. 3 Feedback analysis as per the data gathered in Table 1 (From post-activity survey)

the survey, rating the attainment level of stated objectives using attributes SA, A, D, SD, or Neutral.

Also the performance of previous batch of students undergoing theory and laboratory courses without any additional activity is compared with that of the current batch of students undergoing the courses along with literature survey-based project activity and data are shown in Table 2. It was noted here also that there was an improvement in class average and individual performance for the batch with project activity as compared with previous batch. Thus all teams could get information about the practical models covering all topics of the theory syllabus through demos and presentations of peer teams in order to establish the link between theory and practical effectively [1, 7, 9]. Students could get very good exposure about the interrelation between the different mechatronics elements combined to work as a system [4].

Table 2 Comparison of performance of 2013 and 2014 batch students

Parameters/batch	2013 batch (previous)	2014 batch (current)
Activities planned	Test and assignments only	Tests, project activity, and survey report
Class average	66 %	75 %
Summary of student feedback	Satisfactory	Good

6 Conclusion

The undertaken activity helped in motivating students and staff to be part of such activities in order to boost the teaching learning process. The outcomes of the activity certainly gave some valuable inputs for design changes in the program as a whole in order to accommodate such active learning activities under different subjects [3, 4]. Thus the proposed activity helped students to acquire some of the competencies required by them in the engineering workplace as autonomous and lifelong learners through the process of scientific enquiry-based learning, problem identification, building solution, verification, and validation. Thus the requirements of attaining ABET competencies as stated under objectives in Sect. 2 could be achieved to a good extent. Based on student feedback and experiential learning, there is a need for refining the guidelines to be given to students, implementation and assessment methods in order to elevate the performance of students and the satisfaction level of involved faculty members, in turn making the activity more effective.

References

1. Davis C. Learning and teaching in laboratories. An engineering subject centre guide. Loughborough: Higher Education Academy Engineering Subject Centre; 2008.
2. Bailie C, Moore I, editors. Effective learning and teaching in engineering. Oxon: Routledge Falmer; 2004.
3. Wahab HFA, Ayob A, Zaki WMDW, Hussain H, Hussain A, Mokri SS. Program outcomes measurement and assessment processes. Faculty of Engineering and Built Environment, Universiti Kebangsaan Malaysia, Bangi.
4. Passow HJ. What competencies should engineering programs emphasize? A meta-analysis of practitioners opinions informs curricular design. In: Proceedings of the 3rd international CDIO conference. MIT, Cambridge, MA, 11–14 Jun 2007.
5. Kolb DA. Experiential learning – experience as a source of learning and development. Englewood Cliffs: Prentice Hall; 1984.
6. Felder RM, Forrest KD, Baker-Ward L, Jacquelin Dietz E, Mohr PH. A longitudinal study of engineering student performance and retention. I. Success and failure in the introductory course. J Eng Educ. 1993;82(1):15–21.
7. Houghton W. Learning and teaching theory for engineering academics. Available online at: www.engsc.ac.uk. 2004.

8. Boylan H. What works: research-based best practices in developmental education. Boone: Appalachian State University; 2002.
9. Gabriel KF. Teaching unprepared students: strategies for promoting success and retention in higher education. Sterling: Stylus Publishing; 2008.
10. Dahm K, Rowan University. Efficient and effective strategies for collecting assessment data. In: Proceedings of the Fall 2010 mid-Atlantic ASEE conference, Villanova University, 15–16 Oct 2010.
11. Angelo TA, Partricia Cross K. Classroom assessment techniques. San Francisco: Jossey-Bass; 1993.
12. Standahl J. Student outcomes assessment plan, course assessments and program reviews. Office of Assessment and Institutional Research Paul D Camp Community College, Franklin, November 2008.

Attainment of Professional Outcome by Active Learning Method

Shrinivas D. Desai, Savitha S. Desai, and Linganagouda Kulkarni

Abstract The program outcomes defined in outcome-based education shall be broadly categorized into technical outcome and professional outcome. In engineering education, the professional outcomes are addressed to some extent by introducing humanity science course. Professional outcomes are addressed by mini and capstone projects. However in core courses the activities need to be designed to address the professional outcome. In this paper, we propose an active learning method called "Let us join" which is intended to address the professional outcomes. These activities are necessary in today's scenario of education as knowledge sharing and group learning among students is seldom observed. The process of conduction of the activity and attainment of POs are presented in this paper. A survey was carried out before and after the activity, looking at the social relationship, self-drive for initiating communication and ability to integrate the information and document the information. The activity assesses students and also grades them according to their contribution in the activity.

Keywords Active learning • Professional outcome • Program outcome • Soft skill

S.D. Desai (✉)
Department of Information Science and Engineering, BVB College of Engineering and Technology, Vidya Nagar, Hubli, Karnataka 580031, India
e-mail: sd_desai@bvb.edu

S.S. Desai
Department of Biotechnology, BVB College of Engineering and Technology, Vidya Nagar, Hubli, Karnataka 580031, India
e-mail: joshisavitha@bvb.edu

L. Kulkarni
Department of Computer Science and Engineering, BVB College of Engineering and Technology, Vidya Nagar, Hubli, Karnataka 580031, India
e-mail: lgk@bvb.edu

R. Natarajan (ed.), *Proceedings of the International Conference on Transformations in Engineering Education: ICTIEE 2014*, DOI 10.1007/978-81-322-1931-6_57,

1 Introduction

In today's scenario the students from this part of Karnataka state are lacking most important skill, i.e., soft skill. This skill fulfills an important role in shaping an individual's traits. It is of high importance for each and every student to acquire adequate skills beyond academic or technical knowledge. The outcome-based education method stresses on this issue [1]. The reported literature presents that the responsibility of who will educate soft skill for today's youth is shifting between teachers and parents [2]. Where the parents, schools, and the community once contributed an equal share to the teaching of our youth, a deficit is now seen. It is clear that school, community, and parents equally share the responsibility of providing soft skill which will play an important role in shaping student's personality. If we look for the most essential soft skills, we need to look at much on the context and one's personal perception. However, the major features that immediately come to one's mind when soft skills are mentioned are communication skills, presentation skills, and interpersonal skills. And indeed, it is these skills which are mostly lacking among graduates from colleges and universities.

Communication skill is all about how effectively one can communicate. Effective communication is about conveying our ideas, feelings, knowledge, or messages to other people clearly and unequivocally. This skill also focus about receiving information that others are sending and understanding with as modest alteration as possible. Through communication, people reach some understanding of each other, learn to appreciate each other, influence one another, and build conviction, and study more about themselves and how others perceive them [3]. People who communicate efficiently know how to interact with others flexibly, dexterously, and responsibly, but without sacrificing their own requirements and integrity. Presentation skill is all about exhibiting our ideas through visual, aural medium using tools or technology [4]. These are one of the first managerial skills which an engineer must acquire. Interpersonal skills include the habits, attitudes, etiquette, manifestation, and behaviors that we exhibit, which is perceived by others [5]. The acquisition of interpersonal skills begins early in life and is influenced by family culture, friend circle, and our observations of the world around us.

Most of the soft skills shall be addressed through collaborative learning technique. It is a pedagogy tool to teaching and learning process that involves groups of learners working collectively to solve a problem, complete a project, or construct a product [6]. Collaborative learning is a state in which two or more people learn together [7, 8]. More particularly, it is based on the model that knowledge can be acquired within a population where members vigorously interact by sharing experiences and take on various roles [9]. It also refers to methodologies and environment in which learners connect in a common task where each individual depends on and is answerable to each other. It is learnt that there exists an inherent social nature of learning for acquisition of knowledge [10]. Often, collaborative learning is used as an umbrella term for a variety of pedagogies that involve joint scholarly effort by students or students and teachers [11, 12]. The list of advantage of collaborative learning is infinite [12]. As per literature there are 44 benefits of collaborative learning. We can summarize the benefits of Collaborative learning (CL) as the element of benefit as listed in [11, 12]. (Johnson, Johnson, and Smith, 1991): There are different strategies, tools, and methods for collaborative learning method. The detailed list shall be found in literature [13].

2 Methodology

The strategy for the "Let us join" a collaborative learning experiment is as presented below. This activity is intended for Computer Organization course at 2nd-year level of UG course. But it can be redesigned for any course at appropriate level. Instructor designs the page consisting of an "Image of computer part" on one side, and an "alphabet" on the other side. Number of papers will be equal to number of groups, where each group consists of either four or five students. Then each page is cut into four parts. Instructor provides each one part of page to each one of student.

When all members of a group (students) join all appropriate parts, they can identify "image" and "alphabet." After identifying image and alphabet, the group should write technical report of that computer part. If all the alphabets are arranged properly it portray name of a person. Groups should identify the name. Assessment is based on identification of image/alphabet, meaningful word, and technical report.

2.1 Assessment of Soft Skill

Before carrying out collaborative learning method "let us join" activity, survey is done to evaluate the presentation, communication, and interpersonal skills of students. For this the survey questions are gathered from various sources and survey was done when students are free from academic load, so that they think effectively and provide valid input toward the survey.

2.1.1 Communication Skill

Questions for communication skills test are originally designed by Robin Jacobs, Portland Community College, Portland, Oregon [14]. The survey done with these questions will help students to evaluate own communication skills and style, and provide them with helpful tips for becoming a good communicator and a team player. Questions are designed in such a way that student has to mark the one (out of three options A, B, or C) that best describes their communication style. Totally there are 24 questions. The best answer is not necessarily the correct one. In fact there is no Right or Wrong answer.

Table 1 Evaluation rubrics

Sl. no.	Evaluation parameter	PO addressed	Program outcome
1.	Analyzing the problem	a	Ability to apply the knowledge of science, mathematics, and engineering concepts
2.	Executing the activity	a	
3.	Team work	d	Ability to function on multidisciplinary teams
4.	Preparation of report	g	Ability to communicate effectively with a range of audiences

2.1.2 Presentation Skill

For presentation skill questions designed by Mandel, S is used [15]. The question provides the student to evaluate their own presentation skills. It helps them to identify areas where they should try to improve. Student has to read the question and tick the number that best describes him/her. Every question is evaluated for 5 marks. And there are 20 questions. If the student scores:

- Between 80 and 100, he is an accomplished speaker who simply needs to maintain basic skills through practice.
- Between 60 and 80, he has the potential to become a highly effective presenter.
- Between 40 and 60, he requires major improvement in presentation skill.
- Between 30 and 40, he should show dramatic improvement with practice.
- If the total is below 30, he requires lot of efforts to gain presentation skill.

2.1.3 Interpersonal Skill

For interpersonal skill the questions designed by R B Hill is used [5]. For each statement listed in the question paper, student should rate himself on a scale of 1–10 for each of the items. A rating of 10 would indicate that the statement is always true and a rating of 1 would indicate that it is never true.

2.2 Assessment of Program Outcome

The average of earned score of students is calculated for the different evaluation parameters such as analyzing the problem, executing the activity, team work, and preparation of report. These activities are mapped on to following POs as in Table 1.

3 Results and Discussion

Survey was carried out prior to "Let us Join" activity and survey after this activity reveals a major change in interpersonal skill and communication skill, while not much change in presentation skill. Figure 1a–c shows the comparative statistics of

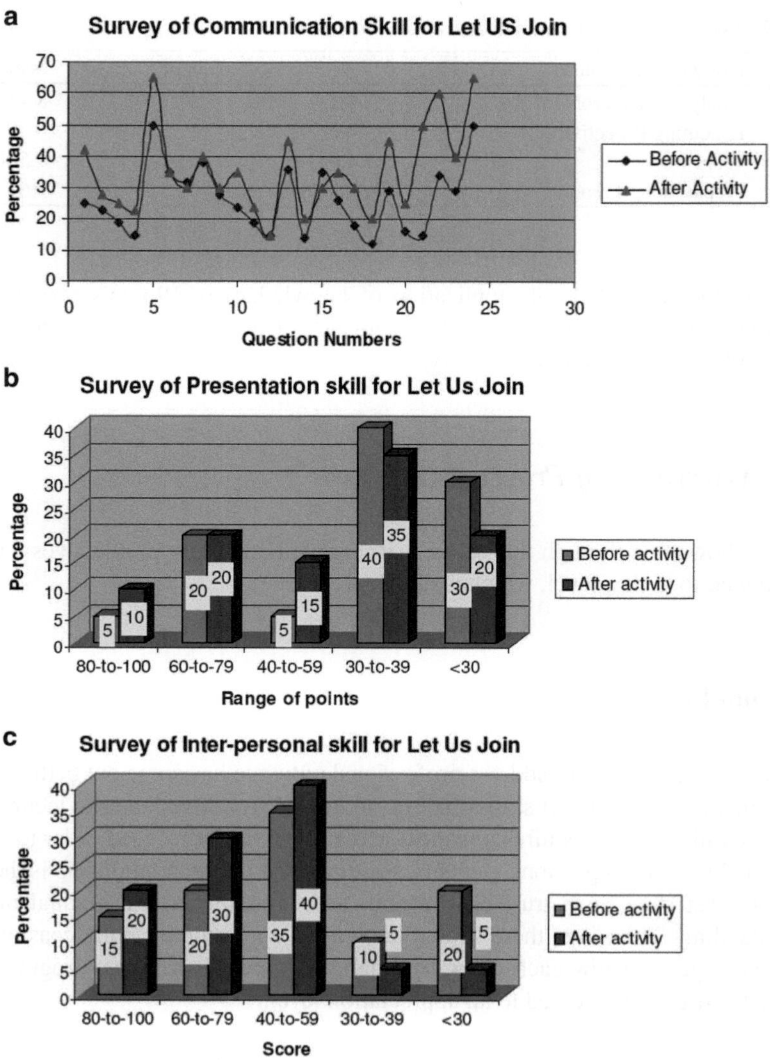

Fig. 1 Survey of (**a**) communication skill, (**b**) presentation skill, and (**c**) interpersonal skill, before and after "Let us Join" activity

survey done prior and after "Let us Join" activity. When this score is compared with earlier scores, students were pleased with the progress that they have made.

Over all "Let us Join" proved that collaborative learning improves learning where students enjoy the course and were able to recognize their classmates by their names. Majority of the students met the expected outcome. However, some percentages of students were reluctant to work in team. Observation from the Fig. 1a–c shows that there is around 20 % of improvement in presentation skill, 50 % of students were found to be improved in communication skill, and 30 % of improvements in inter personal skill. As an outcome the students were able to meet following intended

Table 2 Assessment of program outcomes

Sl. no.	Evaluation parameter	PO addressed	Average score (out of 5 marks)	Scale to 10
1.	Analyzing the problem	a	4.33	8.41
2.	Executing the activity	a	4.08	
3.	Team work	d	4.78	9.56
4.	Preparation of report	g	4.81	9.62

learning objective. L3 level (application of knowledge) of Bloom's taxonomy is achieved and Logical, Interpersonal, Linguistic – Gardner's multiple intelligence level is addressed through this activity.

3.1 Assessment of Program Outcome

Table 2 shows that through this active learning method program objectives d and g are attained to higher level, which is obviously true.

4 Conclusion

The activity carried out to address professional outcome has come out better. Study shows that majority of soft skills shall be addressed by active learning techniques. Active learning intern requires openness and knowledge sharing. In order to facilitate knowledge sharing among students, building trustworthy relationship is the first and foremost step. Such trust can be built and strengthened via gradual mutual understanding. Therefore, there should be various opportunities and occasions for students to get to know each other. Through this type of activities, program outcomes d and g are addressed to an appreciation level.

References

1. Felder RM, Brent R. Designing and teaching courses to satisfy the ABET engineering criteria. J Eng Educ. 2003;92(1):7–25.
2. Kitty-Sue Schlink. Addressing educational needs of youth in today's society. 38(4), Commentary, 4COM1. August 2000, p. 1–4.
3. Handbook on communication skills. http://www.cgg.gov.in/Handbook%20on%20Communication%20Skills.pdf
4. Presentation skills by Dr. Pooja Purang. http://www.hss.iitb.ac.in/courses/purang06.ppt
5. Hill RB. On line lesson. http://www.coe.uga.edu/workethic/less3.htm
6. Susan Fountain, UNICEF. Teaching and learning methodology of global education. In: Education in the changing world, Hungary. 1991.

7. http://en.wikipedia.org/wiki/Collaborative_learning#head
8. Dillenbourg P. Collaborative learning: cognitive and computational approaches, Advances in learning and instruction series. New York: Elsevier Science, Inc.; 1999.
9. Mitnik R, Recabarren M, Nussbaum M, Soto A. Collaborative robotic instruction: a graph teaching experience. Comput Educ. 2009;53(2):330–42.
10. Lee CD, Smagorinsky P, editors. Vygotskian perspectives on literacy research: constructing meaning through collaborative inquiry. Cambridge: Cambridge University Press; 2000.
11. Smith BL, MacGregor JT. What is collaborative learning? National Center on Postsecondary Teaching, Learning, and Assessment at Pennsylvania State University. 1992.
12. Harding-Smith T. Learning together: an introduction to collaborative learning. New York: HarperCollins College Publishers; 1993.
13. http://www.gdrc.org/kmgmt/c-learn/44.html
14. Jacobs R. Effective communication skills. http://spot.pcc.edu/~rjacobs/career/effective_communication_skills.htm
15. http://www.communication-skills-4confidence.com/communication-skills-test.html

A Case Study on Improving Students' Conceptual Understanding in Engineering Courses

S.A. Halkude and Sunita B. Aher

Abstract Globally, teaching engineering courses using PowerPoint lecturing and assessing through objective tests is not a new practice. However, from the Indian context, these practices are still getting explored in teaching. The current study attempts to improve students' conceptual understanding of engineering courses, using objective tests followed by immediate group discussion practice. Objective test encourages students to explain answer to questions that seem tricky or confusing. The present study showed improved performance of students in theory as well as in practical of complex course like System Programming. This study provides evidence on improvement in performance of students by using this teaching practice.

Keywords Objective test • Feedback • University result • Likert's scale

1 Introduction

Teaching and learning engineering courses requires proper guidance and effective teaching methods. The multiple-choice test is a staple of higher education because it provides an efficient and effective measure of student learning (McKeachie 1999). The practice of conducting objective test and arranging the discussion after it helps the students to improve the knowledge of particular course. Students also get their doubts cleared about course during the discussion. This is the optimal method of testing

S.A. Halkude (✉)
Department of Civil Engineering, Walchand Institute of Technology,
Solapur, Maharashtra, India
e-mail: halkude60@gmail.com

S.B. Aher
Department of Computer Science and Engineering, Walchand Institute of Technology,
Solapur, Maharashtra, India
e-mail: sunitaaher@gmail.com

R. Natarajan (ed.), *Proceedings of the International Conference on Transformations in Engineering Education: ICTIEE 2014*, DOI 10.1007/978-81-322-1931-6_58,

students' knowledge about the course. Guessing a wrong answer and selecting a wrong answer through a reasoning process are negative consequence of multiple-choice tests. Students may learn false facts from multiple-choice tests, the positive effects of testing outweigh this cost [1].

The current study uses objective type tests and group discussion technique to improve students' understanding in engineering courses. In the present study, we have considered one of the complex course namely Systems Programming (SP) of Third Year Computer Science & Engineering, U.G. program. Direct and Indirect method is used to test the effectiveness of objective type tests and group discussion techniques.

2 Literature Survey

Various types of teaching methods exist, some applied descriptive and inferential statistical techniques to test the effectiveness of five teaching methods, namely, lecture, lecture/discussion combination, jigsaw, case study, and team project in a large class setting [2]. Experiments conducted by [3] show how teacher have used both multiple-choice and constructed response questions to evaluate students' understanding of course materials and principles.

The use of multiple-choice questions (MCQ) has both positive and negative consequences [4]. This study also shows that feedbacks provide positive effects and reduce the negative effect of multiple-choice testing. Study conducted by [5] provided verification feedback for learners to correct errors, which improved their understanding and performance in the subject. Study [6] shows the effectiveness of active learning methods, in which students do activities other than listen to lecture, such as read, write, pose questions, respond to questions, solve problems, and so on. Burton et al. [7] discussed various methods to formulate multiple-choice tests along with their advantages and limitations. The advantages of conducting multiple-choice tests are discussed by various researchers [8, 9].

By employing active learning strategies, students not only learn content, but also the process [10]. The use of multiple-choice questions as a formative assessment technique is explained in [11]. The research [12] implemented a combination of periodic short objective tests, followed by immediate group discussion and feedback, in a Systems Programming course. The present study is the extension of the research in [12].

3 Methodology Used

The research questions examined in this study are as follows:

- How to improve students' conceptual understanding theoretically as well as practically of System Programming course?
- How to improve student's performance in various competitive exams for higher studies and for employment?

Table 1 Formation of group

Sl. no.	Number of students	Description of performance
1.	10	Distinction
2.	10	First class
3.	10	Second and pass class
Total	**30**	

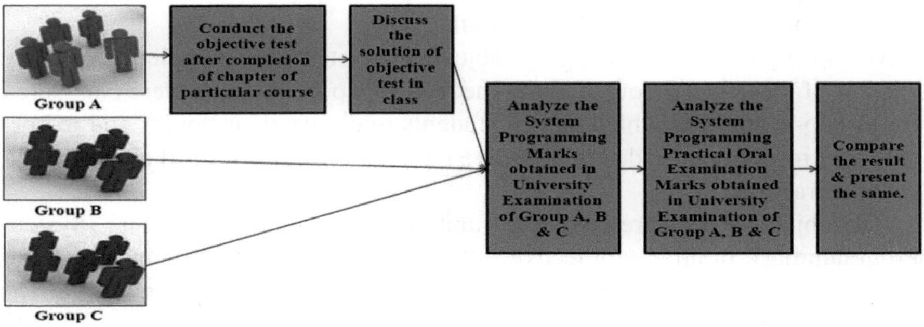

Fig. 1 Organization of case study

Objective test followed by immediate group discussion technique is used in the present study.

3.1 *Sample Used*

Three groups were formed for this study: Third Year Computer Science & Engineering (TECSE) (Group A – Experimental Group) Year 2012–2013, TECSE (Group B – Control Group) Year 2012–2013, and TECSE (Group C – Control Group) Year 2011–2012. Each group consisted of 30 students. Formation of each group is shown in Table 1.

For this study, students' subject marks in university examination, practical oral examination (POE), and Aptitude Test (AT) conducted as a part of "Value Added Program" marks were considered.

3.2 *Organization of Case Study*

The organization of case study in detail is shown in Fig. 1. The objectives of this study are as follows:

- To improve the result of complex courses
- To make concept more clear and give more practical knowledge to students
- To improve students' performance in various competitive exams for higher studies and for employment

3.3 Procedure

Three-group experimental studies to test the effectiveness of objective test followed by immediate group discussion were conducted. The sample consists of 90 students from Third Year Computer Science Engineering (TECSE) of Solapur University. A randomized sample of 30 students were selected and equally distributed among the three groups: Group A – Experimental Group, Group B and C – Control Group based on their merit in the university result.

Group-A participants were given objective type test after completion of every chapter of a particular course. After the test, group discussions were conducted for Group-A students which helped students to clarify their doubts and to correct their response provided in objective type tests; while Group-B and C had no such interaction.

The objective tests are conducted unit wise of Course Curriculum (System Programming). In all, six objective tests were conducted for six units and one objective test was conducted for the entire course. Group discussion was carried out with regard to the answers of the objective questions. During the group discussion, first, the students were made to understand the approach of the questions in the test. Then, the various options for each question were considered and the right answer was brought to the fore by explaining the reasons. Students were also encouraged to raise questions regarding the solution of the questions. This process helped the students in their conceptual understanding regarding various topics of a particular unit. Objective tests containing multiple-choice questions were used as direct assessment tool.

In our institute, an aptitude test for TECSE students is conducted to prepare students for Technical interview and competitive exams like GATE. In semester I of TECSE, non-technical aptitude test and in semester II, technical aptitude tests based on the topics like C-Programming, Java Programming, Microprocessor, Formal System & Automata, SP, etc., are conducted. Table 2 shows the objective, data set, and performance metric used to achieve these objectives.

Table 2 Data set and performance metric to achieve objectives

Sl. no.	Objective (Obj.)	Data set used	Performance metric used
1.	Obj-1	Solapur University TE(CSE) Part-I November–December 2011–2012 & 2012–2013 result	Marks obtained in SP, % and class
2.	Obj-2	Solapur University TE(CSE) Part-I POE November–December 2011–2012 & 2012–2013 result	Marks obtained in SP POE, % and class
3.	Obj-3	AT result conducted in program "Value Added Program"	Aptitude test marks in SP

Table 3 Topics covered and their duration for objective tests [12]

Sl. no.	Name of topic (duration of test in minutes)	Number of objective questions in test	Discussion time (in minute)
1.	Language processors (60)	65	60
2.	Assemblers (45)	55	45
3.	Macro (20) (small chapter)	25	15
4.	Compilers (45)	45	30
5.	Linker (30)	40	30
6.	Loader (30)	40	30
7.	Grand objective test (45)	50	30
8.	Aptitude test (45)	50	–

3.4 *Objective Test Questions Classification*

Table 3 shows objective test conducted and their duration in this study.

Objective test conducted after completion of chapter consists of following types of questions:

- True/false questions
- Questions with multiple-choice options
- Questions based on arranging the activities
- Questions based on assembly language programming
- Questions to improve the programming skill in assembly language
- Questions on a given diagram

Table 4 shows objective test questions classification with example related to System Programming subject.

3.5 *Bloom's Taxonomy*

Bloom's taxonomy divides educational objectives into three "domains": Cognitive, Affective, and Psychomotor [13]. Objective test questions conducted in present study are covering cognitive category of Bloom's taxonomy that includes remember, understand, apply, and analyze. Figure 2 shows the cognitive category of Bloom's taxonomy, while Table 5 shows the cognitive level analysis of objective test question types.

3.6 *Feedback*

In this study, indirect assessment tool like survey questionnaire with Likert's scale of five values (strongly agree (SA), agree (A), neutral (N), disagree (D), and strongly disagree (SD)) is used to know about student's perception about this case study. The

Table 4 Objective test question classification with example

Type 1.True/ False Questions Semantic gap is the gap between programming language domain & execution domain. True/ False
Type 2.Questions with multiple options given Address assigned by the loader to symbol is called a) Load origin b) Load time adder c) Link origin d) Link time origin
Type 3.Questions based on arranging the activities Following activities are performed during which pass: i) Isolate the label, mnemonic opcode and operand fields of statement ii) Obtain the machine opcode corresponding to mnemonic from the mnemonic table iii) Check validity of mnemonic opcode through lookup in mnemonics table iv) Obtain address of memory operand from symbol table a)Pass-I, Pass-II, Pass-I, Pass-II b)Pass-II, Pass-II, Pass-I, Pass-I c)Pass-I, Pass-I, Pass-II, Pass-II d) None of the above
Type 4. Questions based on assembly language START 101 READ N PRINT N N DS 01 What is equivalent machine code? a) 101) + 09 0, 103 b)101) + 08 0, 103 102) + 10 0, 103 102) + 10 0, 103 103) 103) c) 101) + 09 0, 104 d) 101) + 07 0, 103 102) + 11 0,103 102) + 10 0,103 103) 103)
Type 5. Questions to improve the programming skill in assembly language Consider the following macro INCR &MEM_VAL, &INCR_VAL , & REG MOVER &REG, & MEM_VAL ADD &REG & INOR_VAL MOVEM &REG, &MEM_VAL MEND Macro call for above definition will be a) INCR A,B ,AREG b) INCR_M MEM_VAL=A,INCR_VAL=B,REG=AREG c) INCR_M INCR_VAL=B,REG=AREG,MEM_VAL =A d) All of above
Type 6.Questions based on given diagram For a*b, Avail_in = true for which blocks: 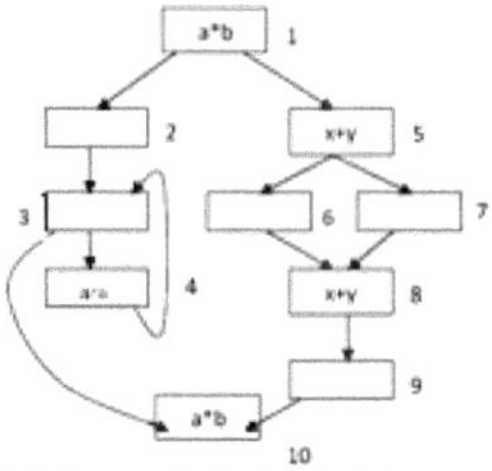 a) 2,5,8,9 b)5,6,8,9 c)1,7,8,9 d) 2,5,6,7,8,9

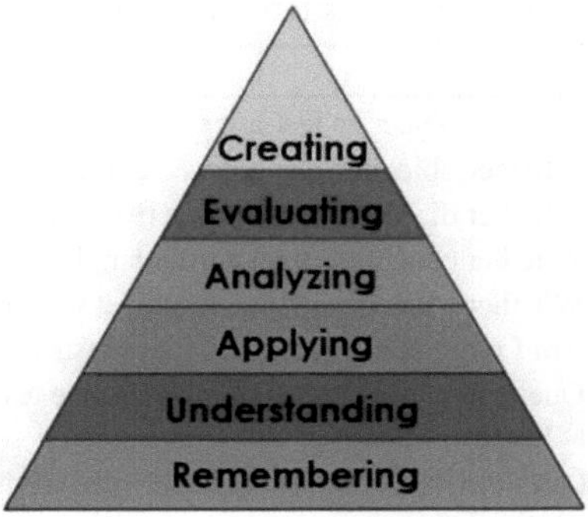

Fig. 2 Bloom's taxonomy: cognitive category [14]

Table 5 Bloom's taxonomy: cognitive level analysis

Sl. no.	Objective test questions type	Bloom's taxonomy – cognitive level
1.	Type 1	Remember
2.	Type 2	Remember
3.	Type 3	Understand
4.	Type 4	Apply
5.	Type 5	Apply
6.	Type 6	Analyze

feedback was conducted twice – at the end of the semester and after the University examination. Tables 6 and 7 show these feedback questions and percentage of feedback. From Table 6 it is found that 100 % students agreed that objective test after completion of chapter is useful and 93 % students agreed that discussion after objective tests is helpful.

4 Experimental Result

The format of question paper of Solapur University Exam consists of both subjective and objective questions. The results obtained for System Programming course in University examination, POE, and Aptitude test marks are analyzed and are shown in Fig. 3.

Data on x-axis represents marks while y-axis represents students. Group-A students showed better performance in all three assessments compared to Group B & C students who were not subjected to any kind of interaction. The *t*-test results also show significant statistical difference among the groups for three interventions as shown in Figs. 4 and 5. Group-A students showed better understanding of concepts as compared to Group B & C.

Table 6 Feedback form (at the end of the semester)

	SA	A	N	D	SD
Whether objective test after completion of chapter was useful?	60	40	–	–	–
Whether objective test covered complete chapter?	43.33	53.33	3.33	–	–
Whether discussion after objective test was useful?	46.67	46.67	6.67	–	–
Whether concepts about particular chapter were cleared?	43.33	43.33	13.33	–	–
Whether concepts about SP subject were cleared?	36.67	50	13.33	–	–
Am I more confident due to these tests for examination?	53.33	26.67	20	–	–
Question bank containing all objective tests will be useful at the time of examination.	50	43.33	6.67	–	–
Question bank containing all objective tests will be useful for practical oral examination.	26.67	46.67	23.33	3.33	–
Question bank containing all objective tests will be useful in exam like Aptitude test, GATE examination, etc.	50	43.33	6.67	–	–

Table 7 Feedback form (after the university examination)

	SA	A	N	D	SD
I got more interest in studying for examination of SP	36.67	40	23.33	–	–
Question bank containing all objective tests was useful at the time of examination	43.33	56.67	–	–	–
Question bank containing all objective tests was useful practical oral examination	26.67	56.67	16.67	–	–
Question bank containing all objective tests was useful in exam like Aptitude test	56.67	33.33	10	–	–
Question bank containing all objective tests was useful in exam like GATE examination	60	33.33	6.67	–	–

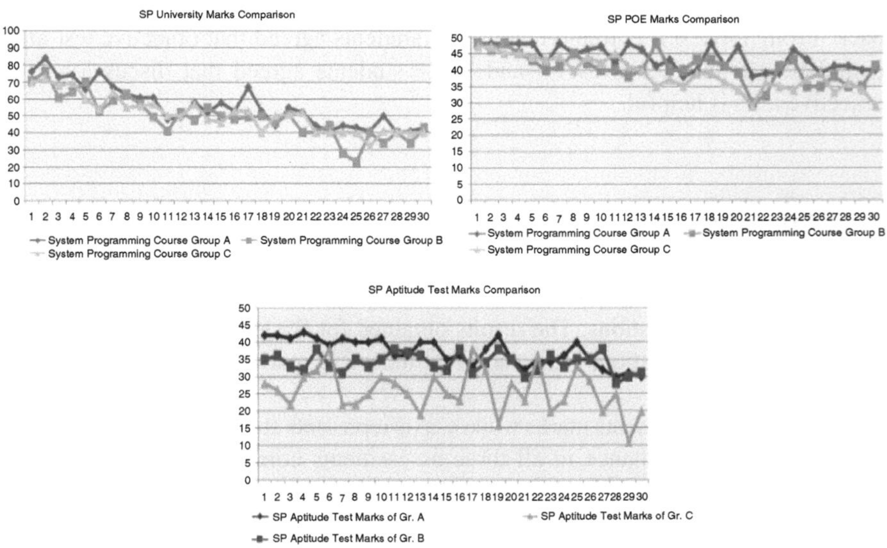

Fig. 3 Performance of students in SP

Test	p=0.05	t- value	df
Objective	p=.01	3.14 (significant)	78
POE	P=.05	4.10 (significant)	78
AT	P=.05	4.73 (significant)	78

Fig. 4 *T*-test result for Group A and Group B

Test	p=0.05	t-value	df
Objective	0.05	4.10(significant)	58
POE	0.04	4.21(significant)	58
AT	<0.01	8.30(significant)	58

Fig. 5 *T*-test result for Group A and Group C

5 Conclusion

This paper presents a case study on improving students' conceptual understanding in System Programming course using objective type tests and group discussion technique. It is observed that these teaching techniques are useful to improve students' interaction, in addition to improvement in conceptual and practical knowledge of the course. The intervention used in the case study is very popular in other parts of the world but it is still not used on a large scale as a part of pedagogy in several parts of India. Therefore, this paper emphasizes the application and importance of this teaching technique for various engineering courses conducted in Indian context.

References

1. Marsh EJ et al. The memorial consequences of multiple-choice testing. Psychon Bull Rev. 2007;14(2):194–9.
2. Carpenter Jason M. Effective teaching methods for large classes. J Fam Consum Sci Educ. 2006;24(2):15–9.
3. Simkin MG, Kuechler WL. Multiple choice tests and student understanding: what is the connection? Decis Sci J Innov Educ. 2005;3:73–97.
4. Roediger III HL, Marsh EJ. The positive and negative consequences of multiple-choice testing. J Exp Psychol Learn Mem Cognit. 2005;3(5):1155–9.
5. Marsh EJ, Lozito JP, Umanath S, Bjork EL, Bjork RA. Using verification feedback to correct errors made on a multiple-choice test. Memory. 2012;20(6):645–53. doi:10.1080/09658211.2012.684882. Epub 2012 May 29.
6. Bonwell CC, Eison JA. Active learning: creating excitement in the classroom. Washington, DC: School of Education and Human Development, George Washington University; 1991.

7. Burton SJ, et al. How to prepare better multiple-choice test items: guidelines for university faculty. Accessed 31 Dec 2013.
8. Little JL, Bjork EL. The persisting benefits of using multiple-choice tests as learning events. Accessed 31 Dec 2013.
9. McCoubrie P. Improving the fairness of multiple-choice questions: a literature review. Med Teac. 2004;26(8):709–12.
10. McConnell JJ. Active learning and its use in computer science. ACM SIGCUE Outlook. 1996;24(1–3):52–4.
11. Black P, Wiliam D. Assessment and classroom learning. Assess Educ. 1998;5:7.
12. Aher SB. Effect of periodic sTiGD activity on long-term performance in system programming course, Technology for Education (T4E) 2013. In: Proceedings of the 5th IEEE international conference on Technology for Education (T4E), Indian Institute of Technology, Kharagpur, 18–20 Dec 2013, p. 91–2.
13. http://en.wikipedia.org/wiki/Bloom%27s_taxonomy. Accessed 1 Jan 2014.
14. http://ww2.odu.edu/educ/roverbau/Bloom/blooms_taxonomy.htm. Accessed 1 Jan 2014

Twenty-First Century Classroom Engineering – Designing Effective Learning Environments: A Conceptual Case Study

Arunkumar C. Giriyapur and B.B. Kotturshettar

Abstract A lot of emphasis is being placed on the development of the global engineer with the recommended twenty-first century skills. The shift to outcomes based engineering education and accreditation is already complete in the developed countries while the shift is still in progress in India. Rapid changes in technologies, a diversified classroom populated with students having different epistemological beliefs and thinking and rising cost of education are putting a lot of pressure on engineering education in the developing countries. This calls for innovations in the engineering curricula. The challenge of the twenty-first century skills is applicable to both the engineering teachers and the students. This paper makes an attempt to show that by adopting proper instructional design principles and a right balance of both objectivist and constructivist pedagogies combined with a technology enriched classroom a true transformation of the engineering teacher to a teaching engineer can happen in the context of classroom engineering or re-engineering. This should enable the teachers to move from a purely didactic approach to teaching to a more constructivist based teaching environment. This paper draws upon various research papers, reports and recommendations and outlines a conceptual design approach to create effective learning environments for classroom engineering and enable the teaching engineer. A typical course in engineering design has been taken as an example.

Keywords Effective learning environment • Classroom engineering • Teaching engineer • Objectivist approach • Constructivist approach • Courseware design

A.C. Giriyapur (✉)
Department of Automation and Robotics, BVB College of Engineering and Technology, Vidya Nagar, Hubli, Karnataka 580031, India
e-mail: aaron@bvb.edu

B.B. Kotturshettar
Department of Industrial and Production Engineering, BVB College of Engineering and Technology, Vidya Nagar, Hubli, Karnataka 580031, India
e-mail: bbkshettar@gmail.com

R. Natarajan (ed.), *Proceedings of the International Conference on Transformations in Engineering Education: ICTIEE 2014*, DOI 10.1007/978-81-322-1931-6_59,

1 Introduction

In India 75 % of the engineering graduates are taught at the private engineering colleges and universities. A lot of changes have happened in the field of engineering education in India and more so in the developed countries. Froyd et al. [1], have identified five major shifts that have reshaped or are currently reshaping, engineering education over the past 100 years. The first shift was moving from a practice based approach to a science based approach which happened after the World War II. The second was outcomes based approach based largely on the actions of the Accreditation Board of Engineering and Technology (ABET). The first two shifts have been completed. The three shifts that are still in progress are: (1) a renewed emphasis on design; (2) the application of research in education, learning, and social-behavioural sciences to curricula design and teaching methods; and (3) the slowly increasing prevalence of information, communication, and computational technologies in engineering education. Rajala [2] has suggested that there is a need for a sixth major shift in engineering education i.e. the integration of attributes of a global engineer. In India the shift towards outcomes based accreditation is still in progress. This shift requires the engineering teacher to grow from an effective teacher to a scholarly teacher and beyond. Sanghi [3] in his article "The twenty-first century engineering crisis" states that there are two problems with our existing University education setup:

- Lack of innovation because of standardization of curriculum structure and delivery with hundreds of engineering colleges under a single university
- The students are being trained to solve problems of the developed world.

Added to these are a number of other challenges such as:

- A highly diversified student population with different epistemological beliefs and thinking
- Rapid changes in technologies
- Aging professors and new inexperienced faculty
- Faculty lacking in experience with real life engineering projects as well as engineering education research
- Rising cost of education
- Slowing job market
- Uncertain economic scenarios

Hence twenty-first century has lots of challenges and opportunities for engineering education in India. The awareness of outcomes based education is slowly spreading and a lot of emphasis is also being laid on developing the engineering graduates with twenty-first century skills and enabling the scholarly teacher to enhance the teaching-learning environment in engineering education. This requires changes in the curricula where the course design and delivery has a proper balance of breadth and depth. The outcomes based education is an objectivist approach and it becomes imperative to move beyond and add a constructivist approach to enhance engineering classrooms for the twenty-first century. Time has now come for the engineering teacher to get involved in classroom engineering to enhance the teaching-learning activity.

2 Need for Twenty-First Century Classroom Engineering

Duderstadt [4] has stated that "It could well be that faculty members of the twenty-first century college or university will find it necessary to set aside their roles as teachers and instead become designers of learning experiences, processes, and environments". Ramo [5] also essentially predicted the rise of engineering education as a separate discipline, "There is probably a new profession known as 'teaching engineer,' that kind of engineering which is concerned with the educational process and with the design of the machines, as well as the design of the material." Chan and Fishbein [6] have noted, "The engineer's role in society is no longer, if it ever was, limited to solving technical issues. Today, broader issues with society-wide and global implications require the profession's attention". Engineering education and the profession are confronting a challenging crossroad. The engineering teacher now has a greater role to play. The teacher should be able to design the learning experiences, the teaching-learning processes and an effective learning environment in the classroom. This can be termed as classroom engineering. To reengineer the existing classroom, the outcome based approach can be extended by including a constructivist approach. Here the teacher can act as a consultant, and also coach and provide required scaffolding for active learning. As the students become more proficient the coaching can be diminished and the scaffolding gradually removed so that the students become more autonomous. The reengineering can include factors like proper breadth and depth in the course design and classroom engineering to build the right kind of environment for active learning. Classroom engineering can play a dual role of enabling the student as well as the teacher as shown in Fig. 1. Hence the key to twenty-first century classroom engineering is to design effective learning environments. Effective learning environments have the ability to empower the learner towards constructivism and at the same time enable the teachers to move from a purely didactic approach to teaching to a more constructivist based teaching environment. The use of technologies in the design of the learning environments has the potential to expand choices about how we teach and learn.

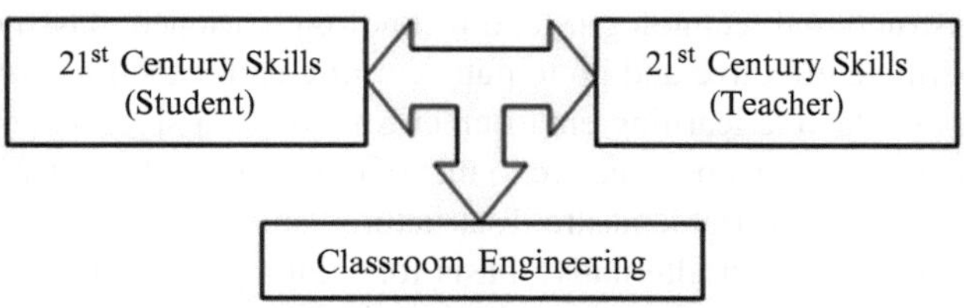

Fig. 1 Classroom engineering

3 Effective Learning Environments

"Learning itself cannot be *designed.* It can only be *designed for* through the design of learning environments." While re-engineering the classroom, the teaching engineer, has to lay more emphasis on design, learning, social-behavioural science research and the role of technology. An effective learning environment is the intersection of good design approach and the application of sciences of learning. Some of the basic requirements of effective learning environments can be written as follows:

- Capture learners' attention (intrigue, inspire, confound, personalize)
- Build on prior knowledge/experience (or lack thereof)
- Facilitate active, social, collaborative engagement
- Convey a sense of presence/immediacy among participants
- Support and challenge diverse learners by providing multiple, accessible methods of content presentation, student expression, and student engagement
- Demand practice and provide for ongoing assessment and feedback
- Engage in authentic representations of real situations
- Facilitate development of process skills and metacognition

Some of the benefits of effective learning environments can be:

- Problem/Inquiry based approach brought into teaching-learning process
- Faculty move from lower level to higher level i.e. as scholars/researchers
- Students as scholars/researchers
- Scholarly endeavors enrich classroom lecture/presentation
- Scholarly endeavors incorporated into classroom experience

4 Conceptual Design of Effective Learning Environment

A lot of emphasis has been laid by many researchers on the use of higher order thinking skills in the construction of knowledge and understanding and the requirement of the learners to be engaged with associated activities in their learning process. Technology itself does not bring about the learning but it has been observed that technology-enriched classroom environment has a positive effect on students and leads to acquisition of higher order thinking skills. An appropriate blend of learning theory, instructional technologies and technology-enriched classroom environments can facilitate effective and up to date learning and teaching. The design and development of effective learning environments involves a significant shift from a behavioural and objectivist perspective to that of cognitive and constructivist. Since most engineering design problems are ill-structured and open ended constructivism provides designers with an alternative basis for thinking. Didactic instruction has always been the paradigm in traditional education where the learner is passively receiving information. The challenge is now to shift the emphasis on constructivism approach to learning at the same time maintaining the outcomes based objectivist approach. Illeris [7] has stated that learning includes three dimensions namely, the cognitive dimension of knowledge and skills, the emotional dimension of feelings and motivation, and the social dimension of communication and cooperation – "all of which are embedded in a societally situated context".

According to Gance [8] the main pedagogical components commonly associated with constructivist learning model are:

- A cognitively engaged learner who actively seeks to explore his environment for new information.
- A pedagogy that often includes a hands-on, dialogic interaction with the learning environment.
- A pedagogy that often requires a learning context that creates a problem-solving situation that is realistic.
- An environment that typically includes a social component often interpreted as actual interaction with other learners and with mentors in the actual context of learning.

The proposed design of effective learning environment is centered on a focus problem and involves the student in a learning cycle that consists of four distinct phases:

- Activation of prior experience
- Demonstration of skills
- Application of skills
- Integration of these skills into real-world activities.

The proposed design uses the 4C/ID model as given by Van Merrienboer [9] and first principles of instruction by Merrill [10] as reference. The elements in the conceptual design of an effective learning environment are shown in Fig. 2.

The learning environment is made up of instructional activities which are instructor centered, learning activities which are student centered, and feedback and assessment which are organized so that there is a progression of knowledge and skill components and the student gets to demonstrate metacognitive skills at the end of the course. The design objective is to achieve a balance between the breadth and the depth of the course design and delivery and at the same time to achieve all the targeted outcomes. The course design starts with the course learning goals which are used to identify the learning and instructional activities. The instructional activities are derived from the first principles of instruction and are aligned to the learning activities. The course learning outcomes are identified in alignment with the course learning goals and are used for assessment and feedback. The instructional activities can be adjusted based on the feedback. The elements of the course design are shown in Fig. 3.

The prescriptive design process for the effective learning environment design consists of five stages as shown in Table 1.

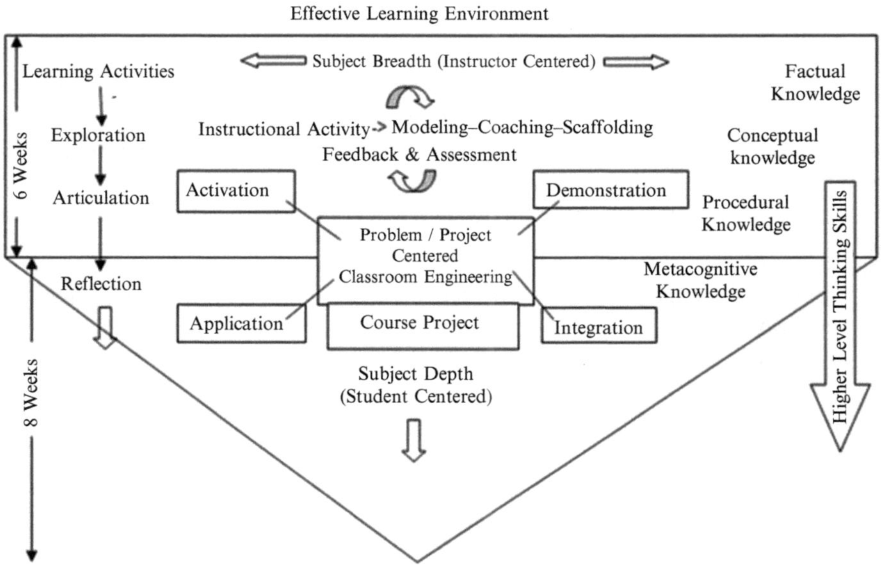

Fig. 2 Conceptual design of classroom engineering for effective learning environment

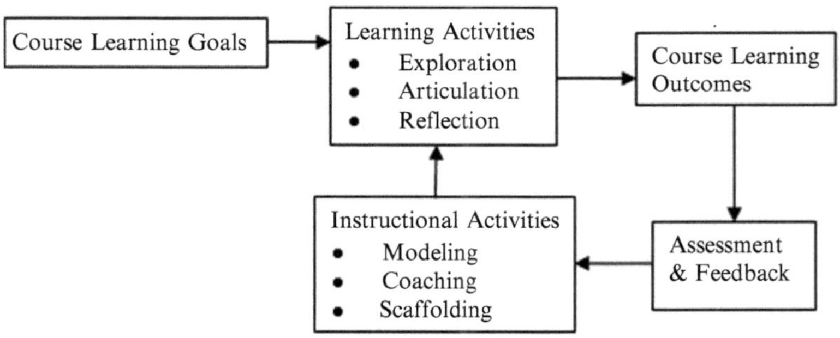

Fig. 3 Elements of course design

5 Courseware Design Example: Engineering Design

A course on engineering design is taken as an example to demonstrate the courseware design. The courseware design is done keeping in mind the first principles of instruction given by Merrill [10] so that a relationship between the instructional objectives and instructional activities are established as shown in the Table 2.

The complete course is divided into two phases. In phase I the student teams are formed and the problem is stated. The teacher then demonstrates the whole process of engineering design using a real world problem. The whole task is broken into sub-tasks and arranged in progression and then demonstrated. Various

Table 1 Stages in prescriptive design process

Stage	Activities
1	Identify desired learning goals and learning outcomes.
2	Identify the whole task or the focus problem(s) of the type that students will be taught to solve by the teacher.
	Identify a progression of problems of increasing difficulty or complexity or breakup the whole task into sub-tasks or sub-problems and arrange them in progression.
	Identify the component knowledge and skills required to complete each task or solve each sub-problem in the progression.
	Select the instructional strategy that will be used to engage learners in solving the problems and help them to acquire the component knowledge and skills required to complete the sub-tasks or solve the sub-problems.
	Plan learning experiences and instructions in terms of breadth (instructor centered) and depth (student centered).
3	Select appropriate tools and techniques
4	Establish feedback and assessment
5	Student course project

Table 2 Instructional strategy

Instructional objectives	First principles of instruction	Courseware design
Problem-centered instructions	Learning is promoted when learners are engaged in solving real-world problems	The courseware is presented in the context of real-world problems. Students are shown the problem and engaged at both the task and operation level and also involved in a progression of tasks or problems.
Activation of prior experience	Learning is promoted when existing knowledge is activated as a foundation for new knowledge	The courseware is designed to activate prior knowledge or experience. The students are directed to recall past experience and provided relevant experience. A prescriptive based structure is used.
Demonstration of skills	Learning is promoted when new knowledge is demonstrated to the learner	The courseware demonstrates what is to be learned. The demonstrations are consistent with the instructional goals.
Application of skills	Learning is promoted when new knowledge is applied by the learner	The students will apply their newly acquired knowledge and skills to complete a set of tasks or problems. The application is consistent with the instructional goals and involves a progression of tasks or problems with feedback. Students are provided with gradually diminishing coaching.

(continued)

Table 2 (continued)

Instructional objectives	First principles of instruction	Courseware design
Integration of these skills into real-world activities	Learning is promoted when new knowledge is integrated into the learner's world.	The courseware provides techniques that encourage students to integrate the new knowledge or skill to complete a course project. The students have to publicly demonstrate their new knowledge through their course project, reflect on their new knowledge and create new ways to use their new knowledge.

resources, tools and techniques are used during the instructional activities which consist of modeling, coaching and scaffolding. The students are coached to undergo learning activities like exploration, articulation and reflection. The teacher provides scaffolding wherever required and gradually removes it so that the students become more autonomous. Phase I involves all the activities of from stage 1–4 as shown in Table 1.

The learning goals are the drivers and they can be mapped to the course learning outcomes which in turn can be used for assessment and feedback. The outcomes based accreditation criteria (ABET), a–k, are used for outcomes assessment. The course learning goals are framed as follows:

- Perform a systematic need analysis based on contemporary issues and formulate an open ended problem statement. (ABET criteria – a, e, j)
- Apply the engineering design process to the indentified problem and develop a working model or a prototype. (ABET criteria – a, b, e)
- Work in a team to complete the above two activities. (ABET criteria d, g)
- Apply critical thinking and creative problem solving techniques to design systems, subsystems and components, individually and in a team. (ABET criteria a, c, e, i)
- Apply the necessary mathematical and computer based tools to solving complex design problems. (ABET criteria a, k)
- Apply appropriate research methods including design, data analysis, and interpretation to a research project. (ABET criteria a, c)
- Practice ethical behavior while engaging in course learning activities. (ABET criteria f)
- Communicate effectively in an oral presentation and written project report. (ABET criteria g)

In phase II the student teams are autonomous and using the metacognitive knowledge and skills developed has to complete the course project. During the course project stage the teacher will act as a consultant and a mentor. Stage 5 is part of phase II. The student course project will be of 8 weeks duration. The students work in autonomous mode most of the time with the teacher acting as a consultant. The teacher may provide scaffolding if required and gradually remove the scaffolding. The various phases in the course project and the time durations are shown in Table 3.

Table 3 Student course project

Student centered activity – *course: engineering design*	
Team formation, project proposal and objectives	2 weeks
Generating different alternative solutions for selected problem using morphological charts and choose best alternatives	2 weeks
Design, model and analyze the selected alternative and fabricate the model	2 weeks
Testing of model, presentation and report submission	2 weeks

Assessment: The assessment of the various outcomes is done as per the Table 4.

Table 4 Assessment

Assessment	Learning outcomes	Assess what?
Student work samples – prototypes	b, c, j, k	Integrating learning experience
Project-embedded assessment	e	Skill development
Course-embedded assessment	a	Critical thinking skills
Internal juried review of student projects	d	Student progress
External evaluations of student performance	g	Oral & written presentation
Document analysis, e.g., meeting minutes, policies, design notebook	f, i	Team discipline, work-breakdown structure
Performance on a case study/problem	h	Connectivity with the real world
Performance on problem and analysis	a	Problem solving skills

6 Conclusion

It can be seen that a proper instructional designs and a right balance of outcomes based objectivist and constructivist pedagogies based approach enhances teaching-learning experiences by providing proper breadth and depth to the course design and delivery. Since the approach is that of problem based learning, most of the criterions of the outcomes based accreditation will be naturally satisfied at the same time enhancing metacognitive knowledge of both the student and the teacher. Hence the above approach not only helps in shifting to outcomes based design and delivery but also enables the teacher to move to a higher level and enable him/her to carry out educational research.

References

1. Froyd JE, et al. Five major shifts in 100 years of engineering education. In: Proceedings of the IEEE, vol. 100, 13 May, 2012.
2. Rajala SA. Beyond 2020: preparing engineers for the future. In: Proceedings of the IEEE, vol. 100, 13 May, 2012.
3. Sanghi D. The 21st century engineering crisis. EDuTECh, July 2010.
4. Duderstadt JJ. The challenges and opportunities of the digital age for higher education. Seminar on academic computing, Colorado. Available at: http://www.mvac.org/Snowmass.html. Accessed 1997.

5. Ramo S. A new technique of education. IRE Trans Educ. 1958;1(2):37–42.
6. Chan ADC, Fishbein J. A global engineer for the global community. J Policy Engagem. 2009;1(2):4–9.
7. Illeris K. A comprehensive understanding of human learning. In: Human learning: an holistic approach. London: Routledge, 2005. p. 87–100.
8. Gance S. Are constructivism and computer-based learning environments incompatible? J Assoc Hist Comput 2002;5(1).
9. Van Merriënboer JJG, Clark RE, De Croock MBM. Blueprints for complex learning: the 4C/ID-model. Educ Technol Res Dev. 2002;50(2):39–61.
10. Merrill MD. First principles of instruction. Educ Technol Res Dev. 2002;50(3):43–59.
11. Hopson MH, Simms RL, Knezek GA. Using a technology-enriched environment to improve higher-order thinking skills. J Res Technol Educ. 2002;34(2):109–20.
12. Fink LD. A self-directed guide to designing courses for significant learning. San Francisco: Jossey-Bass, 2003. p.1–35.
13. Entwistle NJ, Peterson ER. Conceptions of learning and knowledge in higher education: relationships with study behaviour and influences of learning environments. Int J Educ Res. 2004;41(6):407–28.
14. Goel S. Competency focused engineering education with reference to IT related disciplines: is the Indian system ready for transformation? J Inf Technol Educ. 2006;5(27):27–52.
15. Tryggvason G, Apelian D. Re-engineering engineering education for the challenges of the 21st century. JOM J Miner Met Mater Soc. 2006;58(10):14–7.

Project Based Learning in Laboratories Using Open Source Technologies: Case Study of a Frugal Approach

Shridhar T. Doddamani and Arunkumar C. Giriyapur

Abstract A course on Microcontrollers is being offered in the Department of Automation and Robotics at BVBCET since last 2 years. The students enrolled in this department come from different disciplines such as Mechanical, Mechatronics, and Electronics. Teaching microcontrollers for such students is a challenging task because of the diverse background of the students. In this paper we propose a course design which will hold the interest of the students and at the same time be able to teach about the current microcontrollers. The course is so designed that it has three objectives. One, the student should be able to program both in assembly and C language. Two, use real-world concepts and techniques to identify the real world problems and solve to automate them. Third and the important objective is to make students capable of using this course in mini, minor and major projects and in their industrial career. The laboratory was developed using a frugal approach in which low cost open source kits were identified and acquired in more numbers so that the students could easily take the kits and work wherever they want. The experiments and projects were built around these open source inexpensive kits. The results are very encouraging and they show that all the students took interest and were comfortable with microcontroller kits and were able to build good projects at the end of the laboratory.

Keywords Microcontrollers • Assembly and C language • PIC microcontroller development board and Arduino

---

S.T. Doddamani (✉) • A.C. Giriyapur
Department of Automation and Robotics, BVB College of Engineering and Technology, Vidya Nagar, Hubli, Karnataka 580031, India
e-mail: shridhar_d@bvb.edu; aaron@bvb.edu

R. Natarajan (ed.), *Proceedings of the International Conference on Transformations in Engineering Education: ICTIEE 2014*, DOI 10.1007/978-81-322-1931-6_60,

1 Introduction

A course on embedded systems is included in most of the undergraduate programs in Electrical and Computer science disciplines, while other programmes like mechanical and production sciences are now realizing the importance of the embedded microcontroller. The microcontrollers have become ubiquitous and are found in a variety of applications starting from a washing machine to an industrial robot. It is important for students working in multidisciplinary teams to learn embedded systems so that they can handle embedded systems efficiently to solve the problems related to their discipline [1]. This paper proposes a method to design and implement laboratory experiments that allow students to study and understand the real world problems in the context of embedded systems and solve them.

Microcontroller courses are usually difficult to teach and learn because of rapid growth in technology, increased complexity and application in real-time systems. Teaching the application aspect of microcontrollers requires students to gain extra knowledge in the application domains. Teaching and learning the microcontrollers can be even more difficult when students come from diverse backgrounds. But a well organized, application oriented and project based [2] teaching will help in motivating the students to gain the required knowledge. To meet these objectives, the following steps have been taken during the course delivery:

1. Co-relate theory and laboratory.
2. Identify real-world problems.
3. Play with different laboratory equipments like sensors and actuators.
4. Collect and analyze the data and
5. Be motivated to learn new things.

Section 2 discusses the development strategies and platforms selected and also explain about the approaches followed in the laboratory. The methods used to motivate the students towards the course and use of modern engineering tools is explained in Sect. 3. The Sect. 4 reports student's feedback on the course and discusses about course learning objective.

2 Development Strategies and Platforms

The microcontroller theory is taught using PIC microcontroller, where students learn about the internal architecture along with I/O port programming, timers, serial communication, interrupt programming, interfacing of hardware, ADC and DACs [3]. While in the laboratory students will work on two different microcontrollers i.e. PIC microcontroller and Arduino [4]. This helps the student to understand the behavior of different microcontrollers, analyze results and compare their performance.

Figure 1 shows the steps followed for each laboratory experiment. Teams of four are setup in the laboratory and each team is provided with a common goal (problem

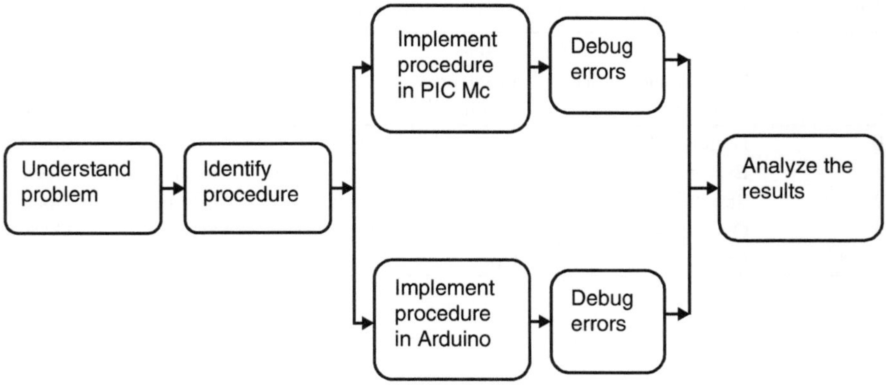

Fig. 1 Steps followed for each laboratory experiment

statement) and components. Students will analyze the given problem and identify the procedure required to implement. Once the procedure is developed each team is divided into sub-teams of two students in which one sub-team work on PIC microcontroller and other will work on Arduino and implement the obtained procedure to get the results and analyze the results. After the completion of the given task the teams will then interchange the components and repeat the process. Each sub-team will teach the other sub-team about their experience. This method will help in understanding the procedure; boosts in achieving the objectives in short time and improve their communication skills. The teams not only learn problem solving skills but also learn how to identify the suitable microcontroller and components based on the needs and requirements.

2.1 *Structured Approach to Laboratory Using Categorization*

A structured approach is followed where laboratory experiments are categorized into five different levels [5] according to the requirements and the extent to which they are open. The laboratory progresses from teacher-led demonstrations, where all aspects of the laboratory are given by the teacher, to student-led projects in which all aspects of the design are open to the student. The categorization is shown in Table 1.

2.2 *Configuring the Experiments and Projects*

The introduction to the laboratory is started with real world interfacing problems. In the first level as shown in Table 1 the students are demonstrated interfacing of various input and output devices to the microcontroller. Examples of input and output

Table 1 Levels of autonomy for different exercises in microcontroller laboratory

		Givens			
Level of autonomy	Types of laboratory exercise	Problem statement	Procedure	Components required	Results
0	Demonstration	Given	Given	Given	Given
1	Exercise	Given	Given	Given	Open
2	Structured enquiry	Given	Given	Open in part	Open
3	Open-ended enquiry	Given	Open in part	Open	Open
4	Project	Open	Open	Open	Open

devices are LED, LCD, seven segment displays, sensors and dc motors. In the second level, students have to complete exercises on interrupts, 4×4 keypad matrixes, serial communication and controlling the speed and direction of a DC motor. In the third level, students have to apply a structured enquiry process and develop the required procedure. They also need to identify the components and analyze the results at the end. Here students work on controlling speed and direction of DC, stepper and servo motors and interface different sensors. In level four students work on open ended enquiry wherein only the problem statement is given and students need to understand the problem statement and solve to get the procedure and identify the components required and analyze the results. Here students implement real world applications like controlling the fan speed using temperature sensor, obstacle detection and indication to car driver. Finally the students will develop a system considering the real-world problem as a laboratory project.

2.3 *Student Projects*

Once the students complete the structured enquiry and open-ended enquiry exercises they are capable of developing small projects in the laboratory. Some of the projects implemented during the class and laboratory activity are:

Traffic light control based on density.
Water level indicator.
Battery voltage indicator.
Temperature based fan speed control.
Obstacle detection robot.
Line following robot.
Automated door open and close.
Automated railway gate using sensors, motors, buzzer and LED.
Detect the violation of traffic signal using sensor, RGB LED and buzzer.
Security buzzer in automobiles using sensors and buzzer.

Table 2 Cost of laboratory components

Components	Quantity	Cost per component in INR	Total cost in INR
PIC microcontroller development board	10	600	6,000
Arduino board	10	550	550
Arduino shield	5	350	1,750
Motors	20	150	3,000
16/2 LCD display	5	10	500
Wire/connectors and LEDs	200	1	2,000
Sensors and keypad	25	80	2,000
Seven segment LED.	5	70	350
MPLAB, Mikro C and PICkit2	45	0	0
	Total:		15,300

2.4 *Overall Cost of Setting Up the Experimental Resources*

A frugal approach has been followed in microcontroller laboratory i.e. using the low cost components that gives the good performance. Table 2 shows the cost of each component and total investment for the microcontroller laboratory during the course.

3 Methodology and Contributions

During the course different methodologies have been adopted that helped students to stay motivated towards subject and get more knowledge in a short time. The methods adopted are:

Presentations on current trends. To motivate students towards theory as well as laboratory, teaching of present developments and applications of microcontroller in embedded world is done by showing the videos of microcontroller application in different areas like home, industrial, commercial application and projects developed by hobbyist. As a presentation in class students need to choose one important application of microcontroller that is applicable to modern world and give presentation for 10 min.

Circuit Simulation. Hardware is costly and damage to it leads to a security issue. So simulating using Proteus [6] virtually before implementing in hardware will help students to identify the problem and solve it and then implement physically.

On-spot project event using Arduino with shield. After completion of structured enquiry a workshop kind of activity is performed to teach and make students capable of using the low-cost hardware efficiently for various real-world applications. A 4 h laboratory workshop is divided into two parts:

(1) *Learning and understanding*. This is first and important part in which students get exposed to new components and make comfortable them self by

doing hands on exercises that are given by the instructor and try to modify given program and analyze the behavior of all the features present in the given hardware.

(2) *Identification of real-world problems and implementation.* Once students get comfortable with the components and their working, they are asked to identify the real world applications and implement the same with available components.

This activity has helped students in understanding how to use the available resources effectively within a given time and build a component or product.

Interfacing Arduino with modern engineering tools. The study of microcontroller would not be complete until it's been integrated with modern engineering tools like NI LabVIEW [7, 8]. Students have learnt how to interface Arduino with NI LabVIEW and perform different operations like motor control, reading sensor values etc.

4 Results and Analysis

Initially when the course was started students were a little apprehensive. After the initial laboratory classes' students started showing interest in the subject which resulted in an improved performance in their continuous internal evaluation. Figures 2 and 3 show some of the projects developed by student during the course.

Fig. 2 (**a**) Line following robot (**b**) Automatic door open and close (**c**) Obstacle avoiding robot

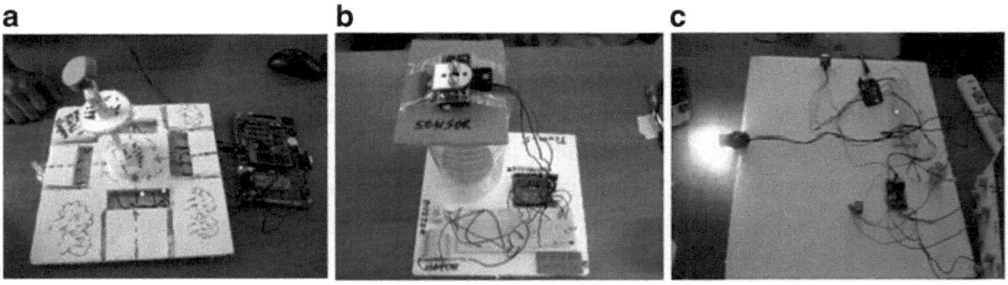

Fig. 3 (**a**) Traffic signal indicator (**b**) Automatic water level indicator (**c**) Automatic light on/off control

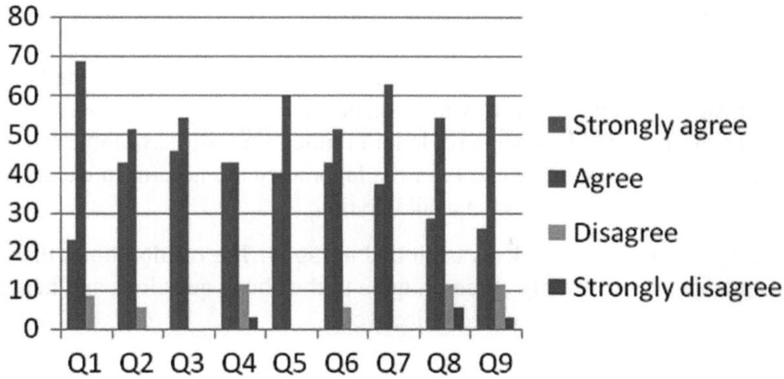

Fig. 4 Feedback analysis

The line following robot is shown in Fig. 2a which follows white line on a black surface. Figure 2b shows the simulation of automatic door open and close in which door will open by sensing the human and closes automatically after some delay. Figure 2c shows the robot that avoids the obstacle, where after sensing any of the obstacles the robot takes left or right turn randomly. Figure 3a shows the simulation of traffic signal where signal is controlled based on the density. Figure 3b shows the simulation of water tank indicator and control, in which when water gets empty buzzer will give a beep, a led will blink and motor starts automatically to fill the tank and once the tank is full the sensor senses the fullness and microcontroller sends stop signal to motor and motor gets off to avoid the overflow. Automatic light control is demonstrated in Fig. 3c in which the system senses the presence of human movement and automatically put on the light and gives a buzzer and automatically switch-off the light when there is no human movement.

4.1 Feedback Analysis

A survey was conducted at the end of the semester putting the nine questions to students:

Q1. The microcontroller laboratory has succeeded in convincing the concept which is taught in theory.
Q2. The concepts which are taught in laboratory will help in handling the mini-project/major projects.
Q3. The exercises which are done in laboratory will help to identify the real world problems and solve them.
Q4. The laboratory practices will help to work in group and communicate effectively.

Q5. The presentations which are given by students have helped to motivate towards subject.
Q6. The activities which are performed in laboratory will help in lifelong learning.
Q7. The instructor in the laboratory has succeeded in solving problems.
Q8. Got ability to solve the given problem in limited time with available resources?
Q9. The components (kits) available in the laboratory are helpful in executing the programs with less effort and in available time.

The response from the students is plotted in Fig. 4. The results shows that many of the students either strongly agree or agree to the above questions and hence feel that the course has met the required objectives.

4.2 Course Learning Objective

At the end of the course students are able to meet the following objectives:

Demonstrate knowledge of the basic building blocks of PIC MICROCONTROLLER and Atmega328 microcontrollers.
Ability to differentiate between wide varieties of microcontrollers.
Ability to write the program both in assembly and C code using open-source software like: MPlab, MiKroC, processor etc.
Ability to simulate in virtual environment and test on real hardware
Interface different devices/components with the microcontroller
Identify and develop autonomous systems that are applicable to the real world

5 Conclusion and Future Goals

The teaching practices indicate that if students are motivated using the above methods, they can easily achieve the desired goal/objectives in a very short time. This paper discussed methods like class presentations, quiz's, peer learning, exposure to new technologies and project based learning and how these methods have helped in understanding the basic concepts of microcontrollers and their application to real world problems. The utilization of low-cost and open source software are the added advantage where students can think of doing real-world and low-cost projects. Results and feedback shows that students are happy with the way they have learnt the microcontroller course. In future, the laboratories can be more open so that students with limited background can be motivated to gain the knowledge while self-motivated students can be more creative and productive.

References

1. Pasca S. The challenge of teaching electronic hardware for today's medical engineering students. Advanced topics in electrical engineering (ATEE), 2013 8th international symposium on IEEE, 2013.
2. Chu RH, Lu DD-C, Sathiakumar S. Project-based lab teaching for power electronics and drives. IEEE Trans Educ. 2008;51(1):108–13.
3. Mazidi MA et al. Pic microcontroller and embedded systems. Upper Saddle River: Prentice Hall; 2005.
4. Margolis M. Arduino cookbook. Sebastopol: O'Reilly Media; 2011.
5. Davies C. Learning and teaching in laboratories. © Higher Education Academy Engineering Subject Centre, Loughborough: Loughborough University; 2008.
6. Brewer EA et al. Proteus: a high-performance parallel-architecture simulator. Cambridge, MA: Massachusetts Institute of Technology. Laboratory for Computer Science; 1991.
7. Johnson GW. LabVIEW graphical programming: practical applications in instrumentation and control. New York: McGraw-Hill School Education Group; 1997.
8. Travis J, Kring J. LabVIEW for everyone: graphical programming made easy and fun, National instruments virtual instrumentation series. Upper Saddle River: Prentice Hall PTR; 2006.

A Study on Methodology and Implementation of Flipped Classroom Teaching for Engineering Courses

Aswin Karthik Ramachandran Venkatapathy

Abstract The aim of this paper was to study the flipped classroom teaching method and its implementation. It describes various technologies which can be easily adapted to the flipped classroom teaching method. Flipped classroom is a pedagogical tool where the learning is transformed from passive learning to active by exposing the student to the course content before the lecture sessions and those conventional sessions are converted to act as a meeting session with the instructor where various activities such as solving worksheets, discussions with the instructor, a hands-on session on the current topic or a debate can be held thereby creating a inclusive learning experience for all the students in the classroom environment. This method also transforms the lecturer from being a pedagogical tool as described in the Victorian classroom styles to a source of knowledge and inspiration for the students. This also provides the time for fostering the purpose on learning the course by citing examples, various real time use cases of different concepts learnt during the course. The course itself is analyzed for its performance and generalized to create a strategy to select courses in different fields of engineering so that it can be a tool to diverge from traditional teaching practices implementing flipped classroom technique.

Keywords Flipped classroom • Methodology • Implementation • Inclusive learning • Technology

A.K.R. Venkatapathy (✉)
Student Platform for Engineering Education Development,
Technical University of Dortmund, Dortmund, Germany
e-mail: aswinkarthik.rv@worldspeed.org

R. Natarajan (ed.), *Proceedings of the International Conference on Transformations in Engineering Education: ICTIEE 2014*, DOI 10.1007/978-81-322-1931-6_61,

1 Introduction

At the forefront of technological innovation and societal modernization, the highly used pedagogical model is still the Victorian style classrooms where the instructor itself acts as the tool and he prepares the lecture from the assumption of what would be helpful for the students to learn the course. As per Sugata Mitra's[1] hypothesis [1] of why such classroom setup is obsolete is because it was organized to create students with identical knowledge to be applied further in various fields of work but as technology progressively expands in folds with a high degree of divergence that the lines of different fields is vanishing. There is high amount of research in the education community to rightly evolve the education system to correctly cater to the progressing technological innovation which in turn will keep the pace of advancements in technology. There are many diverse schools of thoughts on education and one of the successful practices is considered to be online learning of free and open courseware practiced by Massachusetts Institute of Technology. This paper provides insight of how this learning is a subset of flipped classroom method.

2 Inclusive Learning

A wide range of educational, behavioral, and societal research has concluded that there have been distinct as well as individual styles of learning. The learning style has been broadly categorized after Garsha Reichmann Learning Styles Questionnaire (GRLSQ) was analyzed [2, 3]. The three different styles of learning were concluded to be independent, collaborative, and dependent-learning. An independent learner performs best when he or she is left to their own devices. They learn by themselves and the right direction is to be provided by the lecturer during the course of study. Collaborative learners perform when they are in a group while dependent learners require the help of the lecturer to assimilate what is being taught at the lecture [3]. A change in the traditional system could increase the performance of all classes of learners that are described which will increase the performance of the whole participation of the course. The proposed technique to facilitate this kind of performance improvement is the flipped classroom teaching method.

3 About Flipped Classroom Teaching Method

The flipped classroom teaching method describes a reversal of traditional teaching where students gain first exposure to new material outside of class, usually via reading or lecture videos, and then class time is used to do the harder work of

[1] The Victorians were great engineers. They engineered a [schooling] system that was so robust that it is still with us today, continuously producing identical people for a machine that no longer exists.

Table 1 Comparison between traditional and flipped classroom teaching method [5]

Timeline	Conventional classroom	Flipped classroom
Before class	Students assigned material to read	Students guided through learning module and questions are generated
	Instructor prepares lecture	Instructor prepares learning opportunities
Beginning of class	Students have limited information about what to expect	Students have specific questions in mind to guide their learning
	Instructor makes general assumption about what is helpful	Instructor can anticipate where students need the most help
During class	Students try to follow along	Students practice performing the skills they are expected to learn
	Instructor tries to get through all the material	Instructor guides the process with feedback and mini-lectures
After class	Students attempt the homework, usually with delayed feedback	Students continue applying their knowledge skills after clarification and feedback
	Instructor grades past work	Instructor posts any additional explanations and resources as necessary and grades higher quality work
Office hours	Students want confirmation about what to study	Students are equipped to seek help where they know they need it
	Instructor often repeats what was in lecture	Instructor continues guiding students toward deeper understanding

assimilating that knowledge through strategies such as problem-solving, discussion or debates [4]. In this paper, a lecture where flipped classroom technique was employed in an engineering course is studied and all the effects through this teaching method are discussed. The implementation of this teaching method without taking overhead time of the instructor is also described on the later part of this paper. Table 1 gives a brief overview of flipped classroom teaching method and also a comparative study between the conventional classroom and the flipped classroom with a timeline where the instructor and student interaction is expected [5]. There is need for preparation of the learning opportunities by the instructor and also the learning module so that the students when start the course can also start with their learning of the theoretical concepts for the course and generate questions that arise when guided through the learning module. When the students are at the lecture sessions, they have an idea of the session and are more specific about their learning. The session is accompanied with many different tools like debate, quiz, solving worksheets, and practical demonstration as demanded by the course content. This type of teaching method improves the time for interaction between the instructor and the student which can also foster a way to provide motivation on learning the course which lacks in most of the traditional teaching due to the focus on the learning material and not the practical use cases. After the class the students apply the skills acquired through the learning module and the sessions to further apply them on practical applications to deepen their understanding of the learnt concepts. The later part of the paper provides information about identifying the right tools to be used for the course and a methodology followed to successfully implement flipped classroom teaching method for an engineering course that demands a high amount of fundamental concepts to be understood by the student and also requires practical training. It also comments about the inclusive learning and how different styles of learning are influenced with the different aspects of organization of the course.

4 Implementation and Methodologies

A case where flipped classroom teaching method was implemented is considered in this paper for studying the implementation of the method and how this method was adapted diverging from the existing traditional teaching method without creating considerable overhead on the instructor. The course is Cyber Physical Systems Fundamental taught at Technical University of Dortmund. This course demands lab sessions to provide better understanding of the fundamental concepts that are learnt and also requires various demonstrations using the traditional concepts and its evolution to become the state of the art. Preparation of organizing the course started a term ahead of the pedagogical tool discussed here is to be implemented. The traditional lectures are recorded in the previous term as a preparation for the learning module. The course content is refreshed for the following term and the videos that are not in the scope of the course are removed from the learning module that composes the course content but held in the repository to provide additional information. The repository used for videos is YouTube™ since it provides a video streaming platform with capabilities to edit the videos online and compose them into playlists for better accessibility and flow of the course. It is chosen also for the reliability it provides for creating, editing, and publishing the videos online. Figure 1 provides an idea of the video frame composition which was identified as the ideal format as

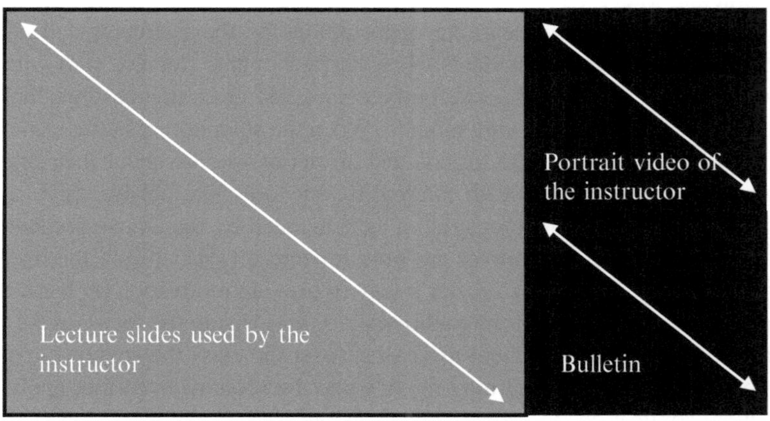

Fig. 1 Video frame showing the composition of the video lectures

Table 2 Methodology used in implementing flipped classroom teaching method

Timeline	Methodology
Before class	Students are provided with a plan for the learning modules. YouTube™ Videos are regarded as the learning module
	Creating worksheets and planning the activities
Beginning of class	Questions are handed over to the instructor
	Students work on the worksheets which resonate with the concepts from video
	Instructor answers all the doubts from the video
During class	Students have the opportunity to have practical training, hands on demonstration or a discussion with an expert from the field of concept that was discussed
	Instructor can facilitate the students with learning activities
	Instructor provides motivation for learning the course, real time uses cases and provide exposure on state-of-the-art research
After class	Students continue with the video after having more motivation on the further topics and work on training themselves with the various tools that were exposed to them during the practical sessions
	Instructor posts answers for the worksheets and plans activities for the next session such as a debate, discussion, or a presentation from a industry expert
Office hours	Students use these hours rarely as most spent time in the sessions interacting with the instructor
	Instructor creates more resources for further learning and creating a profound knowledge on the course concepts

it shows the content of the lecture, the instructor lecturing and a bulletin to share implying information.

This way of recording the videos during the previous lecture sessions reduces the overhead of redoing all the lectures again for the preparation of implementing this pedagogical tool. Adapting this methodology is more feasible for courses which have content that teaches basic concepts of that field of engineering or fundamentals about various practical applications. The classroom setup as called by the traditional teaching is coined here as meeting sessions with the instructor where the questions from the videos are discussed with a mini lecture. The meeting session proceeds with an activity that in this case is worksheets. There are many other recommendations for activities such as debate, quizzing, demonstrations [4, 5]. After the meeting session is concluded, the answers for the worksheets are posted with the mini lecture slides and the actual lecture content from the videos. Now the student has the lecture that is taught by the instructor at his disposal which helps independent learners to learn the concepts by themselves. The meeting session serves as a place where independent learners are motivated to do more whereas dependent learners have the time to interact with the instructor to assimilate information and collaborative learners can learn from the activities all of which stress on deep understanding of the concept that is taught in the video [6, 7] (Table 2).

5 Conclusion

Flipped classroom teaching method, discussed in this paper, provides insights about how such a course can be implemented which followed the traditional teaching method. It discusses the various tools used and the methodology that was employed in implementing such a course. Additional information about positively influencing all the three classes of learning styles together by improving the teaching method is also discussed. It also provides basic information about transforming the learning module into an online learning method. The performance analysis of this pedagogical tool is not discussed in this paper. Further research is to be carried out to create a rubric for analyzing both quantitatively and qualitatively the different parameters of this technique, to prove its effectiveness in improving the performance of the students.

References

1. http://www.ted.com/talks/sugata_mitra_build_a_school_in_the_cloud.html
2. Kolb DA, Fry R. Toward an applied theory of experiential learning. In: Cooper C, editor. Theories of group process. London: Wiley; 1975.
3. Riechmann SW, Grasha AF. A rational approach to developing and assessing the construct validity of a student learning style scales instrument. J Psychol Interdiscip Appl. 1974;87(2):213–23.
4. Center for Teaching, Vanderbuilt University. http://cft.vanderbilt.edu/teaching-guides/teaching-activities/flipping-the-classroom/
5. Center for Teaching and Learning, University of Texas at Austin. http://ctl.utexas.edu/teaching/flipping_a_class/what_is_flipped
6. Lage MJ, Platt GJ, Treglia M. Inverting the classroom: a gateway to creating an inclusive learning. J Econ Educ. 2000;31(1):30–43.
7. Veenema S, Gardner H. Multimedia and multiple intelligences. Am Prospect. 1996;29:69–75.

Curriculum Design and Instructional Practices for Experiential Learning

V.S.V. Laxmi Ramana, A. Padmaja, and P. Rajeshwar Reddy

Abstract In developing countries, many higher education institutions experience a growing gap between their curricula and the demands from society, business, and industry for a more flexible workforce with high skills in problem solving, teamwork, and project management. Many universities want to put a strong emphasis on traditional, disciplinary knowledge production because timeless, universal knowledge is important in a world where everything is in flow. Chronic industry complaints about skill deficiencies in engineering graduates, the gradual dwindling rates of engineering students with good academic performance records, the worldwide adoption of outcome-based engineering program accreditation, and findings from both cognitive science and thousands of educational research studies showing serious deficiencies in traditional teaching methods force the universities for changes in how engineering curricula are structured, delivered, and assessed. This article tries to discuss the various ways suggested by Kolb and Fry's theory of experiential learning and instructional strategies in designing the curriculum. Effective teaching and learning throughout the study was conceptualized as that situated within a social constructivist framework. This is a descriptive study in design. The literature component of the study utilized a content analysis methodology with a view to identifying strategies of practical application and potential to facilitate learning in classes, such as active learning, collaborative learning, cooperative learning, inductive teaching methods, and novel practices for change are suggested.

Keywords Competence-based education • Outcome-based education • Instructional strategies • Experiential learning

V.S.V. Laxmi Ramana • A. Padmaja • P. Rajeshwar Reddy (✉)
Vidya Jyothi Institute of Technology (Anurag Group of Institutions), Hyderabad, India
e-mail: rajeshwar_palla@yahoo.com

R. Natarajan (ed.), *Proceedings of the International Conference on Transformations in Engineering Education: ICTIEE 2014*, DOI 10.1007/978-81-322-1931-6_62,

1 Introduction

Traditionally, the curriculum provides a plan of instruction that indicates structured learning experiences and outcomes for students. It specifies the details of student learning, instructional strategies, the teachers' roles, and the context in which teaching and learning take place. Students can benefit from learning through broad, contextual experiences. In fact, many individuals become more engaged in contextualized learning, thereby improving their preparation for future studies and entry into the professional world. Unfortunately, many institutions have continued to operate in the traditional mode, replicating the early automobile factory model. Students are sent down a narrow, predetermined path, stopping along the way for periodic inspection, and to gather all the required courses for eventual graduation. If students continue to be mass produced, will today's graduates be adequately prepared for life and living? Will they be able to compete in the world class workplace? If educators hope to prepare graduates who can compete head-to-head with graduates from other nations, "product" quality must greatly improve. Here we must note the suggestions provided by Goertz et al.:

> If all students are to learn to new standards, not only teachers, but administrators, teacher educators, and other participants in the education system must change their roles and expectations. Educators, researchers, and policymakers are beginning to explore different ways to enhance the ability of the system and its teachers to improve student learning. But before they can design effective policies, policymakers must determine what capacities are needed and what mechanisms and strategies might foster their development. Most capacity-building strategies in education today are targeted on individual teachers and are designed to enhance their knowledge and improve their instructional skills through the provision of workshops and university courses. Yet, our data and that of other researchers suggest that the traditional model of professional development reflects a limited conception of the dimensions of teacher capacity necessary to support and sustain instruction reform and ignores the role of the school and other communities of practice in teacher learning and educational improvement.

This may conflict with existing practices that are in favor of a centralistic approach. New ways of teaching and learning, creating rich learning environments, designing new forms of assessment also require intensive training and coaching of academic staff and a continuously applied monitoring and evaluation of the curriculum in action. However, global developments in science, society, and economy also affect the developing countries and their higher education institutions are closing the gap between "classical" disciplinary knowledge and know-how required for the new job market. Hence, a new kind of curriculum has to be designed to focus on the *quality* rather than on the *quantity*.

The three areas that the curriculum designers to concentrate are as follows:

1. Identifying curriculum practices and processes that emphasize "all aspects" in the long term
2. Preparing a document that describes these long-term curriculum practices and processes and ways they may be designed
3. Assisting educators at selected school sites with their curriculum implementation

2 The Thematic Curriculum and All Aspects

It is a set of organized learning experiences that provide students with exposure to a broad, predominant content theme. This broad theme (or multiple themes) can serve as an organizer for an entire institution, provide a focal point or furnish students with content where they can engage in self-directed learning. As with any approach, value and benefits are dependent on educator and student learning objectives. Some possible benefits include the following:

- In-depth exposure to a broad field of interest and a wide range of occupations
- Opportunities to learn through more contextualized and authentic learning experiences
- Exposure to linkages between school-based learning and work-based learning
- Greater potential to prepare for both higher education and employment
- Opportunities for teachers to work together on self-directed teams
- Opportunities for counselors and administrators to influence students' career development and school experiences

Listed below are several factors that distinguish "all aspects." The first five factors are considered essential to "all aspects," whereas, factors six through eight are highly recommended since they have been shown to enhance the benefits "all aspects" offers to students.

1. "All aspects" encompasses a broad "industry" or field within the workplace. Illustrative fields include agriculture, business, construction, communication, community services, health, manufacturing, marketing, and transportation.
2. Curriculum content and focus includes instruction in a wide range of industry or field-wide functions, concerns, issues, and technological knowledge and skills. Illustrative of the areas are community issues, environment issues, economic issues, finance, health, labor issues, leadership, management, planning, safety, technological knowledge and skills, and underlying principles of technology.
3. Instruction includes strong emphasis on developing problem solving skills and basic and applied academic skills in technological settings. Academic studies focus on mathematics, oral communication, reading, writing, science, and social and behavioral sciences instruction.
4. Students are provided with extensive experience in and an understanding of those aspects of the broad industry of field they are exploring and/or preparing to enter.
5. Students completing the instruction are able to link their school studies and related experiences directly to a broad industry or field including the functions, concerns, issues, and technological knowledge and skills associated with that industry or field.
6. Teachers actively collaborate with each other in determining content and experiences, planning instructional strategies, and teaching in a manner that integrates instruction and emphasizes contextualized, applied learning.
7. "All aspects" is available to all students regardless of their future education and employment plan and aspirations.
8. "All aspects" is creatively and effectively combined with educational reforms to produce a comprehensive school-to-work system.

3 Exploring Curriculum Options

The thematic curriculum may be organized in a variety of ways. Since the ways themes are arranged can affect *what* is taught and *how* it is taught, implementation options should be considered early in the development process. When considering how thematic curricula can look in a particular counsel, four basic questions need to be answered:

1. What is the context for the thematic curriculum?
2. How may the curriculum be organized within the institution and the workplace?
3. In what different ways can the curriculum be offered to students?
4. What is included in the curriculum content?

The four questions address the contextual, organizational, delivery, and content options that must be explored, both individually and collectively, by persons creating the thematic curriculum. A competence-based curriculum is therefore dependent on the context of the institution offering the curriculum. The technical institutes should have a dialog with the industry to include their need in the curriculum.

4 Types of Experiential Learning

4.1 Competency-Based Education

Competency-based strategies provide flexibility in the way that credit can be earned or awarded, and provide students with personalized learning opportunities. This type of learning leads to better student engagement because the content is relevant to each student and tailored to their unique needs. It also leads to better student outcomes because the pace of learning is customized to each student.

By enabling students to master skills at their own pace, competency-based learning systems help to save both time and money. Each of these presents an opportunity to achieve greater efficiency and increase productivity.

4.2 Outcome-Based Education

OBE is a method of curriculum design and teaching that focuses on what students can actually do after they are taught. OBE goes beyond "structured tasks" (e.g. memorization) by demanding that students demonstrate his/her skills through more challenging tasks like writing project proposals and completing the projects, analyzing case studies and giving case presentations, etc. Such exercises require students to practice and demonstrate their ability to think, question, and research; make decisions; and give presentations. OBE also identifies higher levels of thinking (e.g., creativity, ability to analyze and synthesize information, ability to plan and organize tasks). Such skills are emphasized especially when students are assigned to organize and work as a community or entrepreneurial service teams to propose solutions to problems and market their solutions.

5 Experiential Learning

Experiential learning is widely recognized as a high-impact educational practice that occurs outside the classroom through experiences such as internships, study abroad, and service-learning. In experiential learning, the teacher takes the role of the guide on the side rather than the teacher on the stage. Experiential learning addresses the needs and wants of the learner. Learning is facilitated when:

1. The student participates completely in the learning process and has control over its nature and direction.
2. It is primarily based upon direct confrontation with practical, social, personal, or research problems.
3. Self-evaluation is the principal method of assessing progress or success. It also emphasizes the importance of learning to learn and an openness to change.

Rogers also emphasizes the importance of learning to learn and an openness to change. He was discouraged by the emphasis on cognitive theory of education. He believed this was responsible for the loss of excitement and enthusiasm for learning. His 1983 book, *Freedom to Learn for the 80's* presented his full theory of experiential learning. He believed that the highest levels of significant learning included personal involvement at both the affective and cognitive levels, were self-initiated, were so pervasive they could change attitudes, behavior, and in some cases, even the personality of the learner. Learning needed to be evaluated by the learner and take on meaning as part of the total experience.

5.1 Steps for Design and Implementation of Experiential Learning

To get started with an experiential learning activity in the course, Jacoby offers the following guidelines:

1. Identify learning outcomes
2. Create a safe environment
3. Select an activity that pushes students to their learning edge
4. Introduce students to the concept and practice; cover basic material
5. Engage students; provide guidance and support
6. Discuss the process and result
7. Provide structure for critical reflection

8. Obtain feedback throughout the process
9. Assess learning

6 Instructional Practices

The process of learning has been viewed as a largely passive experience in which knowledge is received and stored for future use. Teachers who use flexible grouping strategies often employ several organizational patterns for instruction. Students are grouped and regrouped according to specific goals, activities, and individual needs. When making grouping decisions, the dynamics and advantages inherent in each type of group must be considered. Both teacher-led and student-led groups can contribute to learning.

6.1 Teacher-Led Groups

Teacher-led groups are an effective and efficient way of introducing material, summing-up the conclusions made by individual groups, meeting the common needs of a large or small group, and providing individual attention or instruction.

6.2 Student-Led Groups

Student-led groups provide opportunities for divergent thinking and encourage students to take responsibility for their own learning. Students working in groups learn to work with people from varying backgrounds and with different experiences, sharpening social skills and developing a sense of confidence in their own abilities. A variety of group types and a sampling of activities that may be appropriate for each are described below.

6.3 Collaborative Groups

The essence of collaborative learning is the team spirit that motivates students to contribute to the learning of others on the team. Because team success depends on individual learning, members share ideas and reinterpret instructions to help each other.

6.4 Performance-Based Groups

Performance-based groups are most effective when formed on the basis of a particular need rather than in response to predetermined performance levels. Performance-based groups provide a means for increasing students' access to a particular concept or skill.

6.5 *Think, Pair, Share*

After whole-class instruction, have individuals think about what strategies they would use for approaching the investigation. Students should write down their ideas. After a time, have pairs meet to share their ideas and strategies. This approach helps encourage divergent thinking and provides students with immediate feedback on their approaches to problem solving.

7 Assessment Criterion

A Rubric is a coherent set of criteria for student work that describes levels of performance quality. When rubrics are created and used correctly, they are strong tools that support and enhance classroom instruction and student learning. The two essential components of effective rubrics are as follows:

- Criteria that relate to the *learning* (not the "tasks") that students are being asked to demonstrate
- Clear descriptions of performance across a continuum of quality

8 Assessment Criteria

Rubrics are designed to help educators and evaluators to:

- Develop a consistent, shared understanding of what proficient performance looks like in practice
- Develop a common terminology and structure to organize evidence
- Prepare informed professional judgments about formative and summative performance ratings on each standard and overall.

9 The ESE Model Teacher Rubric

9.1 *Standards*

Standards are the broad categories of knowledge, skills, and performance of effective practice detailed in the regulations. There are four standards for teachers: Curriculum, Planning, and Assessment; Teaching All Students; Family and Community Engagement; and Professional Culture.

9.2 *Indicators*

Indicators, also detailed in the regulations, describe specific knowledge, skills, and performance for each standard. For example, there are three indicators in standard I of the teacher rubric: Curriculum and Planning; Assessment; and Analysis.

9.3 *Elements*

The elements are more specific descriptions of actions and behaviors related to each indicator. The elements further break down the indicators into more specific aspects of educator practice and provide an opportunity for evaluators to offer detailed feedback that serves as a roadmap for improvement.

9.4 *Descriptors*

Performance descriptors are observable and measurable statements of educator actions and behaviors aligned to each element and serve as the basis for identifying the level of teaching or administrative performance in one of four categories: Unsatisfactory, Needs Improvement, Proficient, or Exemplary.

While assessing the language proficiency, we make use of the performance-based rubrics. It helps us to interpret information that what our students *know and can do.*

Example:

Speaking (interpersonal skills)	Generic	Role plays
		Debates
		Group discussions
Presentations	Generic/task based	Situational role plays
		Presentations on different topics: technical or general
		Narrations
Writing	Generic	General and technical reports, writing a slogan, bulletin board, magazine contributions
		E-mails, note-making, note-taking, precis-writing, developing the ideas into paragraphs

This has been designed to assess and to ensure that students were able to demonstrate their ability to use the language of study at the appropriate level.

A rubric with too many dimensions may be unworkable in classroom assessment. Some common descriptive terms are listed in the following chart.

	4	3	2	1
Task requirements	All	Most	Some	Very few or none
Frequency	Always	Usually	Some of the time	Rarely or not at all
Accuracy	No errors	Few errors	Some errors	Frequent errors
Comprehensibility	Always comprehensible	Almost always comprehensible	Gist and main ideas are comprehensible	Isolated bits are comprehensible
Content coverage	Fully developed, fully supported	Adequately developed, adequately supported	Partially developed, partially supported	Minimally developed, minimally supported
Vocabulary				
Range	Broad	Adequate	Limited	Very limited
Variety	Highly varied, non-repetitive	Varied, occasionally repetitive	Lacks variety, repetitive	Basic, memorized, highly repetitive

There are many rubric formats. In the grid format shown here, which is one of the possible ways to lay out a rubric, we illustrate a few common, frequently recommended, features of multiple trait rubrics:

9.5 *High to Low Scale*

In the graphic, the highest level of performance is described at the left. Students read first the description of an exemplary performance in each criterion. A few labels for a four-point scale include the following:

4	3	2	1
Exemplary	Excellent	Acceptable	Unacceptable
Superior	Good	Fair	Needs work

9.6 *Bloom's Taxonomy*

In Bloom's taxonomy, the evaluation level is where students make judgments about the value of ideas, items, materials, and more. Evaluation is the final level of the Bloom's taxonomy pyramid. It is at this level where students are expected bring in all they have learned to make informed and sound evaluations of material. *Bloom's taxonomy* is also tied to the *learning pyramid* which displays that the deeper one descends on the pyramid the higher the retention rate of knowledge.

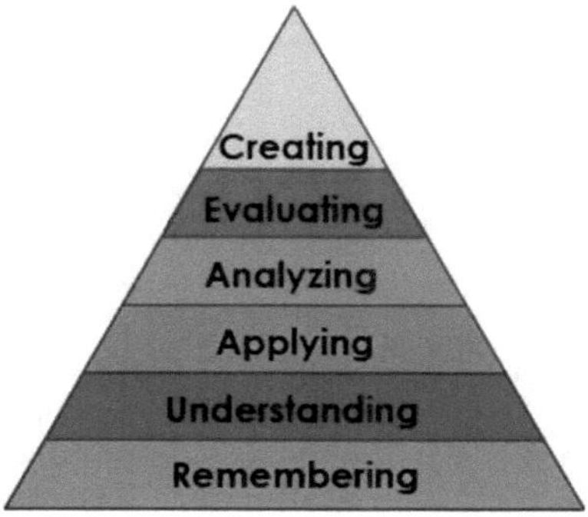

Bloom's taxonomy can be used across grade levels and content areas. By using Bloom's taxonomy in the classroom, teachers can assess students on multiple learning outcomes that are aligned to local, state, and national standards and objectives. Within each level of the taxonomy, there are various tasks that move students through the thought process.

10 Conclusion

New ways of teaching and learning, creating rich learning environments, designing new forms of (authentic) assessment also require intensive training and coaching of academic staff, monitoring and evaluation of the curriculum in action. Active learning increases the effectiveness and efficiency of the teaching and learning process. Teachers want students to leave a class with knowledge and/or skills they did not have when they began the class. The Revised Curriculum aims to empower our young people to develop their potential as individuals and to make informed and responsible decisions for living and working in the twenty-first century. Our society today needs young people who are flexible, creative, and proactive – young people, who can solve problems, make decisions, think critically, communicate ideas effectively, and work cogently within teams and groups.

References

1. Boud D, Cohen R, Walker D. Introduction: understanding learning from experience. In: Boud D, Cohen R, Walker D, editors. Using experience for learning. Buckingham: Society for Higher Education and Open University Press; 1993. p. 1–17.
2. Kolb DA. Experiential learning: experience as the source of learning and development. Englewood Cliffs: Prentice-Hall; 1984.
3. Kolb DA, Boyatzis RE, Mainemelis C. Experiential learning theory: previous research and new directions. In: Sternberg RJ, Zhang LF, editors. Perspectives on cognitive, learning, and thinking styles. New York: Lawrence Erlbaum; 2000. p. 227–47.

4. Rogers CR, Freiberg HJ. Freedom to learn. 3rd ed. Columbus: Merrill/Macmillan; 1994.
5. Chi MTH, de Leeuw N, Chiu MH, La Vancher C. Eliciting self-explanations improves understanding. Cognit Sci. 1994;18(3):439–77.
6. Felder RM, Henriques ER. Learning and teaching styles in foreign and second language education. Foreign Lang Ann. 1995;28(1):21–31.
7. Massachusetts model system for educator evaluation Part III: guide to rubrics and model rubrics.
8. Bloom BS. Taxonomy of educational objectives: the classification of educational goals. Handbook 1; cognitive domain. New York: David McKay Co. Inc; 1956. p. 7–8 [return].
9. Overbaugh R, Schultz L. Bloom's taxonomy. http://www.odu.edu/educ/roverbau/Bloom/blooms_taxonomy.htm. Accessed 3 Nov 2008 [return]
10. Krathwohl DR. A revision of bloom's taxonomy: an overview. Theory Pract. 2002;41(4):212–18 [return].

Part III
Presented Posters

Moving Towards Experiential Learning

Dhananjay S. Patil and Prakash M. Mohite

Abstract Engineering education scenario today is failing to meet the industrial requirements, and one major reason for this may be exam-oriented students, but again a question can be asked, *why are students exam oriented?* Many reasons right from their background to classroom teaching may give us the answer. It has been found that around 30 % of students are not involved in the teaching-learning process. The way a subject is posed by the faculty in the class and the way the faculty delivers methods adopted for communicating and interaction with students will be the basis of how students will take up interest in the subject. Hundred percent involvement of students in the classroom teaching-learning process *(not physical)* may help changing their attitude as well as enabling them to understand the importance and application of what they are studying. Conventional means of teaching and learning should be kept aside and new innovative ways should be developed and a faculty should divert himself or herself from being a teacher to a facilitator; this will encourage students to be participative and improve their involvement leading to proper understanding and grasping. Through this paper, the authors have tried to convey means through which involvement of students can be increased in the class. Concepts like learning by doing, model-based study and freehand drawing are used to penetrate the concept deeply and improve the involvement of students. Results have shown that learning by doing is one of the best ways to make students understand and learn the subject more clearly than other conventional teaching methods. Some limitations do exist, but through some more skills, efforts and time, experiential learning will enhance the technical background of budding engineers. Case based, model based and involving maximum number of students in performing/doing will take conventional blackboard and even content delivered through multimedia to an experiential learning and take technical education to new heights.

Keywords Experiential learning • Learning pyramid • Learning by doing

D.S. Patil (✉) • P.M. Mohite
Rajarambapu Institute of Technology, Rajaramnagar, Urun Islampur, Maharashtra 415414, India
e-mail: dhananjay.patil@ritindia.edu

R. Natarajan (ed.), *Proceedings of the International Conference on Transformations in Engineering Education: ICTIEE 2014*, DOI 10.1007/978-81-322-1931-6_63,

Smart Assessment of Program and Course Outcome with Students' Objectives

Jayashree Awati and Sudhir Awati

Abstract Most of the institutes concentrate to achieve program and course outcome based on result analysis, students' placements, students' feedback to faculty, and the National Board of Accreditation period of the departments. This paper concentrates on students' learning objectives of particular courses as well as program. Paper helps to achieve the vision, mission, and goals of institutes in coordination with the program and course outcome of students' learning objectives. This facilitates smart assessment of the program and course outcome. It concentrates on smart faculty feedback system as well as students' feedback system by the faculty.

Keywords Feedback • Assessment • Outcome • Teacher • Student • Program • Course

J. Awati (✉)
ETC Department, RIT, Sakharale, Maharashtra, India
e-mail: jayashree.samai@ritindia.edu

S. Awati
Computer Department, D.K.T.E's Textile and Engineering Institute,
Ichalkaranji, Kolhapur, Maharashtra, India
e-mail: sbawati@gmail.com

R. Natarajan (ed.), *Proceedings of the International Conference on Transformations in Engineering Education: ICTIEE 2014*, DOI 10.1007/978-81-322-1931-6_64,

Technology-Enhanced Teaching Learning Using Palm Devices

Shivananda R. Poojara, Khyamling A. Parane, Naveenkumar C. Patil, and Sachin S. Patil

Abstract In today's ever-increasing world of information technology, it is not enough to say that teaching learning process in engineering education is not considerably efficient, since the use of traditional teaching methods like blackboard and presentations will not cope up with the new generation. Nowadays hand-held and PDA devices are becoming part of human life. As today's generation are more interested in hand-held devices, teaching learning method should focus to gather point of interest of students to enhance practical learning. The proposed work explains the technology enhanced methods implemented for UG/PG courses using palm device like Aakash Tablet. Techniques used in making the students enjoyable learning in class room Clicker app for MCQs make allowing students to solve instant queries, quotes, phrases using tabs and PDAs, Cloud apps for data sharing within the class, Use of intranet in handled devices (Moodle), Virtual labs, and Classroom presentation using android apps. These techniques were used in teaching both UG and PG subjects like soft computing and advance database systems respectively. The observations were made while implementing the abovementioned methods and some of the key factors were identified. The students found enthusiastic in learning interactively using palmtops and result analysis of both the have been made and it was observed that the student's performance was good compared to last year.

Keywords Hand-held devices • Cloud apps • Engineering education

---

S.R. Poojara (✉) • K.A. Parane • N.C. Patil • S.S. Patil
Department of CSE, Rajarambapu Institute of Technology, Rajaramnagar, Urun Islampur, Maharashtra 415414, India
e-mail: shivananda.poojara@ritindia.edu; khyamling.parane@ritindia.edu; naveenkumar.patil@ritindia.edu; sachin.patil@ritindia.edu

R. Natarajan (ed.), *Proceedings of the International Conference on Transformations in Engineering Education: ICTIEE 2014*, DOI 10.1007/978-81-322-1931-6_65,

Transformation of Engineering Education Using Quality Circle Approach

Sandeep R. Desai, Sushma S. Kulkarni, and Martand T. Telsang

Abstract Quality circle is a group of people that come together and work voluntarily to solve work-related problems. It has the proven and systematic 12-step methodology that can be applied to solve almost all types of problems. It involves application of several quality circle tools and techniques such as Pareto diagram, cause and effect diagram, why-why analysis, 5W-1H principle, etc. to solve the problems of interest. The present work underlines the concept, methodology and applications of quality circles in engineering education with special reference to its implementation at Rajarambapu Institute of Technology (R.I.T.). The concept of 'quality circle' has received wide acceptance in industries due to the systematic problem-solving approach and monetary savings it offers after solving the problem. In case of educational institutes, though the concept has been used all over the world to tackle the problems such as students' projects, development of laboratory setups, etc., still the use is not widespread when compared to its implementation and outputs in industries. Since 1983, R.I.T. is well known for its continuous experiments and leadership in academic practices, thus striving for excellence in engineering education that is being delivered to its stakeholders. The institute is devising and implementing a number of practices with an objective to empower academics. One of

S.R. Desai (✉)
Automobile Engineering, Rajarambapu Institute of Technology, Rajaramnagar,
Urun Islampur, Maharashtra 415414, India
e-mail: sandeep.desai@ritindia.edu

S.S. Kulkarni
Director, Civil Department, Rajarambapu Institute of Technology, Rajaramnagar,
Urun Islampur, Maharashtra 415414, India
e-mail: sushma.kulkarni@ritindia.edu

M.T. Telsang
Mechanical Department, Rajarambapu Institute of Technology, Rajaramnagar,
Urun Islampur, Maharashtra 415414, India
e-mail: martand.telsang@ritindia.edu

R. Natarajan (ed.), *Proceedings of the International Conference on Transformations in Engineering Education: ICTIEE 2014*, DOI 10.1007/978-81-322-1931-6_66,

such best practices which is consistently implemented from the last 15 years is 'quality circles'. The institute has taken sincere efforts and proved over the years with the number of problems that are solved and the appreciation it has received that quality circle is equally good in education, though the benefits are not of monetary nature. In this paper, the process of formation of the groups, types of problems addressed, methodology to solve the problems, awards received and benefits of the activity are presented in sequence. Also, details about the financial assistance provided to quality circle groups and challenges that are faced by award-winning group along with its case study are reported in the paper at the end.

Keywords Academics • Best practices • Industries • Methodology • Quality circle

Personal Transformation from a Teacher to a Learning Facilitator: A Case Study

Martand T. Telsang

Abstract Transformation from a traditional teacher to a learning facilitator is a major change. This change is resisted quite often because of two reasons: lack of strong need to change and loosing the class control. In student-centred instruction (SCI), the teacher still has traditional functions like lecturing, designing assignments and tests, and grading and providing students with opportunities to learn independently and from one another and coaching them in the skills they need to acquire. Good professors may feel awkward when they start using student-centred methods and their course-end ratings initially may drop. It is tempting for instructors to give up in the face of all that, and many unfortunately do. Attempt is being made in this paper to demonstrate the personal transformation of a traditional engineering teacher to a learning facilitator, and various issues concerning the change process are discussed. Faculty Leadership Institute (FLI) of Indo-US Collaboration for Engineering Education (IUCEE) on Effective Teaching during July 2008 is a turning point for change. This workshop had enough influence and strength to dispel all the preconceived ideas about teaching and create an urge to embrace student-centric outcome-based teaching learning. A strong belief was developed in IUCEE mantra "*I am teaching, are they (students) learning*". The challenge lies in preparing both self and the students to accept the new learner-centric teaching methodology and facilitate the process of learning. The author experimented the outcome-based teaching for the courses taught and convinced about the positive response and change in classroom dynamics. Once convinced, the author started conducting workshops on the outcome-based education to bring in a change in the process of teaching and learning.

Keywords Student-centred instruction • Learning styles • Outcome-based education • Course outcomes • Learning facilitator

---

M.T. Telsang (✉)
Mechanical Department, Rajarambapu Institute of Technology, Rajaramnagar, Urun Islampur, Maharashtra 415414, India
e-mail: martand.telsang@ritindia.edu

R. Natarajan (ed.), *Proceedings of the International Conference on Transformations in Engineering Education: ICTIEE 2014*, DOI 10.1007/978-81-322-1931-6_67,

Strategic Planning for an Engineering Institute: A Case Study of Rajarambapu Institute of Technology, Sakharale

Sushma S. Kulkarni, A.C. Attar, and Binod Sinha

Abstract Engineering education has taken a paradigm shift in post liberalization era. A number of colleges imparting undergraduate, postgraduate, and Ph.D. programmes in engineering and technology increased, and hence availability of seats for engineering aspirants is changed. But it is observed that in the last 3 years, seats are remaining vacant and students are selecting only reputed institutes. There is a fierce competition in every aspect to attract admission and build image with respect to academic and placements. This demand leads to the change in the working of engineering colleges. Many of them are not in a position to sustain, either they are closed or on the verge of winding up. This clearly indicates that institutions did not focus proper strategic planning which shows the path of future direction and action to cope with the changes and complexes in the environment. Generally, corporates are making strategic plan and implementing properly. A few foreign universities and big institutions are using strategic plan, and they are getting edge in the competitive environment locally and globally. Strategic plan helps in planned development and growth of an engineering institution. This leads to achieve and to impart quality education. It is a road map, lightly filled in so that it gives you plenty of room to

S.S. Kulkarni (✉)
Director, Civil Department, Rajarambapu Institute of Technology,
Rajaramnagar, Urun Islampur, Maharashtra 415414, India
e-mail: sushma.kulkarni@ritindia.edu

A.C. Attar
Professor, Department of Civil Engineering, Rajarambapu Institute of Technology,
Rajaramnagar, Urun Islampur, Maharashtra 415414, India
e-mail: abdulrashid.attar@ritindia.edu

B. Sinha
Associate Professor, Department of Management Studies, Rajarambapu Institute
of Technology, Rajaramnagar, Urun Islampur, Maharashtra 415414, India
e-mail: bkumar.sinha@ritindia.edu

R. Natarajan (ed.), *Proceedings of the International Conference on Transformations in Engineering Education: ICTIEE 2014*, DOI 10.1007/978-81-322-1931-6_68,

manoeuvre. You get specific when you are deciding the action part of the plan, where you link it with people and operations. This paper outlines a one-page strategic plan of Rajarambapu Institute of Technology, Sakharale, and gives insight on planning and implementing strategy of RIT to impart quality engineering education in diverse, dynamic, and complex environment.

Keywords Paradigm • Strategy • Strategic plan • One-page strategic plan

Prominent Assessment of Students Learning and Statistical Analysis of Quizzes

Prakashgoud Patil, Deepa Mulimani, and B.L. Desai

Abstract Assessment of student learning has become indispensable for determining if students meet the desired standards of course content. An innovative way of carrying out student learning assessment, which is not the typical pen-and-paper method, is by using a tool like Moodle. Moodle facilitates us to build different types of assessments. Moodle, which stands for modular object-oriented dynamic learning environment, is an enormously versatile system for course and learning management. It is a course management system (CMS), also known as learning management system (LMS) or a virtual learning environment (VLE). It is a free web application that offers a wide variety of teaching tools, and educators can use it to create effective online learning sites. One of these tools is the quiz module that represents an alternative to traditional face-to-face courses and paper-based testing. This paper presents reports on experience of 6 months (one semester) with e-learning using open-source software "Moodle" for the course Wireless and Mobile Computing (MCAE608) of IV semester MCA. As part of continuous internal assessment, online quiz was conducted using Moodle to check the level of understanding at the end of each chapter of the course. Moodle-generated quiz reports that include Grades report, Responses report, Statistics report, and Manual grading are presented in the paper. Quiz questions and students' answers can be analyzed to carry out a psychometric analysis to identify the appropriateness of the questions stated in the quizzes and to assess student ratings on this activity as a guide for improving the teaching and learning process. The feedback from the students suggests that students were very positive about the use of Quizzes for Formative

P. Patil (✉) • D. Mulimani
Department of MCA, BVB College of Engineering
and Technology, Vidya Nagar, Hubli, Karnataka 580031, India
e-mail: prpatilji@gmail.com; deepamulimani@bvb.edu

B.L. Desai
Department of Electronics and Communication Engineering, BVB College of Engineering
and Technology, Vidya Nagar, Hubli, Karnataka 580031, India
e-mail: bldesai@bvb.edu

R. Natarajan (ed.), *Proceedings of the International Conference on Transformations in Engineering Education: ICTIEE 2014*, DOI 10.1007/978-81-322-1931-6_69,

Assessment of the course. From the teacher's perspective this method can be called as Green Assessment of Students' Performance, since it is a paperless one. The teacher can also hook into the critical areas of poor performance of students and plan improvements accordingly.

Keywords Learning management system (LMS) • Formative assessment • Quiz • Moodle • Assessment • Psychometric analysis

Engineering Education and Employability: Our Commitment

Anupama Deshpande, Pragya Jain, Sonali Singh, Priyanka Tripathi, and Suvarna More

Abstract The goal of Atharva College of Engineering (ACE) is to create a new generation of engineers with the spirit of innovation and creativity. The institute cherishes a dream of becoming the leading 'Centre of Excellence' in Engineering and Technology.

This paper suggests various steps taken by ACE to improve the skill set of graduates; emphasize soft skills; refocus the assessments, teaching-learning process and curricula towards analysis and creativity; and interact more with employers to understand the particular demand for skills in that region and sector. The details of the application of IUCEE learning at Atharva College may be viewed as a case study.

The outcomes that have been successfully achieved or are in progress to ensure enhanced use of technology to improve the teaching learning process are as follows: (1) a virtual class room to get advantage of expert lectures through Webinars, Distance Education Centre of IIT Bombay, NPTEL Lectures, etc.; (2) a virtual laboratory under the guidance of the College of Engineering, Pune for practicing simulations to understand the basic concepts thoroughly; (3) a hobby centre where students work on small electronic circuits on breadboard to get the output leading to introduction of mini projects in each subject at least for 2nd and 3rd year students to develop awareness of trying something with their own hands; and (4) Ember, an entrepreneurship development cell for relevant awareness and skills amongst the students. Within a span of 2008–2011, the students started commercial ventures and technical ventures for the benefit of Atharva students only.

The gains from external sources are as follows: (1) ACE succeeded in getting Innovation Entrepreneurship Development Centre and funding for projects leading to multidisciplinary engineering approach in our students. IEDC funding

A. Deshpande (✉) • P. Jain • S. Singh • P. Tripathi • S. More
Atharva Engineering College, Atharva Educational Complex Malad
Marve Road, Malad (W), Mumbai, India
e-mail: mangla_d_2000@yahoo.com; jainpragya123@rediffmail.com; sonalibkn@gmail.com; pooja_geco@yahoo.com; suvilas@gmail.com

R. Natarajan (ed.), *Proceedings of the International Conference on Transformations in Engineering Education: ICTIEE 2014*, DOI 10.1007/978-81-322-1931-6_70,

scheme is started by the Department of Science and Technology, Govt. of India, (2) A Robotic Cell established in ACE gets mentoring from IIT-Bombay, and (3) Atharva students started a Receiving Station for soon to be launched satellite 'Pratham' of IITB students.

Now that we have become a member of IUCEE, we look forward to have more guidance on research methodologies, project-based learning and direct collaboration with experts for the latest technologies. Partnering with a foreign university for mutual benefit remains yet another focus.

Keywords NBA • IEDC • IUCEE • IEEE • E-CELL • VLAB • Virtual class

Application of Curriculum Innovation for Production Management Subject

K.N. Nandurkar and A.S. Kamble

Abstract In the affiliated system of higher education, implementing the curriculum decided by the university becomes a challenging task for an individual faculty member. It becomes essential to apply the curriculum innovation in response to stakeholder needs. One such attempt done in one of the unaided engineering colleges in Maharashtra is presented in this paper. The syllabus for Production Management subject at the final year Production Engineering course was developed by the University of Pune. However, the faculty members developed a plan for curriculum innovation based on the inputs received during the IUCEE FLI programme. Some of the active learning techniques adopted provided an opportunity for the students to work in teams, collect information, visit industries, prepare reports and present their findings. This has resulted not only in better understanding of the subject as reflected in the university results but also in developing institute-industry partnerships. This exercise has been carried out for the last 6 years and over 400 students have been benefitted. The feedback received clearly indicates that if the curriculum innovation is applied properly, it is possible to transform the traditional course syllabus into a competency-based syllabus resulting in overall improvement in the quality of technical education.

Keywords Curriculum innovation • Active learning • Production management

K.N. Nandurkar (✉) • A.S. Kamble
Department of Production Engineering, K. K. Wagh Institute of Engineering Education and Research, Nashik 422003, India
e-mail: Keshav1965@gmail.com

R. Natarajan (ed.), *Proceedings of the International Conference on Transformations in Engineering Education: ICTIEE 2014*, DOI 10.1007/978-81-322-1931-6_71,

Soft Skills Training Through Cooperative Learning: A Case Study

P.K. Shahabadkar, P.S. Vispute, and K.N. Nandurkar

Abstract In this competitive world, teamwork, learning ability and soft skills along with academic excellence are prerequisites for the career growth of a student. In the global work culture, engineering students have to work in groups. It is the responsibility of the institute to restructure the learning situations in the changing scenario. A few changes in college teaching are required for bringing in the desired change in which students learn in groups. This paper is about structuring the appropriate teaching-learning environment for learning the soft skills. Herein, we describe the experiment conducted at K. K. Wagh Institute of Engineering Education & Research, Nashik, India, for teaching the characteristics of a master student and developing soft skills through cooperative learning. In cooperative learning, students work in groups to maximize their own and each other's learning. An attempt has been done in this study to demonstrate the application of cooperative learning for developing the soft skills through the master student programme.

Keywords Cooperative learning • Master student • Soft skills

P.K. Shahabadkar (✉)
Training and Placement Cell, K. K. Wagh Institute of Engineering Education and Research, Nashik 422003, India
e-mail: pramodshahabadkar@gmail.com

P.S. Vispute
Department of Information Technology, K. K. Wagh Institute of Engineering Education and Research, Nashik 422003, India

K.N. Nandurkar
Department of Production Engineering, K. K. Wagh Institute of Engineering Education and Research, Nashik 422003, India

R. Natarajan (ed.), *Proceedings of the International Conference on Transformations in Engineering Education: ICTIEE 2014*, DOI 10.1007/978-81-322-1931-6_72,

Blog-Based Student Lab Assignment and Assessment

Prakashgoud Patil, Shivanand Seeri, and Deepa Mulimani

Abstract Blogs and wikis are some of the new and exciting Web 2.0 technologies used in classrooms. These technologies of wikis and blogs can help the faculty for direct online learning activities that encourage peer support, collaboration, and dialogue. Well-constructed blog assignments may provide a safe environment and encourage a collaborative learning culture. This study explores the use of Web 2.0 tools in classroom and laboratory activities. In this work the authors (teachers) used blogs for lab assignments, submission and maintenance of online lab journals, and continuous assessment of students' progress in the Visual Basic Programming Practices course (MCAL509). The purpose of this work is to make use of public social software Web 2.0 tools like Web wikis and blogs for classroom activities and for continuous internal assessment of lab assignments of this course. As part of this work, the blog was created at the blogger called http://prpatilji.blogspot.com. The blogger is a blog-publishing service that allows private or multiuser blogs with time-stamped entries. The feedback from the students provided evidence that the blog assignments had a positive impact on in-class participation rate and laboratory environment. Key advantages identified include green assessment of student performance, ability to retrieve feedback while away from home, confirmation of receipt of assignments, more convenient storage of past assignments, and the ability to obtain a replacement copy of lost assignments.

Keywords Web 2.0 • Wikis • Blogs • Online assessment

P. Patil (✉) • S. Seeri • D. Mulimani
Department of MCA, BVB College of Engineering and Technology,
Vidya Nagar, Hubli, Karnataka 580031, India
e-mail: prakashpatil@bvb.edu; seeri@bvb.edu; deepamulimani@bvb.edu

R. Natarajan (ed.), *Proceedings of the International Conference on Transformations in Engineering Education: ICTIEE 2014*, DOI 10.1007/978-81-322-1931-6_73,

Overview of Effective and Efficient Learning Model Project-Based Learning (PBL)

M. Narayana Moorthi and J. Vaideeswaran

Abstract Teaching and learning are most important for teachers and students in any educational institution. The students need to perform well in their academic to gain knowledge, skill, and competency for their better career. Teachers play an important role in making the students to learn the fundamental and advanced concepts, various domain expertise and skills and competency. The most challenging process in any education institute is the evaluation of student performance. There are many ways by which the student performance can be evaluated. Traditionally we follow class-based learning (CBL) where we conduct an examination to evaluate the student performance. This paper discusses about the more effective and efficient learning model named as project-based learning (PBL).

Keywords Education • Evaluation methods • Learning • Teaching • PBL

M.N. Moorthi (✉) • J. Vaideeswaran
School of Computing Science and Engineering, VIT University,
Vellore 632014, Tamil Nadu, India
e-mail: Mnarayanamoorthy@vit.ac.in; jvaideeswaran@vit.ac.in

R. Natarajan (ed.), *Proceedings of the International Conference on Transformations in Engineering Education: ICTIEE 2014*, DOI 10.1007/978-81-322-1931-6_74,

A Global Knowledge Mining Hub: Technology-Enhanced Learning

Srinivas Aluvala and Nagendar Yamsani

Abstract In the past decade, the swift growth of computer and communication technologies opens up many opportunities for developing innovative learning environments with rich resources. Technology-enhanced learning shifted the focus from technology to support factual learning and the reinforcement of very basic skills to stimulate students to engage in meaningful learning and situated learning. The basic objective of education is to prepare the students for life. Technology-enhanced learning is technology-based learning and instructional systems through which students acquire skills or knowledge, usually with the help of teachers or facilitators, learning support tools, and technological resources. With the support of technological development, students are able to gain higher-order skills, such as rational thinking and problem-solving skills individually or collaboratively. Technology-enhanced learning has become an interdisciplinary issue attracting researchers from various disciplines to work together. In the recent years, the technological development of new technologies has generated considerable enthusiasm towards TEL. In this paper, we define the importance of TEL and different tools available with their classification and discuss TEL environments which enable access to a range of materials, learning tools, and communication facilities that enable the students to become more actively involved in developing their understandings, increased learner responsibility, and a focus on realistic tasks with passion.

Keywords Technology-enhanced learning • Situated learning • Rational thinking

S. Aluvala (✉) • N. Yamsani
Department of Computer Science and Engineering, SR Engineering College, Warangal, Andhra Pradesh, India
e-mail: srinu.aluvala@gmail.com; nagendar.y@srecwarangal.ac.in

R. Natarajan (ed.), *Proceedings of the International Conference on Transformations in Engineering Education: ICTIEE 2014*, DOI 10.1007/978-81-322-1931-6_75,

Project Based Learning: An Enhanced Approach for Learning in Engineering

Lipika Gupta, Pooja Arora, Rajnish Sharma, and Madhu Chitkara

Abstract In engineering studies, the undergraduates usually take up project as a major subject in the final year of their degree programme. However, the incorporation of project-based learning (PBL) and integrated projects in each year can improve the learning of undergraduates. This paper provides a comparative study of project-based learning and traditional classroom learning. The desired learning outcomes and effects of PBL on undergraduates are also discussed. The study is based on integrated PBL which is incorporated in the course curriculum of Chitkara University, Himachal Pradesh. Two groups of undergraduate students of different batches have been considered for the analysis. The results show that the students who had PBL in the 1st year and integrated project in the 2nd year show better performance than the students who were not involved in PBL or integrated projects till the 3rd year.

Keywords Project-based learning • Integrated project • Multidisciplinary approach • Learning outcomes • Classroom learning • Effects and challenges

---

L. Gupta (✉) • P. Arora • R. Sharma • M. Chitkara
Electronics and Communication Engineering, Chitkara University,
Barotiwala, Himachal Pradesh, India
e-mail: lipika.gupta@chitkarauniversity.edu.in; pooja.arora@chitkarauniversity.edu.in; rajnish.sharma@chitkarauniversity.edu.in; director@chitkara.edu.in

R. Natarajan (ed.), *Proceedings of the International Conference on Transformations in Engineering Education: ICTIEE 2014*, DOI 10.1007/978-81-322-1931-6_76,

An Effective Lab in Digital Signal Processing

P. Meena

Abstract In the era where digital technology reigns supreme over most engineering applications, digital signal processing is the heart of smart engineering solutions. Therefore, there is a need to grasp the fundamental techniques involved. A fundamental course in digital signal processing can turn out be a nightmare in the absence of an inspirational method of course delivery. The availability of simulation tools and their widespread use have facilitated quick and easy implementation of ideas and concepts. This also encourages one to develop prototype models on seeing the simulated results. This paper presents an initiative to ameliorate the learning experience of students belonging to the electrical and electronics discipline in a course as important as digital signal processing. Hitherto the emphasis for most course instructors was on developing good standard material for teaching a course. The transformation brought about in engineering education demands a paradigm shift that calls for effective strategies contributing to experiential learning. A sound mathematical foundation is a pre-requisite for the course. It serves well to have Matlab/Simulink usage to enhance the learning experience. The learning outcome attainment results presented justify the effectiveness and improvement in learning brought in by the introduction of the lab in place of a tutorial in the curriculum.

Keywords DSP • Laboratory • Simulation tools • Learning • Attainment

---

P. Meena (✉)
Department of Electrical and Electronics, BMS College of Engineering, Bangalore, India
e-mail: meenabms@gmail.com

R. Natarajan (ed.), *Proceedings of the International Conference on Transformations in Engineering Education: ICTIEE 2014*, DOI 10.1007/978-81-322-1931-6_77,

Open Assessment Method for Better Understanding of Student's Learnability to Create Personalised Recommendations

Shivanagowda G.M., R.H. Goudar, and U.P. Kulkarni

Abstract It is well accepted that learners even at academic institutions come with different abilities and flavours. Personalising the education is one of the 16 grand challenges as per the National Academy of Engineering (NAE). Measuring the outcomes of an education, progress of learning, etc., plays an important role in generating personalised feedbacks and recommendations in personalised learning environments facilitated by either an intelligent tutoring system (ITS) or an expert academician in conventional institutions. Any assessment would require a medium of expression to measure the achievements/outcomes of learning. Most common among them are writing a set of reasoning statements, writing algorithms, diagrams/sketches, occasional viva voce, etc.. The intention of this paper is to share one of our assessment practices during 2011–2013 where the medium of expression is as chosen by the students; we call this method of assessment as "open assessment". We also share the techniques of recording assessment data, policies to ensure student's participation in assessments and grading philosophy for such assessment methods. Finally, we present a short analysis to highlight advantages and disadvantages of this method when practiced in institutions with academic autonomy in India.

Keywords Open assessments • Teaching practices • Google docs • Students modelling • Personalised education

S. G.M. (✉) • U.P. Kulkarni
Department of Computer Science and Engineering, SDMCET,
Dharwad 580002, Karnataka, India
e-mail: shivana.gowda@gmail.com; upkulkarni@yahoo.com

R.H. Goudar
Department of Computer Network Engineering, Visvesvaraya Technological University,
Belgaum, 590018, Karnataka, India
e-mail: rhgoudar@vtu.ac.in

R. Natarajan (ed.), *Proceedings of the International Conference on Transformations in Engineering Education: ICTIEE 2014*, DOI 10.1007/978-81-322-1931-6_78,

Learning Outcomes in Engineering Education: A Review of Experimentation at Walchand Institute of Technology, Solapur

S.A. Halkude, D.D. Awasekar, M.A. Nirgude, and S.B. Aher

Abstract In Indian context, outcome-based education is gradually being adopted. Our institute has switched over to outcome-based education in the recent past. This methodology leads to a quality-based attainment of the program educational objectives (PEOs). However, the program outcomes (POs), based on the graduate attributes, need to be evaluated during their matriculation. These POs can be achieved, to a great extent, through the various course outcomes, which are the results of pre-defined learning objectives that were set for the course. Higher level of attainment of course outcome entirely depends upon the course content and its effective content delivery. At our institute, certain experiments were carried out to assess the course outcomes. The experiments conducted used a blended instructional strategy. Two experiments were conducted for the computer programming "C" language, while the other was for the System Programming course. Instructional strategies include program visualization, debugger tool, and short-test immediate group discussion used in these experiments.

Program visualization used visual cues to illustrate the flow and execution of a computer program as scaffold in resource-constrained classroom to teach "C" language compared with the traditional approach. This simplified the understanding of dynamic and usually hidden processes during program execution. Debugger tool

S.A. Halkude (✉)
Department of Civil Engineering, Walchand Institute of Technology, Solapur, Maharashtra, India
e-mail: halkude60@gmail.com

D.D. Awasekar • M.A. Nirgude
Department of Information Technology, Walchand Institute of Technology, Solapur, Maharashtra, India
e-mail: dipali.awasekar@gmail.com; nirgudemanisha@gmail.com

S.B. Aher
Department of Computer Science and Engineering, Walchand Institute of Technology, Solapur, Maharashtra, India
e-mail: sunitaaher@gmail.com

R. Natarajan (ed.), *Proceedings of the International Conference on Transformations in Engineering Education: ICTIEE 2014*, DOI 10.1007/978-81-322-1931-6_79,

demonstrated a step-by-step execution of the program and showed the presence of errors and values of memory in the watch window. This helped students to minimize errors and predict the output of the program correctly. Short test followed by immediate feedback and group discussion enhanced learning outcome of the System Programming course. From the above experiments conducted, it is inferred that the students' learning outcome depends on instructional strategy being used by an instructor. Therefore, while planning for course delivery, on the basis of predefined learning outcome, the teacher needs to develop the assessment measures to be used and then accordingly select instructional strategies for engaging students and assist students in attaining the identified learning outcome. There is no single best strategy available that a teacher should choose, but rather a combination of various instruction strategies will assist students in developing interest, understanding the content, and eventually achieving learning outcomes. Hence, subject teachers need to facilitate the students' learning process by using instructional strategies relevant to the learning objectives.

Keywords Learning outcome • Instructional strategy • Visualization • Debugger • Objective test

Overview of Accreditation System and Investigations of Assessment Methods/Techniques for Quality Assurance of Engineering Education

Kota Chandra Bhushana Rao

Extended Abstract Higher education system in India and across the world is experiencing sea change with the information, opportunities, and knowledge available. Engineering education, in particular, benefitted immensely with modern tools available. Quality assurances through assessment by the Washington Accord (WA) of International Engineering Alliance (IEA), Accreditation Board of Engineering and Technology (ABET-USA), and National Board of Accreditation (NBA-India) across the world have set standards to excel to meet set outcomes!

The Washington Accord Agreement recognizes that "Accreditation of engineering academic programs is a key foundation for the practice of engineering at the professional level in each of the countries or territories covered by the Accord." Engineering education in India has been transformed from input-output mode to outcome-based mode recently to meet global challenges. In this regard, all stakeholders in the society shall come together by keeping a positive will with unified thought process to achieve the outcome.

Quality products shall be attained by robust accreditation systems, assessment procedures, curriculum design, and coordination between regulatory bodies like AICTE, UGC, MHRD, State Council for Higher Education (SCHE), NAAC, and NBA in the system.

This paper aims to highlight the key issues related to outcome-based education through standard accreditation procedures and assessment methods to be adopted to get the desired outcomes.

Keywords WA • IEA • ABET • NBA • NAAC • UGC • SCHE • MHRD • AICTE

K.C.B. Rao (✉)
Department of ECE, University College of Engineering,
JNTUK-Vizianagaram, Andhra Pradesh 535 003, India
e-mail: cbraokota@yahoo.com

R. Natarajan (ed.), *Proceedings of the International Conference on Transformations in Engineering Education: ICTIEE 2014*, DOI 10.1007/978-81-322-1931-6_80,

The Institutional Leadership of JNTUK System in Embracing New Paradigms in Engineering Education

G. Tulasi Ram Das, G. Abbaiah, and Madhavi Gudavalli

Abstract Jawaharlal Nehru Technological University Kakinada (JNTUK), Andhra Pradesh, India, is one of the largest providers and facilitators of engineering education, having 289 affiliated colleges of engineering and technology, management, and pharmacy under its fold. The rapid spread and proliferation of engineering education and the entry of private sector have brought in quality concerns in their wake. The hands-on experience reveals that the notable gap between expectations and realizations is the cumulative symptom of ineffectiveness of what is being practiced hitherto, in addressing the quality concerns. JNTUK system has left no stones unturned to channelize its resources and deliberate, research, and innovate the mechanisms and practices to reduce this gap, in tune with the new paradigms showing up in engineering and higher education. It revises the curriculum of UG and PG for different disciplines in affiliated and constituent colleges incorporating outcome-based education (OBE) and is the first government university to take up the task of OBE approach in the country. As JNTUK has to manage 289 colleges, it has evolved a quality assurance tool for uniform application across all its affiliated institutions. JNTUK has demonstrated its leadership to lead and contribute for quality

G.T.R. Das
Department of Electrical & Electronics Engineering, Jawaharlal Nehru Technological University Kakinada, Kakinada, Andhra Pradesh 533 003, India
e-mail: das_tulasiram@yahoo.co.in

G. Abbaiah (✉)
Department of Civil Engineering, Jawaharlal Nehru Technological University Kakinada, Kakinada, Andhra Pradesh 533 003, India
e-mail: abbaiah@yahoo.com

M. Gudavalli
Department of Information Technology, JNTUK UCE Vizianagaram, Vizianagaram, Andhra Pradesh 535 003, India
e-mail: madhavi.researchinfo@gmail.com

R. Natarajan (ed.), *Proceedings of the International Conference on Transformations in Engineering Education: ICTIEE 2014*, DOI 10.1007/978-81-322-1931-6_81,

assurance in technical education at regional and national levels, in collaboration with apex bodies of UGC, NBA, APQN, NPTEL, IUCEE, Mission 10× of Wipro, and Campus Connect of Infosys. To bridge the gap between expected and delivered technical education, JNTUK has planned to establish incubation centers for trans-disciplinary product development and research promotion in the focused areas of research with private and public partnership. This paper outlines how the JNTUK system sustains the quality through concerted efforts of empowerment of faculty, use of ICT for better management, delegation of ownership, and developing models of quality assurance. The university intends to share what it has achieved and leaves the scope for achieving further by learning from others in the arena of engineering and higher education.

Keywords Transformation • Enablers • E-resources • Paradigm shift • Empowerment • Expectations • Quality assurance • Success criteria • Areas of impact

Developing Effective Industry Partnerships to Promote Learning and Entrepreneurship

Ajay D. Kale, Kalyan Sundaram, and Ratnakar R. Kulkarni

Abstract Owing to globalization and liberalization, technology transfer from resourceful countries to all developing countries has become inevitable. Many Indian industries neither opt to invest and establish their own R & D Centers to develop indigenous technology nor do they network systematically with knowledge generators such as academic institutes. The first option requires large funds to create R & D infrastructure and is financially not viable for small-scale sector, whereas for the second option, there is only a theoretical response. Critical analysis is necessary to address the reluctance of both industry and upcoming educational institutes to cooperate effectively to confront their technical, operational, and financial hurdles. At present, both prefer to go for foreign collaborations at individual level instead of creating any symbiotic relationship.

This paper attempts to discuss the need of relationship between industries and educational institutes and broadly states the inherent means of cooperation, the objectives, and the nature of various forms of interactions. The paper also highlights the criticality of the industry-institute interaction.

Keywords Effective learning • Industry-institute cooperation • Technology transfer • Success criteria

A.D. Kale (✉) • K. Sundaram • R.R. Kulkarni
Mechanical Engineering, Vishwakarma Institute of Information Technology,
Pune, Maharashtra, India
e-mail: jaiganeshkale@gmail.com

R. Natarajan (ed.), *Proceedings of the International Conference on Transformations in Engineering Education: ICTIEE 2014*, DOI 10.1007/978-81-322-1931-6_82,

Role of Industry-Institute Interaction to Promote Education and Entrepreneurship

M.C. Padma and V. Sridhar

Abstract The goal of any technical institution is to produce skilled, globally competent professionals through quality technical education and to prepare them for immediate employment. Industries engross these knowledgeable professionals and enhance its production capabilities by contributing the latest technologies. To produce proficient graduates ready for the industry, it is necessary to know the requirements of the industries through industry-institute interaction. Hence, a good and vibrant industry institute interaction to promote education and entrepreneurship is definitely required. To build good rapport between the industry and the institute, institutes should have Memorandum of Understanding (MoU) with the industries.

Industry-institute interaction (I-I-I) is the most preferred activity for mutual benefit and growth of industries as well as institutions. I-I-I provides the best platform for showcasing the best practices, latest technological advancements, and their implementation and impact on the industry. Also, I-I-I promotes industry experts to participate in curriculum design which plays a significant role in preparing the students ready for the industry. Through I-I-I, industries can participate in technical education programs and cross-fertilize ideas for systems improvement. Teaching-learning processes can be improved by integrating industrial training to the students which also provides an exposure of the corporate world. Students should be encouraged to undertake the final year projects in the industry with a joint supervisor from the industry. I-I-I promotes development of entrepreneurs

M.C. Padma (✉) • V. Sridhar
Department of Computer Science and Engineering, PES College of Engineering, Mandya 571401, Karnataka, India
e-mail: padmapes@gmail.com; venusridhar@yahoo.com

R. Natarajan (ed.), *Proceedings of the International Conference on Transformations in Engineering Education: ICTIEE 2014*, DOI 10.1007/978-81-322-1931-6_83,

which further leads to rapid industrialization and hence improved well-being of a country. I-I-I can also increase the research and development activities in both industries as well as institutions which further leads the nation to grow technologically and socioeconomically.

Keywords Industry-institute interaction • Technical education • Modes of interaction • Entrepreneurship • Memorandum of understanding (MoU)

Technology: A Source of Supplement to the Teaching and the Process of Learning in Higher Education

Bandi Surendra Reddy, M. Rajeshwar, and L. Naveen Kumar

Abstract In the contemporary world, technology is inevitable irrespective of the profession, and the realm of education is no exception to this phenomenon. Education, and the teacher's role in the transformation of knowledge, has been challenged to address the contemporary needs of a learner. The advancement of technology has led to changes in the mind-set of learners, as well as better comprehension of information. To make the process of learning better and more effective, certain techniques or methods are employed (i.e., LCD projectors, PowerPoint presentations, and Cloud-based tools). These new technologies have replaced the traditional method of using transparent sheets and projectors in classrooms. This improvement in audiovisual equipment has grabbed the attention of the learners. The combination of new learning styles, teaching pedagogy, new technology, and good infrastructure leads to a better education. If these latest trends are not used in an active manner, then the purpose of the teaching and any effectiveness of delivery will not be useful. This paper is an attempt to study the positive effects of using technology-based education tools in the technical arena.

Keywords Technology • Learning style • Pedagogy • Cloud

B.S. Reddy (✉) • M. Rajeshwar • L.N. Kumar
Department of Computer Science Engineering, Hyderabad Institute of Technology and Management (HITAM), Hyderabad, India
e-mail: surendra.mca@hitam.org; rajeshwarm.cse@hitam.org; naveenkumarl.cse@hitam.org

R. Natarajan (ed.), *Proceedings of the International Conference on Transformations in Engineering Education: ICTIEE 2014*, DOI 10.1007/978-81-322-1931-6_84,

Indian Technical Teaching Service (ITTS): A Proposal to Improve Quality of Engineering Education

V.G. Renumol, G. Krishna Kumar, and R. Gopeekrishnan

Abstract At present All India Services have mainly three sectors such as the Indian Administrative Service (IAS), Indian Police Service (IPS), and Indian Foreign Service (IFS). However, there is no such challenging examination for recruiting teachers in the academic sector in India. In this paper, we suggest the relevance of conducting such an examination to recruit teachers in engineering education from the graduate level, in order to bring a positive change in the quality of engineering education in India. We name it as the Indian Technical Teaching Service (ITTS). Our views to implement this idea have been given through a set of guidelines in this paper. However, actual implementation is a policy decision, which is beyond the scope of this paper.

Keywords Engineering education • Quality • Teacher eligibility test • Teacher training

---

V.G. Renumol (✉)
School of Engineering, Cochin University of Science and Technology, Kochi, Kerala, India
e-mail: renumolvg@gmail.com

G. Krishna Kumar
Open Source Techno Labs, Kochi, Kerala, India
e-mail: kkgkurup@gmail.com

R. Gopeekrishnan
Department of Computer Science, Saintgits College of Engineering, Kottayam, Kerala, India
e-mail: gopankedaram@gmail.com

R. Natarajan (ed.), *Proceedings of the International Conference on Transformations in Engineering Education: ICTIEE 2014*, DOI 10.1007/978-81-322-1931-6_85,

Electronic System Design, Manufacturing, and Human Resource

S. Jayakumar

Abstract The goal is to make younger generation in India, globally competitive design engineers in Electronic System Design and Manufacturing (ESDM). Karnataka government and IESA (India Electronics and Semiconductor Association) are taking major initiative to bring out working policy in ESDM. According to the mentioned ESDM policy, the plan is to bring up 28 million and make them ready for the global product design market. K-Pages provides methods and apparatus to train engineers for the ESDM industry. K-Pages is built on the 4 Quadrant learning method (used in NPTEL) and extends from there on.

- Quadrant I of K-Pages provides synchronous and asynchronous lectures on OSCAD tools which are free and available in http://www.oscad.in/. It is carried out by using online resources and access device like Aakash tablet PC (from MHRD) with OSCAD tools (runs in Ubuntu 13.04).
- Quadrant II of K-Pages is using learning and collaboration tools such that engineers are having option of asking questions on a given online content (via spoken-tutorials http://spoken-tutorial.org/) or lectures. It is providing collaborative learning option by integrating others questions and answers with respective technique of board design and analysis.
- Quadrant III of K-Pages provides training on "product design and product making" to learning students by using hardware and associated software infrastructure.
- Quadrant IV of K-Pages provides test environment and also certification options to learning engineers.

For example, if engineers had finished all three modules and associated test then these engineers are qualified to take up work in the design and development of board design. K-Pages offers low cost option for engineers to get trained in ESDM. K-Pages is expected to deliver very high quality engineers to ESDM industry in

S. Jayakumar (✉)
Piitech Systems Private Limited, Bangalore, India
e-mail: jk@piitechsystems.com

R. Natarajan (ed.), *Proceedings of the International Conference on Transformations in Engineering Education: ICTIEE 2014*, DOI 10.1007/978-81-322-1931-6_86,

India. K-Pages provides opportunity for learning engineers to trade their products via online store. K-pages provides platform to create innovation on human resource development for ESDM industries. The unique features of K Pages are, (1) hybrid delivery model (2) affordable and scalable (3) meets current electronic industry needs and constantly keeps pace with it, and (4) collaboration with students, industry and academia.

Keywords ESDM • 4 Quadrant learning • NPTEL • K-Pages • OSCAD • Aakash IV tablet PC • MHRD • Spoken-tutorials • Karnataka government • Human resource • IESA

Efficient Teaching Aid: Self-Learning Models

S.C. Potnis

Abstract Industry is one of the primary customers of the engineering institutes and universities. Those customers are constantly complaining that fresh engineering graduates lack conceptual understanding, critical thinking, and creativity. According to a recent report published by NASSCOM, only 25 % of the engineering graduates produced in India are readily employable in the industry because of lack of deep technical knowledge and creativity. In this paper, we focus on using self-learning models as an innovative teaching and learning strategy to enhance conceptual understanding and foster creative thinking in engineering students.

The student survey shows that these models help engineering students in many ways. These models not only enhance conceptual understanding of difficult and complicated terms in highly conceptual courses like Engineering Mechanics, Strength of Materials, Structural Engineering, etc., but also motivate students to think on their own. Hence, the use of these models in teaching produces critical and innovative thinking skills required to solve challenging real-world problems. The hands-on experience on such self-learning models helps students to learn from experiences and develops the ability of lifelong learning.

Keywords Conceptual understanding • Creativity • Critical thinking • Hands on experience • Lifelong learning • Self-learning models

---

S.C. Potnis (✉)
Department of Civil Engineering, Maharashtra Institute of Technology, Paud Road, Pune 411038, Maharashtra, India
e-mail: potnissc@yahoo.co.in

R. Natarajan (ed.), *Proceedings of the International Conference on Transformations in Engineering Education: ICTIEE 2014*, DOI 10.1007/978-81-322-1931-6_87,

Developing Innovation Among Undergraduate Students

P. Martin Jebaraj, K. Mallikharjuna Babu, and D. Ajay Kumar

Abstract Our country is going through rapid technological changes and economic growth. Its sustenance will mainly depend on the innovations that must happen in all quarters of the industry and in business enterprises. Though innovation is now talked about, no systematic efforts are undertaken in the engineering colleges to develop and foster skills among interested budding engineers and transform them as inventors. In this paper, the efforts of BMS R & D center for developing and supporting innovative thinking with the first year students. The methodology consists of spreading awareness and continually encouraging innovative skills among interested students. The BMS College of Engineering is pioneering a systematic effort including both academy and industry, to build an environment that will produce young innovators. The actual implementation of the initial stage is presented in this paper. The outcome of the efforts has started to bear fruits. Incentives like student patent filings, paper presentations, awards, etc., have increased dramatically.

Keywords Student innovators • Approaches to grow in innovative environment • Freshman innovates

P.M. Jebaraj (✉) • K.M. Babu • D.A. Kumar
BMS College of Engineering, Bull Temple Road, Bangalore 560019, Karnataka, India
e-mail: kmartin8181@yahoo.co.in; drkmbabu@gmail.com; d.ajay402@gmail.com

R. Natarajan (ed.), *Proceedings of the International Conference on Transformations in Engineering Education: ICTIEE 2014*, DOI 10.1007/978-81-322-1931-6_88,

The Innovative Cloud-Based Solution for Classroom Transformation

Ramesh Pudale, S.S. Mantha, and R.P. Sujata

Abstract Many new interactive learning techniques are emerging as students started showing interest in new digital tools. Students are collaborating and discussing new ideas, figuring out possible solutions. As a result, educational institutes inclined towards project-based leaning exploring real-world challenges. Design, simulations, and social networking technologies have much to offer in immersive learning. We consider here the cloud-based services like Autodesk Fusion 360 as a learning tool to quickly develop new ideas and to create detail designs with the latest industrial and mechanical design workflows. In the classroom, next-generation students can work faster and share their ideas and project assignments with faculties from anywhere. Fusion can bring innovation faster and accelerate the learning process. These cloud services transform classroom teaching into immersive learning to make learning effective and intuitive.

Keywords Classroom transformation • Interactive learning • Cloud services in education • Design • CAD • Collaboration • PBL

R. Pudale (✉)
Autodesk India Pvt. Ltd, Unit A4, 'A' Wing, 2nd Floor,
Divyasree Chambers, Langford Road, Bangalore 560025, India
e-mail: pudale@gmail.com

S.S. Mantha
Sardar Patel College of Engineering, Andheri (W) Mumbai, India

R.P. Sujata
Sambhram Institute of Technology, Bangalore, India
e-mail: sujataramesh22@gmail.com

R. Natarajan (ed.), *Proceedings of the International Conference on Transformations in Engineering Education: ICTIEE 2014*, DOI 10.1007/978-81-322-1931-6_89,

Group Assignment Writing in Engineering: Some Preliminary Findings

Gururaj B. Tennalli and Laxmikant R. Patil

Abstract Generic skills play an important role in increasing employability and credibility of a student. As for the engineering students, these skills are imparted in the learning activities given to the students. In this newer "group assignment" writing method, students are first asked to form compatible groups among them well in advance in the beginning of the semester. Later, the predefined assignment topics were randomly allotted to the students. The activity demands that students working in a group of 3–4 need to understand the broad statement of the given chapter-end assignment topics and collect the information from various sources including textbooks, Internet sources such as blogs, research journals, etc. Later, students need to prepare a report and submit in a stipulated period of time. Reports collected from students were evaluated for different criteria such as the depth and technical significance of the information collected, logical order of content preparation and presentation, and skills and creativities related to document preparation. Finally, the reports were discussed with the students in the class and awarded with marks. Skills such as writing, collection of information, depth of collected information, organization, teamwork, innovation, and creativity were evaluated for this practice. In our preliminary study, we implemented this newer method of "group assignment" practice to the Molecular Biology and Immunology theory courses of the second and third year B.E. Biotechnology students, respectively. Students were actively involved in this newer activity and found it as an interesting activity. Students' perceptions about the assignment work were collected at the end of the semester using questionnaires. Results from the questionnaires indicated the following observations: 38 % of students strongly agreed for the opportunity for creativity in group assignment writing, more than 70 % of students agreed that the group assignment writing instigated them to read more and work actively for a concept, 60 % of students believe that the assignment writing work made them understand better the topics, around

G.B. Tennalli (✉) • L.R. Patil
Department of Biotechnology, BVB College of Engineering and Technology, Vidya Nagar, Hubli, Karnataka 580031, India
e-mail: gururaj_tennalli@bvb.edu

R. Natarajan (ed.), *Proceedings of the International Conference on Transformations in Engineering Education: ICTIEE 2014*, DOI 10.1007/978-81-322-1931-6_90,

38 % of students strongly accepted that the evaluation method adopted was better and judicious over the evaluation for traditional assignment writing, and beyond all about 20 % of students strongly found it a bit laborious. Overall, as a teacher, we personally believe that through "group assignment" writing activity, the generic and professional skills among students can be nurtured by carefully and cautiously looking at the student's changing perceptions about the writing activities inside and outside the classroom for the betterment of the student's future.

Keywords Assignment writing • Generic skills • Teamwork • Innovation and creativity

IPython Notebook for Teaching and Learning

A.B. Raju

Abstract In all branches of engineering, computational work/simulation is now rightly seen as the third vertex of a triangle, complementing observation and theory. This requirement necessitates an engineering student should know computational concepts as well as a whole new language to express these concepts. These are challenging tasks and students might face difficulties in learning the finer details of the language and its usage. While these are important, they are not strict prerequisites for using Python for their computational needs. Minimizing the students mental loads, allow them to concentrate on the core computational ideas. IPython Notebook provides a programming environment that offers many advantages for students as well as for instructors as these are free and open source software (FOSS). Nowadays reproducible research is gaining importance and this idea in scientific computing is to archive and make publicly available all the codes used to create a paper's figures or tables, preferably in such a manner that readers can download the codes and run them to reproduce the results. IPython Notebook feature of Python advocates this philosophy and in this paper, author shares his experience how it can be used effectively to promote teaching/learning and reproducible research.

Keywords Python • IPython • IPython Notebook • Teaching and learning

A.B. Raju (✉)
Department of Electrical and Electronics Engineering, BVB College of Engineering and Technology, Vidya Nagar, Hubli, Karnataka 580031, India
e-mail: abraju@bvb.edu

R. Natarajan (ed.), *Proceedings of the International Conference on Transformations in Engineering Education: ICTIEE 2014*, DOI 10.1007/978-81-322-1931-6_91,

A LabVIEW-Based Laboratory Experience of Real-Time Embedded Concepts

Manjula P. Pandarikar

Abstract The real-time embedded systems (RTES) find wide ranging utility in numerous applications that range from trivial devices to high-end sophisticated systems that include applications in flight control, railway signaling, robotics, medical devices and telephony. The RTES concepts therefore constitute vital know-how segment of engineering graduates dealing with process control and automation. The knowledge of process control systems is very essential for undergraduate students of Automation and Robotics discipline for development of safety-critical applications.

This paper presents the pedagogical approach taken in the course delivery of RTES Laboratory through LabVIEW-based assignments to facilitate learning of necessary competencies in implementation of real-time embedded solutions. This goal is achieved by an introduction to the basic concepts of real-time embedded system, such as state machine diagrams, deterministic and nondeterministic processes, semaphores and interrupts-driven programming techniques.

The re-structured RTES laboratory was conducted using NI CRIO 9075, NI Robotic Starter kit 2.0 and Stellaris ARM cortex M3 boards. The RTES laboratory effectively addressed the (a), (b), (c), (d), (e) and (k) ABET criteria through specific focus on experiential, hands-on learning using industry standard tools like LabVIEW and Keil. The motivation behind this was to create an environment for students to apply the theoretical aspects to real-world problems. The different strategies or approaches used in the present lab setup provide effective solutions to overcome the drawbacks of conventional teaching of the course that lacked the integrating experience. The approach facilitated the development of competency in students in building real-time control algorithms on tightly integrated hardware using LabVIEW Real-time and FPGA modules.

M.P. Pandarikar (✉)
BVB College of Engineering and Technology, Vidya Nagar,
Hubli, Karnataka 580031, India
e-mail: manjula_pp@bvb.edu

R. Natarajan (ed.), *Proceedings of the International Conference on Transformations in Engineering Education: ICTIEE 2014*, DOI 10.1007/978-81-322-1931-6_92,

A Framework for Curriculum Design of Data Structures and Design of Algorithms Course (DSDA)

C. Sujatha, Jayalakshmi G. Naragund, and Suvarna G. Kanakaraddi

Abstract The curriculum design is the process of collecting, organizing and sequencing the lessons or content to study the course in orderly manner. In this paper, authors propose a framework to design the curriculum for the courses: Data Structures and Design and Analysis of Algorithms. The proposed framework integrates these two courses into a new course as Data Structures and Design of Algorithms (DSDA). In this framework authors divide the entire content of these courses into two themes: One is static allocation and another is dynamic allocation. The aim is to enhance the relationship between these two courses by teaching the algorithm design techniques through the applications of data structures, which also improve the skills of teaching and learning the course. The curriculum includes data structures, design techniques and some flavor of parallel computing. The curriculum design addresses the program outcomes (POs) such as apply basic knowledge, identify and formulate a problem, analyze and design the solution and implement the process or component to meet the desired needs.

Keywords Algorithm techniques • Data structures • Static and dynamic memory allocation

C. Sujatha (✉) • J.G. Naragund • S.G. Kanakaraddi
Department of Computer Science and Engineering, BVB College of Engineering and Technology, Vidya Nagar, Hubli, Karnataka 580031, India
e-mail: sujata_c@bvb.edu; jaya_gn@bvb.edu; suvarna_gk@bvb.edu

R. Natarajan (ed.), *Proceedings of the International Conference on Transformations in Engineering Education: ICTIEE 2014*, DOI 10.1007/978-81-322-1931-6_93,

Assessment of Program Outcome by Open-Ended Experiment in Enzyme Technology Laboratory Course

Savitha S. Desai, Basavaraj S. Hungund, and Shrinivas D. Desai

Abstract The laboratory component of Enzyme Technology course is designed to provide hands-on experience to inculcate the experimental skill during isolation, purification and assay of various enzyme preparations. To improve the attainment of program outcomes through Enzyme Technology laboratory course, the experiments were categorized as demonstration, exercise, structured enquiry and open-ended experiment. The primary objective of open-ended experiments is to enable the student to design and conduct experiments, as well as to analyze and interpret data. The experiments were carried out in a group of two so that students shall develop the ability to function effectively as an individual and in a group with the capacity to be a leader as well as an effective team member. The present paper describes the design of the open-ended laboratory experiment and its impact on attainment of program outcome and the overall students' learning experience. Student teams were given only with problem statement, keeping the study approaches open. The students were expected to carry-out literature review and design the experiment, with due consideration to resources and feasibility. All the students were graded by

S.S. Desai (✉)
Department of Biotechnology, BVB College of Engineering and Technology, Vidya Nagar, Hubli, Karnataka 580031, India
e-mail: joshisavitha@bvb.edu

B.S. Hungund
BVB College of Engineering and Technology, Vidya Nagar, Hubli, Karnataka 580031, India
e-mail: bshungund@bvb.edu

S.D. Desai
Department of Information Science and Engineering, BVB College of Engineering and Technology, Vidya Nagar, Hubli, Karnataka 580031, India
e-mail: sd_desai@bvb.edu

R. Natarajan (ed.), *Proceedings of the International Conference on Transformations in Engineering Education: ICTIEE 2014*, DOI 10.1007/978-81-322-1931-6_94,

adopting evaluation rubrics. The evaluation results show significant improvement in the achievement of program outcomes. It is observed that design of experiments as open-ended one improves the learning experiences and nurtures the innovativeness and creativity among students.

Keywords Open-ended experiments • Enzyme Technology lab • Program outcomes • Self-learning

Introduction of Chapter "How Stuff Works" and Course Seminar in Elements of Mechanical Engineering

Sanjeev M. Kavale

Abstract A survey of Freshman students at BVB College of Engineering and Technology revealed that more than 20 % have not given any kind of seminars until their pre-university (K12) education, and more than 50 % have given only thrice. Many engineering graduates fail to explain how some of the Engineering products/gadgets work even though they see and use them in routine life. In this regard, a new chapter "How Stuff Works" was introduced in Elements of Mechanical Engineering course taught at the first-year level. Students in the team were asked to study a product/gadget in detail and were asked to present a seminar to the whole of the class. This paper deals with necessity, implementation and usage of the new chapter and course seminar. This new initiative is in practice for one full academic year and the results are encouraging.

Keywords Course seminar • How stuff works • Mechanical engineering products/gadgets

S.M. Kavale (✉)
Department of Mechanical Engineering, BVB College of Engineering and Technology, Vidya Nagar, Hubli, Karnataka 580031, India
e-mail: sanjeev_kavale@bvb.edu

R. Natarajan (ed.), *Proceedings of the International Conference on Transformations in Engineering Education: ICTIEE 2014*, DOI 10.1007/978-81-322-1931-6_95,

Outcome-Based Pedagogical Approach for Energy Conversion Laboratory Course of Mechanical Engineering UG Programme

N.R. Banapurmath, P.P. Revankar, and R.S. Hosamath

Abstract The Energy conversion course constitutes the application of knowledge acquired in different thermal engineering courses related to classical sciences of thermodynamics, fluid mechanics, turbo-machinery and heat transfer. The abstract ingredients in these courses demand an inquisitive approach of delivery and assimilation of the concepts due to which large group of learners lose focus on the course. This consequently ruins the attentiveness of learners towards these courses and hence demands for modified pedagogical practices that promote effective learning. The majority of learners have an apprehension to learn these courses owing to the phobia developed towards fundamental physics and mathematics. In view of this, the Energy conversion laboratory course that integrates the assimilated knowledge of thermal engineering courses demands special attention to keep the student fraternity informed about the latest innovations happening in this field of engineering.

The students undertaking this laboratory course will be given hands-on experience on the intricate concepts that was earlier restricted to mere reading of concepts. This approach will introduce the concept of categorizing the course content into different category of modules that includes demonstration, exercises, structured enquiry and open-ended projects. This practice will benefit the students to a great extent thereby promoting them to undertake capstone projects and research activities.

Keywords Energy conversion • IC engines • Structured enquiry and open-ended problems

---

N.R. Banapurmath (✉) • P.P. Revankar • R.S. Hosamath
Department of Mechanical Engineering, BVB College of Engineering and Technology, Vidya Nagar, Hubli, Karnataka 580031, India
e-mail: nr_banapurmath@bvb.edu

R. Natarajan (ed.), *Proceedings of the International Conference on Transformations in Engineering Education: ICTIEE 2014*, DOI 10.1007/978-81-322-1931-6_96,

A Comprehensive Method for Defining and Assessing Programme Outcome – A Lifelong Learning Through Direct Assessment Techniques

Padmashree Desai, M. Vijayalakshmi, and G.H. Joshi

Abstract Lifelong learning is the "ongoing, voluntary, and self-motivated" pursuit of knowledge for either personal or professional reasons. It enhances not only social inclusion, active citizenship and personal development, but also competitiveness and employability. It enables students to learn at different times, in different ways, for different purposes at various stages of their lives and careers. Lifelong learning is concerned with providing learning opportunities throughout life, while developing lifelong learners. As per Accredation Board for Engineering and Technology (ABET) 3a-3k Programme Outcomes (POs), a PO-i states the expectation that engineering graduates must have "recognition of the need for, and an ability to engage in lifelong learning". The engineering education must help the students to develop skills, which are necessary for students to develop as lifelong learners and offer courses which help to acquire lifelong learning skills. This paper discusses the elements of lifelong learning, course design through activities and assessment of outcome for undergraduate Computer Science and Engineering programme to meet the Programme Outcome.

Keywords Lifelong learning • Self-motivated • Skills • Activity • Programme outcome • ABET

P. Desai (✉) • M. Vijayalakshmi • G.H. Joshi
Computer Science and Engineering, BVB College of Engineering and Technology, Vidya Nagar, Hubli, Karnataka 580031, India
e-mail: padmashri@bvb.edu; viju11@bvb.edu; ghjoshi@bvb.edu

R. Natarajan (ed.), *Proceedings of the International Conference on Transformations in Engineering Education: ICTIEE 2014*, DOI 10.1007/978-81-322-1931-6_97,

An Attempt to Bridge the Gap Between Industry and Academia – An Experience

M. Vijayalakshmi, Padmashree D. Desai, and G.H. Joshi

Abstract As we know that industry and academia are operating as two different domains, due to paradigms shifts, the growing complexity of the business environment made these two domains to collaborate each other. The universities are not only intended to prepare to produce the skilled human resources to corporate world, but also in various intangible ways. The intersecting requirements and jointly interreliant relationship requires identifying means of further corroboration academia–industry partnerships. Being the college in a tier-2 city it is difficult to attract industry but, here we made an attempt to bridge the gap between the industry and academia.

This paper attempts to explore how we can work closely with industry, study the extent of academia–industry partnership, and identify possible area where industry's contribution to academia would be most valuable in the Indian scenario. We are sharing the challenges faced, solutions adopted for Computer Science & Engineering programme to bring industry–academia together to make the students industry ready.

Keywords Academia • Industry • Incubation • Training • Projects

M. Vijayalakshmi (✉) • P.D. Desai • G.H. Joshi
Computer Science and Engineering, BVB College of Engineering and Technology, Vidya Nagar, Hubli, Karnataka 580031, India
e-mail: viju11@bvb.edu; padmashri@bvb.edu; ghjoshi@bvb.edu

R. Natarajan (ed.), *Proceedings of the International Conference on Transformations in Engineering Education: ICTIEE 2014*, DOI 10.1007/978-81-322-1931-6_98,

Effective Teaching of Digital Electronics for Undergraduate Students Using a Free Circuit Simulation Software—SEQUEL

Anupama R. Itagi and Varsha Tatti

Abstract Recent research on the learning process has shown that conventional approach of teaching fails to address all categories of students. Hence changes should be done in the traditional teaching methods as holding concentration of the students in the classroom for complete 1 h of the lecture is a big challenge. The teachers have to attract the attention of students in a classroom, to enable clear translation of the content of lessons and concepts. Many researchers agree on the fact that learning materials should not just reflect the teacher's teaching style, but should be designed for all kinds of students and all kind of learning styles.

A wise selection of strategies capable of ensuring the smooth and effective delivery of concepts is required. One such strategy that can be adopted is teaching through simulation. Among the available simulation software, Solver for EQuations with User defined ELements (SEQUEL) is a free software with many solved examples, which help beginners to learn it easily. This paper presents the approach followed by the authors to teach Digital Electronics course using SEQUEL. Using SEQUEL as an aid in teaching–learning process has increased student ability to comprehend the analysis of several digital circuits, as it can be used by the students even outside the college campus. This has reflected in the end semester exam results. With this the authors could effectively address level 1 to level 4 of Blooms taxonomy. In this paper, few Digital Electronics examples are presented to demonstrate how free circuit simulation software can assist in teaching the Principles of Digital Electronics to undergraduate students.

Keywords Conventional teaching methodology • Teaching by simulation • Free software for digital electronics

---

A.R. Itagi (✉) • V. Tatti
Electrical and Electronics, BVB College of Engineering and Technology,
Vidya Nagar, Hubli, Karnataka 580031, India
e-mail: anupama_itagi@bvb.edu; varshat@bvb.edu

R. Natarajan (ed.), *Proceedings of the International Conference on Transformations in Engineering Education: ICTIEE 2014*, DOI 10.1007/978-81-322-1931-6_99,

Theme-Based Capstone Projects and Programme Outcome Evaluation: A Case Study

D.G. Narayan and M. Meena

Abstract In this paper, we propose the process of theme-based capstone project implementation and attainment of program outcomes using rubrics. Capstone projects provide the opportunity for student teams to experience real-world software development by applying the knowledge gained through the curriculum. The project work provides a platform for acquiring transferrable skills viz. team work, communication skills and skills for lifelong learning. The regular curriculum course teaching learning experiences do not provide enough space and opportunity to inculcate these transferrable skills in the students. The typical capstone project process includes problem identification, system analysis and design, testing, performance analysis, documentation and report. Software engineering concepts are practiced during this process. However, process of capstone project faces lot of challenges like practicing industry practices, less emphasis on design aspects, inadequate testing, meeting program outcomes etc. Apart from these, evaluating student's teams in each phase of capstone course is always a challenge for the guides and review committee members. To address these issues, capstone course requires good planning, execution and evaluation methodology. Towards this, we have designed a process which can be used to evaluate the teams using common templates, rubrics and guidelines based on themes and measure programme outcomes.

Keywords Rubrics • SDLC • Capstone • Program outcome • Themes

D.G. Narayan (✉) • M. Meena
Department of Information Science and Engineering,
BVB College of Engineering and Technology, Vidya Nagar,
Hubli, Karnataka 580031, India
e-mail: narayan_d_g@rediffmail.com; msm@bvb.edu

R. Natarajan (ed.), *Proceedings of the International Conference on Transformations in Engineering Education: ICTIEE 2014*, DOI 10.1007/978-81-322-1931-6_100,

Importance of Research at Undergraduate Level

A. Padmaja, V.S.V. Laxmi Ramana, and P. Rajeshwar Reddy

Abstract The undergraduate research experience is one of several experiences that can impact the future career choices of our undergraduate students. In academia, both the faculty and the students are challenged to embrace engaged learning experiences and evidence-based education through undergraduate research. This has been the transition in higher education over the last 20 years: moving towards creating powerful educational environments that improve learning, rather than adding more courses that merely transfer knowledge. However, undergraduate research as a retention strategy does not go far enough. A report from the Council on Undergraduate Research succinctly summarizes that undergraduate research should be: faculty-driven, student-centered, and institutionally supported and provides the combination of factors necessary for: pedagogical effectiveness, enhanced learning outcomes, research productivity, and research program sustainability. Research should be at the core and must be instrumental in generating a major interface with the academic and business world. It empowers the faculty for an in-depth approach in teaching. It has the potential to enhance the consultancy capabilities of the researcher. Research can be internally driven or projects can be commissioned by national and international organizations such as the UNO, World Bank, OECD, Asian Development Bank, NCERT, Planning Commission, ISRO, DRDO, Central & State ministries and industrial agencies. Students need to be mentored in the entire research process. The best way for this to happen is to put students in a position to become a research assistant and be truly useful to the research program. Undergraduate research allows students to develop professionally and personally in ways not possible through traditional lecture and laboratory courses. Research is an important theme that threads its way through the undergraduate experience from the first year through to graduation. Weaving together the threads of what is currently underway provides a powerful basis from which to build an enriched, comprehensive

A. Padmaja • V.S.V. Laxmi Ramana (✉) • P.R. Reddy
Vidya Jyothi Institute of Technology (Anurag Group of Institutions), Hyderabad, India
e-mail: principalvjit@vjit.ac.in; laxmiramana@vjit.ac.in; rajeshwar_palla@yahoo.com

R. Natarajan (ed.), *Proceedings of the International Conference on Transformations in Engineering Education: ICTIEE 2014*, DOI 10.1007/978-81-322-1931-6_101,

learning environment for undergraduate students and encourage engineering graduates towards pursuing research. Undergraduate research allows students to develop professionally and personally. Research experiences give students an opportunity to gain a deeper knowledge of research techniques and processes, apply classroom learning in real-world contexts, explore academic literature, and form meaningful relationships with faculty members and professional researchers. In India the technical education institutes realize the immediate need to bridge the gap between the institute and industry needs and the students must be aware of the latest technology. Therefore, it is essential to establish labs of current technologies in the Engineering departments of colleges. This paper looks forward offering the students perspective undergraduate internships available for undergraduate research. This paper tries to present some of the websites which encourage taking up the research projects at undergraduate level.

Keywords Professional education • Perspective internships • Research experience • Research techniques

Index

R. Natarajan (ed.), *Proceedings of the International Conference on Transformations in Engineering Education: ICTIEE 2014*, DOI 10.1007/978-81-322-1931-6,

MIX
Papier aus verantwortungsvollen Quellen
Paper from responsible sources
FSC® C105338

If you have any concerns about our products,
you can contact us on
ProductSafety@springernature.com

In case Publisher is established outside the EU,
the EU authorized representative is:
Springer Nature Customer Service Center GmbH
Europaplatz 3, 69115 Heidelberg, Germany

Printed by Libri Plureos GmbH
in Hamburg, Germany